D0151293

Algebra in Context

Algebra in Context

Introductory Algebra from Origins to Applications

Amy Shell-Gellasch
Montgomery College

J. B. Thoo
Yuba College

Johns Hopkins University Press
Baltimore

Johns Hopkins University Press

2715 North Charles Street

Baltimore, Maryland 21218-4363

www.press.jhu.edu

ISBN-13: 978-1-4214-1728-8 (hardcover : alk. paper)

ISBN-10: 1-4214-1728-6 (hardcover: alk. paper)

ISBN-13: 978-1-4214-1729-5 (electronic)

ISBN-10: 1-4214-1729-4 (electronic)

Library of Congress Control Number: 2014954169

A catalog record for this book is available from the British Library.

Special discounts are available for bulk purchases of this book. For more information, please contact Special Sales at 410-516-6936 or specialsales@press.jhu.edu.

Johns Hopkins University Press uses environmentally friendly book materials, including recycled text paper that is composed of at least 30 percent post-consumer waste, whenever possible.

Contents

Preface

The history of mathematics is a rich and vibrant area of study that has been drawing increased interest in the mathematical and educational communities over the past few decades. There are two main areas of focus in the history of mathematics: historical research and the uses of the history of mathematics in teaching. Publications in both of these areas have grown every year. An outgrowth of these activities is that there are now many books available on the history of mathematics, ranging from popular books for the general public to textbooks for history of mathematics courses.

There are many ways to partition the books on the history of mathematics. One way is into those that assume that the reader has a calculus background, and those that do not. Textbooks designed primarily for history of mathematics courses (which are required for mathematics and mathematics education majors in many states) generally assume that the reader has a calculus background, and popular books, such as William Dunham's *Journey through Genius* [47] and Eli Maor's *e: The Story of a Number* [80], generally do not. Nevertheless, even popular books on the history of mathematics typically assume that the reader has some mathematics background, usually up through high school algebra. As with history of mathematics textbooks, many resource books also exist to aid college teachers in incorporating the history of mathematics into the mathematics classroom, but to date there is no textbook or resource book that is designed to use history as the vehicle for presenting the mathematical content. Our aim is to provide such a textbook.

You will find many of the topics that are covered in algebra (plus a few other topics such as logic and set theory, including infinite sets); however, unlike a traditional algebra textbook, the topics are presented here in their historical and cultural settings. A fair amount of history that does not pertain directly to mathematics is also presented to give a backdrop to the history of mathematics that is presented. Consequently, this book may be used as a textbook for a course in the history of mathematics that does not assume that the students have a calculus background. It also provides a gateway to appreciating many of the popular books on the history of mathematics if one has never had or is rusty in high school algebra.

This book is suitable for a variety of mathematics courses.

- *General education.* As a textbook for any course that is designed for students who are not science, technology, engineering, or mathematics majors (non-STEM students) but who need to fulfill a quantitative reasoning GE requirement or a multicultural GE requirement. For example, it would work very well for a college algebra for non-STEM majors course or for a liberal arts mathematics course.

- *Mathematics education.* As a textbook for a history of mathematics course that does not have a calculus prerequisite. Many states now require mathematics education majors to take a course in the history of mathematics. Such a course would be especially suitable for future elementary and middle school teachers. Moreover, education majors at any level, including the secondary level, will find the material covered to be useful in their understanding of the underpinnings and the development of mathematics.

- *Any mathematics course.* As a supplementary text for any mathematics course to inject doses of the history of mathematics to bring the course to life. Indeed, after teaching the history of mathematics for many years, both as stand-alone courses and embedded in mathematics courses, we have found that the history of mathematics is a great motivator that encourages students to become engaged in the mathematical topic and to see its uses and beauty. Furthermore, although this book was not designed for use by students who have a calculus background or beyond, more advanced students who have used preliminary versions of it have found that they enjoyed learning the back story, so to speak, of how and why the mathematics they have learned was developed.

- *High school.* The material is also appropriate and accessible to high school students.

ABOUT THIS BOOK

This book is organized into four parts:

Part I   Numeration Systems
Part II   Arithmetic Snapshots
Part III   Foundations
Part IV   Solving Equations

with each part broken into several chapters. The chapters are then broken into sections and possibly subsections. The chapters, sections, and subsections are numbered using Indo-Arabic numerals (1, 2, 3, and so on). For example,

9.2   Grating or Lattice Method   is SECTION 2 of CHAPTER 9
22.3.2   Descartes's Rule of Signs   is SUBSECTION 2 of CHAPTER 22, section 3

Figures and tables are numbered with both the chapter number and the figure or table number. For example,

Figure 2.4: Plimpton 322   is FIGURE 4 in CHAPTER 2
Table 2.1: Babylonian number characters   is TABLE 1 in CHAPTER 2

Throughout the book, we provide exercises in a "just in time" manner to reinforce the material presented. The chapters are sprinkled with two types of exercises: *Now You Try* exercises and *Think About It* exercises. The *Now You Try* exercises give you an opportunity to become familiar with the mathematics that was just discussed, and the *Think About It* exercises ask questions that may require you to ponder. These exercises are integral to your appreciating and understanding the mathematical concepts or the history of mathematics that is presented. There are also additional exercises that are collected in a chapter at the end of each part. Some of these exercises are routine, some are nonstandard to add depth and variety, and some (marked with an ∗) may offer a little more challenge or require you to do a little research. The exercises that require a little research can be used as quick Internet research projects or as ideas for larger projects; they are also designed to be used to motivate class discussions.

The book uses the symbol □ to mark the end of certain blocks. For example,

Remark 1.1 The numbers from one to nine hundred ninety-nine may be considered *fractions of 1000* so that 1000 may be considered the *unit* or basic quantity. For example, eight hundred eleven is $\frac{811}{1000} \times 1000$. As another example, 231,811 is $231\frac{811}{1000} \times 1000$.   □

The following table shows the different blocks that end with the □ symbol.

Remark	Think About It	Now You Try	Rule
Example	Definition	Theorem	Corollary

All of these blocks are numbered sequentially within themselves beginning with the chapter number.

To appreciate and to understand any history of mathematics beyond knowing some biographies and anecdotes and sequences of events require understanding some mathematics. This book presents snapshots of the history of mathematics that do not extend beyond high school algebra, from the early beginnings to the eighteenth century. The book assumes that you are already familiar with elementary algebra, specifically, that you are familiar with

- arithmetic with signed numbers (positive and negative numbers)

- the order of operations

- simplifying algebraic expressions (for example, using the distributive law and combining like terms)

- evaluating algebraic expressions when given values of the variables

- solving linear equations in one variable

- solving 2×2 systems of linear equations

- graphing equations in two variables, particularly, graphing linear equations in two variables

that are commonly taught in a traditional high school Algebra I course or in a college remedial algebra course. If you need a refresher or an introduction to these topics, search for "elementary algebra" on the World Wide Web (the Web) to find a slew of materials on the subject. There are also many good videos on the topics on the Web, for example, at <http://www.mathtv.com>.

OTHER SOURCES FOR THE HISTORY OF MATHEMATICS

The bibliography lists many references that you may pursue. The references range from covering very specific topics, to specific time periods or geographical regions, to specific or broad themes, to broader surveys and commentaries. Listed among the references are some of the standard textbooks in the history of mathematics, namely, [15, 21, 22, 35, 41, 73, 74, 75]. Although these textbooks assume that the reader has a calculus background, you may still glean a lot from them even if you have to skip over some of the mathematics.

For a very good overview of the history of mathematics that does not assume that the reader has a calculus background, we recommend the introductory chapter, "The History of Mathematics in a Large Nutshell," of Berlinghoff and Gouvêa's excellent book, *Math through the Ages* [11]. Their survey complements the material in this book very nicely.

ACKNOWLEDGMENTS

This book could not have been written and produced without the help of many people.

We thank our colleagues, students, and family for their help, support, and patience while we created what we hope will be a new type of book, and a new approach to teaching basic mathematics.

For reading drafts, early and late, and providing constructive feedback, we thank Erick Gremlich, Christopher Goff, and Travis Smith, and also the anonymous reviewers from Johns Hopkins University Press (JHUP). We also thank the members of the SIGMAA-HOM listserv (the Special Interest Group of the Mathematical Association of America on the History of Mathematics), who very patiently helped clear up some tricky history.

We thank Vincent J. Burke, Catherine Goldstead, and Kathryn Marguy at JHUP for their guidance and support during the production process. Andre M. Barnett did a terrific job of copy-editing. We are, of course, solely responsible for any shortcomings or mistakes.

This book was typeset using the LaTeX memoir documentclass, and that would not have been possible without the encouragement and help from the members of the MacTeX users group. We especially thank William Adams, Nestor E. Aguilera, John Burt, David Derbes, Paul Dulaney, Murray Eisenberg, Gary L. Gray, Martin Wilhelm Leidig, Themis Matsoukas, Scot Mcphee, M. Tamer Özsu, and Axel E. Retif. (We apologize if we left off anyone.) The font family used for the text is TeX Gyre Termes, with the mathematics fonts provided by the qtxmath package.

Algebra in Context

Introduction

An interest in history marks us for life. How we see ourselves and others is shaped by the history we absorb, not only in the classroom but from films, newspapers, television programmes, novels and even strip cartoons. From the time we first become aware of the past, it can fire our imagination and excite our curiosity: we ask questions and then seek answers from history. As our knowledge develops, differences in historical perspectives emerge. And, to the extent that different views of the past affect our perception of ourselves and of the outside world, history becomes an important point of reference in understanding the clash of cultures and of ideas. Not surprisingly, rulers throughout history have recognized that to control the past is to master the present and thereby consolidate their power.

—George Gheverghese Joseph, *The Crest of the Peacock* [72]

Philosophy is written in this grand book—I mean the Universe—which stands continually open to our gaze, but it cannot be understood unless one first learns to comprehend the language and interpret the characters in which it is written. It is written in the language of mathematics, and its characters are triangles, circles and other geometrical figures, without which it is humanly impossible to understand a single word of it.

—Galileo Galilei, *Assayer* [20]

WHY STUDY THE HISTORY OF MATHEMATICS

These two quotations highlight the complementary aspects of this book. As Joseph notes, "An interest in history marks us for life," for "it can fire our imagination and excite our curiosity: we ask questions and then seek answers from history." Mathematics, as Galileo tells us, is a language. Indeed, mathematics has symbols; it has parts that are equivalent to nouns, verbs, and prepositions; it relies on context and connotation to be understood; it borrows from different cultures and has evolved over time; it even follows fads. The history of mathematics, then, is the history of this language that allows us to understand the workings of our universe.

History can be broadly separated into two parts: prehistory and recorded history. In general, prehistory is the period in which no written records of any kind were left behind. History becomes much more exact when written records can be examined. The earliest written remains are those of tally marks found on animal bones or horns. The most famous is the Ishango bone found in the Congo in 1960. This baboon bone shows groups of parallel gashes that many scholars believe are a form of tally system, perhaps even a lunar calendar. The bone also has a quartz crystal placed in the end for making the markings.

So the earliest written record left to us by man is in the form of a mathematical artifact. One of the earliest acts Homo sapiens performed that started us on the road on which we are still traveling is the combining of abstraction and language into the science and art of mathematics.

Unlike any other area of study, mathematics is the study of truths. Mathematical styles and methods may change, but mathematical truths are eternal, and how these truths are discovered is the story of the history of mathematics. Seltman reflects upon this in her apology for Harriot's *Praxis* [105]:

First, I am going to ask the (perhaps surprising) questions: in what sense does mathematics have a history? I ask this question because there is one sense in which it does not have a real history, and that resides in its consisting of permanent truths such as the theorems of Euclid, under the conditions of Euclidean axioms, which express unchanging relationships. Correct mathematics remains correct however out of date it is. Archimedes' methods for finding volumes and areas may be obsolete but are still valid relative to the axioms of his system. This is the unchanging aspect of mathematics. In this sense, mathematics has no history—it just persists through time. It is not a changing subject. Remember that mathematics itself consists only of *ideas* (relationships between elements)—it is not the symbols, or the paper they are written on or the mathematician who has constructed the mathematics. What does have a history is the total process, consisting of the mathematical ideas together with the mathematician who constructed them, together with the notation, symbolism, methodology, and so on and all the other connections to the social and individual milieu that constitutes the developing mathematical process. And this certainly does have a history. But the mathematical ideas *per se* do not. They are pure, noetic relationships, which are permanently true. We must, therefore, distinguish the mathematics in itself from the historical process as a whole, which includes all the external connections that are contingent to it. Such contingent relationships are 'external' to the ideas to which they are related and may or may not be absorbed into the process, thus altering it. The thought processes of mathematicians, or other people connecting in some way to the mathematics, are contingencies in relation to it and may change the existing mathematics or not in some way, thus being the engine for the historical (or developmental) process of mathematical growth. It is the 'may or may not be' property of contingency which renders history of any sort unpredictable. In the light of all this, there is a second general theoretical question to be asked and that is: Are there revolutions or fundamental changes in mathematics as there are said to be in science? ...There are surely no revolutions in mathematics in itself, since mathematics, consisting of pure relational ideas, may become obsolete but is not replaced (in the sense of being overthrown) by changes. It remains correct, no matter what. There are, however, revolutions in the process of mathematical development, which includes the mathematician and his/her thinking and all the external relations of the mathematics in itself. And such revolutions may occur with regard to, say, symbolism, axioms, methodology, and so forth. All this was stated above.

Studying the history of mathematics is an intriguing and inspiring journey. Along the way you will discover people and places and ways of thinking about numbers that are different from the mathematical concepts you have encountered thus far. You will find that there is no "one way" to do mathematics, but in fact there are many ways: each way is different and creative, evolved from the needs of mankind and the social and historical context of the time. The history of mathematics is an important story to tell and to learn because mathematics is the most "human" of all the human endeavors.

We wish you well on your journey through time as you explore the history of mathematics.

Amy Shell-Gellasch and J. B. Thoo
July 2015

Part I

Numeration Systems

Chapter 1

Number Bases

What does the symbol "3" represent? You may say, "The number three, of course!" Yes, but three *what*? Three ones (3); three tens (30); three hundreds (300)? Now, you may say, "Wait a minute: 3 is different from 30, and both are different from 300." Very well. In that case, would 3000 300 30 3 represent three thousand, three hundred thirty-three? "No, no, no!" you say?

What if we were to write

3 for three,   3T for 30 (three tens),   3H for 300,   and   3Th for 3000?

Would it then be reasonable that 3Th 3H 3T 3 represents three thousand, three hundred thirty-three? It turns out that 3Th 3H 3T 3 is precisely the manner in which three thousand, three hundred thirty-three is expressed using Chinese characters. So, referring to TABLE 1.1, 三千三百三十三 $(= 3 \times 1000 + 3 \times 100 + 3 \times 10 + 3)$ represents three thousand, three hundred thirty-three using Chinese characters. (Today the Chinese are just as apt to write 3333 instead of 三千三百三十三 .)

一	二	三	四	五	六	七	八	九	十
one	two	three	four	five	six	seven	eight	nine	ten

百	hundred	千	thousand	萬 or 万	ten thousand

Table 1.1: Chinese number characters.

Now, to take another example, because 4379 and 3479 do not both express four thousand, three hundred seventy-nine, we say that the number system we commonly use is a *positional number system* or a *place-value number system*, whereby the position of a symbol like 3 within a numeral matters because every position has a different place value. There are other positional number systems besides the one we commonly use, as well as *nonpositional* systems.

Current research indicates that there were four times in history that the principle of place value was discovered [72]: around 2000 BC by the Babylonians; around the beginning of the Christian era by the Chinese; also around the beginning of the Christian era by the Maya; and between the third and fifth centuries AD by the Indians.[1]

The number system that we commonly use is called the *Indo-Arabic* number system.[2] The key features of the Indo-Arabic system we use today are

1. it is a positional system;

2. it uses ten symbols (1, 2, 3, 4, 5, 6, 7, 8, 9, and 0), corresponding in number to the *base* of the system (base 10);

---

[1]The inhabitants of the Indian subcontinent.

[2]The Indo-Arabic number system is also commonly called the *Hindu-Arabic* number system.

3. it has a place holder symbol, 0, that is also a number; and

4. it has a separatrix (the decimal point) that separates the integer part of a numeral from the fraction part.

This was not always the case.

In this part, we will explore a few of the different number systems that have been used throughout history. We will see that not all number systems enjoy the four key features of today's Indo-Arabic number system. Before we discuss a few of the different number systems, however, we take a look at different *number bases*.

The fact that the Indo-Arabic number system is called a base-10 or *decimal* system has nothing to do with the fact that it uses ten symbols to write numerals, although it is very felicitous that the number of symbols used is the same as the base. (Indeed, as we will see, the Babylonian number system is a base-60 number system, but it uses only two symbols for numerals: a symbol for 1 and a symbol for 10.) By the *base* of a number system, we mean the way in which the numbers are grouped. For instance, the American *word name* for numbers is grouped in powers of 1000, that is to say, we have a different word for each new power of 1000. Here are the first few.

one	$1000^0 = 1$	(one)
thousand	$1000^1 = 1000$	(thousand ones)
million	$1000^2 = 1,000,000$	(thousand thousands)
billion	$1000^3 = 1,000,000,000$	(thousand millions)
trillion	$1000^4 = 1,000,000,000,000$	(thousand billions)

So, the American word name for numbers can be said to be base 1000. (See SECTION 12.4.)

Remark 1.1 The numbers from 1 to 999 may be considered *fractions of 1000* so that 1000 may be considered the *unit* or basic quantity. For example, eight hundred eleven is $\frac{811}{1000} \times 1000$. As another example, 231,811 is $231\frac{811}{1000} \times 1000$. □

On the other hand, every position or *place value* in a numeral is a power of 10. For instance, in 4379,

$$4000 = 4 \times 1000 = 4 \times 10^3,$$
$$300 = 3 \times 100 \ = 3 \times 10^2,$$
$$70 = 7 \times 10 \ \ = 7 \times 10^1,$$
$$9 = 9 \times 1 \ \ \ = 9 \times 10^0.$$

So, the place value we commonly use in writing a numeral is base 10, but the American word name is base 1000.

This is different in the Chinese language, where both the Chinese word name for numbers and place value for numerals are base 10. For example, the word name for 四千三百七十九 is literally, "four thousand, three hundred, seven ten, nine," that is to say, there is a different word for each new power of 10.

Think About It 1.1 What is the difference between the traditional Chinese number system and the Indo-Arabic number system? Which would you say is more efficient? □

To help us better understand number bases in general, we will explore bases other than base 10. To avoid confusion, we will use different symbols. These examples are not historical but were made up in order to illustrate the nature of different number systems.

1.1 BASE 6

In base 6, the place values are powers of 6 instead of powers of 10. For example, the first few powers of 6 are

$$6^0 = 1, \quad 6^1 = 6, \quad 6^2 = 36, \quad 6^3 = 216, \quad 6^4 = 1296.$$

We will use the symbols $\alpha, \beta, \gamma, \delta$, and ϵ to represent the numbers 1 to 5, respectively, and the symbol $\cdot$ to represent a place holder.[3]

α	β	γ	δ	ϵ	$\cdot$
one	two	three	four	five	place holder

Remark 1.2 In general a place holder symbol only denotes the absence of a number and is, itself, not a number. In particular, a place holder symbol is not necessarily the number zero, although today it almost always is. It took many, many years for the conception of zero as a number to enter mathematics. The credit for the conception of zero as a number seems to belong to the Indians by the time of Brahmagupta (AD 598–670) [41]. □

We choose to write our base-6 numerals with place values increasing from right to left, just as we do in the Indo-Arabic number system, because we are accustomed to that. Then, for example, the base-6 place-value numeral $\alpha \cdot \epsilon\delta$ represents the number two hundred fifty.[4] This is because

$$\alpha \cdots = 1 \times 6^3 = 1 \times 216 = 216,$$
$$\cdots = 0 \times 6^2 = 0 \times 36 = 0,$$
$$\epsilon \cdot = 5 \times 6^1 = 5 \times 6 = 30,$$
$$\delta = 4 \times 6^0 = 4 \times 1 = 4,$$

so that

$$\alpha \cdot \epsilon\delta = \underline{1} \times 216 + \underline{0} \times 36 + \underline{5} \times 6 + \underline{4} \times 1 = 250.$$

This can be a little confusing because $\alpha \cdot \epsilon\delta$ is a base-6 numeral, but two hundred fifty is a base-10 or decimal word name. We really should invent a base-6 word name.

Now You Try 1.1 Write the decimal numeral and the decimal word name for each of the following base-6 place-value numerals.

1. $\alpha \cdot$

2. $\alpha\beta\epsilon\gamma\gamma\beta\delta$

□

To express a number as a base-6 place-value numeral using these symbols, we need to remember only that the value in any place may not exceed five (just as the value in any place in base 10 may not exceed nine). This is a general principle.

Rule 1.1 The value in any place in base b may not exceed $b - 1$. □

---

[3]These are the first five Greek lowercase letters: *alpha* α, *beta* β, *gamma* γ, *delta* δ, and *epsilon* ϵ. We do not, however, pronounce the symbols as letters here; instead, we pronounce the symbols as numerals. Thus, instead of reading α as "alpha," we read it as "one," and so on.

[4]When we write a numeral such as 250 using the Indo-Arabic symbols, without specifying the base for the numeral, it shall be understood that it is a base-10 numeral. In other words, 250 will be understood to be $2 \times 100 + 5 \times 10 + 0 \times 1$.

So, for example, because we can "fit in" one 6 into 9 with a remainder of 3, nine written as a base-6 place-value numeral is $\alpha\gamma$.

$$\alpha\gamma = \underline{1} \times 6 + \underline{3} \times 1 = 9$$

Similarly, because we can fit in four 6's into twenty-nine with a remainder of five, twenty-nine written as a base-6 place-value numeral is $\delta\epsilon$.

$$\delta\epsilon = \underline{4} \times 6 + \underline{5} \times 1 = 29$$

In general, to express a number as a base-6 place-value numeral, we need to find how many of each place value fits into the number. It is easiest if we begin with the greatest place value that fits into the number and then proceed with each smaller place value until we reach the 6^0 or the unit place.

Example 1.1 Express four thousand, three hundred seventy-nine as a base-6 place-value numeral.

Note that the greatest base-6 place value that fits into the number is $6^4 = 1296$, and that the next smaller place values in sequence are $6^3 = 216$, $6^2 = 36$, $6^1 = 6$, and $6^0 = 1$. Hence, we divide 4379 by 1296, and then divide the remainder of that division by 216, and so on.

$$
\begin{array}{ccccc}
\overset{\textstyle 3}{1296\,)\overline{4379}} & \overset{\textstyle 2}{216\,)\overline{491}} & \overset{\textstyle 1}{36\,)\overline{59}} & \overset{\textstyle 3}{6\,)\overline{23}} & \overset{\textstyle 5}{1\,)\overline{5}} \\
\underline{3888} & \underline{432} & \underline{36} & \underline{18} & \underline{5} \\
491 & 59 & 23 & 5 & 0
\end{array}
$$

Therefore, four thousand, three hundred seventy-nine as a base-6 place-value numeral is $\gamma\beta\alpha\gamma\epsilon$. Note that it is important not to skip any place value when doing the divisions or else we may miss where a place holder should be.

□

Example 1.2 Express ten thousand, five hundred sixty-nine as a base-6 place-value numeral.

Note that the greatest base-6 place value that fits into the number is $6^5 = 7776$, and that the next smaller place values in sequence are $6^4 = 1296$, $6^3 = 216$, $6^2 = 36$, $6^1 = 6$, and $6^0 = 1$.

$$
\begin{array}{cccccc}
\overset{\textstyle 1}{7776\,)\overline{10569}} & \overset{\textstyle 2}{1296\,)\overline{2793}} & \overset{\textstyle 0}{216\,)\overline{201}} & \overset{\textstyle 5}{36\,)\overline{201}} & \overset{\textstyle 3}{6\,)\overline{21}} & \overset{\textstyle 3}{1\,)\overline{3}} \\
\underline{7776} & \underline{2592} & & \underline{180} & \underline{18} & \underline{3} \\
2793 & 201 & & 21 & 3 & 0
\end{array}
$$

Therefore, ten thousand, five hundred sixty-nine as a base-6 place-value numeral is $\alpha\beta \cdot \epsilon\gamma\gamma$.   □

Now You Try 1.2 Express the following numbers as base-6 place-value numerals using the base-6 numerals $\alpha, \beta, \gamma, \delta$, and ϵ, and place holder $\cdot$.

1. One hundred thirty-four.   2. Nine hundred forty-one.

□

Think About It 1.2 To express a number as a base-6 numeral, can we begin with the *least* place value that fits into the number, instead of with the greatest place value as we did in the examples, and then proceed with each greater place value until the number is exhausted?   □

Now that we can express a number in base 6, let us do some arithmetic in base 6. Remember that the value in any place may not exceed five. To begin, TABLE 1.2 shows an abbreviated addition table and an abbreviated multiplication table.

+	·	α	β	γ	δ	ϵ
·	·	α	β	γ	δ	ϵ
α	α	β	γ	δ	ϵ	$\alpha\cdot$
β	β	γ	δ	ϵ	$\alpha\cdot$	$\alpha\alpha$
γ	γ	δ	ϵ	$\alpha\cdot$	$\alpha\alpha$	$\alpha\beta$
δ	δ	ϵ	$\alpha\cdot$	$\alpha\alpha$	$\alpha\beta$	$\alpha\gamma$
ϵ	ϵ	$\alpha\cdot$	$\alpha\alpha$	$\alpha\beta$	$\alpha\gamma$	$\alpha\delta$

$\times$	α	β	γ	δ	ϵ
α	α	β	γ	δ	ϵ
β	β	δ	$\alpha\cdot$	$\alpha\beta$	$\alpha\gamma$
γ	γ	$\alpha\cdot$	$\alpha\gamma$	$\beta\cdot$	$\beta\gamma$
δ	δ	$\alpha\beta$	$\beta\cdot$	$\beta\delta$	$\gamma\beta$
ϵ	ϵ	$\alpha\gamma$	$\beta\gamma$	$\gamma\beta$	$\delta\alpha$

Table 1.2: Abbreviated addition and multiplication tables in base 6.

Now You Try 1.3 Extend the addition table to $\alpha\delta + \alpha\delta$, and the multiplication table to $\alpha\delta \times \alpha\delta$.   □

Example 1.3

1. $\epsilon + \delta = \alpha\gamma$: five plus four equals nine

2. $\beta\epsilon + \epsilon = \gamma\delta$: seventeen plus five equals twenty-two

3. $\alpha\gamma - \delta = \epsilon$: nine minus four equals five

4. $\gamma\delta - \epsilon = \beta\epsilon$: twenty-two minus five equals seventeen

   □

Example 1.4 Find the sum of 23 and 9 in base-10 and base-6 place values.

$$\text{Base 10} \qquad\qquad \text{Base 6}$$

$$\begin{array}{cc} \overset{1}{2} & 3 \\ & 9 \\ \hline 3 & 2 \end{array} + \qquad\qquad \begin{array}{cc} \overset{\alpha}{\gamma} & \epsilon \\ \alpha & \gamma \\ \hline \epsilon & \beta \end{array} +$$

So, $23 + 9 = 32$ or $\gamma\epsilon + \alpha\gamma = \epsilon\beta$.   □

Example 1.5 Find the product of 23 and 4 in base-10 and base-6 place values.

$$\text{Base 10} \qquad\qquad \text{Base 6}$$

$$\begin{array}{cc} \overset{1}{2} & 3 \\ & 4 \\ \hline 9 & 2 \end{array} \times \qquad\qquad \begin{array}{ccc} & \overset{\gamma}{\gamma} & \epsilon \\ & & \delta \\ \hline \beta & \gamma & \beta \end{array} \times$$

So, $23 \times 4 = 92$ or $\gamma\epsilon \times \delta = \beta\gamma\beta$.   □

Now You Try 1.4 Perform the following operations in base-6 place values using the base-6 numerals α, β, γ, δ, and ϵ, and place holder ·.

1. Fifty-eight plus thirty-two.
2. Twenty plus thirty.
3. Fifteen minus six.
4. Forty minus twelve.
5. Twenty-four times nine.

   □

1.2   BASE 4

In base 4, the place values are powers of 4. The first few powers of 4 are

$$4^0 = 1, \quad 4^1 = 4, \quad 4^2 = 16, \quad 4^3 = 64.$$

We will use the symbols $|$, $\wedge$, and $\vee$, to represent the numbers 1, 2, and 3, respectively, and the symbol $\square$ to represent a place holder.

$|$	$\wedge$	$\vee$	$\square$
one	two	three	place holder

Thus, for example, the base-4 place-value numeral $|\square \vee \wedge$ represents the number seventy-eight. This is because

$$|\square\square\square = 1 \times 4^3 = 1 \times 64 = 64,$$
$$\square\square\square = 0 \times 4^2 = 0 \times 16 = 0,$$
$$\vee\square = 3 \times 4^1 = 3 \times 4 \; = 12, \text{ and}$$
$$\wedge = 2 \times 4^0 = 2 \times 1 \; = 2,$$

so that

$$|\square \vee \wedge = \underline{1} \times 64 + \underline{0} \times 16 + \underline{3} \times 4 + \underline{2} \times 1 = 78.$$

Now You Try 1.5 Write the decimal numeral and word name for each of the following base-4 place-value numerals.

1. $|\square$

2. $| \vee \wedge \wedge | \wedge$

$\square$

Example 1.6 Express six hundred forty-three as a base-4 place-value numeral.

Note that the greatest base-4 place value that fits into the number is $4^4 = 256$, and the next smaller place values in sequence are $4^3 = 64$, $4^2 = 16$, $4^1 = 4$, and $4^0 = 1$.

$$
\begin{array}{ccccc}
\quad\;\;2 & \quad\;\;2 & \;\;0 & \;\;0 & \;\;3 \\
256\overline{)643} & 64\overline{)131} & 16\overline{)3} & 4\overline{)3} & 1\overline{)3} \\
\;\;\underline{512} & \;\;\underline{128} & & & \;\;\underline{3} \\
\;\;131 & \;\;3 & & & \;\;0
\end{array}
$$

Therefore, six hundred forty-three as a base-4 place-value numeral is $\wedge \wedge \square\square \vee$. $\square$

Now You Try 1.6

1. Express the following numbers as base-4 place-value numerals using the base-6 numerals $|$, $\wedge$, and $\vee$, and place holder $\square$.

    a) One hundred thirty-four.

    b) Nine hundred forty-one.

2. Create an abbreviated addition table up to $\wedge \wedge \; + \; \wedge \wedge$, and an abbreviated multiplication table up to $\wedge \wedge \times \wedge \wedge$.

$\square$

Example 1.7

1. $\vee$ + $\wedge$ = ||: three plus two equals five

2. $\wedge$| + $\vee$ = $\vee$□: nine plus three equals twelve

3. || − $\wedge$ = $\vee$: five minus two equals three

4. $\vee$□ − $\vee$ = $\wedge$|: twelve minus three equals nine

□

When subtracting, remember that the value in any place value is a power of 4. This is especially important when one needs to "borrow" in the subtraction.

Example 1.8 Find the difference between 61 and 22 in base-10 and base-4 place values.

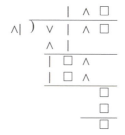

Base 10

$$
\begin{array}{cc}
\overset{5}{\cancel{6}} & \overset{10}{1} \\
2 & 2 \\
\hline
3 & 9
\end{array} \;-
$$

So, $61 - 22 = 39$ or $\vee\,\vee\,| - ||\wedge = \wedge|\vee$.

□

Example 1.9 Find the quotient of 216 and 9 in base-4 place values.

Base 10

$$
\begin{array}{r}
2\;4 \\
9\,)\,\overline{2\;1\;6} \\
1\;8 \\
\hline
3\;6 \\
3\;6 \\
\hline
0
\end{array}
$$

So, $216 \div 9 = 24$ or $\vee\,|\,\wedge\,□ \div \vee| = |\wedge□$.

□

Now You Try 1.7 Perform the following operations in base-4 place values using base-4 numerals symbols |, $\wedge$, and $\vee$, and place holder □.

1. Fifteen plus nine.

2. Twenty-one plus four.

3. Thirty minus two.

4. Thirteen minus six.

5. Eighty-four divided by seven.

□

Chapter 2

Babylonian Number System

When we speak of Babylonian mathematics, we really speak of the mathematics that was developed in the region between the Tigris and Euphrates Rivers known as Mesopotamia, and which is modern-day Iraq and Kuwait, and northeastern Syria, southeastern Turkey, and southwestern Iran. Mesopotamia is commonly known as the "cradle of civilization" because one of the earliest civilizations in the West emerged in Sumer in southern Mesopotamia between 3500 and 3000 BC. Moreover, the world's first cities appeared in Mesopotamia some five hundred years before that. It is also the land of the Old Testament Bible: Noah's ark is said to have come to rest on the mountains of Ararat, part of modern Armenia (Gn 8:4); the story of the Tower of Babel is based on the ziggurats of Babylonia (Gn 11:1–9); Abraham (Abram) is said to be from the Sumerian city of Ur (Gn 11:27–28); the people of Jerusalem, except for the poor, were taken captive by King Nebuchadnezzar II (2 Kgs 24:10–16). It was because Mesopotamia was familiar to Europeans through Old Testament stories that British and French archaeologists in the 1840s who were excavating the ancient Assyrian city of Nineveh near Mosul, today in northern Iraq, "claimed it [the ruins] as part of their own, European heritage, and were little interested in its place in Middle Eastern history and tradition per se. Thus unwittingly the tone was set for interpreting ancient Assyrian—and later, Babylonian and Sumerian—remains" [102]. This attitude, Eleanor Robson [102] tells us, has been changing slowly since the 1970s. As an example of this change, she mentions that Jens Høyrup has "reevaluated Old Babylonian 'algebra' by analysing its language rather than forcing it to fit modern algebraic models."

The hills and high plains of northern Mesopotamia (FIGURE 2.1) received sufficient rainfall annually for growing crops and raising domesticated animals. In the south, however, although the open plains were fertile, a system of irrigation canals had to be constructed to bring water from the rivers to support agriculture. Over time, the people of the region settled down from their nomadic lifestyles, and up sprang urban communities, administrative bureaucracies, and codes of law. Organized agriculture made possible by the two rivers allowed some people to have specialized functions such as blacksmiths, cooks, merchants, priests, tavern owners, and architects [35]. Daily activities, from the pricing of goods to the building of irrigation canals and the construction of temples and palaces to the creation of calendars, required more sophisticated mathematics that was the purview of the palace and temple scribes. Unfortunately, although the Tigris and Euphrates allowed urbanization in the Mesopotamian valley, the open plains of the region also allowed the cities to be attacked easily by other cities, as well as by foreign invaders. Indeed, the history of Mesopotamia is marked by many dynasties.

These are some of the notable Mesopotamian dynasties. Sargon I (ca. 2276–2221 BC), an Akkadian, established a very large empire that stretched from the Black Sea to the Persian Gulf, and from the Mediterranean Sea to the steppes of Persia [15]. Hammurabi (ca. 1792–1750 BC), an Amorite, wrote the first codes of law in recorded history; these laws are reputed to be very strict, for example, containing the dictum "an eye for an eye and a tooth for a tooth" [35]. Nebuchadnezzar II (605–562 BC), a Babylonian, established the Chaldean empire and built the great ziggurat of Babylon. Darius I (521–486 BC) and Xerxes I (486–465 BC) oversaw the height of Persian power, spanning from Thrace in Europe to the Hindu Kush and from the Caspian Sea to the Egyptian desert, that kept Greek civilization in check to live, as Socrates

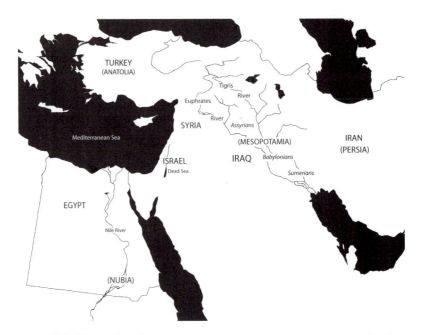

Figure 2.1: The ancient Near East. (Source: Image courtesy of Tom Urban, The Oriental Institute of the University of Chicago.)

put it, "round the [Mediterranean] sea—like frogs round a pond" [85]. Alexander the Great (356–323 BC), a Macedonian, wrestled control from the Persians and conquered much of the known civilized world, from the Mediterranean Sea to as far away as the Indian subcontinent. (He entered Egypt in 331 BC, and there he commissioned the founding of a city to be named after himself near the mouth of the river Nile. Alexandria became home to the great Museum, the cultural and academic center of the Hellenistic Era. The Museum of Alexandria was an important center for Greek mathematics.) Yet, with each new dynasty, the culture and traditions of ancient Mesopotamia not only continued substantially unchanged over the years because it was adopted in great part by the conquerors but also spread to neighboring regions along the trade routes [35].

2.1   CUNEIFORM

Sumerian was one of the first languages in the world to be written. Eventually, all the writings of the Mesopotamian region, which were influenced by the Sumerians, came to be called *cuneiform* script. The term cuneiform comes from he Latin *cuneus* which means "wedge or wedge-shaped." Cuneiform script was written by making impressions in moist clay tablets using a reed pen or stylus. A reed pen could make wedge-shaped impressions with the sharp end and round or semicircular-shaped impressions with the other end. If the writing was to be kept, then the clay tablet would be baked; otherwise, the tablet could be washed and reused. Cuneiform script was written from left to right. Here is a fifth century BC example of cuneiform script. Xerxes had this inscription carved above the doorways of his palace at Persepolis:

This transliterates as:

Xerxes, the great king, the king of kings, the son of Darius the king, an Achaemenian.

That the ancient Mesopotamians wrote on clay tablets (FIGURE 2.2) is both good and not so good for archaeologists and Assyriologists. The good is quite obvious: clay tablets can survive the ravages of time much better than other materials, such as papyrus, leather, wood, and so on. As a result, over 400,000 cuneiform tablets have been discovered, with several hundred of mathematical interest. About two-thirds of these are Old Babylonian (1800–1600 BC; the Hammurabi dynasty) [21]. The drawbacks are at least two. First, unlike materials like papyrus or leather that can be made into a scroll or a codex, clay tablets are not suited for creating a comprehensive work; thus, we do not have any systematic treatises to study [41]. Second, many clay tablets ended up being reused as building rubble [102].

Figure 2.2: The *Epic of Gilgamesh*, an epic poem from Mesopotamia, written in cuneiform. The *Epic of Gilgamesh* is considered to be the first great work of literature, and cuneiform portions of it date to 1200 BC. (Source: © Thinkstock.)

Even though numerous cuneiform tablets had been discovered, deciphering the cuneiform script was a practically impossible task at first because the language had long fallen out of use. What jump-started the decipherment of cuneiform script was the complete translation of carvings along the Behistun Cliffs in Iran by Englishman Henry Rawlinson in 1846. The carvings were an account of Darius I's crushing of a revolt in Persia in 516 BC written in Elamatic (Old Persian), Babylonian, and Persian for all the travelers and passersby to see. According to Calinger [35], "Bilingual inscriptions in Sumerian and Akkadian, a Semitic language similar to Babylonian, permitted extrapolations and further deciphering. Rawlinson had cracked the cuneiform enigma and essentially opened for study the records of ancient Mesopotamia."

2.2 MATHEMATICAL TEXTS

Mathematical cuneiform tablets came to the fore in the late 1800s. In 1906, Hermann Hilprecht, who was the director of the University of Pennsylvania's Nippur excavations, published a book in which he discussed multiplication and reciprocal tables, as well as metrological lists and tables. (Vincent Scheil and François Thureau-Dangin then made significant corrections and additions to Hilprecht's work.) But it was to wait until the 1920s before Otto Neugebauer began the most prodigious work on cuneiform mathematics to date. His work culminated in the publication of his 1940s public lecture series as *The Exact Sciences in Antiquity* [34, 88, 102]. The translation and interpretation of cuneiform mathematics continues to this day. Indeed, Robson [102] tells us that "modern scholarly understanding of Mesopotamian mathematics is changing and improving at a rate unparalleled since the 1930s. Not only are dozens of new sources published every year, causing constant reevaluation of the historical corpus, but recent developments in the scripts, languages, history and archaeology of the region have stimulated exciting new lines of research that contextualise ancient mathematical practice in a way that was unimaginable even ten years ago."

The mathematical cuneiform tablets that have been found come from the Old Babylonian period and the Seleucid period (the last three centuries BC; Seleucus was a Greek general who controlled the region after the death of Alexander the Great). The texts themselves fall into two broad categories, namely, table texts and problem texts. Table texts are tablets that list, for example, multiplication tables, reciprocal tables, and square tables (tables of squares, n^2) to assist in computation. Problem texts, on the other hand, have problems that are much like those in mathematics textbooks today. These texts were often used as "school texts" for training scribes.

The most famous mathematical cuneiform tablet, indeed perhaps the most famous mathematical artifact altogether, is *Plimpton 322*, so called because it is item number 322 of the Plimpton collection at Columbia University. (See FIGURE 2.3.) George A. Plimpton was a publisher from New York who purchased the tablet about 1922 from Edgar J. Banks, a well-known dealer, for the tidy sum of $10 [101]. Plimpton left his entire collection of historical mathematical books and artifacts to Columbia upon his death.

Figure 2.3: Plimpton 322. This clay tablet, believed to have been written about 1800 BC, shows a table of numbers written in cuneiform script. (Source: Rare Book and Manuscript Library, Columbia University.)

According to Banks, the tablet was from ancient Larsa, what is today Tell Senkereh in southern Iraq, just north of the ancient Sumerian city of Ur. The tablet, believed to have been written about 1800 BC, itself is quite small, measuring only about $3\frac{1}{2}$ inches by 5 inches by 1 inch, but it has been the subject of much speculation [17, 101]. It has a table of numbers laid out in fifteen rows and four columns, together with a heading for each column, although the top left-hand corner and a small bit along the right-hand edge of the tablet had chipped off, so the numbers there had to be reconstructed following the pattern of the other numbers. (Lines 2, 9, 11, 13, and 15 do not quite conform to the overall pattern of the numbers, but these discrepancies can be explained as copying mistakes by the scribe [17].) Once the numbers in the table had been deciphered, different theories for the origin of the numbers, as well as the purpose of the tablet, emerged.

The only forthright agreement has been that the right-most column, which contains the numbers 1 to 15 in succession, labels the rows. Beyond that, there have been three major interpretations of the contents of the tablet. Observing the pattern of the numbers through modern mathematical lenses, it has been proposed that [101]

1. The tablet is a form of a trigonometric table in which the second and third columns list, respectively, the short side and the hypotenuse of a right triangle, and the first column lists the square of the tangent or the square of the secant of the included angle.

2. The numbers were generated using positive integers p and q, with $p > q$ and having no common divisor besides 1. The numbers in the table were then found by computing $p^2 - q^2$, $2pq$, and $p^2 + q^2$.

3. The numbers in table were derived from reciprocal pairs of numbers, x and $\frac{1}{x}$, by finding the *average* and the *semidifference* of the pairs, respectively,

$$\frac{x + \frac{1}{x}}{2} \quad \text{and} \quad \frac{x - \frac{1}{x}}{2}.$$

Mathematically, each of these theories is plausible. They all relate to what we would now refer to as solving right triangles, of which there is ample evidence that the ancient Babylonians had knowledge. As Robson [101] points out, however, the theory that is to be believed "should not only be mathematically valid but historically, archaeologically, and linguistically sensitive too." We remark that Robson makes a good case for the third theory.

2.3 NUMBER SYSTEM

From about 3500 BC to the time of Christ, three peoples of the Mesopotamian valley—the Sumerians, the Akkadians, and the Babylonians—contributed to the beginnings of arithmetic and algebra [35]. Regarding their number system, apparently the Sumerians had developed the *sexagesimal* (base 60) system by 2350 BC, but it was not a positional system at that point. Calinger [35] offers several possible reasons for this development. Very early on, the Greek mathematician Theon of Alexandria[1] believed that the Sumerians eventually settled on base 60 because of the many numbers (positive integers) that divide into 60 evenly, thus making it easier to reckon with fractions. (To compare, observe that 10 is properly divisible by only 2 and 5, whereas 60 is properly divisible by 2, 3, 4, 5, 6, 10, 12, 15, 20, and 30.) In the last century, the German Assyriologist G. Kewitsch[2] posited that base 60 was a compromise between two peoples using two different number bases, namely, base 6 and base 10, but later the French Assyriologist François Thureau-Dangin[3] disputed Kewitsch, believing instead that base 60 may have been derived from base 12 instead of base 6.

After Sumer had fallen to the Akkadians in about 2270 BC, we find that the Akkadians adopted the Sumerian script for official or sacred writings (much in the way that Latin was used in medieval Europe). Over time, however, the symbols used in the Sumerian script lost their original meanings and became abstract symbols. These abstract number symbols, which evolved into Akkadian cuneiform, permitted more sophisticated computations.

Finally, in the Amorite period in the early second millennium BC, the Babylonians invented a positional or place-value system of numbers. They also reduced the number of numerals to basically two: a vertical wedge ⌐ for 1 and ⟨ for 10 [35].[4] In addition, there was a symbol for 100, namely, ⌐⊢. The Babylonians also used a circle for 10 and a semicircle for 1.

---

[1] *Commentaire de Théon d'Alexandrie sur le premier livre de la Composition Mathématique de Ptolémée* (éd. Halma), chap. IX.

[2] G. Kewitsch, "Zweifel an der astronomischen und geometrischen Grundlage des 60-Systems," *Z. f. Assyr. Arch.*, Vol. 18, No. 1 (1905), pp. 73–95.

[3] F. Thureau-Dangin, "Sketch of a history of the sexagesimal system," *Osiris*, Vol. 7 (1939), pp. 95–141.

[4] According to Florian Cajori [31, p. 2], "Grotefend believes the character for 10 originally to have been the picture of two hands, as held in prayer, the palms being pressed together, the finger close to each other, but the thumbs thrust out."

Ɨ	⟨	Ɨ⤙
one	ten	one hundred

Table 2.1: Babylonian number characters.

The Babylonian number system is basically additive. For example, referring to TABLE 2.1, Ɨ⤙⣿Ɉ is 143 by adding the values of all the symbols. Further, every ten Ɨ would be replaced by one ⟨. Thus, we may say that the Babylonian number system is a sexagesimal (base-60) positional system on top of a decimal (base-10) numeral system.

In base 60, the place values are powers of 60. The first few powers of 60 are

$$60^0 = 1, \quad 60^1 = 60, \quad 60^2 = 3600, \quad 60^3 = 216{,}000.$$

A space was used to separate place values. So, for example, ⟨ɈɈ may be thirteen (13×1), but ⟨ ɈɈ may be six hundred three ($10 \times 60 + 3 \times 1$), and ⟨ ɈɈ may be thirty-six thousand three ($10 \times 3600 + 0 \times 60 + 3 \times 1$). We say "may be" because early on the Babylonian number system lacked a symbol for a place holder, and it can be hard to tell how many spaces have been left between groups of symbols. (After all, the scribes were probably pressing quickly into the moist clay tablets, so there was not always uniformity in the writing.) So, not only can it be difficult to tell the difference between

$$⟨ \ ɈɈ \ (= 10 \times 60 + 3 \times 1) \text{ and } ⟨ \ ɈɈ \ (= 10 \times 3600 + 0 \times 60 + 3 \times 1),$$

but also ⟨ ɈɈ could be

- thirty-six thousand, one hundred eighty ($10 \times 3600 + 3 \times 60 + 0 \times 1 = 36{,}180$);

- two million, one hundred seventy thousand, eight hundred ($10 \times 216{,}000 + 3 \times 3600 + 0 \times 60 + 0 \times 1 = 2{,}170{,}800$);

- ten and one-twentieth ($10 \times 1 + 3 \times \frac{1}{60} = 10\frac{1}{20}$).

Indeed, the possibilities go on, to include the fractional versions. Here are a few more examples.

Example 2.1

one	Ɨ	two	ɈɈ
three	ɈɈɈ	four	ꜛɈ
five	ꜛꜜ	six	ꜛꜛ
seven	ꜛꜜɨ	eight	ꜛꜛꜛ
nine	ꜛꜜꜛ	ten	⟨
twelve	⟨ɈɈ	twenty-one	⟨⟨Ɨ
twenty-seven	⟨⟨ꜛꜜ	sixty	Ɨ
one hundred	Ɨ ⣿ or Ɨ⤙	one hundred one	Ɨ ⣿Ɨ or Ɨ⤙Ɨ
eight hundred fifty	⟨ꜛ ⟨	nine hundred ninety-nine	⟨ꜛ ⟨ꜛ

Finally, four thousand, three hundred seventy-nine is Ɨ ⟨ɈɈ ⟨⟨⟨ꜛ .   □

Now You Try 2.1 Write two possible decimal numbers and their word names for each of the following Babylonian numerals. Assume that there is at most one space between each grouping of symbols.

1. ⦁

2. ⦁

3. ⦁

4. ⦁

We see that the value of a Babylonian numeral is ambiguous without a symbol for a place holder. It turns out that the only way to tell what is the value of a particular Babylonian numeral is by the context in which it was stated. For instance, the numeral ⦁⦁⦁ written in the context of the number of children in a family probably is 3, whereas written in the context of the number of people in a village may be 180 (3×60).

Remark 2.1 We still sometimes have to tell the value of a numeral from its context. Consider the following excerpt about John Fixx, the son of running great Jim Fixx, from a 2008 article in the magazine *Runner's World*.[5]

> In college, in the heady period following the 1977 publication of his dad's unexpected best-seller, *The Complete Book of Running*, Fixx ran on the cross-country and track teams at Wesleyan University, and continued racing after graduation, achieving bests of 54:10 for 10 miles and 2:51 for the marathon.

In the first place, 54:10 is 54 minutes 10 seconds, and in the second place 2:51 is 2 hours 51 minutes (even though it was not written 2:51:00 with a place holder for the seconds). How were we to know that? By the context: Presumably it would take more time to run a marathon (26.2 miles) than it would to run 10 miles. □

Example 2.2 Express ten thousand, five hundred sixty-nine as a Babylonian numeral.

Note that the greatest base-60 place value that fits into the number is $60^2 = 3600$, and that the next smaller place values in sequence are $60^1 = 60$ and $60^0 = 1$. Hence, we divide 10,569 by 3600, and then divide the remainder of that division by 60 and, last, by 1.

$$
\begin{array}{r}
2 \\
3600\,\overline{)\,10569} \\
7200 \\
\hline
3369
\end{array}
\qquad
\begin{array}{r}
56 \\
60\,\overline{)\,3369} \\
3000 \\
\hline
369 \\
360 \\
\hline
9
\end{array}
\qquad
\begin{array}{r}
9 \\
1\,\overline{)\,9} \\
9 \\
\hline
0
\end{array}
$$

Therefore, ten thousand, five hundred sixty-nine is ⦁⦁ ⦁⦁⦁ ⦁⦁⦁⦁ as a Babylonian numeral. □

Now You Try 2.2 Express the following numbers as Babylonian numerals.

1. Eighty-three.

2. Two hundred nineteen.

3. Six thousand, seven hundred nine.

4. Three million, four hundred seventy-two thousand, six hundred thirteen.

□

---

[5] Amby Burfoot, "Grave concerns," *Runner's World*, Dec 2008, p. 116.

Beyond being additive, Cajori [31] tells of a multiplicative principle for numbers greater than 200 and gives as an example that ⟨⊤⊢ is 1000 (= 10 × 100). He also tells of a subtractive principle, and cites examples where 19 is written as ⟨⟨⊤⊤ (= 20 − 1), which, Cajori explains, uses the symbol "⊤⊢", for *LAL* or 'minus'...."

Regarding fractions, the Babylonians simply continued with sexagesimal place values just as we normally do with decimals. See FIGURE 2.4.

In base 10, the fraction place values are reciprocals of powers of 10. The first few reciprocals of powers of 10 are

$$10^{-1} = \frac{1}{10^1} \text{ or } \frac{1}{10}, \quad 10^{-2} = \frac{1}{10^2} \text{ or } \frac{1}{100}, \quad 10^{-3} = \frac{1}{10^3} \text{ or } \frac{1}{1000}.$$

In this way, for example, 3.14 is three and fourteen-hundredths because

$$3.14 = 3 \times 1 + 1 \times \frac{1}{10} + 4 \times \frac{1}{100} = 3\frac{14}{100}.$$

In base 60, the fraction place values are reciprocals of powers of 60. The first few reciprocals of powers of 60 are

$$60^{-1} = \frac{1}{60^1} \text{ or } \frac{1}{60}, \quad 60^{-2} = \frac{1}{60^2} \text{ or } \frac{1}{3600}, \quad 60^{-3} = \frac{1}{60^3} \text{ or } \frac{1}{216\,000}.$$

one-tenth	=	$\frac{1}{10}$	⊤⊤⊤ / ⊤⊤⊤	=	$\frac{6}{60}$
one-half	=	$\frac{5}{10}$	⟨⟨⟨	=	$\frac{30}{60}$
one-fourth	=	$\frac{25}{100}$	⟨⊤⊤⊤/⊤⊤	=	$\frac{15}{60}$

Figure 2.4: Examples of fractions expressed as Babylonian numerals.

Now, the Babylonians not only lacked a place holder symbol, but they also lacked a notation to separate the whole number part of a number from the fraction part. (In the United States, we use a decimal point as in 3.14, but in other parts of the world, such as in Germany and France, people use a decimal comma as in 3,14. So, the choice of notations for a separatrix is not unique and may continue to evolve.) So, for example, ⟨ ⊤⊤⊤ could be (in expanded form)

- $10 \times 1 + 3 \times \frac{1}{60} = 10\frac{3}{60}$ or $10\frac{1}{20}$.

- $10 \times \frac{1}{60} + 3 \times \frac{1}{3600} = \frac{603}{3600}$ or $\frac{67}{400}$.

- $10 \times 60 + 0 \times 1 + 3 \times \frac{1}{60} = 600\frac{3}{60}$ or $600\frac{1}{20}$.

Now You Try 2.3 Write two possible proper fractions or mixed numbers, as in the example above, for each of the following Babylonian numerals. Assume that there is at most one space between each grouping of symbols. You may write the numbers in expanded form.

1. ⟨ ⊤⊤⊤   2. ⟨⟨⟨⊤⊤⊤   3. ⟨⟨ ⊤⊤⊤

□

Regarding reciprocals, whereas we would express the reciprocal of 12, for example, as the fraction $\frac{1}{12}$, because $12 \times \frac{1}{12} = 1$, expressed as Babylonian numerals, the reciprocal of ⟨⊤⊤ is ⊤⊤⊤/⊤⊤. Now, at first glance, this would seem to be saying that the reciprocal of 12 is 5, which, of course, is not true because $12 \times 5 = 60$ and not 1; however, from the context ("the reciprocal of 12"), one would understand ⊤⊤⊤/⊤⊤ to be the fraction $\frac{5}{60}$: $12 \times \frac{5}{60} = 1$. (Note also that $\frac{5}{60}$ reduces to $\frac{1}{12}$.) Thus, we make the following observation.

To express the reciprocal of a nonzero number a as a Babylonian numeral, find a number b such that ab is a power of 60. This is because the decimal reciprocal of a is $\frac{1}{a}$ and

$$\frac{1}{a} = \frac{1}{a} \cdot \frac{b}{b} = \frac{b}{ab} = \frac{b}{60^n}.$$

Thus, we see the reason why Theon of Alexandria believed that the Sumerians eventually settled on base 60: the many positive integers that divide into 60 evenly make it easier to reckon with fractions.

Now You Try 2.4 Express the following numbers as Babylonian numerals.

1. The reciprocal of 2.

2. The reciprocal of 15.

3. The reciprocal of 20.

4. The reciprocal of 45.

$\square$

Think About It 2.1 Is it always possible to express the reciprocal of a number as a sexagesimal fraction? What is the reciprocal of fourteen as a sexagesimal fraction? $\square$

Remark 2.2 Note that the prime factors of 60^n are 2, 3 and 5. Thus, if a number a has any other prime factor, then there is no number b such that ab is a power of 60. For example, $14 = 2 \cdot 7$; thus, there is no number b such that $14b$ is a power of 60. In this case, using only sexagesimal fractions, the reciprocal of 14 would be a nonterminating expression.

We relate this to what is familiar to us: decimal fractions. Suppose that we were to express all fractions as decimal fractions only, that is, as fractions with denominators that are powers of 10 only. For example, $\frac{5}{10}$ for one-half, $\frac{25}{100}$ for one-fourth, and so on. Using only decimal fractions, the reciprocal of a number that has any other prime factor would be a nonterminating expression because the prime factors of 10^n are 2 and 5. For example, $30 = 2 \cdot 3 \cdot 5$; thus, the reciprocal of 30 is a nonterminating expression, namely,

$$\text{one-thirtieth} = \frac{3}{100} + \frac{3}{1000} + \frac{3}{10\,000} + \cdots = 0.0333\ldots,$$

and the fractions $\frac{3}{100}, \frac{33}{1000}, \frac{333}{10\,000}, \ldots$, no matter how far out we go, would only be approximations of one-thirtieth. $\square$

Two place holder symbols, $\nearrow\hspace{-1mm}\nearrow$ and $\smallblacktriangle$, finally appeared during the Seleucid period (323–63 BC). These place holders, however, were used to indicate spaces *between* place values only and were never used at the end of a numeral. Thus, for example, $\langle\nearrow\hspace{-1mm}\nearrow|||$ would definitely denote $10 \times 60^{n+2} + 0 \times 60^{n+1} + 3 \times 60^n$, but there still remained the small problem that the number could be

$$\overline{10 \times 60^2 + 0 \times 60^1 + 3 \times 60^0} \quad \text{or}$$

$$\overline{10 \times 60^3 + 0 \times 60^2 + 3 \times 60^1 + 0 \times 60^0} \quad \text{or}$$

$$\overline{10 \times 60^4 + 0 \times 60^3 + 3 \times 60^2 + 0 \times 60^1 + 0 \times 60^0},$$

and so on. These three examples illustrate numerals for which there is no trailing zero, one trailing zero, or two trailing zeros, respectively. Later, the Alexandrian astronomer Ptolemy in the second century AD began to use the Greek letter *omicron*, o (the first letter of the Greek word $o\upsilon\delta\epsilon\nu$ that means "nothing"), or a small circle, $\circ$, not only as a place holder between place values but also at the end of numerals [21, 41].

A notation to separate the whole number part of a number from the fraction part, that is, a "sexagesimal point" (like the decimal point we use) never appeared in cuneiform numerals.

Chapter 3

Egyptian and Roman Number Systems

3.1 EGYPTIAN

Just as the Euphrates and the Tigris Rivers allowed the regions in the Mesopotamian valley to flourish some 5000 years ago, so did the Nile River allow Egypt to flourish.

3.1.1 History

The land of Egypt (FIGURE 3.1), situated in a parched desert that provided excellent natural protection from foreign invasions, not only depended on the Nile as a source of water and food but also depended on the annual flooding of the Nile to bring fertile black mud onto the land to make it very suitable for farming. This allowed the Egyptians to grow an abundance of crops. After the flood waters would recede and the planting season began, the Egyptians used a system of canals to direct water from the river to irrigate the fields. The Nile also provided a means of transport and communication for the Egyptians. Indeed, the Nile River proved to be so important to Egyptian civilization that the Greek Herodotus, "the father of history," called Egypt "the gift of the Nile."

As in Mesopotamia, having an abundance of crops meant that not everyone had to work the land, so that classes of specialists such as architects, carpenters, and priests emerged [35]. Of particular import to mathematics were the palace and priest scribes, who spent their time not only solving practical problems in mathematics—irrigation, astronomy, calendars, taxes, wages, interests on loans—but also contemplating mathematical problems that were not necessarily intended to address any practical purpose. Documents suggest that it was this "play element" that helped advance their mathematical knowledge. The scribes enjoyed great prestige, and, indeed, Aristotle, in his *Metaphysics*, attributed the origins of the mathematical sciences in Egypt to the leisure available to them [35].

The history of ancient Egypt spans from about 3000 BC, when it was united as one country under the first pharaoh, Menes, to about AD 30, when it was conquered by first the Greeks of Alexander and then ultimately the Romans. The three main eras of ancient Egypt have come to be called the Old Kingdom (about 2600–2100 BC), when the Great Sphinx and the monumental pyramids were built; the Middle Kingdom (about 2000–1700 BC), when pyramid building declined; and the New Kingdom (about 1550–1050 BC), when pyramids were replaced by tombs in the Valley of the Kings. As you may guess, ancient Egyptians were very interested in tending to life after death. Predynastic Egypt existed before the Old Kingdom, and other lesser periods occurred in between the Kingdom periods, as well.

From the building of the Great Sphinx and the pyramids, it is not difficult to imagine that the Egyptians had used some rather sophisticated mathematics. Mathematics was used to mark again the boundaries of plots of land that would be washed away by the annual flooding of the Nile. Of course, mathematics was needed for the accounting of the people, livestock, and so on, and the levying of taxes. But while the Egyptians were able to solve many different mathematical problems, some of which we shall see later,

Figure 3.1: Ancient Egypt. (Source: © Thinkstock.)

their solutions were specific to each problem, and there is nothing to suggest that they generalized the concepts.

3.1.2   Writing and Mathematics

There were three types of Egyptian writing: *hieroglyphic* script, *hieratic* script, and *demotic* script. Hieroglyphic script, which dates as far back as 3300 BC, is the system of pictographs that you may have seen in pictures carved into monuments, like the walls of a pyramid, whereas hieratic script is a cursive writing that was written on papyrus, leather, wood, and so on, using ink and a reed pen. Hieroglyphic script generally would be carved from right to left, although it was carved in any convenient direction. As a consequence, some pictographs appeared reversed in some carvings, but this did not pose a problem because the writing was very standardized. Hieratic script, on the other hand, was always written from right to left, but it could be difficult to read because it very much depended on the handwriting of the scribe. Demotic script was an abbreviated form of hieratic that appeared around 700 BC.

Just as the decipherment of Babylonian cuneiform had to wait until the complete translation of the Behistun Cliff inscriptions, the decipherment of the cryptic Egyptian writing also eluded archaeologists and Egyptologists for a very long time until the discovery of the Rosetta Stone in 1799 in Egypt during the French occupation [35]. The Rosetta Stone (FIGURE 3.2) is a basalt stone with three identical descriptions inscribed in Greek, hieroglyphic, and demotic scripts. With this key, the secrets of the ancient Egyptian scripts were unlocked in 1822 by Jean-François Champollion. The Rosetta Stone currently sits in the British Museum, where it has been since the British relieved the French of it in 1801.

Figure 3.2: The Rosetta Stone. Discovered in Egypt in 1799, it is a basalt stone with three identical descriptions inscribed in Greek, hieroglyphic, and demotic scripts. (Source: © Thinkstock.)

Hieroglyphic carvings did not tell us much about Egyptian mathematics: they mostly described calendars, accounting, and the like. Egyptian mathematics was to be found, instead, in the hieratic scrolls. Unfortunately, unlike the Babylonian clay tablets, Egyptian scrolls tended to decompose over time. Fortunately, some scrolls have survived. Annette Imhausen [70] lists the scrolls that have been found that contain Egyptian mathematics. They are the

- *Rhind Mathematical Papyrus* (BM 10057–10058);

- *Lahun Mathematical Fragments* (7 fragments: UC32114, UC32118B, UC32134, UC32159–32162);

- *Papyrus Berlin* 6619 (2 fragments);

- *Cairo Wooden Boards* (CG 25367 and 25368);

- *Mathematical Leather Roll* (BM 10250);

- *Moscow Mathematical Papyrus* (E4674);

- *Ostracon Senmut* 153;

- *Ostracon Turin* 57170.

Imhausen remarks that, "Most of these texts were bought on the antiquities market, and therefore we do not know their exact provenance. An exception is the group of mathematical fragments from Lahun, which were discovered by William Matthew Flinders Petrie when he excavated the Middle Kingdom pyramid town of Lahun."

The most prominent of the extant mathematical texts from ancient Egypt is the Rhind Mathematical Papyrus [34], named after Scottish antiquary A. Henry Rhind, who bought the scroll in the Nile resort town of Luxor in 1858 [15, 41]. It measures about 18 feet long and 1 foot wide. Historically called the Rhind Papyrus, it is now also referred to as the Ahmes Papyrus or A'hmosè Papyrus after the Egyptian scribe who copied it about 1650 BC. A'hmosè attributed the material to some document from the Middle Kingdom about two hundred years prior. The papyrus begins, "Correct method of reckoning, for grasping the meaning of things and knowing everything that is, obscurities and all secrets," which is often used as its title. The Ahmes contains 87 mathematics problems, mostly of a practical nature, but some, for instance problems 26 and 27, belong to the small group of "aha" problems ("aha" in Egyptian refers to the unknown quantity) that are purely mathematical. (See examples 19.5 and 19.6.) Only part of the Ahmes has been found. The part bought by Rhind has been at the British Museum since 1865, while another section was discovered in the Egyptian collection of the New York Historical Society in 1922 and is currently at the Brooklyn Museum of Art [41].

Another of the more prominent extant Egyptian texts is the Moscow Mathematical Papyrus, which contains 25 problems. (See FIGURE 3.3.) It can be found at the Moscow Museum of Fine Arts. We will see examples from the Ahmes and the Moscow papyri later on.

Figure 3.3: A portion of the Moscow Mathematical Papyrus. (Source: U.S. public domain.)

3.1.3 Number System

The Egyptian number system is a base-10 system, but Cajori [31, p. 11] tells us that "traces of other systems, based on the scales of 5, 12, 20, and 60, are believed to have been discovered." TABLE 3.1 shows the Egyptian hieroglyphs used for the powers of 10 from one to ten million. Unlike the Babylonian number system, however, the Egyptian number system was not a positional system. It was an additive system, so that each symbol was repeated as many times as needed with the rule that ten of any symbol would be replaced by one of the next higher base value. Also, for ease of reading, no more than four of the same symbol would be grouped together, and, in the case when more than four of the same symbol were needed, the larger group would be written to the left of, or above, the smaller group [31]. Moreover, numerals were usually written with the symbols of larger values preceding, or to the right of, those of lower values, since the Egyptians wrote from right to left. Here are a few examples.

| (staff or stroke) one	**∩** (heel bone or loaf) ten	**ℙ** (coiled rope or snake) hundred	**↥** (lotus flower) thousand
∫ (bent or pointed finger) ten thousand		**↰** (burbot or tadpole) hundred thousand	
𓁿 (astonished man) million		**◯** (rising sun) ten million	

Table 3.1: Egyptian hieroglyphic number characters.

Example 3.1

one			two	||
three	|||	four	||||
five	||| ||	six	||| / |||
seven	|||| / |||	eight	|||| / ||||
nine	||| || / ||||	ten	∩
twelve	||∩	twenty-one	|∩∩
twenty-seven	|||| ∩∩ / |||	sixty	∩∩∩ / ∩∩∩
one hundred	ℙ	one hundred one	|ℙ
eight hundred fifty	∩∩∩ℙℙℙℙ / ∩∩ ℙℙℙℙ	nine hundred ninety-nine	||| ||∩∩∩ ∩∩ℙℙℙ ℙℙ / |||| ∩∩∩∩ ℙℙℙℙ

Finally, four thousand, three hundred seventy-nine is ||| ||∩∩∩∩ / |||| ∩∩∩ ℙℙℙ↥↥↥↥ .   ☐

Now You Try 3.1

1. Write the decimal number and the word name for the following Egyptian hieroglyphic numerals.

   a) |||||∩∩ℙℙℙ

   b) |||∩∩↥ ↰ 𓁿 𓁿 𓁿◯

2. Write the following numbers using Egyptian hieroglyphic numerals.

   a) Eighty-three.
   b) Two hundred nineteen.
   c) Six thousand, seven hundred nine.
   d) Three million, four hundred seventy-two thousand, six hundred thirteen.

   ☐

According to Cajori, a multiplicative principle also came into use around 1600–1200 BC. He gives the following two examples.

| one hundred and twenty thousand |  ⌡ ∩∩ᶜ | 120×1000 |
| two million, eight hundred thousand |  ᵼ ∩∩ | $28 \times 100{,}000$ |

In the first example, we see that a nonstandard (to the left) placement of the glyph ⌡ for thousand implies multiplication by 1000. In the second example, we see that the placement of the glyph ᵼ for hundred thousand above the numeral implies multiplication by 100,000.

Now You Try 3.2 Write the following numbers using Egyptian hieroglyphics employing a multiplicative principle.

1. Two hundred forty thousand.

2. One million, six hundred thousand.

☐

Regarding fractions, with the exception of two-thirds, the ancient Egyptians used *unit fractions* almost exclusively, that is to say, fractions with numerator 1. Generally, the hieroglyph for a fraction would be the symbol � placed over the numeral for the denominator. For example, ⌡⌡⌡⌡⌡ is one-fifth, ⌢ is one-tenth, and ⌢⌡⌡ is one-twelfth. One-fourth, one-half, and two-thirds, however, which were among the most commonly used fractions, had their own symbols [70].

×	⊏	⍑
one-fourth	one-half	two-thirds

The fraction two-thirds is the only non-unit fraction used by the Egyptians.

Think About It 3.1 Why would two-thirds be of special use or meaning for the Egyptians?   ☐

Many glyphs changed over time. For example, during the Old Kingdom, two-thirds was written ⍑ but as ⍑ and ⍑ over the course of the next two thousand years [112]. All non-unit fractions other than two-thirds, like two-ninths, were generally written as sums of unit fractions.

 ⌢⌢∩	$\frac{1}{6} + \frac{1}{18} = \frac{2}{9}$
two-ninths	

Now, there are many ways to decompose a fraction into a sum of unit fractions. In general, however, there were four basic rules that were followed to decide how to write non-unit fractions.

1. Use four or fewer fractions.

2. Use even denominators.

3. Use a larger denominator for the first fraction.

4. Do not repeat a fraction.

Now You Try 3.3 Find the value of each fraction.

1.  ⌢⌡⌡⌡ ⌢∩

2.  × ⌢⌡⌡∩

3.  ⌢∩∩∩ ⌢∩∩ ⌢∩∩∩∩∩

4.  ⊏ ⌢∩

☐

3.2   ROMAN

After the Mesopotamians and the Egyptians, the Greeks were the next people in the region of the Mediterranean to advance mathematics. In fact, beginning with Thales of Miletus (ca. 625–547 BC), who is generally regarded as the father of modern mathematics, the subject moved from being one of intuition and empiricism to one of statements and proofs. (It is likely that Thales learned mathematics during his travels to Egypt and Mesopotamia.) This would continue to the "Golden Age of Greek Mathematics" that was dominated by Euclid (323–285 BC), Archimedes (287–212 BC), and Apollonius of Perga (262–190 BC), and centered about the Museum of Alexandria in Egypt. But the impressive advances in mathematics by the Greeks would come to an end with the expansion of the Roman Empire (FIGURE 3.4). It was not until the "Silver Age of Greek Mathematics" that there was a brief resurgence of Greek mathematical thought with advances by Diophantus of Alexandria (fl. AD 250) and Pappus of Alexandria (early fourth century AD).

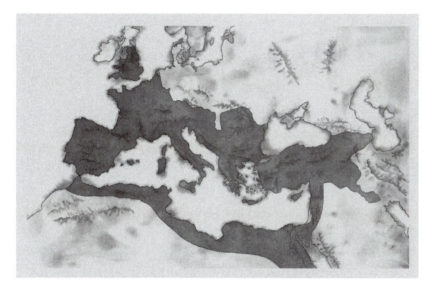

Figure 3.4: The Roman Empire. (Source: © Thinkstock.)

3.2.1   History

The historian Varro (d. 27 BC) put the founding of Ancient Rome in 753 BC [35]. According to legend, the city was founded by the twin brothers Romulus and Remus, who were the grandsons of the Latin King Numitor of Alba Longa. Numitor was overthrown by his brother, Amulius, when the twins were born. Because the twins were conceived when their mother, Rhea Silvia, was raped by the god Mars, the twins were half divine. Fearing that the half-divine twins would take back the throne, Amulius attempted to have the twins drowned. However, the twins were rescued and then raised by a she-wolf, and they ultimately recaptured the throne for Numitor. Subsequently, the twins set off to found their own city that was to become Rome. It was during this time that Romulus killed Remus in a quarrel over who was to be king and after whom the city was to be named. Or so this legend goes.

Rome was captured by the Etruscans in the sixth century BC. In 509 BC, the Romans expelled the Etruscans and formed a republic. Roman society was then composed of patricians (noblemen by birthright), plebeians (commoners who were merchants, small farmers, artisans, and so on, including the urban poor), and slaves. The motto of the republic was *Senatus Populusque Romanus* (Senate and People of the Romans), of which everyone was reminded by its representation on coins by the letters *SPQR*. The Romans valued law and order and patriotism and working for the common good. Over time, the Romans expanded their sphere of influence. In the third century BC, they built a fleet and made their first venture abroad. Between 265 and 146 BC, Rome conquered Carthage in three bloody Punic Wars; Corinth also fell to Rome. In the years that followed, Rome would rule all of the Mediterranean and as far away as Scotland, North Africa, Mesopotamia, and India.

But things were not all well at home. Existing on plunder, taxes, and slavery, Rome began to experience internal turmoil. Long military expeditions meant that many farmers were not able to tend their farms because they served in the military. Wealthy aristocrats began to buy up these farms and work them with slaves that were acquired through military conquests. Slaves were also widely used in the cities by merchants and artisans and households. The use of slaves drove up unemployment among the Romans even as manual labor came to be looked down upon. The rich grew richer and the poor grew poorer. Civil unrest would ensue. In 49 BC, Julius Caesar, in a civil war, restored order and declared himself dictator for life of Rome. This ended the Roman Republic.

It was a short reign, however, for Julius Caesar was assassinated in 44 BC on the Ides of March by a group of senators ("*Et tu, Brute?*"). After this, imperial Rome was ruled by Marc Antony, together with Cleopatra, in the east, and by Octavius Caesar, the grandnephew and adopted son of Julius Caesar, in the west. Before long, however, Marc Antony and Octavius Caesar clashed, resulting ultimately in the defeat of Antony in the Battle of Actium in 31 BC. Following this, Antony and Cleopatra committed suicide, and Octavius Caesar became the sole ruler of imperial Rome. The Roman senate would later bestow upon Octavius Caesar the title Augustus, thereby making him the first emperor of Rome. Thus came about the Roman Empire.

Augustus Caesar ruled from 27 BC to AD 14, and ushered in a period of history that is referred to as the *Pax Romana*, a period of peace that lasted until about AD 180. It was a peace over a very large geographical area that was unprecedented and apparently has not been seen since. It was also a period during which the Christian church was established and began to spread across the Roman Empire. At first largely ignored, Christians were soon persecuted, beginning under Emperor Nero (who ruled from AD 54–63); reached their height of popularity under Emperor Diocletian (who ruled from AD 284–305); and become a legal religion under Constantine the Great (Constantine I, who came to the throne in 312 and himself converted to Christianity). In fact, in 392 Emperor Theodosius prohibited any pagan practices, even in the privacy of one's home, and, by the time of his death in 395, the empire became officially Christian.

Diocletian realized that the empire was too large an area to be governed by one person. He thus divided the empire in two, effectively creating the Western Roman Empire and the Eastern Roman Empire, and made Maximian co-emperor under the title Augustus. The Eastern Roman Empire came also to be called the Byzantine Empire. (This division later gave rise to the Latin Rite or Roman Catholic Church and the Eastern Rite or Byzantine Catholic Church, as well as the Eastern Orthodox Church.) While Latin was the official language of the whole Roman Empire, it was the lingua franca only of the west, while the lingua franca and the administrative language of the east was Greek. In the coming years, Rome would face increasing problems in the west caused by Germanic tribes. The problems eventually led to the Western Roman Empire being overtaken by the Germanic peoples. The fifth century AD would see the decline of the Western Roman Empire.

In 330 Emperor Constantine the Great established the new Christian capital city of Constantinople (current-day Istanbul, Turkey) on the site of the old Greek city of Byzantium. While the Western Roman Empire was being besieged, the Eastern Roman Empire around Constantinople remained isolated from the conflicts in the west. As such, the Eastern Roman Empire lasted for a millennium after the fall in the west, until the sacking of Constantinople in 1453 during the Fourth Crusade against the expansion of the Ottoman Empire. By this time, however, Roman influence had left an indelible mark on western civilizations to follow.

Many of the structures that the Romans built—like their arched bridges, aqueducts, roads, public baths, and the Coliseum—are still around today in great part due to their extensive use of concrete, which was a new building material. (See FIGURE 3.5.) Yet, despite these impressive physical structures, the time of the Roman expansion is one during which there was a dearth of advances in mathematics. Although Greek influence is evident in many aspects of Roman culture, Greek interest in and high regard for mathematics were lost on the Romans. Indeed, Calinger [34] notes that "the Romans did not produce one eminent

Figure 3.5: A Roman aqueduct. Many of the structures that the Romans built—like their arched bridges, aqueducts, roads, public baths, and the Coliseum—are still around today. (Source: © Thinkstock.)

mathematician in their thousand-year history." To the Romans, mathematics was a tool to an end, and spending time on theory was folly, and even unpatriotic, because it did not apparently contribute to the common good. This view is described by the Roman orator Cicero in his work *Tuscalan Disputations* [21]:

> The Greeks held the geometer in the highest honor; accordingly nothing made more brilliant progress among them than mathematics. But we have established as the limits of this art its usefulness in measuring and counting.

(G. H. Hardy, a prominent mathematician of the early twentieth century, held a view diametrically opposite to the Romans', opining in *A Mathematician's Apology* [66] that the only beautiful mathematics is that which is devoid of application. Ironically, the mathematics Hardy most loved, number theory, has turned out to be the basis of data encryption, a field that today is vitally important not only to national security but also to commerce, for example.) The mathematical books that were written were basically technical manuals. The best known technical manual is the ten-book collection *On Architecture* by Vitruvius Pollio (d. ca. 25 BC) that was dedicated to his patron, Caesar Augustus. Yet, *On Architecture* was hardly a mathematical treatise, for, as Calinger [35] remarks, "Although Vitruvius attempts to encompass architectural theory and practice, he presents the theory of only a few Greek architects, whose ideas he had not satisfactorily assimilated." Fortunately, despite the dearth of original research during these times, copyists in the Byzantine Empire preserved many of the scientific and literary texts of antiquity, much to our benefit today.

3.2.2   Number System

The Roman number system gives us another example of a system that is nonpositional. You are probably familiar with the Roman numerals.

I	V	X	L	C
one	five	ten	fifty	hundred
D			M	
five hundred			thousand	

Today we recognize these numerals as letters of the Roman or Latin alphabet: eye, vee, ex, ell, see, dee, and em. We even write them using lowercase letters (especially as page numbers). However, Moritz Cantor tells us that "the later Roman signs, I, V, X, L, C, M ...from their resemblance to letters transformed themselves by popular etymology into these very letters" [31]. In other words, these numerals were not originally chosen from the letters of the Roman or Latin alphabet. More likely the later Roman numerals evolved from the old Roman numerals, the latter of which were similar to the Etruscan numerals that were letters of the Etrurian alphabet. (The Greek and Hebrew numerals, for example, were the actual letters of their respective alphabets.) For instance, the Etrurian V or ∧ and the old Roman V for five; the Etrurian X or + and the old Roman X for ten; the Etrurian ↑ or ↓ and the old Roman ⊤ or ↓ or ⌐ or ⊥ or L for fifty.

The Roman number system generally uses the additive principle. Thus, generally the value of a numeral is the sum of the values of the symbols that compose the numeral. Numbers are written left to right, as our numbers are, with the larger-value numerals proceeding or to the left of lesser value ones. Here are a few examples.

Example 3.2

one	I	six	VI
two	II	seven	VII
three	III	eight	VIII
twelve	XII	twenty-one	XXI
twenty-seven	XXVII	sixty	LX
eight hundred fifty	DCCCL	two thousand eight	MMVIII

□

A subtractive principle is also used. Specifically, if a symbol of lesser value is written before a symbol of greater value, then the value of the numeral is obtained by subtracting the smaller value from the greater. (Recall that the Egyptian system is not subtractive.) Here are a few examples.

Example 3.3 The subtractive principle.

four	IV	$5 - 1$	nine	IX	$10 - 1$
forty	XL	$50 - 10$	ninety	XC	$100 - 10$

Both the additive and the subtractive principles can be used together.

fourteen	XIV	$10 + (5 - 1)$
forty-three	XLIII	$(50 - 10) + 1 + 1 + 1$
sixty-nine	LXIX	$50 + 10 + (10 - 1)$
ninety-four	XCIV	$(100 - 10) + (5 - 1)$

□

Now You Try 3.4

1. Write the following numbers using Roman numerals.

   a) Thirteen   b) Sixteen   c) Eighty-three

   d) Four thousand, three hundred seventy-nine

2. Write word names for the following Roman numerals.

a) MDCCLXXVI	c) MCMXLV	e) XXCIII
b) MCDXCII	d) LXXXIII	f) IIIIX

☐

The subtractive principle is also built into the language. For example, in Latin, eighteen is *duodeviginti* ("two from twenty") and nineteen is *undeviginti* ("one from twenty"). However, the subtractive principle was not always applied uniformly to numerals. For example, Cajori [31, p. 31] quotes the following remarks by Adriano Cappelli.

> The well-known rule that a smaller number, placed to the left of a larger, shall be subtracted from the latter ...was seldom applied by the old Romans and during the entire Middle Ages one finds only a few instances of it. The cases I have found belong to the middle of the fifteenth century and are all cases of IX, never IV, and occurring more especially in French and Piedmontese documents.

So, the subtractive principle that is common to us today was not always commonly used, even by the Romans. Further, Cappelli tells us that

> Walther, in his *Lexicon diplomaticum*, Göttingen, 1745–47, finds the notation LXL = 90 in use in the eighth century. On the other hand, one finds, conversely, the numbers IIIX, VIX with the meaning of 13 and 16, in order to conserve, as Lupi remarks, the Latin terms *tertio decimo* and *sexto decimo*.[1]

We note that *tertio decimo* in Latin is literally "three ten" (thirteen) and *sexto decimo* is "six ten" (sixteen). Finally, Cajori tells us that L. C. Karpinski found on some tombstones and a signboard from 130 BC the numeral XXCIII instead of LXXXIII for eighty-three.

Like the Egyptian number system, the Roman number system also used a multiplicative principle, but using a variety of notations. Cajori [31] gives the following examples from an assortment of sources.

five thousand	VM	5×1000 (not $1000 - 5$)
eighty-three thousand	LXXXIII . M	. M $= \times 1000$
ninety-two thousand	XCII . M	92×1000
ten thousand	$\overline{\text{X}}$	line above $= \times 1000$
one hundred twenty thousand	$\overline{\text{CXX}}$	120×1000

Note that in the first example above, VM does not represent 995 $(1000 - 5)$ because V is not the next lower-value numeral from M; on the other hand, CM would represent 900 $(1000 - 100)$ because C is the next lower-value numeral from M. Cajori gives the following examples of the use of a multiplicative principle.

- $|\overline{\text{X}}|$CLXXXDC:

  one million, one hundred eighty thousand, six hundred

  - lines above and on the sides (or only on the sides) $= \times 100,000$, so $|\overline{\text{X}}|$ (or $|\text{X}|$) $= 10 \times 100,000$ or $\boxed{1,000,000}$

  - $\overline{\text{CLXXX}} = 180 \times 1000 = \boxed{180,000}$

  - DC $= \boxed{600}$

---

[1]*Lexicon Abbreviaturarum* (Leipzig, 1901), p. xlix.

- $\overset{\text{M}}{\text{IIII}}$ $\overset{\text{C}}{\text{IIII}}$ LXXIII:

  four thousand, four hundred seventy-three

- IIII^{xx} :

  eighty; the small xx written like a modern exponent indicates multiplication by 20

- VI^{xx}XI :

  one hundred thirty-one

Now You Try 3.5 Write the following numbers using Roman numerals employing a multiplicative principle.

1. Sixty thousand.

2. Four thousand, three hundred seventy-nine.

3. One million, four hundred seventy-two thousand, six hundred thirteen.

☐

Besides written numerals, the Romans also used finger numerals [35, 119]. Finger numerals were most likely used in the noisy marketplaces as a means for trading over the din. Regarding fractions, the Romans used duodecimal (base-12) fractions. This is very likely because the fractions that were, and are, most commonly used ($\frac{1}{4}$, $\frac{1}{3}$, $\frac{1}{2}$, $\frac{2}{3}$, and $\frac{3}{4}$) are integral parts of 12. Even today we use duodecimal fractions in some instances, for example, to partition one foot into 12 inches, and to partition one day or one night into 12 hours. Their preference is reflected in their language, in which there are special names and symbols for the duodecimal fractions from $\frac{1}{12}$ to $\frac{12}{12}$ or 1 [31]; see TABLE 3.2.

uncia	$\frac{1}{12}$	–	septunx	$\frac{7}{12}$	S –
sextans	$\frac{1}{6}$	=	bes	$\frac{2}{3}$	S =
quadrans	$\frac{1}{4}$	= –	dodrans	$\frac{3}{4}$	S = –
triens	$\frac{1}{3}$	= =	dextans (decunx)	$\frac{5}{6}$	S = =
quincunx	$\frac{5}{12}$	= = –	deunx	$\frac{11}{12}$	S = = –
semis	$\frac{1}{2}$	S	as	1	I

Table 3.2: Word names for Roman duodecimal fractions.

Cajori explains that the Roman *as* was originally a copper coin that weighed one pound, and that was divided into twelve *unciae*. Also, the symbol · was alternatively used for –, and the symbol : alternatively for =. Additionally, even though each duodecimal fraction had its own name, the fraction $\frac{5}{6}$, for instance, would be commonly referred to as *semis et triens* ($\frac{1}{2} + \frac{1}{3}$) instead of *decunx* ($\frac{10}{12}$). Finally, despite the Romans' preference for duodecimal fractions, they did also use other bases for fractions. For example, silver was measured in tenths (decimal) of a *denarius*.

Using mixed bases is not so unusual, actually. Even today, in the United States, while decimal (base-10) fractions are generally preferred, bolts and wrenches, for example, are measured in sixteenths (hexadecimal) of an inch; and, until recently, mortgage rates and stock prices were reported in eighths. In Great Britain, for example, until the introduction of the pound sterling in 1971, from the time of King Henry VII (reigned AD 1485–1509), British currency was both base 12 and base 20, for there were 12 pence in one shilling

(12d = 1s) and 20 shillings in one pound (20s = £1). As another example, in counting time today, it is common for seconds, minutes, and hours to be expressed in base 60 (60 seconds = 1 minute; 60 minutes = 1 hour), but fractions of a second in base 10 (tenths of a second, hundredths of a second, and so on).

Chapter 4

Chinese Number System

4.1 HISTORY AND MATHEMATICS

China represents the mysterious Orient to many, and myths and legends of China certainly abound. China has a long and storied history. To fix some dates, we note that

- Sun Tzu wrote *The Art of War* in the sixth century BC.

- Confucius, whose philosophy has come to be woven into Chinese thought, lived between 554 and 479 BC.

- Laozi (Lao-Tzu), who founded Daoism (Taoism), lived around the fourth century BC.

- The Great Wall of China was built, rebuilt, and maintained between the fifth century BC and the sixteenth century AD.

- Ts'ai Lun is credited with inventing paper around AD 105, although recent archaeological findings place the invention some 200 years earlier.

- Gunpowder was discovered in China in the ninth century AD.

- The Mongol invasion of China occurred in the thirteenth century.

- Marco Polo (1254–1324) left Venice with his father and an uncle in 1271 to travel to China along the Silk Road; they arrived in 1275 and were welcomed by Kublai Khan (1214–1294). Kublai Khan, the grandson of Ghenghis Khan (originally Temujin; 1167–1227), founded the Yuan Dynasty (1279–1368).

- Jesuit missionaries began to arrive during the late Ming period (1368–1644).

- The Tiananmen Square protests took place in 1989.

Much of China's history is marked by dynastic rule, beginning with the Xia Dynasty (2205–1766 BC) until the Qing Dynasty (AD 1644–1911). During this long period, the borders of China were not fixed. (See FIGURE 4.1.) Finally, the fall of the Qing, the last dynasty of China, in the Xinhai Revolution saw the founding of the Republic of China with Sun Yat-sen as the head of the newly established Chinese Nationalist Party or Kuomintang (or Guomindang; 1912–1927); however, China was fragmented during this period. Eventually, the Republic of China was consolidated under the leadership of Chiang Kai-shek, who had succeeded Sun Yat-sen as the head of the Chinese Nationalist Party (1927–present). When under the leadership of Chiang Kai-shek, the Chinese Nationalist Party was forced to retreat to the island of Taiwan (Formosa) by the Chinese Communist Party in the Communist Revolution led by Mao Zedong.

Figure 4.1: Antique map of China. (Source: © Thinkstock.)

This has come to be called the "Kuomintang Debacle of 1949." With this, the People's Republic of China was founded (1949–present).

The Republic of China is today commonly known as Taiwan, while the People's Republic of China is today commonly referred to as Communist China or Mainland China, and is what many today would think of at the mention of "China." The Chinese Nationalist Party is still the ruling political party in Taiwan. Both the Republic of China and the People's Republic of China claim there is one China, and each claims to be the "real China"; in fact, during the early Cold War years (ca. 1945–1991), many Western nations, as well as the United Nations, recognized the Republic of China (Taiwan) as the legitimate China. This resulted in nervous relations between the two governments over the years.

Mathematics in written form in China came about during the late Shang period. Because of natural barriers—the Pacific Ocean to the east, jungles and mountains to the south and west, and the Gobi Desert to the north—China remained isolated for many years. As a result, for most of Chinese history, mathematics developed without any outside influence. Nevertheless, the Chinese made great strides in mathematics, developing techniques in arithmetic and algebra, including root finding methods for solving higher-degree equations. And, according to Calinger [35], Chinese algebra "reached its pinnacle in the thirteenth-century work of Qin Jiushao, Li Zhi, Yang Hui, and Shu Shijie." With a focus on computational techniques, theoretical geometry in China was to wait until the Jesuit Matteo Ricci (1552–1610) and Xu Guanqi (1562–1633) had translated the first six books of Euclid's *Elements* into Chinese in 1607. This was part of a broader introduction of Western mathematics, including trigonometry and logarithms, into China [35]. The remaining seven books of the *Elements* were not translated until much later in 1856 by Li Shanlan (1811–1882) and Alexander Wylie (1815–1887) [108].

The oldest known Chinese text that is exclusively on mathematics is *Jiu zhang suan shu* by an unknown author. Interestingly, there are several translations of the title of this work: *"Nine Chapters on Mathematical Procedures," "Arithmetic in Nine Sections," "Nine Chapters on the Mathematical Art," "Computational Prescriptions in Nine Chapters,"* and *"Nine Categories of Mathematical Methods"* are some examples [44]. However, the work is commonly referred to in short as the *Nine Chapters*, and we do so also. (See page 292 for a description of the contents.)

(a) *Nine Chapters* stamp.   (b) Liu Hui.

Figure 4.2: *Jiu zhang suan shu* (*Nine Chapters*) and Liu Hui. (Source: Courtesy of Jeff Miller.)

The *Nine Chapters* is believed to have been written during the late Qin (221–206 BC) or early Han (206 BC–AD 220) Dynasty [39, 72], and perhaps even completed before the "burning of the books" in 208 BC [44]. There is no surviving original copy of the *Nine Chapters*. What we have is a compilation by Zhang Cang and Geng Shouchang (first century BC) to which was later provided an extensive commentary by Liu Hui (third century AD) [44]. Liu Hui, who is also known for his other classic work, *Hai dao suan jing* (*Sea Island Mathematical Classic*), is often compared to Euclid, and the *Nine Chapters* to Euclid's *Elements* [44, 108]. Liu's version of the *Nine Chapters* was included among the *Shi bu suan jing* (*Ten Books of Mathematics Classics*) with a commentary by Li Chunfeng (AD 604–672) and others. *Ten Books* was the mathematics text at the Imperial College during the Tang Dynasty (AD 618–907). It is to this version of the *Nine Chapters* included in *Ten Books* that current editions may be traced [44].

4.2 ROD NUMERALS

Ancient writings and archeological finds dating back to the second century AD in China have unearthed bone strips used as counting rods. Ancient carvings on tortoiseshells of Chinese rod numerals have also been found. Dating back to at least the 4th century BC, early rod numerals were actual rods made primarily of bone or bamboo. Lam and Ang [77, p. xx] tell us that,

As far back as the Warring States period (475–221 BC), the Chinese used straight rods or sticks to do their calculation. They formed numerals from the rods, and they did their addition, subtraction, multiplication and division with these rod numerals. The performance of a multiplication problem such as the above [3508 × 436] with these rods would be commonly known at a very early time not only among mathematicians, but also among officials, astronomers, traders and others. The rods were carried in bundles and, whenever calculation was required, they were brought out and computation was performed on a flat surface such as a table top or a mat. After the results were obtained, they would probably be recorded and the rods would be put away.

The earliest known written use of rod numerals has been found on wooden artifacts excavated from the Han Dynasty (206 BC–AD 220). For example, on an artifact found in Hubei from that period is the script 當利二月定算丁, where the 丁 at the end is the rod numeral for the number 6.

Rod numerals were later transmitted from China to Japan, Korea, and Vietnam. With the advent and efficiency of the abacus, the use of physical rods for calculating gradually died out during the Ming Dynasty (AD 1368–1644). A number of texts were written that used the abacus as the main method for calculations, with the most influential text being Cheng Dawei's *Suanfa tongzong* (*Systematic Treatise on Arithmetic*; 1592). The written form of rod numerals, however, continued to be used along with traditional written numerals in China for many centuries and can sometimes be found in old markets even today.

According to Lam and Ang [77, p. 10], "The reason for the decline of Chinese mathematics after the 14th century was because it underwent a change of foundations, from mathematics based on rod numerals with its step by step reasoning, to mathematics based on the abacus with its emphasis on learning by rote method." Lam and Ang also advance the thesis that the Indo-Arabic number system (SECTION 6.3) has its origins in the Chinese rod numeral system, and provide their evidence for this.

The ancient Chinese rod numeral system is additive and positional, yet does not have a zero or a place holder. They were placed on a counting board, like a checkerboard, which allowed for a very clear delineation of the place values. The orientation of the rods also alternated between being placed vertically and horizontally from one place value to the next, beginning with being placed vertically in the one place. Over time, the rods eventually became written numerals and were written using very clear spacing. TABLE 4.1 displays the rod numerals as shown in Burton [22, pp. 29, 258]. Furthermore, at first, a vacant space on the counting board represented a zero in that place value, but a circular symbol ◯ later appeared in print in the 1200s. Lam and Ang [77, p. 152], and others, tell us that red rods were used to represent positive numbers (*zheng* 正) and black rods to represent negative numbers (*fu* 負).[1] Moreover, if rods of only one color were available, then an extra rod would be laid across the last nonzero digit to indicate that the number is negative. Lam and Ang give the example 丁 ≡ 𝍭 for −642. As another example, ≡ ||| ⊥ ||||| is 4379.

Used in place value	1	2	3	4	5	6	7	8	9
Unit, hundred,…	|	||	|||	||||	|||||	⊤	⊤|	⊤||	⊤|||
Ten, thousand,…	—	=	≡	≣	≣	⊥	⊥	⊥	⊥

Table 4.1: Chinese rod numerals.

Not only were rods used for recording numbers and for addition and subtraction, but they were also used to express fractions, as well as to carry out fairly advanced operations such as extracting cube roots, solving proportions, and solving systems of linear equations. To express a fraction, which typically appeared as the result of a division problem on a counting board, the numerator would be placed above the denominator just as we do today, but without a fraction line or fraction bar. So, for example [77, pp. 79–80],

$$\frac{||||}{\top} \quad \text{represents} \quad \frac{4}{6}.$$

[1] Burton tells us that black rods were used to represent positive numbers and red rods were used to represent negative numbers.

Like all things, however, the Chinese rod numerals evolved over time. The ones described here represent an early version. As the rod numerals evolved, they became more elaborate and started to resemble more modern Asiatic numerals.

Now You Try 4.1

1. Rewrite the following rod numbers using our numerals.

   a) ≡ 丁   ‖

   b) ‖ — ‖‖‖ ≡ 丁

   c) ‖‖ ≡ 朮

   d) ≡ ○ ○ ‖‖‖‖

2. Write the following numbers in Chinese rod numerals.

   a) 2345   b) 7089   c) −506   d) −888

   □

Think About It 4.1 Explain why a place holder is not needed in Chinese rod numerals. Can you think of some numbers for which Chinese rod numerals would not show place values as clearly as the examples above?   □

Chapter 5

Mayan Number System

Mesoamerica is the region that extends roughly from what is today central Mexico to the border of Costa Rica (FIGURE 5.1). Many pre-Columbian[1] civilizations inhabited Mesoamerica, among them the Aztec, Mixtec, Maya, Olmec, Teotihuacán, Toltec, and Zapotec. The Mayan civilization spanned from about 1500 BC to about AD 1500 and is commonly divided into three periods: the Preclassic or Formative Period (about 1500 BC to AD 250); the Classic Period (about AD 250–900); and the Postclassic Period (about AD 900–1500).[2] The apogee of the Mayan civilization was during the Classic Period. The Maya are noted for being the only of the pre-Columbian peoples to have a fully developed written language that dates back to around 250 BC. They are also noted for their architecture, art, and calendrics.

Figure 5.1: Mesoamerica. (Source: © Thinkstock.)

The Maya did not belong to one nation. Instead, the Maya were ruled by individual kings that were believed to have been divinely ordained. The kingdoms or city-states were often composed of a capital city

---

[1]The time prior to the arrival of the Europeans, most prominently the Spanish, that is generally marked by the arrival of Christopher Columbus in 1492.

[2]There is a very nice interactive map and timeline at the University of Texas at Austin "Precolumbian Art and Art History" Web site <http://www.utexas.edu/cofa/art/347/347m_map.html>.

and a few surrounding towns. Some of the major archaeological sites are Tikal, Palenque, and Chichen Itza. The structure of Mayan civilization was hierarchical, with the royal and priestly class at the top, followed by warrior nobles, merchants, craftsman, and peasant farmers.

The Maya built large palaces and temples made of stucco and stone. It is probable that the most recognized of Mayan architecture are their stepped pyramids, that are reminiscent of the pyramids of ancient Egypt (FIGURE 5.2), and their ball courts. Among what is remarkable about the construction of these monuments is that the Maya lacked draft animals and had no metal tools. Therefore, it must have required an enormous amount of manpower to build these impressive structures. It is believed that the building of the many public monuments depended on slave labor [35].

Figure 5.2: The Castillo Pyramid in Chichen Itza in the Yucatán Peninsula of Mexico. The pyramid was built between the eleventh and thirteenth centuries AD as a temple to the Yucatec Maya god Kukulcan that is closely related to the Aztec god Quetzalcoatl. (Source: © Thinkstock.)

As with many civilizations and cultures, religion was an integral part of the Mayan fabric, and the priests played an important role in interacting with the gods. Natural disasters and diseases, for example, were thought to have been brought about by angry gods. Some gods, like the Sun and the Moon, were thought to be good and would be called upon for protection from the evil gods. The planet Venus, which was recognized as the morning and evening star, was a particularly important god to the Maya and is depicted as a menacing spear thrower. Ritual human sacrifice to appease the gods was practiced, with the one to be sacrificed assured of great reward in the afterlife. In all of this, as well as in agricultural and other important civil and political activities, astronomy and astrology played a vital role [35, p. 300].

5.1   CALENDAR

The Maya made very precise observations of the stars and the planets, particularly of the planet Venus, and devised very sophisticated calendars. In fact, the Maya used three calendars and "were able to secure precision in dating their events which is not surpassed" according to Morley [84]. The first is the *tzolkin*, which was a 260-day calendar composed of thirteen vigesimal months (20 days in a month). (Morley tells us that this calendar was also long referred to by its Aztec name, *tonalamatl*, or so-called Sacred Year.) The second is known as the *haab* (the Solar Year [84]) that had 365 days composed of 360 "regular" days plus another five days that Cooke [41] says "were apparently regarded as unlucky (and so, best not included in the count)." The tzolkin was used in conjunction with the haab in what is called the *Calendar Round*, a period of 52 years or 18,980 days that is the length of time that it would take the two calendars to coincide. But, because the Calendar Round would repeat itself every 52 years, so that one cannot distinguish any day (name) in one Calendar Round from the same day in another Calendar Round, the Maya cleverly devised another way to record long intervals of time that marked the number of days from a fixed starting point.[3] This is the third calendar known as the *Long Count* that records the absolute number of days since the start of the current epoch. Cooke [41, p. 125] tells us that this date is uncertain, but it is believed to be August 12, 3113, BC on the proleptic Julian calendar,[4] and so ended on December 21, 2012. (Morley [84] remarks that the Aztec, for example, did not have an equivalent of the Mayan Long Count, and so there is much confusion over exactly when some of their recorded events occurred.) With their calendars and astronomical records, the Maya not only were able to record and order historical and current civil and political events, but they also were able to predict such events as solar and lunar eclipses, which were then used in their astrology to forewarn of particular impending disasters. (See FIGURE 5.3.) Even Mayan architecture corresponded to their astronomy and astrology. For example, Calinger [35] informs us that "[the] plaza of seven temples in Tikal, one of the largest classical Mayan cities in what is now Guatemala, has notable geometric features and astronomical orientation."

Figure 5.3: A Mayan calendar. (Source: © Thinkstock.)

The Mayan civilization went into decline in the Postclassic Period before the arrival of the Spanish conquistadors [35]. The reasons for their decline remain a mystery. It is not known if their decline was due to widespread disease or a depletion of the soil for agriculture or revolt or foreign invasion, or perhaps some combination of these, just to name a few conjectures. However, descendants of the pre-Columbian Maya continue to inhabit present-day Mesoamerica with a culture that is the result of a melding of pre- and post-Columbian traditions and beliefs. Unfortunately, although the spoken language has been handed down, the written language has not. Consequently, we have only a limited knowledge of the ancient Maya from the relatively few written records that have survived.

---

[3]This is no different, in principle, from a modern calendar that takes the birth of Christ as a fixed starting point, with the years following the birth of Christ annotated AD (*anno Domini*, Latin for "in the year of the Lord") or CE (current era), and the years preceding annotated BC (before Christ) or BCE (before the current era).

[4]The proleptic Julian calendar is created by applying the Julian calendar backward to dates prior to AD 4, when quadrennial leap years were instituted.

5.2 CODICES

The Maya had a sophisticated written language and, as such, much of what we know about the Mayan number system has resulted from deciphering their written records. This is not to imply, however, that stores of Mayan written records have been uncovered. The fact is that much of what we know about Mayan mathematics has been gleaned from the small number of surviving codices, namely, the Dresden Codex, the Madrid Codex, the Paris Codex, and the lesser Grolier Codex. Further, only about five hundred of about eight hundred known hieroglyphs found on stone monuments, ceramic vessels, and the codices have been deciphered since they were first studied about 1940 [35]. The paucity of evidence with which archaeologists and others have had to work means that although much is known about Mayan mathematics (and Mayan culture in general), very much more is not known.

The Dresden, Madrid, and Paris Codices—named after the cities where the codices are currently kept—were sent to Europe by colonizers before the Franciscan friar Diego de Landa (1524–1579) set out to destroy all Mayan books in the 1550s;[5] the Grolier Codex, so named because it was published by the New York publisher Grolier, was found in Mexico City in 1965 [41]. The authors of the codices are unknown. Of these codices, the most prominent is the Dresden Codex [117], which comprises 74 wooden folios that are folded like a screen or an accordion. As Cooke [41] describes it, the Dresden Codex contains almanacs; astronomical and astrological information; sections on the Moon Goddess; the so-called *Venus Tables* that record 312 years of the appearance of Venus as the morning and evening star; eclipse tables from AD 755–788; records of floods and storms to predict the end of the next world cycle; and a description of the coming end of the current world cycle. The astronomical records have helped researchers place dates on the Mayan civilization.

Unfortunately, as Cooke points out, "The Maya documents that have survived to the present are 'all business' and contain no whimsical or pseudo-practical problems of an algebraic type such as can be found in ancient Chinese, Hindu, Mesopotamian, and Egyptian texts." As a consequence, while we know somewhat the mathematics of which the Maya were capable—for example, they built great monuments, studied astronomy, and engaged in commerce—we do not know at all how they accomplished it; we do not even know how they performed the arithmetic operations of addition, subtraction, multiplication, and division. One thing that we do know, and in what we are interested here, is the Mayan number system.

5.3 NUMBER SYSTEM

Recall that there are different ways of writing ancient Egyptian numerals: hieroglyphic script that was carved into monuments, and hieratic script (and later also demotic script) that was only written on papyrus, leather, wood, and so on, using ink and a reed pen. Similarly, there are two ways of writing Mayan numerals: *head variant form* that was carved into monuments, and *normal form* that was carved into monuments and also written on paper made from tree bark using ink and brushes [84, 100]. (When normal form numerals were used on monuments, they would generally be embellished to make them decorative, as well as functional.) Our main interest here is the normal form numerals, which are composed of dots and bars, and we shall refer to them henceforth as dot-and-bar numerals.

The Mayan number system, using dot-and-bar numerals, bears similarities to the Babylonian number system: both are positional systems and both use two numeral symbols. See TABLE 5.1. Two differences

---

[5]Thompson [117, p. 14] notes that "a great deal of exaggeration has been published on the supposed destruction of quantities of Mayan books by Landa," asserting that more Mayan books have been lost because of neglect or were discarded "after being kept for some time as curiosities than were burned." Moreover, Thompson rightly observes that

> Landa was a product of his century. He had come upon revolting evidence of the sacrifice of children, in some cases in churches, by the supposedly converted Maya, and an end must be put to such practices; Mayan books were an important element in the old pagan structure which had to be eliminated if such sacrifices were to be ended. Setting the book, in which almost no one was then interested, against the lives of innocent children and the souls of their sacrificers, he made the decision which would have won the general approval of his age. We cannot judge him by twentieth-century attitudes.

Nor by twenty-first century attitudes.

Fɪɢ. 51. Head-variant numerals 1 to 7, inclusive.

Figure 5.4: The head variant form numerals for 1 to 7. (Source: Sylvanus Griswold Morley [84, p. 97]. U.S. public domain.)

	Babylonian		Maya	
Place value	Base 60		Base 20	
Symbols	one	ten	one	five

Table 5.1: Babylonian versus Mayan number characters.

between the Babylonian and the Mayan number systems are the bases (base 60 versus base 20) and the values of the symbols (1 and 10 versus 1 and 5). (The Babylonian number system is a sexagesimal positional system on top of a decimal numeral system, whereas the Mayan number system is a *vigesimal* (base-20) positional system on top of a *quinary* (base-5) numeral system.) A big difference between the two number systems, however, is that the Babylonian generally lacks a place holder symbol, whereas the Mayan has a place holder symbol, ⬭ (a cowrie shell that had several variants). See TABLE 5.2, and FIGURES 5.4 and 5.5.

Mayan dot-and-bar numerals are written vertically with the place values increasing from bottom to top, and in any place value every five dots are replaced by one bar. Having a place holder symbol means

•	—	
one	five	place holder

Table 5.2: Mayan number characters.

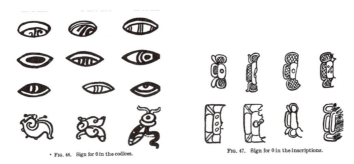

(a) Sign for 0 in the codices.   (b) Sign for 0 in the inscriptions.

Figure 5.5: The Mayan had a variety of ways in which they represented the place holder or zero symbol. On the left is how the symbol appears in the codices, and on the right is how the symbol appears in the inscriptions carved into stone. (Source: Sylvanus Griswold Morley [84, pp. 92–93]. U.S. public domain.)

that, unlike when using Babylonian numerals, a number can be expressed unambiguously using Mayan numerals. This gives the Mayan system a big leg up on the Babylonian [11]. The introduction of a place holder symbol was a significant step toward an improved number system.

In base 20, the place values are powers of 20. However, there seems to be a quirk in the Mayan vigesimal system. In a *true* base-20 system, *every* place value is a power of 20; however, in the Mayan system, the first two place values are powers of 20, but every place value thereafter is the product of a power of 20 and 18.

	Place Values				
True base 20	$20^0 = 1$	$20^1 = 20$	$20^2 = 400$	$20^3 = 8000$	20^n
Mayan base 20	$20^0 = 1$	$20^1 = 20$	$20^1 \times 18 = 360$	$20^2 \times 18 = 7200$	$20^{n-1} \times 18$

To illustrate this, here is the number 12,489,781 written using the Mayan dot-and-bar numerals. This number, which is the greatest number found in the codices, appears in the Dresden Codex among the so-called *Serpent numbers*, so called because serpents are depicted with their coils separating the place values of the numerals [94, 100, 117]. See FIGURE 5.6.

$$
\begin{aligned}
\bullet\bullet\bullet\bullet \quad &= \ 4 \times 20^4 \times 18 = 11{,}520{,}000 \\
\bullet \quad &= \ 6 \times 20^3 \times 18 = 864{,}000 \\
&= 14 \times 20^2 \times 18 = 100{,}800 \\
&= 13 \times 20^1 \times 18 = 4680 \\
&= 15 \times 20^1 \qquad = 300 \\
\bullet \quad &= \ 1 \times 20^0 \qquad = 1
\end{aligned}
\quad \Bigg\} \quad \text{Total} = 12{,}489{,}781
$$

Here are a few more examples of numbers expressed as Mayan dot-and-bar numerals.

Figure 5.6: A transcription of the serpent numbers from Folio 61 of the Dresden Codex. The serpent number 12,489,781 is seen as the numeral within the serpent on the left. (Source: U.S. public domain.)

Example 5.1

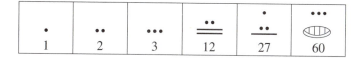

□

Now You Try 5.1

1. Write the decimal number and word names for the following Mayan numerals. Also express the Mayan numerals as Babylonian numerals.

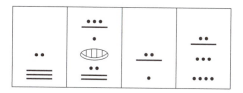

2. Express the following numbers as Mayan dot-and-bar numerals and as Roman numerals.

   a) Twenty-one.

   b) Two hundred nineteen.

c) Six thousand, seven hundred nine.

d) Three million, four hundred seventy-two thousand, six hundred thirteen.

□

Think About It 5.1 Given the close association between Mayan numbers and astronomy, why do you think the Maya developed their number system in this way?   □

It is natural to ask if the Maya indeed used a vigesimal number system because the Mayan number system has this quirky $20^n \times 18$ in all the place values after the second place.

$$20^0 = 1,$$
$$20^1 = 20,$$
$$20^1 \times 18 = 360,$$
$$20^2 \times 18 = 7200,$$
$$20^3 \times 18 = 144,000,$$
$$20^4 \times 18 = 2,880,000.$$

The answer is that those who have studied Mayan calendrics and numeration have advanced differing theories that depend on what is regarded as the basic unit [100].

Examples of Mayan numbers that have been deciphered deal with counting time events and periods (calendrics), and many authorities believe that this was the sole purpose of Mayan numbers [100]. Now, in our calendar we use the time periods *day*, *week*, *month*, *year*, and so on. In the Long Count, the Maya used the time periods *kin*, *uinal*, *tun*, *katun*, and so on.

		1 kin	=	1 day
20 kins	=	1 uinal	=	20 days
18 uinals	=	1 tun	=	360 days
20 tuns	=	1 katun	=	7200 days
20 katuns	=	1 baktun	=	144,000 days
20 baktuns	=	1 pictun	=	2,880,000 days

Observing that the values on the far right in the table above are precisely the place values of the Mayan number system, it becomes clear why they did not use 20 in the third place, but used 18 instead. If they had continued with 20, they would have 20×20 or 400 for the number of days in one tun, which is not a good approximation for a year, whereas 360 is a very good approximation. So, if one takes the unit or basic quantity of time to be the kin, as many have [84], then the Mayan number system is not a true vigesimal system. However, if one takes the unit to be the tun, with the kin and uinal being fractions of the tun [116], namely,

$$1 \text{ kin} = \tfrac{1}{360} \text{ tun} \quad \text{and} \quad 1 \text{ uinal} = \tfrac{1}{18} \text{ tun},$$

then the Mayan number system is a true vigesimal system.

Remark 5.1 The baktun is also referred to as the *cycle* and the pictun as the *great cycle* [84]. The fact is that the Mayan names for these and higher place values are not known, and the names "baktun" and "pictun," and so on, are modern inventions formed by prefixing the Mayan "tun" with the Yucatec terms for 400, 8000, and so on [118].   □

Regarding fractions, recall that the Egyptian hieroglyph for a fraction would be the symbol ∞ placed over the numeral for the denominator—for example, $\overset{\frown}{\text{iii ii}}$ is one-fifth—and the Babylonians leveraged their place values to express fractions—for example, $\langle$ III could be (in expanded form) $10 \times 1 + 3 \times \frac{1}{60}$ or $10 \times \frac{1}{60} + 3 \times \frac{1}{3600}$, and so on. The Maya, however, did not use any notation for fractions.

Think About It 5.2 How does the fact that the Mayan number system is a modified base-20 system affect how numbers are written? In particular, can every number be written uniquely or can some numbers be written in more than one way?	$\square$

## 5.4	NATIVE NORTH AMERICANS

To wrap up, we touch on the many different number systems used by the Native peoples of North America. Generally speaking, Native North American peoples had no written language, and only a few had some system of simple hieroglyphics. As such, much of what we know about their number systems stem from the later recording of their oral number systems, relying on experts in Native North American linguistics. The sources that do exist are widely scattered among a variety of deposits, but Eells [48] has done an excellent job of distilling the vast amount of material; indeed, what follows are highlights of Eells work that was based on an examination of the number systems of over 300 Native North American languages.

We have seen that the prevalent bases used in the number systems of the Old World were the decimal, duodecimal, and sexagesimal bases. In the New World, the prevalent bases used were the decimal, quinary, and vigesimal bases; however, there is also evidence that some Native American tribes used *ternary* (base-3), *quaternary* (base-4), and *octonary* (base-8) number systems, plus traces of evidence that others used binary, *sexanary* (base-6), *nonary* (base-9), *quadragesimal* (base-40), and sexagesimal systems; and very often some mixture of bases (for instance, quinary-decimal, quinary-vigesimal, decimal-vigesimal, and even quinary-decimal-vigesimal).[6] Moreover, while the association of word names for numbers with concrete objects (usually parts of the body, particularly the digits) in the languages of the Old World have mostly been lost, the word names in many of the languages of the New World still bear their association with concrete objects, particularly the digits (the fingers or toes).

Eells notes that there is a remarkable uniformity in finger counting among the many tribes of Native North Americans in that it almost always begins with the little finger of one hand, progresses to the thumb, then to the thumb of the other hand, and continues finally to the other little finger. With this in mind, here are how some numbers are said.

For the number **one**, Eells tells us that the Massachusetts used *pasuk* from *piasuk*, which means "very small"; the Montagnais used *inlare*, which means "end is bent"; and the Zuni used *topinte*, which means "taken to start with." For the number **five**, the Ojibwa used *nanan* meaning "gone" or "spent" and the Hidatsa used *kichu* from the words for "completely [*ki*] turned down [*chu*]," both examples illustrating the idea that one hand has been completed in counting. For the number **ten**, the Zuni used *astemthla* which means "all of the fingers"; the Wintun used *pampa-sempta* from the words for "two [*pampu-ta*]" and "hand [*sem*]"; and the Konkau used *machoko* from the words for "hand [*mar*]" and "double [*choko*]." And, as a last example, for the number **twenty**, the Navaho used *natin* from the word for "man [*tine*]"; the Greenland used *inuk-mavdlugo*, which means "man come to an end" or *inup-avatai-navdlugit* meaning

---

[6]In a pure quinary system, for example, only five "elemental numbers" are required: 1, 2, 3, 4, and 5. Then, for example, after counting from 1 to 5,

$$\text{six} = 5 + 1, \quad \text{seven} = 5 + 2, \quad \text{eight} = 5 + 3, \quad \text{nine} = 5 + 4, \quad \text{ten} = 2 \times 5,$$

and so on. Similarly, in a pure decimal system, only ten "elemental numbers" are required: 1, 2, 3, 4, 5, 6, 7, 8, 9, and 10. But a quinary-decimal system may have five "elemental numbers" and some other qualifier. For example, in a quinary-decimal system, after counting from 1 to 5 on one hand, perhaps

$$6 = \text{"other hand 1,"} \quad 7 = \text{"other hand 2,"} \quad 8 = \text{"other hand 3,"} \quad 9 = \text{"other hand 4,"}$$

and 10 = "two hands complete" (2×5). Of the different mixed bases, Eells tells us that the quinary-vigesimal is the most frequent and gives as an example the Greenland Eskimo [48, p. 89]: "The Greenland Eskimo says 'other hand two' for 7, 'first foot two' for 12, 'other foot two' for 17 and similar combinations to 20, 'man ended.'"

"man's outer members completed"; and the Wintun used *ketet-wintun* from the words for "one [*ketet*]" and "Native American [*wintun*]," all expressing the idea that all the digits (fingers and toes) have been counted.

For naming numbers, there is evidence that different tribes used in some combination the additive, subtractive, multiplicative, duplicative, and divisive principles. The use of the divisive principle is the rarest; the few examples include the Unalit that used *kolin* ("upper half of the body") for **ten** and the Pawnee that used *sihuks* ("half of two hands") for **five**. Yet, despite the ability to count at least into the hundreds, there was almost no conception of fractions. In the very few instances in which fractions have been detected, they have all been unit fractions, the most common by far being one-half; as Eells puts it, "examples of other fractions are almost negligible."

Chapter 6

Indo-Arabic Number System

6.1 INDIA

India, located in South Asia, is situated on what is called a subcontinent that juts into the ocean and seas. Once referred to as *Bharat* in classical Sanskrit sources, the West has used the name *India* since at least 500 BC [49]. The land borders of India are Nepal, backed by the Himalayas (a great mountain range in which sits Mount Everest, the highest peak on Earth), to the northeast; Pakistan (formerly West Pakistan) to the northwest; and Bangladesh (formerly East Pakistan) to the east. The maritime borders of India are the Indian Ocean to the south; the Arabian Sea to the southwest; and the Bay of Bengal to the southeast. (See FIGURE 6.1.)

Figure 6.1: Modern-day map of India. (Source: © Thinkstock.)

6.1.1   History

We have no early history of India in the Western sense of history; the earliest (Western sense) history of the region was written in the fourth century BC by the Greek historian Herodotus. (The early history of India is found in texts known as the *Puranas* that describe the complex relations of men and gods [49].) The history of India, nevertheless, is generally divided into three periods, namely, the Ancient Period (before ca. AD 1200), the Medieval Period (ca. AD 1200–1700), and the Modern Period (from ca. 1700), which were previously referred to as the Hindu Period, the Muslim Period, and the British period, respectively.

Figure 6.2: The Taj Mahal. The construction of the Taj Mahal in Agra, Uttar Pradesh, India, was started in 1632. It is considered to be a masterpiece of Arabic art. (Source: © Thinkstock.)

Indian society traditionally has been based on a caste system that dates back to the period 1000–500 BC. The system was divided into four groups called *varnas*, with the *Brahmans* (teachers, scholars, priests) who controlled ritual practice and religious knowledge at the top; then the *Kshatriyas* (warriors, landowners); the *Vaishyas* (farmers, merchants); and the *Shudras* (artisans, laborers); in addition, excluded from main society were the *parjanyas* or *pariahs* ("untouchables"). Today the caste system has been officially outlawed by the Indian Constitution.

The beginnings of civilization in the Indian subcontinent extend as far back as 3000 BC, with the great Indian civilization developing in the Indus Valley around 2500 BC. But sometime before 1700 BC they went into decline, eventually ceasing to have a common civilization. This demise, however, was to be erased by a group of migrants from the northwest that were known as Aryans. It is with them that we associate the development of the Indo-Aryan languages, including Sanskrit, Hindi, Bengali, and Marathi, and a large body of Vedic texts called *Vedas*, which are among the most ancient works of literature in the world [41].

Alexander the Great (356–323 BC) conquered much of the known Western civilized world during his rule, extending the reach of his empire to the Indian subcontinent—considered then to be the end of the inhabited world—but his death in 323 BC kept him from incorporating India into his empire. Nevertheless, communication had been established with India, and trade brought an exchange of goods and ideas between the worlds. (Greco-Babylonian influence on Indian astronomy is certainly evident, for example.)

The British colonial period, however, is probably the period of India's history that comes to mind most readily, if only because that period is probably the most portrayed in the movies (*Gandhi*; *A Passage to India*; *Gunga Din*; *The Jewel in the Crown*, to name a few). Beginning with Company rule by the British East India Company (AD 1757), and later direct rule by the British Crown under the administration of a British Raj (1858), British influence on Indian society continues to be felt today. (India is a frequent contender in the Cricket World Cup.) As Cooke [41] points out, British rule of India allowed European scholars to acquaint themselves with Indian literature, for many Sanskrit works were translated into English.

Since having gained formal independence from Great Britain on August 15, 1947, India has become a very influential nation: India is not only an important player in today's world economy but also one of a handful of nations with nuclear weapons. Yet, India is considered to be one of the poorest nations in the world today.

6.1.2   Mathematics

Recall that the ancient Greeks were the first to move mathematics from being a subject of intuition and empiricism to one of statements and proofs. There is no denying that the Greeks made enormous contributions toward advancing mathematics, most notably during the Golden Age and, later, the Silver Age. But the Greeks had no conception of negative numbers or even zero because they were constrained by geometry. Indian mathematicians, on the other hand, were not so constrained, and thus dealt freely with negative numbers and zero. (See REMARK 12.1 on page 121.)

Most of the mathematics that we have come to ascribe to the Indians of South Asia were recorded in Sanskrit written on birch bark in the north and palm leaves in the south.[1] As a result, unlike the durable Babylonian clay tablets, nearly all of the extant manuscripts written on these materials are of relatively recent date. According to Kim Plofker [95], "[the] first known texts …are the 'Vedas' (literally 'knowledge'), a canon of hymns, invocations, and procedures for religious rituals. The Vedic texts (generally composed in verse or in short prose sentences called *sūtras* to make them easier to memorize) were carefully learned, recited, and handed down orally." The best preserved and handed down Vedas were the ones that were most important for religious sacrifices. In addition to the Vedic hymns, other texts that were carefully handed down because of their importance were on the six subjects called the "limbs of the Vedas," which Plofker lists as:

1. phonetics, which preserved the correct pronunciations of the archaic Sanskrit invocations;

2. grammar, which explained how its sentences would be understood;

3. metrics, which preserved the structure of its verses;

4. etymology, which explained its vocabulary;

5. astronomy and calendrics, which ordered the timing of the sacrifices; and

6. ritual practice, which preserved the sacrificial tradition.

It is from the Vedas and the limbs of the Vedas that we have obtained glimpses of the earliest Indian mathematics. For example, the limb of ritual practice includes the *Śulbasūtras* ("*Cord-Rules*," which reminds us of the ancient Egyptian *harpedonáptai* or "rope stretchers" [41]) that was written sometime in the first half of the first millennium BC, and that describes rules for finding areas and volumes for building brick altars. Even the limb of metrics contains some mathematics. However, perhaps not too surprisingly, the limb of the Vedas that led to most of ancient Indian mathematics was the one on astronomy and calendrics. (Here we obtain a hint of Babylonian influence on early Indian mathematics: the way linear proportion was used for finding daylight length is similar to a common Babylonian method; also, references to gnomons and water clocks, and the use of similar time units add to the evidence [95].) But Plofker cautions that "these texts can reveal only scattered fragments of this era's wider mathematical interests." Other glimpses of early Indian mathematical interests, particularly in daily life, have been revealed by parts of the culture that have been uncovered by archaeologists.

In 1881 near the village of Bakhshālī, near modern Peshawar in Pakistan, a farmer found a mathematical birch-bark manuscript that contains rules and examples for a variety of arithmetic techniques. The dating of the so-called *Bakhshālī Manuscript* is still uncertain. Some scholars place the manuscript between the eighth and twelfth centuries AD [35, 95], while others assert that it is at the least a copy of a manuscript

---

[1]"Indian mathematics" has also been called "Hindu mathematics" by some historians of mathematics.

that was written no later than the end of the third century [35, 41]. Calinger [35] notes that the Bakhshālī Manuscript, which is "a treasury of information on Jaina arithmetic, is the first surviving Indian text systematically to treat the rule of three and compute square roots,"[2] and he makes the astute observation that dating the manuscript around the third century lets it fill a gap between the *Śulbasūtras* and the Gupta era (AD 320–675; also referred to as the Classical Period).

Three notable mathematicians of the late Ancient Period were Āryabhata I (AD 476–550), Brahmagupta (598–670), and Bhāskara II (1114–1185).[3] Āryabhata is best known for his work *Āryabhatiya*, a collection of mathematical problems and procedures in astronomy, mathematical ideas, and time reckoning [112]. Brahmagupta's major work was *Brāhmasphutasiddhānta* (*Corrected Treatise of Brahmā* or *The Opening of the Universe*), in which he expounds on such things as arithmetic with positive and negative numbers, zero, and fractions; finding square roots and cube roots; and the rule of three [41]. And Bhāskara II, who is better known in India as Bhāskaracharya or Bhāsharacharya (*archarya* meaning "master" or "teacher"), is best remembered for two works: *Līlāvatī* (*The Beautiful*) and *Bījaganita* (*Seed Computation* or *Seed Counting*). The first is a book of mathematical problems in arithmetic, algebra, and geometry that was written as puzzles addressed to a person named Līlāvatī, which was a common name among women;[4] and the second is a more systematic treatment of numbers and algebra [35]. Bhāskara II built his work on that of Brahmagupta, and in several instances corrected or extended the results.

In addition to acquainting Europeans with Indian literature, British rule of India also acquainted Europeans with Indian mathematics. In the eighteenth century, European-style universities were established, but they were geared toward producing government officials and not scholars [41]. In spite of this, India produced one of the most astounding mathematicians of the turn of the twentieth century (and perhaps "one of the greatest mathematical geniuses of all time" [41]), Srinivasa Ramanujan (1887–1920; FIGURE 6.3). Ramanujan was exceptional for having had no formal education in mathematics. Because of the way the system was set up in India, Ramanujan appealed directly to mathematicians in Britain for help. Finally, his raw genius was recognized by G. H. Hardy (see page 29), with whom he collaborated closely.

Figure 6.3: Srinivasa Ramanujan. (Source: Courtesy of Jeff Miller.)

---

[2]The "rule of three" is explained in CHAPTER 23.

[3] Many sources will classify these three as Medieval Indian mathematicians, in which case they would have divided India's history into the Ancient Period, the Classical Period, the Medieval Period, and the Modern Period.

[4] There is a charming story told by Fyzi, the fifteenth century Persian poet who translated this text, about the person Līlāvatī, who was thought to be Bhāskara's daughter. Calinger [35] relates the story like this:

> From her [Līlāvatī's] horoscope Bhāskara discovered the best time for her marriage. He then placed a cup with a small hole at the bottom in a vessel of water, so that it would sink at the propitious time [a water clock]. But when Līlāvatī out of curiosity bent over the vessel, a pearl dropped and blocked the hole. [Some versions of the story say a seed dropped to block the hole; see [112], for example.] The propitious hour passed without the cup's sinking. [And now Līlāvatī would never get married.] To console her, Bhāskara promised to name his first book after her, "which will last to the latest times."

6.2   THE MIDDLE EAST

The Middle East is roughly the part of Southwest Asia from Turkey to Iran and down the Arabian Peninsula, as well as the part of North Africa along the Red Sea. The region is also called the Near East, particularly in reference to the period before the nineteenth century AD. The larger countries geographically in the Middle East today are Turkey, Syria, Iraq, Iran, Saudi Arabia, Egypt, and Sudan, and some of the smaller countries include Kuwait, Qatar, the United Arab Emirates, Oman, Yemen, Jordan, Israel, and Lebanon. (See FIGURE 6.4.)

Figure 6.4: The Persian Empire. (Source: © Thinkstock.)

6.2.1   History

The history of the Middle East stretches back to the times of ancient Mesopotamia (Iraq and Kuwait, and northeastern Syria, southeastern Turkey, and southwestern Iran; CHAPTER 2) and Egypt (CHAPTER 3). Almost from the very beginning, the Mesopotamian valley has been a region of great tumult, marked by wars and ruled by numerous dynasties; Egypt, on the other hand, although it was eventually drawn into the chaos that was centered about Iraq, was relatively stable as long as it remained insulated [1, p. 221].

Alexander the Great died unexpectedly in 323 BC without any apparent heir to his empire. The result of this was a jostling for control among his chief lieutenants, from which two powers emerged: the Ptolemaic Greek Empire founded by Ptolemy I Soter (ca. 367–283 BC) and centered in Egypt, and the Seleucid Greek Empire founded by Seleucus I Nicator (358–281 BC) and centered in Iraq.

Beginning around 250 BC, the Parthians slowly began to displace the Seleucids from Iran, but a counteroffensive by Antiochus the Great (Antiochus III; ca. 242–187 BC) in 209 BC took back much of the Seleucid Empire. Following this great achievement, Antiochus allied himself with Philip V of Macedon (238–179 BC) and made an incursion into Greece, but only to meet defeat by the Romans near Thermopylae

(191 BC) and, after being chased by the Romans, at Magnesia (190 BC). This led to his accepting the Treaty of Apamea in 188 BC that ceded practically all of Seleucid Mediterranean to Rome. The Seleucid Empire was eroded further when Mithridates I (ca. 195–138 BC) ascended to the Parthian throne in 171 BC, and what remained eventually fell to the Romans [85, p. 191].

In the years following the culmination of the Punic Wars (265–146 BC), the Romans would come to rule from Britain, across all of the Mediterranean, parts of the Near East, and all the way to India. In particular, from the first century BC to the seventh century AD, the Near East was divided between the Roman Empire and rulers from Iran, first the Parni and then the Persians, the latter of whom were from the province of Fars that was the homeland of the second Persian Empire (the Sassanid dynasty) in the third century AD [1, p. 223]. During this time, Christianity spread across the Roman Empire. At first largely ignored, and then officially persecuted, Christianity became the official religion of the Roman Empire under Constantine the Great (Constantine I; reigned AD 306–337), who himself converted to Christianity after he claimed to have received a sign from God on the eve of battle (312). (See FIGURE 6.5.)

Figure 6.5: Coinage was an important legacy of Constantine the Great. These gold coins depict Constantine the Great (left) and Diocletian. (Source: © Thinkstock.)

The expansion of the Sassanid Empire brought the Sassanians into conflict with Rome, and the conversion of Constantine brought a new facet to Romano-Persian relations. At the beginning, Roman Christian prisoners from the wars of Shapur I (reigned ca. AD 241–272) were allowed to worship freely in Persia; but, from the time of Bahram I (reigned AD 273–276), Jews, Christians, and Manichaeans found themselves being persecuted by Zoroastrian priests. This led Constantine, who saw himself now as a defender of Christians, to intervene on behalf of the Christians in Persia. Constantine's intervention, however, caused the Christians in Persia to be seen as a fifth column. This perception of the Christians in Persia was exacerbated when Simeon the Catholicos refused to collect a double tax from Christians to support the war, an insubordination that resulted in the execution of Simeon and his closest companions by Shapur II (reigned AD 309–379) [85, p. 55].

The seventh century saw the beginning of Islam,[5] a new religion in the Near East [1, 85, 34]. Islam was founded by Muhammad ibn Abdullah who was born into the aristocratic tribe of Quraysh in the city of Mecca in Arabia around AD 570. He was orphaned at the age of six and grew up in Mecca. By the age of twenty, Muhammad had joined a caravan service along the trade route, which exposed him to Jews and Christians. It is believed that this exposure led Muhammad to develop a monotheistic perspective; prior to Islam, Arabs who were neither Jews nor Christians were polytheistic. Muhammad is believed to have been called to prophethood at the age of 40 when he began to receive a number of messages from

---

[5]The name Islam derives from the root Arabic word for "submission," as does the name Muslim, and therefore Islam is the religion of total submission to Allah. Islam is the third of the three major monotheistic religions to be born in the Near East, following Judaism and Christianity. In fact, Muslims believe that Islam is the completion of the earlier religions. According to Islamic jurists or legal scholars, the world is divided into the "domain of Islam" (*dar al-Islam*) and the "domain of war" (*dar al-harb*), and that there may exist truces, but no permanent peace, between the two. Moreover, every able-bodied Muslim has a duty to defend the domain of Islam from aggressors.

Allah (God) through Gabriel the archangel that were eventually compiled into the *Qur'an* (or *Koran*).[6] In the meantime, Mecca was becoming an important destination for pilgrimage. Thousands of people visited Mecca every year to see the Ka'ba, a shrine for about 360 local deities. The Ka'ba, a sacred black meteorite, is believed to have been given by Gabriel to Abraham, and his elder son, Ishmael (the founder of the Arabs). The pilgrimages meant a booming trade for Mecca and, for this reason, Muhammad's first efforts to convert the people in Mecca to Islam met with stiff opposition from the city's leaders. Word of a planned assassination caused Muhammad to flee Mecca in 622. Muhammad's flight from Mecca led him to Yathrib, which later came to be called Medina (city of the Prophet), where he met with much greater success in his mission. From this point on, Islam spread like wildfire.

Muhammad's flight from Mecca to Medina, called the *Hijra* (or *Hegira*; departure or exodus) was so pivotal in the history of Islam that the year 622 was later taken as the beginning of the Muslim lunar calendar: the first date MuHarram 1, 1 AH[7] corresponds to July 16, 622. Within ten years, Muhammad unified all of Arabia for the first time in history. From then, the Arabs went on to conquer all the lands from the Indus Valley to Syria and Egypt, along the coast of North Africa, and even across the Strait of Gibraltar into most of Spain by 711. Today, converts to Islam are found all around the world, yet throughout its history until now, Arabia has been the center of the Muslim world, with Arabic being the lingua sacra. (See FIGURE 6.6.)

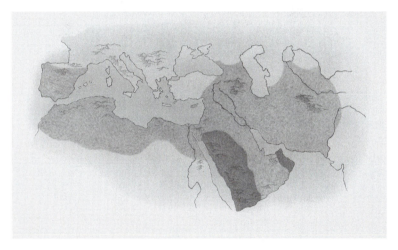

Figure 6.6: The influence of Islam in the Middle East, Africa, and Europe.
(Source: © Thinkstock.)

The death of Muhammad in 632 brought on the problem of selecting a successor to maintain the Medinan hegemony. What emerged was the era of the caliphates or Muslim dynasties that lasted until the twentieth century. Some of the notable caliphates are the Rashidun (632–662; the Orthodox or Rightly-Guided Caliphs), the Umayyads (661–750; claimed universal authority from its capital in Baghdad), the 'Abbasids (750–1258; claimed universal authority from its capital in Damascus), the Seljuks (1038–1194; Iran, Iraq), the Safavids (1501–1786; Iran), and the Ottomans (1281–1922; Turkey, Syria, Iraq, Egypt, Cyprus, Tunisia, Algeria, western Arabia). What also emerged were two principal groups of Muslims, namely, the Sunni and the Shiite. The split occurred with the fourth Rashidun caliph, 'Ali, with the chief dispute being how the temporal authority of Muhammad was to be transmitted. Today, the Sunni account for a majority of Muslims, with the Shiite being the second largest group.

---

[6]Muslims believe that the Qur'an is the *literal* word of Allah. Moreover, the belief that Muhammad was illiterate makes the writing of the Qur'an miraculous. It is believed that, following a long line of prophets, including Abraham, Moses, and Jesus of Nazareth, Muhammad is the "Seal of the Prophets," and that no other religion after Islam will be revealed. Thus, after the Qur'an, the most important sources for Islam are the *Hadith*, the sayings of Muhammad, and the *Sunna*, his traditions and practices.

[7]The designation "AH" (*anno Hegirae*, Latin for "in the year of the Hijra") indicates the number of years in the Islamic Hijri calendar, a lunar calendar.

In an effort to repel the invasion of Seljuk Turks, Blessed Pope Urban II announced the first Crusade in 1095. The next centuries would see several more Crusades, but, eventually, with the fall of Constantinople in 1453—and its subsequent renaming to Istanbul—the Ottoman Empire essentially successfully replaced the Byzantine Empire. During the same time, Mongols, first led by Genghis Khan, were pressing westward, and, by 1256, Hülagü Khan, a grandson of Genghis, had conquered much of southwest Asia and ruled Iraq and Iran as the first Ilkhan (1256–1265). But the Mongols' advance toward Syria and Egypt was met and checked by the Mamluks. After that, the Mongol Ilkhans loosely controlled Iraq and Iran from their base in Azerbaijan. Syria and Egypt would be ruled by the Mamluk Sultanate. A peace between the Mongols and the Mamluks was arrived at in 1322. By the sixteenth century, the Ottoman Empire entered into decline, and between 1800 and 1939, most of the Middle East would find itself under a variety of European colonial rule.

Figure 6.7: The Dome of the Rock, located on the Temple Mount in Jerusalem. The Dome of the Rock was built by Muslims, and was completed in AD 458 upon a site that is sacred to Judaism, Christianity, and Islam. (Source: © Thinkstock.)

Two reasons for European interest in the Middle East at this time were, first, interest in improving communications (the British were interested in bettering communications between British India and Britain overland; the French were interested in cutting the Suez Canal in Egypt to provide a passage between the Mediterranean Sea and the Red Sea; the Germans were interested in building a Bosporus-to-Baghdad railroad); and, second, the discovery of a rich deposit of oil. The Arabian Peninsula, however, was spared European imperialism because of Europeans' stinging experience with the Turks, and because oil had not yet been discovered there. During this time, European influence percolated into the Middle East. Indeed, the Ottoman Empire had diplomats reside in European capital cities, and sent students to Europe to attend universities, as well as to study their languages. Printing presses and translation bureaus opened in Istanbul (Turkey) and Bulaq (Egypt) to translate into Turkish and Arabic, among other things, many European technical and scientific textbooks. Because of the Middle East's ties to Europeans' Biblical heritage (FIGURE 6.7), many Europeans pictured the people of the Middle East to coincide with their version of the people during the time of Abraham. But, also, a goodly number of Europeans held a romantic vision of the Middle East, imagining it to be exotic and mysterious, much like they imagined India—a place of free sensuality, where the usually forbidden was not—an image that was suggested at least in part by the Arabic tales *One Thousand and One Nights*[8] (with stories such as "Ali Baba and the Forty Thieves"), and that inspired numerous novels, operas, paintings, and films.

---

[8]Often known in English as the *Arabian Nights*.

6.2.2 Mathematics

Whereas most of the mathematics we have come to ascribe to the Indians were recorded on birch bark and palm leaves, what we know of Arabic mathematics was recorded using pen and ink on paper; Muslims had learned to make paper from Chinese prisoners who were captured at the Battle of Atlakh around 750 [10].

By Arabic mathematics,[9] we mean the corpus of mathematics that was transmitted in Arabic from about the eighth to the fifteenth century AD. Arabic mathematicians include non-Arabs, as well as non-Muslims. (Qusṭā ibn Lūqā (820–912) was Christian and Ibn Yaḥyā al-Maghribī al-Samaw'al (ca. 1130–1180) was Jewish, to name but two.) Nevertheless, Islam played a role in shaping Arabic mathematics. For instance, the requirement to face Mecca when praying motivated the study of astronomy, spherical geometry, and geography; complicated Islamic inheritance laws and legacy rules also posed mathematical problems to which algebra and unit fractions were applied [41, p. 54]. Moreover, religion was woven into the fabric of the Muslim way of life; for example, we find the following confession and exhortation at the beginning of al-Khwārizmī's book, *Hisab al-jabr wa'l muqābala*, on solving equations [2]:

> IN THE NAME OF GOD, GRACIOUS AND MERCIFUL!
>
> This work was written by MOHAMMED BEN MUSA, of KHOWAREZM. He commences it thus:
>
> Praised be God for his bounty towards those who deserve it by their virtuous acts: in performing which, as by him prescribed to his adoring creatures, we express our thanks, and render ourselves worthy of the continuance (of his mercy), and preserve ourselves from change: acknowledging his might, bending before his power, and revering his greatness! He sent MOHAMMED (on whom may the blessing of God repose!) with the mission of a prophet, long after any messenger from above had appeared, when justice had fallen into neglect, and when the true way of life was sought for in vain.... may his benediction rest on MOHAMMED the Prophet and on his descendants!
>
> [...]
>
> The fondness for science, by which God has distinguished the IMAM AL MAMUN, the Commander of the Faithful ... has encouraged me to compose a short work on Calculating by (the rules of) Completion and Reduction.... My confidence rests with God, in this as in every thing, and in Him I put my trust. He is the Lord of the Sublime Throne. May His blessing descend upon all the prophets and heavenly messengers!

Al-Khwārizmī (ca. 780–850; see page 312) is only one of many notable Arabic mathematicians; Cooke [41], for example, recounts several others: Thabit ibn-Qurra (826–901), Abu-Kamil (ca. 850–893), Mohammed Abu'l Wafa (940–998), Abu Sal al-Kuhi (ca. 940–ca. 1000), Abu Ali ibn al-Haytham (965–1040), Abu Arrayhan al-Biruni (973–1048), Omar Khayyam (1044–1123), Sharaf al-Din al-Muzaffar al-Tusi (ca. 1135–1213), and Nasir al-Din al-Tusi (1201–1274).

The era of the caliphates following the death of Muhammad in 632 saw the spread of Islam from Spain to the borders of India and China. A consequence of this expansion was an increased interest in the ancient learning of the newly conquered or bordering lands which led to a massive acquisition of scientific works, notably Greek, Hellenistic, and Indian. Indeed, according to Berggren [10], the mathematics that Muslims acquired came from three main traditions: Greek mathematics (principally the works of Euclid, Archimedes, and Apollonius in geometry and geometric algebra, Diophantus in numerical solutions of indeterminate problems, and Heron in practical manuals); Indian mathematics (the number system that we use today, algebraic methods, a budding trigonometry, and methods from solid geometry to solve problems in astronomy); and the practitioners' mathematics (the practical mathematics used by "surveyors, builders, artisans in geometric design, tax and treasury officials, and some merchants" [10, p. 516]).

The acquisition of scientific works required a great amount of resources: time and money to travel and secure these works, and the people required to translate the works into Arabic. In 830, Caliph al-Ma'mūn (reigned 813–833) endowed the House of Wisdom in Baghdad, which is often compared to the Museum of Alexandria in Egypt, that included a library and an astronomical observatory [35, p. 314]. As Calinger [35] tells us, "The House of Wisdom gave new impetus to the study of theoretical and natural knowledge in Islam. Leading scholars were brought together in this complex and financially supported by the caliphal

---

[9]"Arabic mathematics" is also called "Islamic mathematics" by some historians of mathematics.

treasury. Besides agriculture, alchemy, botany, and medicine, mathematics was a field of study along with subjects closely related to it: astronomy, astrology, cosmology, geography, natural philosophy, and optics. The House of Wisdom quickly became the foremost center in Islam dedicated to the encouragement of such studies."

However, the flow of knowledge was not only one way—"from the outside in"—for the rest of the world has also benefited tremendously from these Arabic scholars: first, Arabic translations of mathematical texts facilitated the transmission of mathematics between different cultures at that time, particularly the transmission of Indian mathematics to Europe; second, Arabic mathematicians, after having absorbed the works they had been translating, contributed original works of their own that allowed mathematics to progress in Europe through the Middle Ages and, especially, during the Renaissance; and, third, many original works that have since been lost are available to us only through their Arabic translations, or through Latin translations of subsequently lost Arabic translations [10, 41]. To that last point, we mention as one example *Conics* [5, 6], the seminal work of Apollonius (ca. 262–190 BC) that is a collection of eight books on conic sections: only Books I–IV exist in the original Greek, Books V–VII exist only in their Arabic translations, and Book VIII has been lost altogether. For all intents and purposes, the Arab culture of the Middle East was the keeper of the scientific and mathematical legacy of the Hellenistic world until it was passed on to Europe when she finally awakened from the so-called Dark Ages. Calinger [35, p. 317] provides the following partial list of Arabic translations of Greek mathematical texts.

Author	Title	Translator	Date/Comments
Euclid	*The Elements*	Al-hajjāj b. Matar	Time of Harun al-Rashid and al-Mamūm.
		Ishāq b. Hunayn Thābit b. Qurra	Late ninth century Died in 901.
	The Data	Ishāq b. Hunayn	
	The Optics		
Archimedes	*Sphere and Cylinder*	Ishāq b. Hunayn Thābit b. Qurra	Revised a poor early ninth-century translation.
	Measure of the Circle	Thābit b. Qurra	Used commentary of Eutocius.
	Heptagon in the Circle	Thābit b. Qurra	Unknown in Greek
	The Lemmas	Thābit b. Qurra	
Apollonius	*The Conics*	Hilāl al-Himsi Ahmad b. Mūsā Thābit b. Qurra	
Diophantus	*Arithmetic*	Qustā b. Lūqā	Died 912
Menelaus	*Spherica*	Hunayn b. Ishāq	Born 809

6.3   NUMBER SYSTEM

The Indo-Arabic number system that is commonly used around the world today uses the following familiar symbols.

1	2	3	4	5	6	7	8	9	0
one	two	three	four	five	six	seven	eight	nine	zero

The number system entered Europe from India through Spain in the middle ages by way of the Arab countries. The key features of the Indo-Arabic number system today are

- it is positional;

- it uses ten symbols of numeration, corresponding in number to the base (ten) of the system;

- it has a place-holder symbol that is also a number (zero);

- it has a separatrix (decimal point) that separates the integer part of a numeral from the fraction part.

Now while any proper subset of these features may make for a remarkable number system, like the Babylonian and the Maya, we have a truly superior number system when all four features are taken together.

Example 6.1 The numeral 3.141 59 denotes a different number from the numeral 3141.59. Note the base-10 place value in play. In the first instance,

$$3.141\,59 = 3 \times 10^0 + 1 \times 10^{-1} + 4 \times 10^{-2} + 1 \times 10^{-3} + 5 \times 10^{-4} + 9 \times 10^{-6}$$
$$= 3\frac{14\,159}{100\,000}$$

is three and fourteen thousand, one hundred fifty-nine-hundred-thousandths. In the second instance,

$$3141.59 = 3 \times 10^3 + 1 \times 10^2 + 4 \times 10^1 + 1 \times 10^0 + 5 \times 10^{-1} + 9 \times 10^{-2}$$
$$= 3141\frac{59}{100}$$

is three thousand, one hundred forty-one and fifty-nine-hundredths. □

Remark 6.1 The *American* English word names for some place values differ from the British English word names for the same place values in other parts of the world. (See SECTION 12.4.) Also, as we noted earlier, in the United States as well as in many other parts of the world, we use a point or dot to separate the whole number part of a numeral from the fraction part; however, in some parts of the world, a comma is used for the separatrix—for instance, 3,14 for 3.14—and in yet other parts a *raised* dot is used—for instance, 3·14 for 3.14. Finally, in the United States, the product of 3 and 14, for example, may be written

$$3 \cdot 14,$$

but in countries that use a raised dot for the separatrix, the product may be written

$$3 . 14$$

instead. All in all, there is definitely room for confusion. □

According to Calinger [35], the positional feature of the Indo-Arabic number system and the fact that it is base 10 (without a zero) are very ancient. Some scholars hold that this was an indigenous development in India, while others hold that it was due to Greco-Babylonian or Chinese influence. The symbols 0 to 9 for the digits that we use today can be traced back at least to the Brahmi symbols for the digits 1, 4 to 9, multiples of 10, and multiples of 100 to 900 that appeared during India's Ashoka's reign (d. 232 BC). So, we see that early Indians used more than ten symbols to express a number. The work of Indian mathematician Āryabhata (476–550) helped move this system forward. Calinger states that in Āryabhata's work *Āryabhatiya*, a three-book collection of mathematical problems, he "constructed his Brahmi numeral system so that 'from place to place each [numeral] is ten times the preceding.' In this system, the number 'eight hundred fifty six' [*sic*] was written no longer as 800′50′6 [that is, using the symbol for 800, the symbol for 50, and the symbol for six], as in the old Brahmi manner, but as the Brahmi equivalent of 856 [eight-five-six]" [35, p. 276]. But it would take nearly three-quarters of a millennium until the symbols for numbers beyond nine were discarded.

Cooke [41, p. 118] writes that the "idea of using a symbol for an empty place was the final capstone on the creation of a system of counting and calculation that is in all essential aspects the one still in use."[10] While a symbol for an empty place may have been used in India well over 1500 years ago, no such symbol appears in Āryabhata's work. It would not be until in the work of Brahmagupta (598–670) that such a symbol appears [41]. Also, in Brahmagupta's major work, *Brāhmasphuṭasiddhānta*, he explains zero as the result of subtracting a number from itself [112]. And he goes on to explain addition, subtraction, and multiplication with zero correctly, but explains that zero divided by zero is zero, and that any nonzero number divided by zero is a fraction with zero in the denominator. Nevertheless, we see that by this time the conception of zero was not only as a place holder, but also as a number. (See REMARK 1.2 on page 5.)

The symbol "0" that we use today can be traced back to an inscription at Gwalior, India, dated about 876 [112].[11] Otherwise, a dot or small circle was used. The Indians called the symbol *sunya* (void), which the Arabs translated as *as-sifr*, and which became *zephirim* in medieval Europe. It is from the word *zephirim* that we get the words zero and cipher. Interestingly, the Indians also used the dot to represent an unknown quantity, in that the unknown quantity was empty until found [30].

One way that the Indo-Arabic number system entered Europe was through Latin translations of the book *On the Indian Numbers*[12] by the Arabic mathematician al-Khwārizmī (ca. 780–850) that was probably based on the work of Brahmagupta [112]. By this time, the Brahmi numerals had evolved into two forms, the West Arabic or Gobar (or Gubar) numerals and the East Arabic numerals. It was the Gobar numerals that entered Europe. They were often referred to as "dust numerals" probably to describe the dust or sand boards often used in calculations [30]. Meanwhile, the rest of the Arab world used what evolved into the East Arabic numerals. So, the symbols of numeration that Europeans of the middle ages used were not the symbols of numeration that Arabs of the middle ages used!

The earliest appearance of Indian numerals outside of India is in a reference in 662 by Severus Sebokht, a Syrian priest, in a fragment of a manuscript (MS Syriac [Paris], No. 346) in which he mentions the *nine* signs for numeration [19, 31]. Katz [73] speculates that perhaps Sebokht did not consider a dot to be a sign. Since then, as Cajori [31, pp. 48–50] tells us,

> The earliest Arabic manuscripts containing the [Indian] numerals are of 874[13] and 888 A.D. They appear again in a work written at Shiraz in Persia[14] in 970 A.D. A church pillar[15] not far from the Jeremias Monastery in Egypt has the date 349 AH (= 961 AD).

It is not until around 976 in Spain that we find the earliest appearance of the Indo-Arabic numerals in Europe in the *Codex Vigilanus*, written in the Albelda Cloister in Spain [19, 31]. (See FIGURE 6.8.) After that several people helped to spread the new system [112]. One was Gerbert Aurrilac (ca. 940–1003), later Pope Sylvester II, who wrote a treatise to introduce the system, but he had left out "0" because it was not yet known in western Europe. Another was Leonardo of Pisa (ca. 1170–1240), later known as Fibonacci, who wrote *Liber Abaci* (1202) to familiarize merchants with the Indo-Arabic number system [112] (see page 107). Leonardo was one of the first Italian mathematicians. As the Indo-Arabic number system spread

---

[10]For a different perspective on the significance of the introduction of a symbol for zero, we turn to Biggs, who opines [12, p. 73] after giving a brief history of the use of abacuses in Europe for calculating: "Even on the simplest abacus, zero could be indicated by the absence of a counter. Consequently, the zero should be seen as an important part of the algorithmic process, rather than a ground-breaking concept."

[11]The Web site <http://www.ams.org/featurecolumn/archive/india-zero.html> describes the fort at Gwalior within which is the ninth-century Hindu temple where the oldest recorded "0" is inscribed on a dedication tablet, along with photographs of the inscription.

[12]*Kitāb al-jam'wal tafrīq bi ḥisāb al-Hind*, that is also translated variously as *Book on Addition and Subtraction after the Method of the Indians* [73, p. 240] and *Book of Addition and Subtraction According to the Hindu Calculation* [34, p. 182], for example.

[13]Karabacek, *Wiener Zeitschrift für die Kunde des Morgenlandes*, Vol. II (1897), p. 56.

[14]L. C. Karpinski, *Bibliotheca mathematica* (3d ser., 1910–11), p. 122.

[15]Smith and Karpinski, *op. cit.*, p. 138–43.

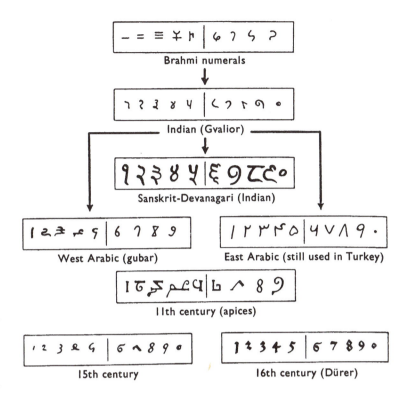

Figure 6.8: The evolution of the Indo-Arabic numerals from India to Europe.
(Source: Karl Menninger's *Number Words and Number Symbols: A Cultural History
of Numbers* published by Dover, November 2011.)

in Europe, the ease with which a number written with these numerals can be altered did not escape the
attention of the establishments of the time. Indeed, in 1299, the bankers guild in Florence, Italy, issued an
edict that "forbade the use of Indo-Arabic numerals in account books 'or any part thereof where payments
and receipts are recorded', and insisted that the numbers be written 'openly and at length, using letters' "
[12, p. 73];[16] and the University of Padua required that book prices be marked " '*non per cifras, sed per
literas clara*' (not by figures, but by clear numbers)" [99]. (Even today, when writing a check, we are
required to write the word name for the amount in addition to writing the amount using numerals, so there
still seems to be a distrust of Indo-Arabic numerals, at least in this instance.) But Europe would eventually
embrace this new number system. Indeed, Cajori [31, p. 70] provides us with the following quotation
by Pierre Simon de Laplace (1749–1827),[17] who was perhaps "the greatest mathematical physicist since
Newton" [112], that expresses his admiration of the Indo-Arabic number system:

> It is from the Indians that there has come to us the ingenious method of expressing all numbers,
> in ten characters, by giving them, at the same time, an absolute and a place value; an idea fine and
> important, which appears indeed so simple, that for this very reason we do not sufficiently recognize its
> merit. But this very simplicity, and the extreme facility which this method imparts to all calculation,
> place our system of arithmetic in the first rank of the useful inventions. How difficult it was to invent
> such a method one can infer from the fact that it escaped the genius of Archimedes and of Apollonius
> of Perga, two of the greatest men of antiquity.

---

[16]Biggs [12, p. 74] points out that this edict did not prevent bankers in Florence from using Indo-Arabic numerals in performing
numerical calculations and that, "indeed they had probably been doing so for at least a century, and it would have been impossible
for them to carry on their businesses in any other way."

[17]*Exposition du système du monde* (6th ed.; Paris, 1835), p. 376.

It is interesting to note that even as late as the fifteenth century, the Indo-Arabic numerals were still not widely used in Germany, France, and England.

In China, it was even only more recently that Indo-Arabic numerals were introduced. As Liping Ma tells us:[18]

> Today it might be hard to believe this, but a hundred years ago most Chinese people had not even seen the Hindu-Arabic numeral system, let alone learned how to calculate with it. During the late 1880s, a U.S. missionary, Calvin Wilson Mateer, wrote the first Chinese textbook on Western arithmetic and used it to teach his Chinese students. The title of his book, "pen-calculation arithmetic," suggests how it differed from traditional Chinese arithmetic—"abacus-calculation arithmetic." At that time, writing numbers horizontally was unknown to the Chinese.[19] The numeral system used in China did not have place value and was written vertically.[20] In his textbook, Mateer carefully and thoughtfully introduced to Chinese people the Hindu-Arabic numeral system and algorithms in this system for addition, subtraction, multiplication, and division. With great care, he composed all kinds of "practical problems" that seamlessly fit everyday Chinese social and economic life, so that learners could easily benefit from the content of the book. That was the dawn of modern Chinese mathematics education.

Finally, we remark that Lam and Ang advance the thesis that the Indo-Arabic number system has its origins in the Chinese rod numeral system (SECTION 4.2). They summarize their evidence for this claim as follows [77, p. 22]:

1. The earliest known methods of multiplication and division were identical with the earliest recorded Chinese methods.

2. Both numeral systems have a place value notation of base ten. They are the only known numeral systems with this notation. The Indian numeral system from which the first nine sysmbols were derived does not have this notational property.

3. The Arab schorlars never claimed they originated the numeral system, while the Western and Indian scholars were unable to provide convincing evidence of its Indian origin.

4. The Chinese use the rod numeral system continuously for almost 2000 years. The system and the accompanying arithmetic so essential for the progress of a nation were easily available to those foreigners, who were able to grasp their significance and adapted them to suit their own needs.

6.3.1 Whole Numbers

Suggestions that the Indians used a decimal positional or place-value system for most of their history can be seen from some of the Vedic texts. For example, Plofker [95] points out that,

> One sacrificial ritual in the *Yajur Veda* [which is among the oldest known Vedic compositions] invokes not only gods and various creatures, but also sets of numbers: "Praise to one, praise to two, praise to three …" and so on (*Yajur Veda* VII 2.11–20). One of these sets lists successive powers of ten from a hundred up to a trillion!"

Further evidence comes to us from the *Yavanajātaka* (*Greek Horoscope*) of Sphujidhvaja, a third century AD Sanskrit version of a Greek astrological text [95]. In it we find the use of "concrete numbers" or "object numerals," that is to say, the use of common word names to represent particular numbers. (This is like our saying "goose egg" to mean "zero.") As Plofker describes it [95, p. 395],

---

[18]Liping Ma, *Knowing and Teaching Elementary Mathematics: Teachers' Understanding of Fundamental Mathematics in China and the United States*, Routledge, New York (2010), pp. xii–xiii.

[19]Daniel W. Fisher, 1911, *Calvin Wilson Mateer, Forty-Five Years A Missionary in Shantung, China: A Biography*, Westminster Press, p. 162.

[20]There are three essentially different methods of arranging numeral characters to form a numeral system: additive, multiplicative, and positional notation. Roman numeration belongs to the category of additive notation. Chinese numeration belongs to the category of multiplicative notation, and it used to be written only vertically. Hindu-Arabic numeration belongs to the category of positional notation.

any number word may be represented by the name of any object or being that is naturally or traditionally found in sets of that number—"moon" and "earth" both represent "one," for example, while "eye" and "twin" mean "two," "limb" is "six" (for the six limbs of the Veda), "tooth" is "thirty-two," and so on. Words for "sky," "void," and "dot" mean "zero," referring to the general concept of emptiness and probably also to the circular shape of its symbol.

Numbers that were then constructed using this means would be put together in a decimal positional way. The example that Plofker gives is "moon-eye-limb-moon" for the digits "one-two-six-one." Were we to write this as a numeral, 1261, we would obtain the number one thousand, two hundred sixty-one:

$$1261 = \mathbf{1} \times 1000 + \mathbf{2} \times 100 + \mathbf{6} \times 10 + \mathbf{1}.$$

And we would be wrong! This is because, while numerals were ordered with the place values beginning with the most significant digit on the left, this *verbal encoding of numbers begins with the* least *significant digit* [95]. Thus, "moon-eye-limb-moon" ("one-two-six-one") is correctly interpreted as

$$\mathbf{1} + \mathbf{2} \times 10 + \mathbf{6} \times 100 + \mathbf{1} \times 1000$$

or one thousand, six hundred twenty-one; but *numerals were written beginning with the most significant digit*: 1621. Still, despite all of the circumstantial evidence that the Indians used a decimal positional system for a very long time, the earliest physical documents that have been found in India that record numbers using a decimal positional system are inscriptions in stone and metal deed-plates from only the first half of the first millennium AD.

A question that arises is how did we come upon the ordering of the place values in the Indo-Arabic system we use today, from larger to smaller, left to right? After all, the Indian numerals were transmitted to Europe via the Arabs, and

Indian script is written left to right; but
Arabic script is written right to left; and
Latin (and Greek) script is written left to right.

The fact is that there was some confusion at the beginning over the order in which Indo-Arabic numerals should be written when the numerals were brought into Latin, to say nothing of the Greek, so this is worth a little discussion.

For example, Burnett [18] tells us that one late thirteenth-century manuscript of an algorismic[21] text explains, "The rule is that every numeral, when it is placed before another numeral, signifies itself only, but when placed after another numeral, signifies ten times its value." Burnett notes that "it is clear he understands the lower value to be on the left, the higher on the right: '12' is twenty-one, '21' is twelve, '09' is ninety, '001' is one hundred, '002' is two hundred, and '003' is three hundred, etc." However, as another example, Burnett points out that in *Liber Alchorismi de practica arismetice*, another early algorism, the author writes in Latin that "the ten is in the 'second position *towards the left of the writer*', whereas the zero is in the 'first position *towards the right of the writer*' [Burnett's emphasis]," which is just how we write numbers today.

To sort out the confusion, we begin with how the Indians themselves expressed numerals. Plofker tells us[22] that we first need to distinguish among three different expressions for numbers: verbal, numerical, and verbally encoded. By verbal we mean the word names of numbers. Like in other cultures, the Indians used word names that started with the *most significant place*, for example, "forty-three," "five hundred seventy-six," and "one thousand, six hundred twenty-one." Consequently, they wrote their numerals *beginning on the left* (because they wrote from left to right) *with the most significant digit* (MSD), for example, "43" for forty-three, "576" for five hundred seventy-six, and "1621" for one thousand, six

---

[21]The word "algorism" derives from the name of the Arabic mathematician al-Khwārizmī, and is from which we have the word "algorithm."

[22]History of Mathematics Special Interest Group of the Mathematical Association of America (HOM SIGMAA) listserv posting (thread Subj: Hindu-Arabic numerals written in Hindu or Arabic), January 28, 2009. <http://www.homsigmaa.org/>

hundred twenty-one. Their verbal encoding of numbers, however, as we mentioned above, begins with the *least significant digit* (LSD), for example "moon-eye-limb-moon" for the digits "one-two-six-one": $1 + 2 \times 10 + 6 \times 100 + 1 \times 1000$ or one thousand, six hundred twenty-one; but this did not affect how they wrote numerals, to wit, beginning with the MSD on the left: 1621.[23]

Arabic mathematics was influenced by both Greco-Babylonian and Indian mathematics. One reflection of this is the fact that early on the Arabs expressed numbers with a decimal integer part and a sexagesimal fraction part (and without a separatrix) [35]. It is not until about 952 that we find the earliest examples of their use of decimal fractions in *Book of Chapters on Hindu Arithmetic* by Abu'l-Hasan al-Uqlīdisī (920–980) [112]. Now, following the Indians, the Arabs expressed numbers from smaller to larger place value. However, Arabic writing is from right to left and, when Arabic works were translated into Latin, translators transcribed the numbers exactly as they appeared in the Arabic texts. [112].

As we noted above, there was some confusion for the ordering of the place values when the Indo-Arabic numerals entered the West. In the end, as we know, the West settled upon writing numerals with the MSD on the left, in accordance with the Indian and the Arabic practices. Burnett [18] gives two reasons for this. First, numerals in Arabic manuscripts were generally set apart from the text, appearing in examples of computations, or were boxed when placed within text. As such, they were generally copied as they appeared (with the MSD on the left), just as diagrams were also generally copied as they appeared. Second, Latin (and Greek) number word names generally were spoken and written ordered from higher place values to lesser place values (corresponding to the fact that Roman numerals and Greek alphabetical numerals were written with the higher-valued numbers first, that is, on the left).[24]

6.3.2   Fractions

The Indians expressed fractions much like we do, vertically with the numerator above the denominator, but they did not use a fraction line or fraction bar. (Cf. rod numeral fractions on page 38.) For instance, the following example appears in the Bakhshālī Manuscript [31, p. 78]:

$$\left| \begin{matrix} 5 & 32 \\ 8 & 1 \end{matrix} \right| \ \textit{phalaṁ } 20$$

meaning $\frac{5}{8} \times \frac{32}{1} = 20$. We also find expressions like

$$\begin{matrix} 1 \\ 1 \\ 3+ \end{matrix}$$

which means $1 - \frac{1}{3}$ or $\frac{2}{3}$, where the cross "+" is used for subtraction instead of addition; without the "+" sign:

$$\begin{matrix} 1 \\ 1 \\ 3 \end{matrix}$$

would mean $1 + \frac{1}{3}$ or $1\frac{1}{3}$. Thus,

$$\begin{matrix} 1 & 1 & 1 \\ 1 & 1 & 1 \\ 3+ & 3+ & 3+ \end{matrix} \quad \text{means} \quad \frac{2}{3} \times \frac{2}{3} \times \frac{2}{3} = \frac{8}{27}.$$

---

[23]Hebrew numerals are written with the MSD on the left and are said beginning with the MSD, but Hebrew script is written from right to left. Thus, when reading a sentence in Hebrew that contains a numeral, one would have to skip to the "end" of the numeral to read it from left to right, and then resume reading the text from right to left.

[24]Burnett [18, 19] provides a fuller and quite interesting discussion of all of this.

As another example, Cajori [31, p. 82] offers the following from Bhāskara II's *Līlāvatī*

> Example: Tell me the fractions reduced to a common denominator which answer to three and a fifth, and one-third, proposed for addition; and those which correspond to a sixty-third and a fourteenth offered for subtraction. Statement:
>
> $$\begin{array}{ccc} 3 & 1 & 1 \\ 1 & 5 & 3 \end{array}$$

Answer: Reduced to a common denominator

$$\begin{array}{ccc} 45 & 3 & 5 \\ \overline{15} & \overline{15} & \overline{15} \end{array} \cdot \quad \text{Sum}\ \frac{53}{15} \cdot$$

Statement of the second example:

$$\begin{array}{cc} 1 & 1 \\ 63 & 14 \end{array} \cdot$$

Answer: The denominator being abridged, or reduced to least terms, by the common measure seven, the fractions become

$$\begin{array}{cc} 1 & 1 \\ 9 & 2 \end{array} \cdot$$

Numerator and denominator, multiplied by the abridged denominators, give respectively $\frac{2}{126}$ and $\frac{9}{126}$. Subtraction being made, the difference is $\frac{7}{126}$.

We elaborate upon Bhāskara's examples.

In his first example, Bhāskara asks for the sum of three-and-one-fifth and one-third, for which he expands $3\frac{1}{5}$ as $3 + \frac{1}{5}$. Thus, he finds that

Bhāskara

$$3\tfrac{1}{5} + \tfrac{1}{3} = 3 + \frac{1}{5} + \frac{1}{3}$$

$$= \frac{3}{1} + \frac{1}{5} + \frac{1}{3} \qquad \begin{array}{ccc} 3 & 1 & 1 \\ 1 & 5 & 3 \end{array}$$

$$= \frac{45}{15} + \frac{3}{15} + \frac{5}{15} \qquad \begin{array}{ccc} 45 & 3 & 5 \\ 15 & 15 & 15 \end{array}$$

$$= \frac{53}{15} \qquad\qquad \text{Sum}\ \frac{53}{15} \cdot$$

In his second example, Bhāskara asks for the difference between one-sixty-third and one-fourteenth, but later subtracting the smaller from the larger. To begin, he notes that 63 and 14 have a common factor ("common measure") of 7; in fact,

$$63 = 7 \cdot 9 \quad \text{and} \quad 14 = 7 \cdot 2,$$

which he indicates as

$$\begin{array}{cc} 1 & 1 \\ 9 & 2 \end{array} \cdot$$

Then, using the "abridged denominators" of 9 and 2, he expresses the two fractions with a common denominator of 126:

$$\frac{1}{63} = \frac{1}{7 \cdot 9} \cdot \frac{2}{2} = \frac{2}{126},$$

$$\frac{1}{14} = \frac{1}{7 \cdot 2} \cdot \frac{9}{9} = \frac{9}{126}.$$

Thus, Bhāskara finds that

$$\frac{1}{14} - \frac{1}{63} = \frac{9}{126} - \frac{2}{126} = \frac{7}{126}.$$

According to Cajori [31, p. 269], the first known use of the fraction line or fraction bar comes to us from the twelfth-century AD Arabic author al-Ḥaṣṣâr, who directed: "Write the denominator below a [horizontal] line and over each of them the parts belonging to it; for example, if you are told to write three-fifths and a third of a fifth, write thus, $\frac{3\ 1}{5\ 3}$." Later, in Leonardo of Pisa's *Liber Abaci* (1202) [31, p. 89] we find expressions like "$\frac{1}{2}12$" (for twelve-and-a-half) using the fraction line; Leonardo consistently wrote the integer part of a mixed number to the right of the fraction part and read the numeral from right to left. Moreover, Leonardo expressed fractions similarly to the Arabic fashion. For instance, in chapter 5 of *Liber Abaci*, Leonardo explains [90, p. 50]:

> If under a certain fraction line one puts 2 and 7, and over the 2 is 1, and over the 7 is 4, as here is displayed, $\frac{1\ 4}{2\ 7}$, four sevenths plus one half of one seventh are denoted. However, if over the 7 is the zephir, thus $\frac{1\ 0}{2\ 7}$, one half of one seventh is denoted. Also under another fraction line are 2, 6, and 10; and over the 2 is 1, and over the 6 is 5, and over the 10 is 7, as is here displayed, $\frac{1\ 5\ 7}{2\ 6\ 10}$, the seven that is over the 10 at the head of the fraction line represents seven tenths, and the 5 that is over the 6 denotes five sixths of one tenth, and the 1 which is over the 2 denotes one half of one sixth of one tenth, and thus singly, one at a time, they are understood.... And if in the fraction line there will be made several fraction parts, and the fraction line will terminate in a circle, then the fractions of it will be denoted in another way than was said, as in this $\frac{2\ 4\ 6\ 8}{3\ 5\ 7\ 9}o$ in which the line denotes fractions, eight ninths of the whole, and six sevenths of eight ninths, and four fifths of six sevenths of eight ninths, and two thirds of four fifths of six sevenths of eight ninths of the whole. And if this fraction line will terminate from another part in a circle thus, $o\frac{8\ 6\ 4\ 2}{9\ 7\ 5\ 3}$, it will denote two thirds of four fifths of six sevenths of eight ninths of a whole. Also if fraction lines will be drawn above the fraction line in this manner, $\frac{1\ 1\ 1\ 5}{5\ 4\ 3\ 9}$, it denotes fractions of five ninths and a third and a fourth and a fifth of one ninth.

Leonardo's convention may have been influenced by the monetary systems at the time. For example, in the Roman system of lira, soldi, and denari, in which there are 12 denari in 1 soldo and 20 soldi in 1 lira, it would be easy to express 5 lira, 6 soldi, and 4 denari as $\frac{4\ 6}{12\ 20}5$ [61, 73].

Example 6.2 We elaborate on Leonardo's examples above.

$$\frac{1\ 4}{2\ 7} = \frac{4}{7} + \frac{1}{2}\cdot\frac{1}{7} = \frac{9}{14}$$

$$\frac{1\ 0}{2\ 7} = \frac{0}{7} + \frac{1}{2}\cdot\frac{1}{7} = \frac{1}{14}$$

$$\frac{1\ 5\ 7}{2\ 6\ 10} = \frac{7}{10} + \frac{5}{6}\cdot\frac{1}{10} + \frac{1}{2}\cdot\frac{1}{6}\cdot\frac{1}{10} = \frac{95}{120}\quad\text{or}\quad\frac{19}{24}$$

$$\frac{2\ 4\ 6\ 8}{3\ 5\ 7\ 9}o = \frac{8}{9}\cdot 1 + \frac{6}{7}\cdot\frac{8}{9}\cdot 1 + \frac{4}{5}\cdot\frac{6}{7}\cdot\frac{8}{9}\cdot 1 + \frac{2}{3}\cdot\frac{4}{5}\cdot\frac{6}{7}\cdot\frac{8}{9}\cdot 1 = \frac{2520}{945}\quad\text{or}\quad\frac{8}{3}$$

$$o\frac{8\ 6\ 4\ 2}{9\ 7\ 5\ 3} = \frac{2}{3}\cdot\frac{4}{5}\cdot\frac{6}{7}\cdot\frac{8}{9}\cdot 1 = \frac{384}{945}\quad\text{or}\quad\frac{128}{315}$$

$$\frac{1\ 1\ 1\ 5}{5\ 4\ 3\ 9} = \frac{5}{9} + \frac{1}{3}\cdot\frac{1}{9} + \frac{1}{4}\cdot\frac{1}{9} + \frac{1}{5}\cdot\frac{1}{9} = \frac{347}{540}$$

□

Now You Try 6.1 Find the values of the following fractions that are expressed using Leonardo's conventions.

1. $\dfrac{3\ 4}{4\ 5}$

2. $\dfrac{3\ 0}{4\ 5}$

3. $\dfrac{2\ 4\ 4}{3\ 5\ 5}o$

4. $\dfrac{2\ 4\ 4}{3\ 5\ 5}9$
5. $\text{o}\dfrac{2\ 4\ 4}{3\ 5\ 5}$
6. $\dfrac{2\ \underline{4}\ 4}{3\ 5\ 5}$

☐

After the introduction of the printing press, it was found that typesetting fractions in the form $\frac{a}{b}$ was cumbersome, so a move was made to use the *solidus* "/" to typeset fractions, for example, typesetting a/b for $\frac{a}{b}$. The British mathematician Augustus De Morgan (1806–1871), for instance, recommended using the solidus in his article, "The calculus of functions," that was published in the *Encyclopaedia Metropolitana* (1845) [31, p. 313]; yet De Morgan, himself, used a colon, as in $a : b$ for $\frac{a}{b}$ in his subsequent work. Both notations were adopted by others.

Other notations for common fractions[25] were also used by others, but eventually they all fell out of use. Decimal fractions (such as 2.718 that we would write today), on the other hand, were not used in Europe even as late as the Renaissance [73, p. 375]. It would have to wait for Simon Stevin (1548–1620) from the Netherlands, in his work *De Thiende* (*The Art of Tenths*; 1585) to introduce decimal fraction notation in Europe.

The decimal fraction notation that Stevin created appears differently from what we use today but is easily understood nevertheless. Stevin refers to a whole number as a *unit* [39] or *commencement* [73], denoted ⓪; then he describes his notation for decimal fractions thus: "And each tenth part of the unity of the commencement we call the *prime*, whose sign is ①, and each tenth part of the unity of the prime we call the *second*, whose sign is ② and so of the other; each tenth part of the unity of the precedent sign, always in order one further." Stevin calls the numbers written according to his description *decimal numbers*. In this way, in Stevin's notation, 2 ⓪ 7 ① 1 ② 8 ③ denotes what we would write today as 2.718 or $2\frac{718}{1000}$. As another example, 365 ⓪ 2 ① 5 ② denotes 365.25 or $365\frac{25}{100}$. Note that, except for the sign ⓪, only single digits appear to the left of the signs ①, ②, and so on.

But Stevin did not stop at introducing notation for decimal numbers. In *De Thiende*, he also explained how to use the new notation. For example, in addition, subtraction, multiplication, and division, he showed how operating with decimal numbers is exactly the same as operating with whole numbers as long as one kept track of the signs. For example, here is how Stevin explains multiplication with decimal numbers [39, p. 39]. Notice that it is just as we would do it today: multiply the decimal numbers as whole numbers, then count the number of decimal places in the multiplicands to determine the position of the decimal point in the product.

> *La Disme*[26] Proposition III – to multiply decimal numbers
> Given the number 32 ⓪ 5 ① 7 ② and the multiplier 89 ⓪ 4 ① 6 ②. Required to find their product.
>
> Construction
>
> Place the numbers in order and multiply in the ordinary way of multiplying whole numbers (by the third problem of *l'Arithmétique*). This gives the product 29137122. To find what this is, add the last two signs of the given numbers, the one ② and the other ② also, which together are ④. We say then that the sign of the last figure of the product will be ④. Once this is established, all the signs are known on account of their continuous order. Therefore, 2913 ⓪ 7 ① 1 ② 2 ③ 2 ④ is the required product.

Stevin goes on to provide a proof.

Now You Try 6.2 Express the following as decimal fractions using Stevin's notation.

---

[25]Today, a *common fraction* or *vulgar fraction* is an expression of the form $\frac{a}{b}$, where a and b are integers with $b \neq 0$. A common fraction $\frac{a}{b}$ is said to be in *lowest terms* if b is positive and a and b have no common factor greater than one. The fraction $\frac{0}{b}$ in lowest terms is $\frac{0}{1}$.

[26]From *La Disme* [*De Thiende*] in *Œurves mathématiques de Simon Stevin de Bruges*, ed. Girard, pub. Leyden, 1634, pp. 208–209. English translation: Vera Sanford in E. E. Smith, *A Source Book in Mathematics*, McGraw-Hill Book Co., 1929, reprinted Dover, New York, 1959, pp. 24–27.

1. 12.753   2. $8\frac{12}{100}$ 3. $7\frac{9}{12}$ 4. $111\frac{2}{3}$

□

Our use of a point or a comma to separate the integer part of a numeral from the fraction part was first introduced by John Napier (1550–1617) in his *Rabdologia* of 1617 [31, p. 323]. According to Cajori [31], that Napier "demonstrated that the comma was intended to be used in this manner by performing a division, and properly placing the comma in the quotient, is admitted by all historians. But there are still historians inclined to the belief that he was not the first to use the point or comma as a separatrix between units and tenths." Cajori illustrates this division, shown here in FIGURE 6.9. Apparently, Napier's example shows $861{,}094 \div 432 \approx 1993.273$ using the galley method (see page 113). In *Constructio*, Napier's groundbreaking work on logarithms that was written before 1617, but was not published until 1619 (see page 461), he had a need for numbers with up to seven decimal places and explains the use of the decimal point thus: "In the numbers distinguished thus by a period in their midst, whatever is written after the period is a fraction, the denominator of which is unity with as many ciphers after it as there are figures after the period" [75, p. 454]. So, Napier moved back and forth between using a point and a comma as a decimal separatrix, and we find this continues to be the case today.

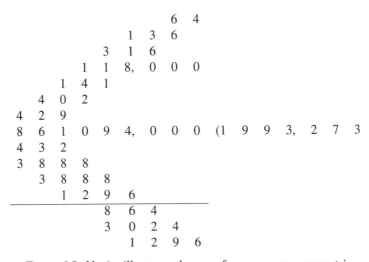

Figure 6.9: Napier illustrates the use of a comma as a separatrix.

In addition to the point and the comma, other forms were used as a decimal separatrix. As Cajori tells us [31, pp. 326–327],

> The great variety of forms for separatrix is commented on by Samuel Jeake in 1696 as follows: "For distinguishing of the Decimal Fraction from Integers, it may truly be said, *Quot Homines, tot Sententiae;* every one fancying severally. For some call the Tenth Parts, the *Primes;* the Hundredth Parts, *Seconds;* the Thousandth Parts, *Thirds,* etc. and mark them with *Indices* equivalent over their heads. As to express 34 integers and $\frac{1426}{10000}$ Parts of an Unit, they do it thus, 34.1′. 4′′. 2.6′′′. Or thus, 34.1′. 4′′. 2′′′. 6′′′′. Others thus, 34,1426′′′′; or thus, 34,1426(4). And some thus, 34.1 . 4 . 2 . 6 . setting the Decimal Parts at little more than ordinary distance one from the other..... Others distinguish the Integers from the Decimal Parts only by placing a Cöma before the Decimal Parts thus, 34,1426; a good way, and very useful. Others draw a Line under the Decimals thus, 34 <u>1426</u>, writing them in smaller Figures than the Integers. And others, though they use the Cöma in the work for the best way of distinguishing them, yet after the work is done, they use a Rectangular Line after the place of the Units, called *Separatrix,* a separating Line, because it separates the the Decimal Parts from the Integers, thus 34|1426. And sometimes the Cöma is inverted thus, 34'1426, contrary to the true Cöma, and set at top. I sometimes use the one, and sometimes the other, as cometh to hand." The author generally used the comma. This detailed statement from this seventeenth-century writer is remarkable for the omission of the point as a decimal separatrix.

So, perhaps surprisingly, we see that even after decimal fractions were introduced, the simplicity of using a single separatrix was not obvious and, like most other advances, needed to evolve.

We close with K. Subramaniam's post on an online discussion board on the Web:[27]

> It might interest members [of the discussion board] to know that at least 5 distinct types of numerals are still in currency in India: the international numerals, the Hindi numerals, the Marathi (Western state of Maharashtra) numerals, the Kannada numerals (Southern state of Karnataka), the Urdu numerals (I believe these are similar to the Arabic). All of these resemble each other to a greater or lesser extent, and of course, derive from older numerals used in India.
>
> The different numeral systems are preserved through the primary education system, where children in the respective regions are taught these numerals in addition to the international numerals. Besides, these numerals appear as page numbers in printed books, in bus tickets, etc.

---

[27] *The Math Forum @ Drexel*, November 29, 2002 <http://mathforum.org/kb/message.jspa?messageID=1184543&tstart=0>.

Chapter 7

Exercises

Ex. 7.1 — Express the following numbers using Chinese characters (TABLE 1.1 on page 3). Note that there is no need for a place holder symbol. For example, 五百九 is five hundred nine.

1. Seven thousand, nine hundred eighty-two.

2. Fifty-six thousand.

* **Ex. 7.2** — TABLE 1.1 on page 3 shows some Chinese number characters. Trace the history of the Chinese numerals.

* **Ex. 7.3** — Research other Chinese numbers and their evolution.

Ex. 7.4 — Rachel Hall [65] gives some word names for numbers in Igbo, a Nigerian language, and asks the following questions.

1 otu	6 isii	11 iri na otu	40 ohu abuo
2 abuo	7 asaa	12 iri na buo	50 ohu abuo na iri
3 ato	8 asato	20 ohu	100 ohu iso
4 ano	9 toolu	21 ohu na otu	300 ohu iri na ohu iso
5 iso	10 iri	30 ohu na iri	400 nnu

1. To what number does *ohu abuo na iri na ano* refer?

2. What is the Igbo word for 16? For 71?

3. What is the base of the Igbo counting system, and how do you know? What is the secondary base? (For example, the numerals we commonly use employ base-10 place values, but the American word names are base 1000.)

Ex. 7.5 — Make up new word names for numbers expressed in base 6 so that it would be easy to grasp the value of the numbers without having to convert to base 10.

Ex. 7.6 — Make up new word names for numbers expressed in base 4 so that it would be easy to grasp the value of the numbers without having to convert to base 10.

* **Ex. 7.7** — Hall, in the same article [65] (see exercise 7.4), writes,

> Among J. R. R. Tolkien's unpublished writings on the lore of his fictitious world, Middle Earth, is a base-12 number system which included both spoken and written numbers. My students create something similar. Each group of students develops an entire number system, complete with spoken and written

numbers, a multiplication table, a "history" explaining the origin of their numbers and significance within the fictional culture that produced them, and a mathematical artifact. Examples of "artifacts" my students have created include calendars, recipes, and even a tombstone. They could not use base 2, 5, 10, 12, 20, or 60.

Either by yourself or with another student, complete the assignment that Hall gave to her students. Exercise your imagination.

Ex. 7.8 — It is common to use six-digit base-16 or *hexadecimal* numerals in many computing applications, such as to specify colors when designing a Web page; see <http://www.computerhope.com/htmcolor.htm>. The following numerals are used to represent the numbers zero to fifteen in base 16; the numeral for zero doubles as a place holder.

0	1	2	3	4	5	6	7	8	9
zero	one	two	three	four	five	six	seven	eight	nine

A	B	C	D	E
ten	eleven	twelve	thirteen	fourteen

F
fifteen

When designing a Web page, each six-digit hexadecimal numeral specifies a combination of red, green, and blue (RGB) to produce a specific color. The first two digits specify the value of red, the second two specify the value of green, and the last two specify the value of blue. White, FFFFFF, is the combination of all three colors fully and black, 000000, is the absence of all three colors. Hence, each color can range in value from 0 to 255 because FF is two hundred fifty-five. That gives a total of 256^3 or 16,777,216 combinations of red, green, and blue, resulting in that many different colors.

1. Using a six-digit hexadecimal numeral we can specify 16,777,216 different colors. We call this 24-bit color because 16,777,216 is 2^{24}.

   a) How many colors are available in 16-bit color? In 32-bit color?

   b) How many colors are available in 8-bit color? (GIF or *Graphics Interchange Format* files use 8-bit color, which is why they sometimes appear to be smeared when viewed on a monitor that supports more than 8-bit color.)

   c) How many colors are available in 1-bit color? What are they?

2. In base 16 the place values are powers of 16. For example, the first few powers of 16 are

$$16^0 = 1, \quad 16^1 = 16, \quad 16^2 = 256, \quad 16^3 = 4096.$$

   Write the decimal word names for the following base-16 place-value numerals.

   a) FFFFFF (white)

   b) FF0000 (red)

   c) 00FF00 (green)

   d) 0000FF (blue)

   e) 810541 (maroon)

   f) 2B60DE (royal blue)

3. Express the following numbers as base-16 place-value numerals. What colors do they represent?

   a) Four thousand, three hundred seventy-nine.

   b) Ten thousand, five hundred sixty-nine.

   c) One hundred thirty-four.

d) Nine hundred forty-one.

Ex. 7.9 — In computer science it is common to use base 2 or *binary*. The following numerals are used to represent the numbers zero and one in base 2; the numeral for zero doubles as a place holder.

0	1
zero	one

A numeral in base 2 can be thought to be a collection of switches, each of which is either off or on. A 0 in a place value can be thought to mean that the switch in that position is *off*, and a 1 can be thought to mean that the switch is *on*. For instance, the numeral 1011 can be thought to mean *on-off-on-on*. This is what makes base 2 particularly suitable in computer science.

1. In base 2 the place values are powers of 2. What are the first five place values, beginning with 2^0?

2. Write the decimal word names for the following base-2 place-value numerals.

    a) 1000   b) 1011   c) 11001   d) 111111

3. Express the following numbers as base-2 place-value numerals.

    a) Ten.   b) Sixteen.   c) Thirty-nine.

4. Find the sum of ten and sixteen in base-2 place values.

5. Find the product of ten and sixteen in base-2 place values.

6. Find the difference between twelve and three in base-2 place values.

7. Find the quotient of twelve and three in base-2 place values.

8. Explain this joke: "There are only 10 types of people in the world—those who understand binary and those who don't."

Ex. 7.10 — We are familiar with the fact that, in base 10, multiplying by a power of 10 that is greater than 1 (that is, 10, 100, 1000, and so on) is easily done by appending an appropriate number of zeros after the other factor. For example,

$$4379 \times 1\underline{0} = 43{,}79\underline{0} \quad \text{and} \quad 2500 \times 10{,}000 = 25{,}000{,}000.$$

In the following, we see that this is true in other bases as well.

1. Perform the following operations in base-6 place values using the the base-6 numerals $\alpha, \beta, \gamma, \delta$, and ϵ, and place holder $\cdot$.

    a) Twenty-three times six (6^1). How many place holders are appended to the base-6 numeral for twenty-three in the product?

    b) Four times one thousand, two hundred ninety-six (6^4). How many place holders are appended to the base-6 numeral for four in the product?

2. Perform the following operations in base-4 place values using the base-4 numerals $|$, $\wedge$, and $\vee$, and place holder $\square$.

    a) Nine times sixty-four (4^3). How many place holders are appended to the base-4 numeral for nine in the product?

b) Sixty-one times sixteen (4^2). How many place holders are appended to the base-4 numeral for sixty-one in the product?

Ex. 7.11 — You are an archaeologist, and you have discovered the ruins of the ancient city of Middle of Nowhere. Among the artifacts you have unearthed are some bone carvings. After a long period of study, you have deciphered many of the carvings. Most interesting among the bone carvings you have found are what appear to be lessons teaching how to solve simple word problems. On four bones, you have decided that the following symbols must mean

⊗ ↗ ⊕ ↝ ⊙ ○	Three plus two is five.
⊕ ⊙ ↘ ⊗ ↝ ⊙ ⊗	Eleven minus three is eight.
⊙ ⊕ ◊ ⊙ ⊗ ↝ ⊕ ⊙ ⊙	Seven times eight is fifty-six.
⊗ ⊗ ⋔ ⊗ ↝ ⊙ ⊙	Eighteen divided by three is six.

Given that you have decided to assume that the people of Middle of Nowhere used a positional number system with a base that is less than 10, decipher the symbols above and name the base of the system.

* Ex. 7.12 — Trace the evolution of counting, from tally marks to "adjectival word names" to word names as nouns or abstract concepts in their own right. Distinguish between *cardinal numbers* and *ordinal numbers*.

* Ex. 7.13 — Ronald Calinger [35, p. 9] writes, "By the late precivilization period, human beings probably signaled numbers by holding up fingers or pointing to body parts.... Most likely, finger and body numbers were widely used because they transcended language differences. They are still used at auctions and in stock exchange signals." Let us call this "body counting." Which were some of the early civilizations that employed body counting? What were some of the different ways of body counting?

* Ex. 7.14 — What is *protomathematics*?

Ex. 7.15 — Our familiar base-10 number system uses the ten symbols 1, 2, 3, 4, 5, 6, 7, 8, 9, and 0, with 0 doubling as a place holder. Suppose, however, that we had only three symbols instead of ten, say, the following three.

|	∋	◇
one	three	place holder

Then, for example, ∋|| is five, | ∋|| is fifteen ($1\times10+5$), | ∋|| ∋ is one hundred fifty-three ($1\times100+5\times10+3$), and | ◇ ∋|| ∋ is one thousand fifty-three ($1 \times 1000 + 0 \times 100 + 5 \times 10 + 3$).

1. Write the following numbers in base-10 place values using the numerals | and ∋, and place holder ◇.

   a) Eighty-three.

   b) Two hundred nineteen.

   c) Six thousand, seven hundred nine.

   d) Three million, four hundred seventy-two thousand, six hundred thirteen.

2. Perform the following operations in base-10 place values using the numerals | and ∋, and place holder ◇.

   a) Two hundred eighteen plus one hundred twenty-three.

   b) Four hundred thirty-five minus one hundred twenty-nine.

c) Eighteen times twenty-three.

d) Two hundred twenty-four divided by seven.

The next exercises refer to the Babylonian number system. For convenience, we use our usual numerals instead of cuneiform script to express numerals in base 60. In addition, many historians use a semicolon or colon for a separatrix (to separate the whole number part of a number from the fraction part). For example, for  we may write

$$14\ 10 \quad \text{for} \quad 14 \times 60 + 10 \times 1, \quad \text{and}$$
$$0;14\ 10 \quad \text{for} \quad 14 \times \frac{1}{60} + 10 \times \frac{1}{3600}.$$

Ex. 7.16 — Find the values of the following sexagesimal numerals.

1. 15 07
2. 01 12 59
3. 08;30
4. 01 12;06

Ex. 7.17 — Decimal fractions and sexagesimal fractions.

Number	Decimal fraction	Sexagesimal fraction
one-fourth	$0.25 = \frac{2}{10} + \frac{5}{100}$	$0;15 = \frac{15}{60}$
one-third	$0.333\ldots = \frac{3}{10} + \frac{3}{100} + \frac{3}{1000} + \cdots$	

1. Verify that the sexagesimal fraction 0;15 is one-fourth.
2. The decimal fraction for one-third is nonterminating (0.333...). Find the sexagesimal fraction for one-third.
3. Compare decimal fractions with sexagesimal fractions with regard to whether they are terminating or nonterminating.

Ex. 7.18 — Perform the following operations in base 60.

1. 04 15 05 + 04 06 20
2. 04 15 05 + 03 50 27
3. 04 15 05 − 04 06 20
4. 04 06 20 − 03 50 27

Ex. 7.19 — Perform the following operations in base 60.

1. 04 15; 05 + 04 06; 20
2. 04 15; 05 + 03; 50 27
3. 04; 15 05 − 04; 06 20
4. 04 06; 20 − 03; 50 27

* **Ex. 7.20 —** There is some research that suggests that the base-60 system arose out of various mensuration systems in use. Find out what the mensuration systems were and how they may have led to the base-60 system of ancient Babylonia.

Ex. 7.21 — The following is a transcription of BM[1] 106444, a cuneiform tablet that was probably from the city of Umma [102].

Sixty: its 2nd part is	30
its 3rd part is	20
its 4th part is	15
its 5th part is	12
its 6th part is	10
its 7 1/2-th part is	8
its 8th part is	7;30
its 10-minus-1th part is	6;40
its 12th part is	5
its 15th part is	4
its 16th part is	3;45
its 18th part is	3;20
its 20th part is	3
its 24th part is	2;30
its 25th part is	2;24
its 27th part is	2;13 20
its 30th part is	2
its 32nd part is	1;52 30
its 36th part is	1;40
its 40th part is	1;30
its 45th part is	1;20
its 48th part is	1;15
its 50th part is	1;12
its 54th part is	1;06 40

1. Write the numbers in the right-hand column of the table shown above using cuneiform script.

2. Write the numbers in the right-hand column of the table using decimal (base-10) numerals.

3. The table has the form, "Sixty: its mth part is n," where n is a sexagesimal numeral. Relate the numbers sixty, m, and n in the table.

4. Why are some parts, such as the 7th, left out?

Ex. 7.22 — The following is a partial transcription of Ash[2] 1924.796. According to Robson [102, p. 156], "[the tablet], from sixth-century Kish, is a beautifully written library copy, with a decorative pattern of firing holes on its surface. It uses three vertically aligned diagonal wedges to indicate empty sexagesimal places, here marked by a colon : ."

[1 times 1	1	the square-side is 1]
[1 30 times 1 30	2 15	the square-side is 1 30]
[2 times 2	4	the square-side is 2]
[2 30 times 2 30	6 15	the square-side is 2 30]
⋮	⋮	⋮
[15] times 15	3 45	the square-side is 15
15 30 times 15 30	4 : 15	the square-side is 15 30

16 times 16	4 16	the square-side is 16
16 30 times 16 30	4 32 15	the square-side is 16 30
17 times 17	4 49	the square-side is 17
⋮	⋮	⋮

Duplicate the table shown above using base-10 numerals. Relate the numbers in the three columns.

Ex. 7.23 — The following is a transcription of the numerical data in Plimpton 322 [101]. (See page 14 for a brief discussion of Plimpton 322, and also FIGURE 2.3.) The empty brackets indicate the missing "digits" where the tablet had been chipped.

Column A	Column B	Column C	Column D
[ ] 00 15	1 59	2 49	1
[ ] 58 14 50 06 15	56 07	1 20 25	2
[ ] 41 15 33 45	1 16 41	1 50 49	3
53 10 29 32 52 16	3 31 49	5 09 01	4
48 54 01 40	1 05	1 37	[ ]
47 06 41 40	5 19	8 01	[ ]
43 11 56 28 26 40	38 11	59 01	7
41 33 45 14 3 45	13 19	20 49	8
38 33 36 36	8 01	12 49	9
35 10 02 28 27 24 26 40	1 22 41	2 16 01	10
33 45	45	1 15	11
29 21 54 2 15	27 59	48 49	12
27 00 03 45	2 41	4 49	13
25 48 51 35 6 40	29 31	53 49	14
23 13 46 40	28	53	15

1. Except for column A and the entries with missing digits, duplicate Plimpton 322 shown above using base-10 numerals. What do you notice?

2. Compare the entries in columns B and C. What is it about the entries in row 13 that is different from the other rows that suggests row 13 is in error?

3. Given two numbers, a and b, Mesopotamian scribes often used their average, $(a + b)/2$, and their semidifference, $(a - b)/2$. Extend the table by two columns; label the columns AVG and SD. Fill in column AVG with the average of columns C and B, and fill in column SD with the semidifference of C and B. What do you notice?

4. Compare the entries in columns AVG and SD. What is it about the entries in rows 13 and 15 that is different from the other rows that suggests rows 13 and 15 are in error?

5. Extend the table again by one column; label the column (p, q). Fill in column (p, q) with pairs of numbers p and q, where $p = \sqrt{AVG}$ and $q = \sqrt{SD}$. What do you notice? What can you say about rows 2, 9, 11, 13, and 15?

6. Extend the table by two more columns; label the columns DS and SS. Using the values in column (p, q), fill in the column DS with the difference of squares $p^2 - q^2$, and the column SS with the

sum of squares $p^2 + q^2$. What do you notice? In particular, compare columns B and DS, and also C and SS.

7. Row 9, among a few others, appears to be in error. Let us try to correct it and discover the source of the error. Remember that the original columns of Plimpton 322 are A, B, C, and D; we focus here on B and C, from which we obtained all the added columns.

   a) From $p = \sqrt{AVG}$ and $q = \sqrt{SD}$ we have the system of equations

   $$\frac{C + B}{2} = p^2,$$
   $$\frac{C - B}{2} = q^2.$$

   Solve the system for B and C in terms of p^2 and q^2.

   b) In row 9 we assume that $B = 541$ is wrong and $C = 749$ is correct. By trial and error, find integers p and q that give C. Express C using sexagesimal numerals.

   c) Using p and q, find B. Express B using sexagesimal numerals.

   d) Write the value of B that you found and the value of B that was inscribed on Plimpton 22 using cuneiform script. What do you think is a plausible reason for the error in the tablet?

Ex. 7.24 — When expressing the amount of time that has elapsed, it is common to use a colon to separate the hours, the minutes, and the seconds. In this way, 3:50:27 is 3 hours, 50 minutes, and 27 seconds. Suppose that four friends, A, B, C, and D, complete a marathon race (26.2 miles) in the following times.

   A: 3:50:27 B: 4:15:05 C: 4:06:20 D: 4:37:00

1. What is the average amount of time that all four friends took to complete the race?

2. How much less time did C take than B?

3. How much more time did C take than A?

* **Ex. 7.25** — "Mesopotamia" is a Greek word. What does "Mesopotamia" mean? Why do you think that region, that is today Iraq and Turkey, was called Mesopotamia?

* **Ex. 7.26** — What is the Epic of Gilgamesh? In particular, what is the story of Utnapishtim, and how does it compare with the Bible story of Noah and the ark?

* **Ex. 7.27** — What is a ziggurat, when were they built and by whom, and for what were they used?

* **Ex. 7.28** — Describe the Seven Wonders of the (Ancient) World? Write a paragraph on each of them.

* **Ex. 7.29** — Write about the Museum of Alexandria and the House of Wisdom. How do they compare?

* **Ex. 7.30** — Write about the Behistun Cliffs and their significance. Find images of the Behistun Cliffs.

* **Ex. 7.31** — Look at a map of the Middle East and Central Asia. What cities can you find that were founded by Alexander the Great?

* **Ex. 7.32** — Research the evolution of using clay tablets for keeping records in ancient Babylonia.

Ex. 7.33 — Write the fraction $\frac{4}{7}$ in two different ways using unit fractions as the Egyptians did.

Ex. 7.34 — Write the fractions $\frac{1}{4}, \frac{1}{3}, \frac{1}{2}, \frac{2}{3}$, and $\frac{3}{4}$ as parts of 12. For example, $\frac{1}{4} = \frac{3}{12}$.

Ex. 7.35 — Write the Roman fractions *sextans* and *dextans* as twelfths.

Ex. 7.36 — In base 12 or duodecimal, the place values are powers of 12.

　　1. What are the first five place values, beginning with 12^0?

　　2. One dozen is 12^1. What are one gross and one great gross?

　　3. What are some other examples of how we count by 12 today, that is to say, of how we have twelve *this* in one *that*?

* **Ex. 7.37** — Give several examples of how Roman numerals are used today.

* **Ex. 7.38** — Find and practice the Roman finger counting method.

* **Ex. 7.39** — Look up the Latin word names for the numbers one to twenty-nine; for the multiples of ten from thirty to ninety; and for the multiples of one hundred up to one thousand.

* **Ex. 7.40** — Look up the German word names for the numbers one to twenty-nine; for the multiples of ten from thirty to ninety; and for the multiples of one hundred up to one thousand.

* **Ex. 7.41** — The French language offers a good example of a vestigial base-20 system. What French numbers show this? What word in English is also remnant of a base-20 system? (Abraham Lincoln used it in the Gettysburg Address.)

* **Ex. 7.42** — "Hieroglyphic" is a Greek word. What does "hieroglyphic" mean? (The term was used to describe this type of writing because this writing was found along temple walls.)

* **Ex. 7.43** — Egyptian hieratic script is a cursive writing that was written on papyrus, leather, wood, and so on, using ink and a reed pen. What is papyrus?

* **Ex. 7.44** — How were numerals written in Egyptian hieratic script? In demotic? When were these alternative number symbols developed and why?

* **Ex. 7.45** — Who were the ancient Egyptian *harpedonáptai* or "rope stretchers"? What did they do, and what mathematics did they use?

* **Ex. 7.46** — Write about the Rosetta Stone and its significance.

* **Ex. 7.47** — Aristotle was a Greek scholar. Describe some of the things for which he is known.

* **Ex. 7.48** — Write about Archimedes and some of his mathematical achievements.

* **Ex. 7.49** — Write about Apollonius of Perga and some of his mathematical achievements.

* **Ex. 7.50** — Calinger [35, p. 188] states, "The Romans rank with the Old Kingdom Egyptians as the most skilled engineers in antiquity." What were some of the engineering achievements of the Romans?

* **Ex. 7.51** — Explain the ancient Greek number system.

* **Ex. 7.52** — Explain the Hebrew number system.

* **Ex. 7.53** — Explain the Julian calendar and the Gregorian calendar. Which calendar, if either, do we commonly use today, and why?

* **Ex. 7.54** — Describe the lunar calendar. (A good source is Helmer Aslaksen's Web site, "The Mathematics of the Chinese Calendar," <http://www.math.nus.edu.sg/aslaksen/calendar/chinese.html>.)

* Ex. 7.55 — Research the origin and development of the English system of measurement (inch, ounce, gallon, and so on).

Ex. 7.56 — According to Calinger [35, p. 304]:

> A sacred round calendar [the Tzolkin] of 260 days in 13 vigesimal months coexisted with a civil or chronological calendar of 360 days consisting of 18 vigesimal months. The civil year with 5 added days formed the vague calendar [the Haab], approximating the sidereal year. No account was taken of a leap year. Correlation of the dates of the sacred and the vague calendar involved passing through a complete cycle of 18,980 paired dates. This number arises from the indexing of the two calendars into a calendar round in which the passage of 52 vague years coincides with 73 sacred years.... Scribal mathematicians had to be able to compute the matching dates forward or backward in time. The approach of a calendar cycle's end point brought anxiety, and magical interventions were invoked against calamities.

Find the least common multiple of 260 and 365. To what does the least common multiple refer in the passage above? Explain how the least common multiple of 260 and 365 is related to 52, 73, 260, and 365.

Ex. 7.57 — According to Cooke [41, p. 125]: "An important aspect of Maya astronomy was a close observation of Venus. The Maya established that the synodic period of Venus (the time between two successive conjunctions with the Sun when Venus is moving from the evening to the morning sky) is 584 days." How many synodic periods are there in two Calendar Rounds? (Recall that there are 365 days in one haab year and 52 haab years in one Calendar Round; see page 42.) Because of this coincidence, Cooke goes on to say that "the Maya calendar bears a particularly close relation to this planet [Venus]."

Ex. 7.58 — Cooke [41, p. 125] tells us that the Long Count date for December 31, 2000, is 1,867,664.

1. Express this Long Count date in the form p haab years (of 360 "regular" days each), q tzolkin months (of 20 days each), and r days.

2. Express this Long Count date as a Mayan dot-and-bar numeral.

Ex. 7.59 — Consider the following ways we use to tell time: second, minute, hour, day, week, month, year, decade, century, and millennium.

1. If we take 1 year to be the unit or basic quantity of time, what is the base for this system of telling time?

2. What fractions of the unit (1 year) is then one second; one minute; one hour; one day; one week; and one month? (You will need to make some assumptions about the number of days in a week, in a month, and in a year.)

* Ex. 7.60 — Describe how the Maya reckoned time. Compare how the Maya reckoned time with how we reckon time.

* Ex. 7.61 — The Mayan Long Count calendar marks the number of days from a fixed starting point to record long intervals of time. How did they choose the fixed starting point and what is its significance?

* Ex. 7.62 — Write about the major pre-Columbian civilizations of Mesoamerica. Who were they? What time periods did they span? For what are they renowned?

* Ex. 7.63 — Write about the Incas of South America. Who were they? What time period did they span? For what are they renowned?

* Ex. 7.64 — Burton [21] tells us, "In the New World, the number string is best illustrated by the knotted cords, called *quipus*, of the Incas of Peru." Describe the quipus and how they were used for recording numbers.

* Ex. 7.65 — Describe the Mayan ball courts and the games that were played there.

* Ex. 7.66 — Describe the plaza of seven temples in the Mayan city of Tikal. Make particular note of their geometric features and astronomical orientation.

* Ex. 7.67 — Describe the Mayan head variant form of writing.

Ex. 7.68 — What is the flaw in Brahmagupta's explanation (see page 62) that zero divided by zero is zero, and that any nonzero number divided by zero is a fraction with zero in the denominator?

Ex. 7.69 — Recall that Plofker [95] explains that in concrete numbers or object numerals, "any number word may be represented by the name of any object or being that is naturally or traditionally found in sets of that number...."

1. Following the early Indian tradition of ordering the place values in this verbal encoding beginning with the least significant digit, find the values of the following:

   a) eye-limb-moon

   b) limb-twin-void-earth

   c) twin-sky-earth

   d) dot-tooth

2. Make up your own list of words to represent the digits 0 to 9, as well as any other numbers you would like, where "any number word may be represented by the name of any object or being that is naturally or traditionally found in sets of that number." For instance, "hand" could mean "five" and "baseball" could mean "nine." Use these words to translate the following into concrete numbers. Order the place values beginning with the least significant digit.

   a) one hundred thirteen

   b) four thousand, eight hundred forty

   c) forty-three thousand, five hundred sixty

   d) one million, two hundred twenty thousand, six hundred seventy-two

Ex. 7.70 — Verify the computations with fractions shown in the example from *Līlāvatī* on page 67.

Ex. 7.71 — A *palindrome* number is one that is the same whether it is read forward or backward. Here are some examples of palindromes:

$$767, \quad 63536, \quad 22, \quad 9, \quad 1881.$$

We may obtain a palindrome from a number that is not a palindrome by reversing the digits and adding, and repeating if necessary. Here are two examples.

Example 1 The number 312 is not a palindrome.

Reverse the digits: 213; now add: $312 + 213 = 525$, which is a palindrome.

Example 2 The number 362 is not a palindrome.

Reverse the digits: 263; now add: $362 + 263 = 625$, which is *not* a palindrome, so we repeat the process.

Reverse the digits: 526; now add: $625 + 526 = 1151$, which still is not a palindrome, so we repeat the process again.

Reverse the digits: 1511; now add: $1151 + 1511 = 2662$, which *is* a palindrome.

1. Create a palindrome from each of the following numbers.

$$32, \quad 17, \quad 375, \quad 9913$$

2. Suppose that ab is a 2-digit number, that is,

$$ab = a \times 10 + b,$$

with $a \neq b$. Explain why the process of "reversing and adding" (and repeating if necessary) will eventually result in a palindrome.

3. Why did the first two steps in the second example above *not* result in a palindrome?

* Ex. 7.72 — What are some of the American English word names for place values that differ from the British English word names for the same place values? What countries differ from the United States in usage?

* Ex. 7.73 — Name some countries that use a comma to separate the integer part of a numeral from its fraction part. Name some countries that use a raised dot.

* Ex. 7.74 — Give a short history of the development of the use of zero, both as a place holder and as a number.

* Ex. 7.75 — Arabic script is written from right to left, but numerals are written with the least significant digit on the right. Are numerals in Arabic read beginning with the least significant digit or with the most significant digit (MSD)? And if a numeral is read beginning with the MSD, how does a person proceed to read Arabic texts that contain numerals?

* Ex. 7.76 — Research the difference between the West Arabic (Gobar) numerals and the East Arabic numerals.

* Ex. 7.77 — Write a history of the decimal separatrix that separates the integer part of a numeral from the fraction part.

* Ex. 7.78 — A region in the Middle East is referred to as the Fertile Crescent. Who coined this moniker and why? Draw a map of the region.

* Ex. 7.79 — What is the significance to Judaism, Christianity, and Islam of the site upon which the Dome of the Rock, Jerusalem, is built?

* Ex. 7.80 — Briefly describe the technical difficulties in finding the direction of Mecca required of the faithful during their five daily prayers.

* Ex. 7.81 — Write about Āryabhata I, Brahmagupta, and Bhāskara II, and about some of their work in mathematics.

* Ex. 7.82 — Write about Srinivasa Ramanujan and about some of his achievements in mathematics.

* Ex. 7.83 — Write about Leonardo Pisano Bigolo, who later became known as Fibonacci, and some of his mathematical achievements.

* Ex. 7.84 — Fibonacci's operations with fractions seem very convoluted compared to today's methods. Research when some of our modern mathematical operation symbols were introduced. What symbols were available to Fibonacci?

* Ex. 7.85 — Write about Pierre Simon de Laplace (page 63) and some of his mathematical achievements.

* Ex. 7.86 — What are the mathematical philosophies of intuitionism and empiricism?

* **Ex. 7.87** — We mentioned that Omar Khayyam is one of the early notable Arabic mathematicians. For what is Omar Khayyam more commonly known to westerners?

* **Ex. 7.88** — Write about Pope Sylvester II and his contributions to mathematics.

Part II

Arithmetic Snapshots

Chapter 8

Addition and Subtraction

Humans began counting very early on in their use of language. The beginnings of computation are likewise lost to the sands of time. Every era and culture had different ways to accomplish the basic arithmetical operations of addition, subtraction, multiplication, and the like, that it needed for commerce and construction. In this chapter, we take a look at some of the ways the basic operations of arithmetic were carried out in various cultures.

Addition and subtraction are generally straightforward operations. To start, we examine how addition and subtraction were performed using Roman numerals.

Figure 8.1: Boethius and Pythagoras in competition. It depicts a competition between Boethius (left, using the Indo-Arabic numerals) and Pythagoras (using a counting board). Supervising the competition is Arithmetica (center). (Source: © Thinkstock.)

Today we generally do not use Roman numerals for computation, but recall that Roman numerals continued to be used widely in Europe even after the Indo-Arabic numerals were introduced in the early thirteenth century [31]. Indeed, in 1299 bankers in Florence, Italy, were required by law to use Roman numerals instead of the new Indo-Arabic numerals; and the University of Padua required that book prices be marked " '*non per cifras, sed per literas clara*' (not by figures, but by clear numbers)" [99]. (Even today, when writing a check, we are required to write the word name for the amount in addition to writing the amount using numerals.) FIGURE 8.1 is a famous image showing the the controversy in Europe over switching to the Indo-Arabic numerals. So, how were some computations done using Roman numerals?

Besides mental computations, the Romans used a counting board or an abacus much like the Chinese *suan pan* or Japanese *soroban* [99, 119]; in fact, the soroban may be descended from the Roman abacus [119]. See FIGURE 8.2.

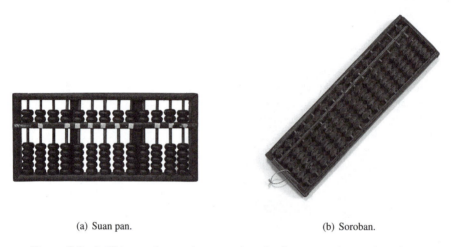

(a) Suan pan.   (b) Soroban.

Figure 8.2: A Chinese abacus (*suan pan*) and a Japanese abacus (*soroban*). (Source: © Thinkstock.)

The Roman abacus or counting board used two sets of counters separated by a horizontal bar; it could be a handheld device, or it could simply be any flat surface with a horizontal line drawn across it to separate the counters. The counters would be lined up in columns marking, from right to left, the one place, the ten place, the hundred place, and so on. One counter would be placed above the bar in each column, and four counters would be placed below the bar.

You will need an abacus to follow along with the following examples, as well as to do the exercises. The easiest may be to use a piece of paper for a surface and some coins for counters. Mark the paper as shown here.

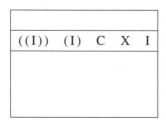

We shall refer to the row with the Roman numerals as the "bar." Note that we follow Turner and use (I) to denote 1000 and ((I)) to denote 10,000.

Remark 8.1 Cajori [31, p. 33] describes the development of the Roman symbols for large numbers:

For 1,000 the Romans had not only the symbol M, but also I, ∞ and (|). According to Priscian, the celebrated Latin grammarian of about 500 AD, the ∞ was the ancient Greek sign X for 1,000, but modified by connecting the sides by curved lines so as to distinguish it from the Roman X for 10....When only the right-hand parenthesis is written, |), the value represented is only half, i.e., 500. According to Priscian[1] ... |)) stood for 5,000, ((|)) for 10,000; also |))) represented 50,000; and ((([|)))), 100,000; (∞, 1,000,000....Through Priscian it is established that this notation is at least as old as 500 AD; probably it was much older, but it was not widely used before the Middle Ages.

□

Now, each counter placed below the bar represents the value of the Roman numeral above it, and a counter placed above the bar represents fives times the value of the Roman numeral below it. Here are two examples.

```
                •                           •           •
 ─────────────────────       ─────────────────────────────
 ((I))   (I)   C   X   I       ((I))   (I)   C   X   I
                   •   •                •         •   •   •
                   •   •                     •    •   •   •
                   •                              •   •
                                                  •   •

            37                               15,492
```

We now turn to the basic arithmetical operations using the abacus. In the following, dots represent the counters that are first placed on the board, and circles represent new counters that are placed on the board.

Example 8.1 Find the sum of sixty-one and twenty-three.

```
                •                           •
 ─────────────────────       ─────────────────────────────
 ((I))   (I)   C   X   I       ((I))   (I)   C   X   I
                   •   •                          •   •
                                                  ○   ○
                                                  ○   ○
                                                      ○

            61                        61 + 23 = 84
```

□

Sometimes we may need to "carry" in an addition.

Example 8.2 Find the sum of sixty-two and twenty-three.

```
                •                           •
 ─────────────────────       ─────────────────────────────
 ((I))   (I)   C   X   I       ((I))   (I)   C   X   I
                   •   •                          •   •
                   •                              ○   •
                                                  ○   ○
                                                      ○
                                                      ○

            62                          62 + 23
```

[1]"De figuris numerorum," *Henrici Keilii Grammatici Latini* (Lipsiae, 1859), Vol. III, 2, p. 407.

But now there are more than four counters below the bar in the I column. Thus, we replace every five counters below the bar with one counter *above* the bar in the I column.

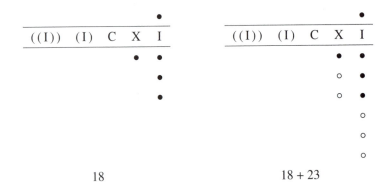

$$62 + 23 = 85$$

☐

Example 8.3 Find the sum of eighteen and twenty-three.

18

18 + 23

As in the last example, there are more than four counters below the bar in the I column, but now there is also one counter above the bar in the I column. Thus, in the I column, we remove the counter above the bar together with five below the bar (totaling 10), and place one counter below the bar in the X column.

$$18 + 23 = 41$$

☐

Subtraction is done by removing counters. Sometimes we may have to "borrow" in a subtraction.

Example 8.4 Find the difference between forty-two and eighteen.

$$
\begin{array}{ccccc}
((I)) & (I) & C & X & I \\
\hline
& & & \bullet & \bullet \\
& & & \bullet & \bullet \\
& & & \bullet & \\
& & & \bullet &
\end{array}
$$

42

Now, $18 = 10 + 8$, and we can easily remove one counter from the X column. But to remove eight from the I column, we have to exchange another counter in the X column for ten counters in the I column as follows.

$$
\begin{array}{ccccc}
& & & & \circ \\
((I)) & (I) & C & X & I \\
\hline
& & & \bullet & \bullet \\
& & & \bullet & \bullet \\
& & & \bullet & \circ \\
& & & & \circ \\
& & & & \circ \\
& & & & \circ \\
& & & & \circ
\end{array}
$$

42

Now, again because $18 = 10 + 8$, we may remove one counter from the X column and eight from the I column.

$$
\begin{array}{ccccc}
((I)) & (I) & C & X & I \\
\hline
& & & \bullet & \bullet \\
& & & \bullet & \bullet \\
& & & & \circ \\
& & & & \circ
\end{array}
$$

$42 - 18 = 24$

□

Now You Try 8.1 Use the abacus to find the following.

1. $315 + 129$ 2. $4739 + 1456$ 3. $520 - 346$ 4. $4739 - 1456$

□

As an interesting note, the counters were often small pebbles, or *calculi* in Latin, from whence came our words to "calculate" and "calculus." Also, in medieval Europe, counting boards were often painted on the table or "counter" at stores, thus our counters in stores, kitchens, and bathrooms.

Chapter 9

Multiplication

Multiplication in a positional number system is very systematic (think of how we start on the right and carry over to the next place value), but this is not the case in number systems that are not positional. Throughout history, different cultures have devised various methods to multiply and divide, some of them quite ingenious.

9.1 ROMAN ABACUS

We begin with a look at multiplication on the Roman abacus. To multiply together "large" numbers, we would usually proceed from the smallest place value to the largest (right to left); however, when using the abacus we proceed from the largest place value to the smallest (left to right). Note that multiplication using the abacus is essentially done by finding the partial products mentally, and using the abacus for bookkeeping along the way.

Example 9.1 Find the product of eighteen and twenty-three.

Because $18 = 10 + 8$ and $23 = 20 + 3$, we find the partial products mentally in the following order and record a running total of the partial products at each step using the abacus.

$$10 \times 20$$
$$10 \times 3$$
$$8 \times 20$$
$$8 \times 3$$

Here, circles represent partial products (shown on the left) and dots represents running totals (shown on the right).

$((I))$	(I)	C	X	I
		∘		
		∘		

$((I))$	(I)	C	X	I
		•		
		•		

$$10 \times 20 = 200 \qquad\qquad 200$$

((I))   (I)   C   X   I
• ○
• ○
○

$$10 \times 3 = 30$$

((I))   (I)   C   X   I
• •
• •
•

$$200 + 30 = 230$$

| ○ |
| ((I))   (I)   C   X   I |
| • • |
| • • |
| ○ • |
| ○ |

$$8 \times 20 = 160$$

| • |
| ((I))   (I)   C   X   I |
| • • |
| • • |
| • • |
| • |

$$230 + 160 = 390$$

| • |
| ((I))   (I)   C   X   I |
| • • ○ |
| • • ○ |
| • • ○ |
| • ○ |
| ○ |
| ○ |

$$8 \times 3 = 24$$

| • |
| ((I))   (I)   C   X   I |
| • • • |
| • • |
| • • |
| • • |

$$390 + 24 = 414$$

Therefore, $18 \times 23 = 414$. □

Remark 9.1 According to Turner [119], "[B]y the time of Archimedes (287–212 BC) and of Heron (date uncertain) the Greeks had evolved a method of written multiplication not unlike ours, except that the operation began with the highest power of 10 instead of with the digits; that is, it proceeded left to right instead of right to left. This, incidentally, has been the standard practice in abacus reckoning from the time of our earliest recorded instructions." □

Now You Try 9.1 Use the abacus to find the following.

1. 32×47 2. 50×112 3. 153×312 4. 68×225

□

Think About It 9.1 Most abacuses do not have column headings. Are they necessary? When may it be helpful not to have column headings? □

There is no surviving example of how the Romans (or the Greeks) may have performed division using the abacus, although some have conjectured how this may have been done. See [119] for an example. We now turn to a method that found much wider use in Europe and is a forerunner of our modern method.

9.2 GRATING OR LATTICE METHOD

Turning now to Indian methods, Bhāskara II, in his work *Līlāvatī*, gave five methods for multiplication, one of which was the *grating* or *lattice method* [35]. The grating method for multiplication appears to have been introduced into Europe from India via the Arabs ([15], [112]), but it was also used in China. This method is referred to by other names, as well, including *jalousie*, "so called because it referred to a type of Venetian blind in the form of a grating, common in Venice, through which nuns or ladies could see out from the inside without being observed from the outside" [39]. Because of this, the grating method also became known as the method of *gelosia*, which means jealousy in Italian [112, p. 233]. According to Boag [13], "The earliest example of the gelosia method is in the Treviso *Arithmetic* of 1478, which is the earliest example of a printed book on arithmetic. The method was used for several centuries until it was replaced by other methods." Scottish mathematician John Napier (1550–1617) devised what has become called Napier's bones or rods, a calculating device to carry out lattice multiplication [15, p. 216], [13, 35].[1]

Example 9.2 We use the grating method to multiply 23×15.

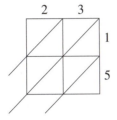

Draw the lattice

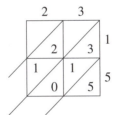

Find the partial products

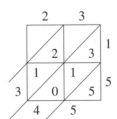

Add down the diagonals

So, we see that $23 \times 15 = 345$. □

Example 9.3 We use the grating method to multiply 34×25.

---

[1] See Chabert et al. [39, pp. 20–28] for figures illustrating the use of this method in *Miftāh al-hisāb* (*The Key to Calculation*; Arab, 1427), *Jiuzhang suanfa bilei daquan* (*Sum of the Methods of Calculation from the Nine Chapters Consisting of Problems Solved by Analogy with Problem Types*; Chinese, 1450), the Treviso *Arithmetic*, and others.

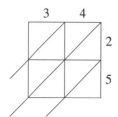

Draw the lattice

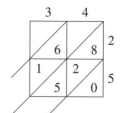

Find the partial products

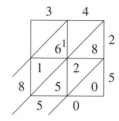

Add down the diagonals

Note the 1 that was "carried" from the middle diagonal, and added to the 6 and the 1, to get 8 down the left-most diagonal.

So, we see that $34 \times 25 = 850$. □

Now You Try 9.2 Use the grating method to find the following products.

1. 80×14 2. 3207×9 3. 253×15

□

Think About It 9.2 Explain the purpose of the diagonals. □

9.3   IBN LABBĀN AND CHINESE COUNTING BOARD

Kūshyār ibn Labbān (fl. ca. 1000), another Indian mathematician, provides us with one procedure for multiplication in his work *Principles of Hindu Reckoning* [112, p. 232]. His method would likely have been carried out on a "dust board" or some other easily erasable surface (it can even be done on the ground). The Chinese also used a method that is very similar to ibn Labbān's method, but using Chinese rod numerals on a counting board [112, p. 207]. We illustrate both methods with the following examples.

Example 9.4 We use the method of ibn Labbān to find 23×18.

First line up the factors as follows.

$$\begin{array}{cc} \mathbf{2} & \mathbf{3} \\ 1 \quad 8 & \end{array}$$

Next, multiply 1×2, add the result to the 0 (empty place) above the 1, and write the result above the 1. Then multiply 8×2 (= 16), and, after erasing the 2 above the 8, write the result above the 8 (and carry so that the 2 above the 1 becomes 3).

$$
\begin{array}{ccc}
2 & \mathbf{2} & 3 \\
1 & 8 &
\end{array}
\quad \rightarrow \quad
\begin{array}{ccc}
3 & 6 & 3 \\
1 & 8 &
\end{array}
$$

Now move the factor of 18 over

$$
\begin{array}{ccc}
3 & 6 & \mathbf{3} \\
& 1 & 8
\end{array}
$$

and repeat the steps by multiplying 1×3 and 8×3. Thus, multiply 1×3, add the result to the 6 above the 1, and, after erasing, write the sum 9; then multiply 8×3, and, after erasing the 3, write the result 24 above the 8 (and carry).

$$
\begin{array}{ccc}
3 & 9 & \mathbf{3} \\
& 1 & 8
\end{array}
\quad \rightarrow \quad
\begin{array}{ccc}
4 & 1 & 4 \\
& 1 & 8
\end{array}
$$

Having finished, we conclude that $18 \times 23 = 414$. □

Think About It 9.3 What is the purpose of aligning the second multiplier on the left and then moving it to the right? □

With all the erasing of the intermediate steps, you can see why ibn Labbān's method is best performed on a dust board or some other easily erasable surface. The intermediate steps are also lost in the Chinese counting board method because rod numerals are removed from the board during the procedure.

Example 9.5 Now we use the Chinese counting board method to find 23×18. For convenience, we write our usual numerals here instead of drawing the Chinese rod numerals (TABLE 4.1 on page 38).

First, place rods for 1, 8, 2, and 3 on a counting board as shown below.

		1	8	
	2	3		

Next multiply 2×1 and 3×1 and place rods for the products on the board as shown below; vertically add the two rows between the factors and remove the 1 rod at the top; and move the rods for the multiplier 23 (on the bottom) to the right.

		1	8	
	2			
		3		
	2	3		

$\xrightarrow{\text{add}}$

			8	
		2	3	
			2	3

Now repeat the steps, that is, multiply 2×8 and 3×8 and place rods for the products on the board; vertically add the three rows between the factors, carry where needed, and remove the 8 rod at the top.

			8
	1	6	
		2	4
	2	3	
		2	3

$\xrightarrow{\text{add}}$

	4	1	4
		2	3

Having finished, we conclude that $18 \times 23 = 414$.   □

Think About It 9.4 Why were three rows left open for use? How many rows would you leave open for a 3-digit by 2-digit multiplication? How do you decide where (from left to right) to place the partial products?   □

Now You Try 9.3 Use ibn Labbān's method and the Chinese counting board method to find the following products.

1. 80×14 2. 3207×9 3. 253×15

□

Think About It 9.5 Although ibn Labbān's method and the Chinese counting board method appear very different at first glance, explain why they are actually very similar.   □

9.4   EGYPTIAN DOUBLING METHOD

The ancient Egyptians used a very intuitive method for multiplying, which many children naturally use to multiply. To multiply, they used the *doubling* method, and, to divide, they inverted the procedure. This is most easily understood by looking at a few examples. The examples here appear in the work by Imhausen [70]. Remember that the problems were written in hieratic script and would have been read from right to left, but we use modern numerals for clarity.

Example 9.6 Rhind Problem 52 [70, p. 14]

In this problem, A'hmosè shows how to multiply 2000 by 5. The solution begins by writing 1 (expressed as a dot) and 2000 in two columns. Then the numbers in both columns are successively doubled until there is a combination in the first column that adds to 5. The sum of the corresponding entries in the second column is then the answer. As in the actual papyrus, a check or backslash mark "\" indicates the entries that are to be added.

$$
\begin{array}{rl}
\backslash\, . & 2000 \\
2 & 4000 \\
\backslash\, 4 & 8000 \\
\text{Total} & 10{,}000
\end{array}
$$

So, we see that $2000 \times 5 = 2000 + 8000 = 10{,}000$.   □

Example 9.7 Rhind Problem 69 [70, p. 15]

This problem multiplies 80 by 14. In the solution of this problem, we see that sometimes a scribe would multiply by 10 before doubling.

$$
\begin{array}{rr}
. & 80 \\
\backslash\ 10 & 800 \\
2 & 160 \\
\backslash\ 4 & 320 \\
\text{Total} & 1120
\end{array}
$$

So, we see that $80 \times 14 = 800 + 320 = 1120$.

☐

Now You Try 9.4 Use the doubling method to find the following products.

1. 35×17
2. 230×15
3. 45×60
4. 3200×9

☐

Think About It 9.6 The scribes apparently took the doubling method for granted. Does it always work? If so, why? ☐

When multiplying with fractions, the doubling method is still utilized. It is easy to double a fraction that has an even denominator by simply halving the denominator (for example, the double of $\frac{1}{12}$ is $\frac{1}{6}$), but to write the double of a fraction with an odd denominator as a sum of unit fractions is not always so easy. In general, Egyptian scribes referred to tables of fractions. Imhausen [70] gives as an example the so-called $2 \div N$ table for odd $N = 3$ to $N = 101$ found in the Rhind Mathematical Papyrus that gives the doubles of $\frac{1}{N}$ as sums of unit fractions; the Mathematical Leather Roll contains a table of 26 sums of unit fractions that equal another unit fraction. (Cooke [41] shares the following interesting bit of trivia about the Leather Roll. He says that the Leather Roll, "[which was] purchased along with the Ahmose Papyrus [Rhind Papyrus], was not unrolled for 60 years after it reached the British Museum because the curators feared it would disintegrate if unrolled. It was some time before suitable techniques were invented for softening the leather, and the document was unrolled in 1927.") Both tables were used in fraction reckoning.

In what follows, a bar over an integer denotes a unit fraction, for instance, $\overline{5}$ is $\frac{1}{5}$ and $\overline{30}$ is $\frac{1}{30}$; as a special case, $\overline{\overline{3}}$ is $\frac{2}{3}$. In TABLE 9.1, the $2 \div N$ table, the column N gives the denominator of the fraction $\frac{2}{N}$. The column $2 \div N$ gives alternatingly a term in the decomposition of $2 \div N$ and its portion of N; the decomposition of $\frac{2}{N}$ is the sum of the unit fractions in bold. For example, in the second row, we see that $\frac{2}{5} = \frac{1}{3} + \frac{1}{15}$ because **3** and **15** are bold; furthermore, $\overline{3}\ 1\ \overline{3}$ tells us that $\frac{1}{3}$ of 5 is $1\frac{2}{3}$, and $\overline{15}\ \overline{3}$ tells us that $\frac{1}{15}$ of 5 is $\frac{1}{3}$. As another example, in the sixth row, we see that $\frac{2}{13} = \frac{1}{8} + \frac{1}{52} + \frac{1}{104}$; further, $\overline{8}\ 1\ \overline{2}\ \overline{8}$ tells us that $\frac{1}{8}$ of 13 is $1 + \frac{1}{2} + \frac{1}{8}$ or $1\frac{5}{8}$, $\overline{52}\ \overline{4}$ tells us that $\frac{1}{52}$ of 13 is $\frac{1}{4}$, and $\overline{104}\ \overline{8}$ tells us that $\frac{1}{104}$ of 13 is $\frac{1}{8}$.

Now You Try 9.5 Explain the fourth and seventh rows of TABLE 9.1, the $2 \div N$ table. ☐

Example 9.8 Rhind Problem 6 [70, p. 16]

Chabert et al. [39, p. 17] provide that this problem is about, "The making of loaves for man 10." This problem multiplies $\frac{2}{3} + \frac{1}{5} + \frac{1}{30}\ (= \frac{9}{10})$ by 10. Notice that, in this instance, multiplication by 10 is not carried out directly.

N	$2 \div N$	N	$2 \div N$
3	$\overline{\overline{3}}$ 2	53	$\overline{30}$ 1 $\overline{\overline{3}}$ $\overline{10}$ **318** 6 $\overline{\mathbf{795}}$ 15
5	$\overline{3}$ 1 $\overline{\overline{3}}$ $\overline{15}$ 3	55	$\overline{30}$ 1 $\overline{\overline{3}}$ 6 $\overline{\mathbf{330}}$ 6
7	$\overline{4}$ 1 $\overline{2}$ $\overline{4}$ $\overline{\mathbf{28}}$ 4	57	$\overline{38}$ 1 $\overline{2}$ $\overline{\mathbf{114}}$ 2
9	$\overline{6}$ 1 $\overline{2}$ $\overline{\mathbf{18}}$ 2	59	$\overline{36}$ 1 $\overline{2}$ 12 $\overline{18}$ **236** 4 $\overline{\mathbf{531}}$ 9
11	$\overline{6}$ 1 $\overline{\overline{3}}$ 6 $\overline{\mathbf{66}}$ 6	61	$\overline{40}$ 1 $\overline{2}$ 40 **244** 4 **488** 8 $\overline{\mathbf{610}}$ 10
13	$\overline{8}$ 1 $\overline{2}$ 8 $\overline{\mathbf{52}}$ 4 $\overline{\mathbf{104}}$ 8	63	$\overline{42}$ 1 $\overline{2}$ $\overline{\mathbf{126}}$ 2
15	$\overline{10}$ 1 $\overline{2}$ $\overline{\mathbf{30}}$ 2	65	$\overline{39}$ 1 $\overline{\overline{3}}$ $\overline{\mathbf{195}}$ 3
17	$\overline{12}$ 1 $\overline{3}$ 12 $\overline{\mathbf{51}}$ 3 $\overline{\mathbf{68}}$ 4	67	$\overline{40}$ 1 $\overline{2}$ 8 20 $\overline{\mathbf{335}}$ 5 $\overline{\mathbf{536}}$ 8
19	$\overline{12}$ 1 $\overline{2}$ 12 $\overline{\mathbf{76}}$ 4 $\overline{\mathbf{114}}$ 6	69	$\overline{46}$ 1 $\overline{2}$ $\overline{\mathbf{138}}$ 2
21	$\overline{14}$ 1 $\overline{2}$ $\overline{\mathbf{42}}$ 2	71	$\overline{40}$ 1 $\overline{2}$ 4 40 **568** 8 $\overline{\mathbf{710}}$ 10
23	$\overline{12}$ 1 $\overline{\overline{3}}$ 4 $\overline{\mathbf{276}}$ 12	73	$\overline{60}$ 1 6 20 **219** 3 **292** 4 $\overline{\mathbf{365}}$ 5
25	$\cdots$	75	$\cdots$
$\vdots$	$\vdots$	$\vdots$	$\vdots$

Table 9.1: A portion of the Rhind Mathematical Papyrus, $2 \div N$ Table, copied from the full table in [70]. The solutions in bold were originally marked in red ink.

	Rhind		Indo-Arabic numerals
.	$\overline{\overline{3}}$ $\overline{5}$ $\overline{30}$	.	$\frac{2}{3}+\frac{1}{5}+\frac{1}{30}=\frac{9}{10}$
\2	1 $\overline{3}$ $\overline{10}$ $\overline{30}$	\2	$1+\frac{2}{3}+\frac{1}{10}+\frac{1}{30}=1\frac{4}{5}$
4	3 $\overline{2}$ $\overline{10}$	4	$3+\frac{1}{2}+\frac{1}{10}=3\frac{1}{5}$
\8	7 $\overline{5}$	\8	$7+\frac{1}{5}=7\frac{1}{5}$
Total	9 loaves of bread. This is it.		$1\frac{4}{5}+7\frac{1}{5}=9$

So, we see that $(\frac{2}{3}+\frac{1}{5}+\frac{1}{30}) \times 10 = 9$. Some explanation is in order.

- Moving from the first row to the second, we see that

  - The double of $\overline{\overline{3}}$ ($=\frac{2}{3}$) is $\boxed{1\ \overline{3}}$ ($=1\frac{1}{3}$ by repeated addition); the double of $\overline{5}$ is $\boxed{\overline{3}\ \overline{15}}$ (from TABLE 9.1, the $2 \div N$ table); and the double of $\overline{30}$ is $\boxed{\overline{15}}$ (halve the denominator).

  - Adding $1\ \overline{3}$ and $\overline{3}\ \overline{15}$ and $\overline{15}$ gives $\boxed{1}$ and $\boxed{\text{the double of } \overline{3}}$ and the $\boxed{\text{double of } \overline{15}}$.

  - From the $2 \div N$ table, the double of $\overline{3}$ is $\boxed{\overline{\overline{3}}}$ and the double of $\overline{15}$ is $\boxed{\overline{10}\ \overline{30}}$, so adding $1\ \overline{3}$ and $\overline{3}\ \overline{15}$ and $\overline{15}$ gives $\boxed{1\ \overline{\overline{3}}\ \overline{10}\ \overline{30}}$.

- Moving from the second row to the third row, we see that

  - The double of 1 is $\boxed{2}$; the double of $\overline{3}$ is $\boxed{1\ \overline{3}}$; The double of $\overline{10}$ is $\boxed{\overline{5}}$; and the double of $\overline{30}$ is $\boxed{\overline{15}}$.

– Adding 2 and $1\,\bar{3}$ and $\bar{5}$ and $\overline{15}$ gives $\boxed{3}$ and $\frac{9}{15}$ $(= \frac{1}{3}+\frac{1}{5}+\frac{1}{15})$ or $\boxed{\bar{2}\ \overline{10}}$ (because $\frac{9}{15} = \frac{3}{5} = \frac{1}{2}+\frac{1}{10}$),

so the double of $1\,\bar{3}\ \overline{10}\ \overline{30}$ is $\boxed{3\,\bar{2}\ \overline{10}}$.

- Moving from the third row to the fourth row, we see that the double of $3\,\bar{2}$ is $\boxed{7}$ and the double of $\overline{10}$ is $\boxed{\bar{5}}$, so the double of $3\,\bar{2}\ \overline{10}$ is $\boxed{7\,\bar{5}}$.

- Finally, adding $1\,\bar{3}\ \overline{10}\ \overline{30}$ and $7\,\bar{5}$ gives 9.

Today, this would have been done as

$$\left(\frac{2}{3} + \frac{1}{5} + \frac{1}{30}\right) \times 10 = \frac{9}{10} \times 10 = 9.$$

□

Now You Try 9.6

1. $5 \times \frac{1}{3}$ 2. $6 \times \frac{1}{5}$ 3. $9 \times \frac{1}{7}$ 4. $2\frac{1}{3} \times 15$

□

Since any whole number, including 1, can be written as a sum of powers of 2, the Egyptian doubling method works well for multiplying whole numbers or whole numbers and fractions. The bulk of multiplication problems found in Egyptian papyri are problems of a useful nature or designed as practice for scribes.

Chapter 10

Division

Division is generally the most difficult of the four basic operations of arithmetic. We first look at how the ancient Egyptians, whom we saw multiplied by doubling, essentially inverted the process to divide. Then we look at two methods that were used in Europe between the twelfth and the seventeenth centuries.

10.1 EGYPTIAN

Multiplying with fractions was no easy task for the ancient Egyptians. Now we see how they inverted this procedure to divide. There are two categories of division problems depending on whether the divisor is less than or greater than the dividend. We start with the former case. Since division is the inverse operation of multiplication, we still use the doubling method, but starting with the divisor.

Example 10.1 Rhind Problem 76 [70, p. 15]

This problem divides 30 by $2\frac{1}{2}$. *If the divisor was smaller than the dividend, doubling of the divisor was employed.* In this example, A'hmosè first puts down 1 and $2\frac{1}{2}$ in two columns, then he multiplies by 10 before successively doubling the columns until there is a combination in the second column (instead of in the first column as when multiplying) that adds to 30. The sum of the corresponding entries in the first column is then the answer.

	Rhind		Indo-Arabic numerals	
.	$2\overline{2}$	.	$2 + \frac{1}{2} = 2\frac{1}{2}$	
\ 10	25	\ 10		25
\ 2	5	\ 2		5
Total	12		$10 + 2 = 12$	

So, we see that $30 \div 2\frac{1}{2} = 12$. Today, this would have been done as

$$30 \div 2\frac{1}{2} = 30 \div \frac{5}{2} = 30 \times \frac{2}{5} = 12.$$

□

Now You Try 10.1 Use the Egyptian method to divide.

1. $42 \div 7$ 2. $8 \div 3$ 3. $20 \div \frac{1}{3}$ 4. $25 \div \frac{1}{2}$

□

Think About It 10.1 Explain how this application of the doubling process is the same as division. □

Next, let us think about how the Egyptian method can be modified for the case of the divisor greater than the dividend. Thinking in modern terms, what would you do if you wanted to divide by a fraction? If you said "invert and multiply," you are right! Let us see how that translates into the Egyptian method.

Example 10.2 Rhind Problem 58 [70, p. 16]

This problem shows how to divide 70 by $93\frac{1}{3}$. *If the divisor is greater than the dividend, successive halving of the divisor is used.* Again, entries in the second column that add to 70 are marked, and the sum of the corresponding entries in the first column is the quotient.

Rhind		Today
. $93\,\bar{3}$		$70 \div 93\frac{1}{3} = 70 \div \dfrac{280}{3}$
\ $\bar{2}$ $46\,\bar{\bar{3}}$		$= 70 \times \dfrac{3}{280}$
\ $\bar{4}$ $23\,\bar{3}$		
Total $\bar{2}\,\bar{4}$		$= \dfrac{3}{4}$

So, we see that $70 \div 93\frac{1}{3} = \frac{3}{4}$ ($\bar{2}\,\bar{4}$ is $\frac{1}{2} + \frac{1}{4} = \frac{3}{4}$). □

Imhausen tells us that, "In more difficult numerical cases the division is first carried out as a division with remainder. The remainder is then handled separately."

Now You Try 10.2

1. Verify the steps in A'hmosè's solution in Rhind Problem 58.

2. Divide the following using the procedure shown in the Rhind problems. You do not have to decompose fractions into sums of unit fractions.

   a) $9 \div 12$ b) $15 \div 30$ c) $26\frac{2}{5} \div 70\frac{2}{5}$ d) $40\frac{5}{6} \div 65\frac{1}{3}$

□

Think About It 10.2

1. Divide 98 by 3 both by modern long division and by using the Egyptian method. How did you deal with the remainder?

2. How can an answer to a division problem be estimated using the Egyptian method? Estimate $10 \div 3$.

3. Can you use the doubling method to multiply fractions, for example, $\frac{1}{4} \times \frac{1}{3}$?

4. Do you think this question of multiplying two fractions would arise very often, if at all, in the context of mathematics needed by the ancient Egyptians?

□

Remark 10.1 As noted earlier, there may be more than one way to express a fraction as a sum of unit fractions. For instance, $\frac{2}{9} = \frac{1}{6} + \frac{1}{18}$ and also $= \frac{1}{8} + \frac{1}{18} + \frac{1}{24}$. So, how did a scribe choose to express a fraction as a sum of unit fractions? Referring to the $2 \div N$ table on page 102, for example, Imhausen [70] says that,

There have been several attempts to explain the choices of representations in the $2 \div N$ table. These attempts were mostly based on modern mathematical formulas, and none of them gives a convincing explication of the values we find in the table. It is probable that the table was constructed based on experiences in handling fractions. Several "guidelines" for the selection of suitable fractions can be discerned. The author tried to keep the number of fractions to represent $2 \div N$ small; we generally find representations composed of two or three fractions only. Another guiding rule seems to be the choice of fractions with a small denominator over a big denominator, and the choice of denominators that can be decomposed into several components.

□

In most systems, multiplication and division are treated separately. But we have seen that in the Egyptian system, they are simply different applications of the same method, namely, that of doubling. We now return to Europe and explore methods of division. Of the methods of division used in Europe, the first that we will look at is by the man credited with bringing the Indo-Arabic numbers to Europe and the second is known for the shape the solution makes.

10.2 LEONARDO OF PISA

In mathematics, the twelfth-century Italian mathematician Leonardo Pisano Bigollo or Leonardo of Pisa (ca. 1170–1240; FIGURE 10.1) is best known for his two works, namely, *Liber Abaci* (*Book of Calculation*)[1]—first written in 1202, with a second version written in 1228 [90]—and *Liber Quadratorum* (*Book of Squares*; 1225). In the nineteenth century, Leonardo became popularly known as Fibonacci [112]. The son of a merchant and public official in the customs house in Bugia, a Pisan colony in North Africa, Leonardo learned about the Indo-Arabic numerals and mathematics during his youth accompanying his father in his travels throughout the Mediterranean. He was influenced by Euclid's *Elements*, the whole Greek mathematical tradition, and the work of Arabic mathematicians [90]. As a young man, Leonardo continued to travel extensively throughout the Mediterranean and "pursued mathematical knowledge from 'whoever was learned in it, from nearby Egypt, Syria, Greece, Sicily and Provence' " [61].

Many of the problems and methods in *Liber Abaci* were taken verbatim or nearly verbatim from the works of Arabic mathematicians such as al-Khwārizmī, Abū Kāmil, and al-Karajī [21, 73], although Leonardo did also introduce his own methods, as well. Goetzmann [61], a professor of finance and management, tells us that *Liber Abaci* is "a book devoted almost entirely to the mathematics of trade, valuation, and commercial arbitrage ... [and] was one of the first works to methodically describe the new [Hindu-Arabic] number system to Europeans and to demonstrate its practical, commercial use with detailed and copious examples"; he goes on to credit Leonardo with the introduction of "present value," an important financial tool today. In fact, Goetzmann describes Leonardo as "arguably the first scholar in world history to develop a detailed and flexible mathematical approach to financial calculation. He was not only a brilliant analyst of the business problems of his day but also an early financial engineer whose work played a major role in the development of Europe's capital market in the late Middle Ages and the Renaissance," and that "[it] may not be purely coincidental that the oldest extant copy of *Liber Abaci* is in the Biblioteca Reccardiana in Florence, the city where modern banking was born."

In mathematics today, Leonardo is widely remembered for a sequence he introduced in chapter 12 of *Liber Abaci* that now bears his name, the *Fibonacci sequence*: 1, 1, 2, 3, 5, 8, 13, 21, 34, 55,.... This sequence, which Leonardo used to describe how fast rabbits could breed in ideal conditions, now has been found to occur in nature in the branching of trees and the spirals of a pine cone, for example. (See SUBSECTION 20.3.3 on page 369.)

It is apparent that *Liber Abaci* was a textbook for teaching the important everyday mathematics of the day. Indeed, Leonardo's work was used in Tuscany's "schools of abaco by boys intending to be merchants

---

[1]In his introduction to the English translation [90], Sigler remarks "that while derived from the word abacus the word *abaci* refers in the thirteenth century paradoxically to calculation without the abacus. Thus *Liber abaci* should not be translated as The Book of the Abacus. A *maestro d'abbaco* was a person who calculated directly with Hindu numerals without using the abacus, and *abaco* is the discipline of doing this."

Figure 10.1: An artist's interpretation of Leonardo of Pisa (Fibonacci). (Source: © Thinkstock.)

or by others desiring to learn mathematics" for over three centuries [90]. Here is the table of contents of *Liber Abaci* to provide an idea of the scope of the work [90].

1. Here Begins the First Chapter

2. On the Multiplication of Whole Numbers

3. On the Addition of Whole Numbers

4. On the Subtraction of Lesser Numbers from Greater Numbers

5. On the Divisions of Integral Numbers

6. On the Multiplication of Integral Numbers with Fractions

7. On the Addition and Subtraction and Division Of Numbers with Fractions and the Reduction of Several Parts to a Single Part

8. On Finding the Value of Merchandise by the Principal Method

9. On the Barter of Merchandise and Similar Things

10. On Companies and Their Members

11. On the Alloying of Monies

12. Here Begins Chapter Twelve

13. On the Method of Elchataym and How with It Nearly All Problems of Mathematics Are Solved

14. On Finding Square and Cubic Roots, and on the Multiplication, Division, and Subtraction of Them, and On the Treatment of Binomials and Apotomes and their Roots

15. On Pertinent Geometric Rules And on Problems of Algebra and Almuchabala

We now turn specifically to Leonardo's presentation of division. In chapter 5 of *Liber Abaci*, Leonardo gives several examples of dividing integers to arrive at a result expressed in his fraction notation (see page 68). He shows how to divide by single-digit numbers; by two-digit prime numbers; by two-digit composite numbers; and by numbers with more than two digits (prime and composite).

Example 10.3 The first concrete example that Leonardo presents is the division of 365 by 2 [90, p. 52]:

And if one will wish to divide 365 by 2, then …2 he puts beneath the 5, and he begins by dividing the 3 by the 2, namely the last figure, saying $\frac{1}{2}$ of 3 is 1, and 1 remains; he writes the 1 beneath the 3, and the 1 which remains he writes above, as is displayed in the first illustration; and the remaining 1 couples with the 6 that is next to the last given figure, making 16; he takes $\frac{1}{2}$ of the 16 which is 8; he therefore puts the 8 beneath the 6 put before, the 1 the 3, as is displayed in the second illustration; and as there is no remainder in the division of the 16, one divides the 5 by the 2; the quotient is 2 and the remainder 1; he writes the 2 under the 5, and the 1 which remains he writes over the whole; and before the $\frac{1}{2}$ he writes the quotient coming from the division, namely 182, as one shows in the last illustration.

These are the steps Leonardo takes:

	3 6 5
"2 he puts beneath the 5"	2

	$\underline{1}$
	3 6 5
"saying $\frac{1}{2}$ of 3 is **1**, and $\underline{1}$ remains"	2

	1
	1 $\underline{0}$
	3 6 5
"he takes $\frac{1}{2}$ of 16 which is **8**" [and $\underline{0}$ remains]	2

	1 **8**
	1 0 $\underline{1}$
	3 6 5
"one divides 5 by 2; the quotient is **2** and the remainder $\underline{1}$"	2

	1 8 **2**
Answer:	$\frac{1}{2}$ 182

Note that
$$\frac{1}{2}\,182 = 182 + \frac{1}{2} \quad \text{or} \quad 182\frac{1}{2}.$$

So, we see that this method of division is not unlike our usual long division. □

Still focusing on dividing by primes, the next example from Leonardo we present is 12,532 divided by 11.

Example 10.4 *Liber Abaci* chapter 5 [90, p. 55]

The Division of Numbers by 11.

The said introductions are indeed noted, and if one will wish to divide 12532 by 11, then he puts 11 under the 32. And he takes $\frac{1}{11}$ of 12 at the head of the dividend which is 1, and there remains 1. Truly $\frac{1}{11}$ of 11 is 1 …therefore $\frac{1}{11}$ of 12 is 1, and there remains 1. One therefore puts the 1 beneath the 2, and the remainder 1 he puts above the 2, and he couples the 1 with the preceding figure, namely with the 5, making 15, of which he takes $\frac{1}{11}$ which is 1, and 4 remains from the said calculation; and he puts the 1 below the 5, and the remainder 4 above the 5; he couples the 4 with the preceding figure, namely with the 3, making 43; of this again he takes $\frac{1}{11}$ which yields 3, and there remains 10, this because $\frac{1}{11}$ of 33 is 3; left in 43 is 10; therefore $\frac{1}{11}$ of 43 is 3, and 10 remains, as we said; he therefore puts the 3 below the 3, and he puts the 10 above the 43; that is, he puts the 1 above the 4 which was put above the 5, and he puts 0 above the 3; and he couples again the 10 with the preceding figure, namely with the 2 that is in the first place; there will be 102, of which again he takes $\frac{1}{11}$; the quotient will be 9, and there remains 3; he puts the 9 below the said 2, and the remainder 3 he puts over the fraction line over the 11, the kept parts; and $\frac{3}{11}$ 1139 will be had for the sought division.

These are the steps Leonardo takes:

"he puts 11 under the 32"

```
1  2  5  3  2
          1  1
```

"he takes $\frac{1}{11}$ of 12 ... which is **1**, and there remains 1"

```
         1
1  2  5  3  2
          1  1

1
```

"he takes $\frac{1}{11}$ [of 15] which is **1**, and 4 remains"

```
      1  4
1  2  5  3  2
          1  1

1  1
```

"he takes $\frac{1}{11}$ [of 43] which yields **3**, and there remains 10"

```
      1
   1  4  0
1  2  5  3  2
          1  1

1  1  3
```

"there will be 102, of which again he takes $\frac{1}{11}$; the quotient will be 9, and there remains 3"

```
   1
1  4  0  3
1  2  5  3  2
          1  1

1  1  3  9
```

Answer: $\frac{3}{11}$ 1139

Note that
$$\frac{3}{11}\,1139 = 1139 + \frac{3}{11} \quad \text{or} \quad 1139\frac{3}{11}.$$

So, again, we see that this method of division is not unlike our usual long division.

Now You Try 10.3 Leonardo follows the example of the division of 365 by 2 by several similar examples, including

1. the division of 1346 by 4.

2. the division of 9000 by 7.

3. the division of 123586 by 13.

4. the division of 13976 by 23.

Carry out these divisions following Leonardo's method and express the results using his fraction notation if necessary.

Before demonstrating the division of numbers by composite numbers with more than two digits, Leonardo discusses how to find the "composition" (factorization) of numbers in a particular way so that he may "find the rule of" (factor) the divisor before carrying out the division. We will not present how Fibonacci "finds the rule of" a number (or factors it), but for the example below, notice that 75 factors into $3 \cdot 5 \cdot 5$.

Example 10.5 *Liber Abaci* chapter 5 [90, p. 70]

The Division of 749 by 75.

Since one will wish to divide 749 by 75, he notes the rule for finding in numbers the factor 5, and he finds the rule for 75, that is $\frac{1\ 0\ 0}{3\ 5\ 5}$. He divides the 749 by 3; the quotient is 249, and there remains 2 which he puts over the 3 in the fraction, and he divides the 249 by 5, namely by that which precedes the 3 in the fraction; the quotient is 49, and there remains 4; this 4 he puts over the 5, and he divides again the 49 by 5, that which is at the end of the fraction; the quotient is 9, and there remains 4; the 4 he puts over the 5, and the 9 he puts before the fraction; and thus one has for the sought division $\frac{2\ 4\ 4}{3\ 5\ 5}9$, as is shown here.

These are the steps Leonardo takes:

"he finds the rule for 75, that is $\frac{1\ 0\ 0}{3\ 5\ 5}$"	$75 = 3 \cdot 5 \cdot 5$ $\frac{1\ 0\ 0}{3\ 5\ 5} = \frac{1}{3 \cdot 5 \cdot 5}$
"He divides the 749 by **3**; the quotient is 249, and there remains **2**"	$\frac{1\ 0\ 0}{3\ 5\ 5} \rightsquigarrow \frac{2\ 0\ 0}{3\ 5\ 5}$
"he divides the 249 by **5** …the quotient is 49, and there remains **4**"	$\frac{2\ 0\ 0}{3\ 5\ 5} \rightsquigarrow \frac{2\ 4\ 0}{3\ 5\ 5}$
"he divides again the 49 by **5** …the quotient is 9, and there remains **4**"	$\frac{2\ 4\ 0}{3\ 5\ 5} \rightsquigarrow \frac{2\ 4\ 4}{3\ 5\ 5}9$

Answer: $\frac{2\ 4\ 4}{3\ 5\ 5}9$

Note that using Leonardo's definition of his fraction (page 68), we have

$$\frac{2\ 4\ 4}{3\ 5\ 5}9 = 9 + \frac{4}{5} \cdot 1 + \frac{4}{5} \cdot \frac{1}{5} \cdot 1 + \frac{2}{3} \cdot \frac{1}{5} \cdot \frac{1}{5} \cdot 1$$
$$= 9 + \frac{74}{75} \quad \text{or} \quad 9\frac{74}{75}.$$

We leave it to you to check by the usual long division that, indeed, $749 \div 75 = 9\frac{74}{75}$. (You are asked in the exercises to explain why this method of division works.) □

Now You Try 10.4

1. Leonardo follows the example of the division of 749 by 75 by an example of the division of 67,898 by 1760. Carry out the division of 67,898 by 1760 following Leonardo's method; in his example, Leonardo uses the factorization $1760 = 2 \cdot 8 \cdot 10 \cdot 11$. Express the result using Leonardo's fraction notation.

2. Use Leonardo's method to divide 325 by 20.

□

Leonardo follows the two examples, the division of 749 by 75 and the division of 67,898 by 1760, by checking the two divisions by "casting out thirteens," a method of checking that is similar to casting out nines that we will discuss in CHAPTER 11. He then notes that one has to take care if, after choosing a factorization of the divisor, one of these factors is also a factor of the dividend. In this case, he instructs that "first the number is divided by the number of the composition which in the fraction of the divisor will have itself in the dividend whether it is greater or less in the fraction because if something will be divided by itself nothing will remain from the division" [90, p. 71]. He then demonstrates this with the following example.

Example 10.6 *Liber Abaci* chapter 5 [90, p. 71]

The Division of 81540 by 8190.

And if one will wish to divide 81540 by 8190, then the composition rule of the divisor is found, which is $\frac{1\ 0\ 0\ 0}{7\ 9\ 10\ 13}$, and as $\frac{1}{10}$ is in the rule of the 81540 by reason of the 0 which is in the first place of it, although $\frac{1}{10}$ is not at the head of the fraction; however, the 81540 is first divided by the 10; that is the 0 is removed from the number; there is left 8154; extracting $\frac{1}{10}$ from the fraction the 8154 remains to be divided with $\frac{1\ 0\ 0}{7\ 9\ 13}$. Also 8154 is divided by 9 because 0 is its residue upon casting out nines. Whence, one divides it by the 9 of the fraction; the quotient is 906 which remains to be divided with $\frac{1\ 0}{7\ 13}$; truly the 906 is divided by the 7; the quotient is 129, and there remains 3; the 3 one puts over the 7. And he divides the 129 by the 13; the quotient is 9, and there remains 12; the 12 he puts over the 13, and the quotient he puts before the fraction; and $\frac{3\ 12}{7\ 13}\,9$ will be had for the sought division.

These are the steps Leonardo takes:

	$8190 = 7 \cdot 9 \cdot 10 \cdot 13$
Factor 8190	$\frac{1\ 0\ 0\ 0}{7\ 9\ 10\ 13} = \frac{1}{7 \cdot 9 \cdot 10 \cdot 13}$
	$81{,}540 \rightsquigarrow 8154$
10 is a common factor of 81,540 and 8190, so remove the common factor	$\frac{1\ 0\ 0\ 0}{7\ 9\ 10\ 13} \rightsquigarrow \frac{1\ 0\ 0}{7\ 9\ 13}$
	$8154 \rightsquigarrow 906$
9 is a common factor of 8154 and 819, so remove the common factor	$\frac{1\ 0\ 0}{7\ 9\ 13} \rightsquigarrow \frac{1\ 0}{7\ 13}$
Now follow EXAMPLE 10.5 $906 \div 7 = 129$ with remainder 3	$\frac{1\ 0}{7\ 13} \rightsquigarrow \frac{3\ 0}{7\ 13}$
$129 \div 13 = 9$ with remainder 12	$\frac{1\ 0}{7\ 13} \rightsquigarrow \frac{3\ 12}{7\ 13}\,9$
	Answer: $\frac{3\ 12}{7\ 13}\,9$

Note that

$$\frac{3\ 12}{7\ 13}\,9 = 9 + \frac{12}{13} \cdot 1 + \frac{3}{7} \cdot \frac{1}{13} \cdot 1$$
$$= 9 + \frac{87}{91} \quad \text{or} \quad 9\frac{87}{91}.$$

We leave it to you to check by the usual long division that, indeed, $81{,}540 \div 8190 = 9\frac{87}{91}$. □

Now You Try 10.5 Perform the following divisions using Leonardo's method. Express the results using Leonardo's fraction notation.

1. Divide 4920 by 420.

2. Divide 6006 by 42.

☐

Leonardo goes on to demonstrate "the division of numbers by prime numbers of three places," and concludes this chapter of *Liber Abaci* by pointing out "by that which was said of the given divisions one can have a full mastery in dividing numbers of IV or more figures; however the said divisions are better understood if they are demonstrated by some numbers of four figures."[2] He then demonstrates the division of 17,849 by 1973 and the division of 1,235,689 by 4007.

10.3 GALLEY OR SCRATCH METHOD

Luca Pacioli (or Paciuolo; ca. 1445–1514), who was also known as Luca di Borgo, worked as a tutor in Venice before he became a friar in the Franciscan order and taught mathematics at Perugia in 1475 [39, p. 505]. (See FIGURE 10.2.) In mathematics, Pacioli is best remembered for his *Summa de arithmetica, geometrica, proportioni et proportionalita* (*Summary of Arithmetic, Geometry, Proportion, and Proportionality*) published in 1494. (A second edition of the *Summa* was published posthumously in 1523, making only changes in the spelling of some of the words [31].) Boyer [15, p. 279] notes that the *Summa* "was more influential than original," being a summary of the main of mathematics known up to that time in "arithmetic, algebra, very elementary Euclidean geometry, and double-entry bookkeeping." Much of the material in the *Summa* was drawn from Leonardo of Pisa's *Liber Abaci*. The *Summa*, however, was not the earliest work in algebra of the Renaissance: that honor goes to Chuquet's *Triparty* (page 131). Though written in the mid-fifteenth century, Chuquet never published *Triparty*. As a result, Boyer [15, p. 278] tells us that the *Summa* "overshadowed the *Triparty* so thoroughly that older historical accounts of algebra leap directly from the *Liber abaci* of 1202 to the *Summa* of 1494 without mentioning the work of Chuquet or other intermediaries." In that sense, Pacioli's *Summa* was much like Euclid's *Elements*.

Figure 10.2: Luca Pacioli. (Source: Courtesy of Jeff Miller.)

Although the *Summa* was a summary of the mathematics known up to that time, Cooke [41, p. 430] tells us that "it did bring the art of abbreviation closer to true symbolic notation. For example, what we now write as $x - \sqrt{x^2 - 36}$ was written by Pacioli as

$$1.co.\tilde{m}Rv.1.ce\ \tilde{m}36.$$

Here *co* means *cosa* (*thing*), the unknown; *ce* means *censo* (*power*), and *Rv* is probably a printed version of *Rx*, from the Latin *radix*, meaning root." See page 342.

---

[2]Note Leonardo's use of a Roman numeral for 4.

As an aside, Pacioli resided with two widely known figures of the Renaissance, namely, Leon Battista Alberti (1404–1472) and Leonardo da Vinci (1542–1519) [35]. In fact, da Vinci provided illustrations for another of Pacioli's works, *De divina proportione* (*The Divine Proportion*; 1508). And the earliest known portrait of a mathematician is of Pacioli.

In the *Summa*, Pacioli presented four methods for division [33, p. 148]. One of these methods was called the *galley* method by Italians because the work can be enclosed within the outline of a ship (FIGURE 10.3). According to Cajori [33], Pacioli "considered this procedure the swiftest [of the four], just as the galley was the swiftest ship." The English called the method the *scratch method*, presumably because digits are scratched off in the work. This method for division finds its origins in India [15, 33], and was employed in Europe until as late as the seventeenth century, having been adopted as well in Spain, Germany, and England [33, p. 149]. The Indians, however, worked on a dust board, and, so, the final appearance of their work did not display scratched off digits.

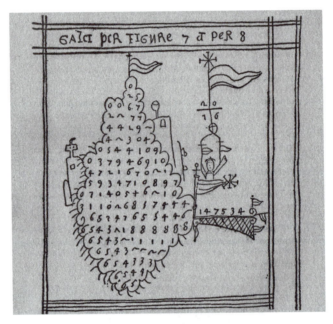

Figure 10.3: The galley method of division. (Source: U.S. public domain.)

We illustrate the galley method in the following examples. Note that the dividend appears in the middle, and that differences are written above instead of below the minuends; that the divisor may or may not be written; that as the process proceeds, "used" digits are scratched off; and that the column placement or vertical alignment of the digits is important (keeping the correct place values), but that the row placement is not.

Example 10.7 Divide 749 by 75.

75| 7   4   9   Write the dividend and the
                                  divisor.

75| 7   4   9 |9   75 goes into 749 at most 9
                                  times.

$$75\,|\quad 7\quad 4\quad 9\quad |\,9$$
$$ 6\quad 7\quad 5$$

$$75 \times 9 = 675$$

$$ 7\quad 4$$
$$75\,|\quad \not7\quad \not4\quad \not9\quad |\,9$$
$$ \not6\quad \not7\quad \not5$$

$$749 - 675 = 74$$

Hence, we conclude that $749 \div 75 = 9\frac{74}{75}$.   □

Example 10.8 Divide 67,892 by 176.

$$176\,|\quad 6\quad 7\quad 8\quad 9\quad 2$$

Write the dividend and the divisor.

$$176\,|\quad 6\quad 7\quad 8\quad 9\quad 2\quad |\,3$$

176 goes into 678 at most 3 times.

$$176\,|\quad 6\quad 7\quad 8\quad 9\quad 2\quad |\,3$$
$$ 5\quad 2\quad 8$$

$$176 \times 3 = 528$$

$$ 1\quad 5\quad 0$$
$$176\,|\quad \not6\quad \not7\quad \not8\quad 9\quad 2\quad |\,3$$
$$ \not5\quad \not2\quad \not8$$

$$678 - 528 = 150$$

$$ 1\quad 5\quad 0$$
$$176\,|\quad \not6\quad \not7\quad \not8\quad 9\quad 2\quad |\,38$$
$$ \not5\quad \not2\quad \not8$$

176 goes into 1509 at most 8 times.

$$ 1\quad 5\quad 0$$
$$176\,|\quad \not6\quad \not7\quad \not8\quad 9\quad 2\quad |\,38$$
$$ \not5\quad \not2\quad \not8\quad 8$$
$$ 1\quad 4\quad 0$$

$$176 \times 8 = 1408$$

$$ 1\quad 0$$
$$ \not1\quad \not5\quad \not0\quad 1$$
$$176\,|\quad \not6\quad \not7\quad \not8\quad \not9\quad 2\quad |\,38$$
$$ \not5\quad \not2\quad \not8\quad \not8$$
$$ \not1\quad \not4\quad \not0$$

$$1509 - 1408 = 101$$

$$
\begin{array}{c}
 1\ \ 0 \\
 \cancel{1}\ \ \cancel{5}\ \ \cancel{0}\ \ 1 \\
176|\ \ \cancel{6}\ \ \cancel{7}\ \ \cancel{8}\ \ \cancel{9}\ \ 2 \quad |\,385 \\
 \cancel{5}\ \ \cancel{2}\ \ \cancel{8}\ \ \cancel{8} \\
 \cancel{1}\ \ \cancel{4}\ \ \cancel{0}
\end{array}
$$

176 goes into 1012 at most 5 times.

$$
\begin{array}{c}
 1\ \ 0 \\
 \cancel{1}\ \ \cancel{5}\ \ \cancel{0}\ \ 1 \\
176|\ \ \cancel{6}\ \ \cancel{7}\ \ \cancel{8}\ \ \cancel{9}\ \ 2 \quad |\,385 \\
 \cancel{5}\ \ \cancel{2}\ \ \cancel{8}\ \ \cancel{8}\ \ 0 \\
 \cancel{1}\ \ \cancel{4}\ \ \cancel{0}\ \ 8 \\
 \phantom{\cancel{1}\ \ \cancel{4}\ \ \cancel{0}}\ 8
\end{array}
$$

$176 \times 5 = 880$

$$
\begin{array}{c}
 1 \\
 \cancel{1}\ \ \cancel{0}\ \ 3 \\
 \cancel{1}\ \ \cancel{5}\ \ \cancel{0}\ \ \cancel{1}\ \ 2 \\
176|\ \ \cancel{6}\ \ \cancel{7}\ \ \cancel{8}\ \ \cancel{9}\ \ \cancel{2} \quad |\,385 \\
 \cancel{5}\ \ \cancel{2}\ \ \cancel{8}\ \ \cancel{8}\ \ \cancel{0} \\
 \cancel{1}\ \ \cancel{4}\ \ \cancel{0}\ \ \cancel{8} \\
 \phantom{\cancel{1}\ \ \cancel{4}\ \ \cancel{0}}\ \cancel{8}
\end{array}
$$

$1012 - 880 = 132$

Hence, we conclude that $67{,}892 \div 176 = 385\frac{132}{176}$.

Now You Try 10.6 Perform the following divisions using the galley method.

1. $67{,}898 \div 1760$

2. $928{,}558 \div 314$

Chapter 11

Casting Out Nines

Whenever we perform calculations on paper, whether it be long addition, subtraction, multiplication, or division, we leave a "calculation trail" that we can use to check our calculation for a mistake. When a written calculation is done on a "dust board," for example, no such trail is left to check for a mistake. One method used heavily in medieval Europe for checking an addition, subtraction, multiplication, or division without needing to see the intermediate steps is called *casting out nines*.

According to Boyer [15, p. 218], "the 'proof by nines,' or the 'casting out of nines,' is a Hindu invention, but it appears that the Greeks knew earlier of this property, without using it extensively, and that the method came into common use only with the Arabs of the eleventh century." It was later used fairly extensively in Europe, up until relatively recent times. We note that this method *can detect several types of mistakes* in a calculation but *cannot confirm the correctness* of a calculation.

FIGURE 11.1 gives the steps for a simplified version of casting out nines. The steps refer to something called a "check digit." In this instance, a check digit is the single-digit number that we obtain by adding the digits of a whole number successively. For example, the check digit of the number 3592 is 1 because

$$3592: \quad 3 + 5 + 9 + 2 = 19 \rightarrow 1 + 9 = 10 \rightarrow 1 + 0 = \boxed{1}.$$

To check if an addition (or a multiplication) has a mistake,

1. Find the check digit of the sum (or the product).

2. Find the check digit of each summand (or each multiplicand).

3. Add the check digits of each summand (or multiply the check digits of each multiplicand), then find the check digit of this sum (or this product).

If the check digit from the first step does *not* match the check digit from the third step, then there was a mistake in the calculation of the sum or product.

Figure 11.1: A simplified version of casting out nines.

Example 11.1 Check if $3416 + 2844 = 6260$ by casting out nines.

1. Find the check digit of the sum.

$$6260: \quad 6 + 2 + 6 + 0 = 14 \rightarrow 1 + 4 = \boxed{5}$$

2. Find the check digit of each summand.

$$3416: \quad 3+4+1+6 = 14 \rightarrow 1+4 = \boxed{5}$$
$$2844: \quad 2+8+4+4 = 18 \rightarrow 1+8 = \boxed{9}$$

3. Add the check digits of each summand, then find the check digit of this sum.

$$5+9 = 14: \quad 1+4 = \boxed{5}$$

Because the *check digits* from the first and the third steps *match*, we have *not detected a mistake* in the calculation of the sum, so it could be (and, if we have practiced addition a lot, it probably is) that $3416 + 2844 = 6260$. □

Example 11.2 Check if $5369 \times 352 = 1{,}879{,}888$ by casting out nines.

1. Find the check digit of the product.

$$1{,}879{,}888: \quad 1+8+7+9+8+8+8 = 49 \rightarrow 4+9 = 13 \rightarrow 1+3 = \boxed{4}$$

2. Find the check digit of each multiplicand.

$$5369: \quad 5+3+6+9 = 23 \rightarrow 2+3 = \boxed{5}$$
$$352: \quad 3+5+2 = 10 \rightarrow 1+0 = \boxed{1}$$

3. Multiply the check digits of each multiplicand, and then find the check digit of this product.

$$5 \times 1 = 5: \quad \boxed{5}$$

Because the *check digits* from the first and the third steps *do not match*, we have *detected a mistake* in the calculation of the product, that is, $5369 \times 352 \neq 1{,}879{,}888$. (The correct answer is $1{,}889{,}888$.) □

Remark 11.1 We can simplify how we find the check digit of a number. We found the check digit of 5369 as follows:

$$5369: \quad 5+3+6+9 = 23 \rightarrow 2+3 = \boxed{5}.$$

However, we could have proceeded in this way instead:

$$\underset{a\,b\,c\,d}{5369}: \quad \underset{a}{5}+\underset{b}{3} = 8, \ 8+\underset{c}{6} = 14 \rightarrow 1+4 = 5, \ 5+\underset{d}{9} = 14 \rightarrow 1+4 = \boxed{5}.$$

In this way, we never have to add anything larger than $9+9 = 18$. □

Now You Try 11.1 For each of the following, check the sum or product using the method of casting out nines; also calculate each sum or product independently. If casting out nines does not detect a mistake, but your independent calculation does, explain the discrepancy.

1. $7384 + 7341 = 14{,}725$

2. $3416 + 2844 = 6620$

3. $930 + 347 + 55 = 1332$

4. $5369 \times 352 = 1{,}889{,}888$

5. $355 \times 726 = 257{,}730$

6. $38 \times 21 = 63{,}062$

Remark 11.2 The term *check digit* is a modern term. Historically, a check digit was called the *residue*. □

Think About It 11.1

1. Why is this method called "casting out nines"? What are we *casting out*? (Hint: What is common about all the numbers that are generated at each stage of finding the check digit?)

2. The process of finding the check digit is equivalent to what mathematical operation?

3. Why would the single digit found be called the "residue"?

□

To use the method of casting out nines to check a difference or quotient, apply the method to the inverse problem. Here is an example for subtraction.

Example 11.3 Check if $6134 - 1621 = 4313$ by casting out nines.

We use casting out nines to check if $6134 - 1621 = 4313$ by checking the inverse problem instead, that is, by checking if $4313 + 1621 = 6134$.

1. Find the check digit of the sum.

$$6134: \quad 6 + 1 + 3 + 4 = 14 \rightarrow 1 + 4 = \boxed{5}$$

2. Find the check digit of each summand.

$$4313: \quad 4 + 3 + 1 + 3 = 11 \rightarrow 1 + 1 = \boxed{2}$$
$$1621: \quad 1 + 6 + 2 + 1 = 10 \rightarrow 1 + 0 = \boxed{1}$$

3. Add the check digits of each summand, and then find the check digit of this sum.

$$2 + 1 = \boxed{3}$$

Because the check digits from the first and the third steps do not match, we have detected a mistake in the addition and, hence, also in the subtraction. (What was the mistake?) □

Think About It 11.2 How would you use the method of casting out nines to check division? □

Now You Try 11.2 For each of the following, check the difference or quotient using the method of casting out nines, and calculate each difference or quotient independently. If casting out nines does not detect a mistake, but your independent calculation does, explain the discrepancy.

1. $7384 - 5569 = 1815$
2. $3416 - 2844 = 752$
3. $63,062 \div 21 = 83$
4. $11,907 \div 441 = 27$

□

Think About It 11.3 Casting out nines does not detect every type of error. What types of errors does casting out nines not detect and why? □

Chapter 12

Finding Square Roots

We start this chapter off with a question: How would you draw a picture of a^2 or of a^3? Can you draw a picture of a^4?

The ancient Greeks had a very geometrical conception of mathematics. For them, a^2 literally meant a *square* of side length a; and a^3 meant a *cube* of side length a. That is why we read a^2 as "*a* squared" and a^3 as "*a* cubed." But being limited to a three-dimensional physical world, that naming scheme ends there, and we continue with "*a* raised to the fourth power," and so on. Reversing this idea, what would be the geometric interpretation of the "square root of a" and the "cube root of a"?

We discuss *n*th roots in greater detail in SUBSECTION 20.2.2. For now, recall that we say that a number b is an *n*th root of a number a if $b^n = a$. For example, the number 2 is a *second* root of 4 because $2^2 = 4$; and 2 is a *third* root of 8 because $2^3 = 8$; and 2 is a *fourth* root of 16 because $2^4 = 16$; and 2 is a *fifth* root of 32 because $2^5 = 32$; and so on. So, a second root is also called a *square root* and a third root is also called a *cube root* because of their geometric significance, assuming they are positive: The square root of a is the side length of a square of area a, and the cube root of a is the side length of a cube of volume a.

Now recall that every positive real number a has two square roots, namely, a positive square root and a negative square root, both with the same magnitude (or absolute value), and that we write

$$\sqrt{a} \text{ or } \sqrt{a} \quad \text{for the positive square root of } a, \text{ and}$$
$$-\sqrt{a} \text{ or } -\sqrt{a} \quad \text{for the negative square root of } a.$$

For example, both 5 and −5 are square roots of 25 because both

$$5^2 = (5)(5) = 25 \quad \text{and} \quad (-5)^2 = (-5)(-5) = 25,$$

and $|5| = |-5|$. For historical reasons, we shall consider only positive square roots here.

Remark 12.1 The concept of negative numbers was not generally accepted by mathematicians until only a few hundred years ago [11]. A big stumbling block was in trying to understand what it would mean for a number to designate "less than nothing." Ancient Greek mathematicians associated numbers with concrete objects. To the Greeks there were counting numbers and magnitudes, and the Greeks even thought of line segments, areas, and volumes as different types of magnitudes. Although Indian mathematicians were using negative numbers by the seventh century AD (treating them as debts), and Chinese mathematicians also seemed to be able to handle negative numbers in solving equations, European mathematicians, on the other hand, grappled with the concept well into the seventeenth century, regarding negative numbers as "fictitious" or "absurd," and negative roots of equations as "false." And these were some of the great mathematicians: Cardano, Viète, Descartes, Wallis. This is not to say that they were not able to operate with negative numbers; they just could not ground the concept in the "real world." Eventually things began to change in the eighteenth century, but it was not until the nineteenth century, when algebra began to be studied as abstract systems, that negative numbers finally gained general acceptance among mathematicians. Then negative numbers were simply additive inverses of positive numbers. Of course, today we find many "real world" applications of negative numbers.   □

The problem of extracting or approximating roots reaches as far back as the ancient Egyptians and the Babylonians. And in the several hundred years that followed, the Greeks, Indians, Arabs, and Chinese all tackled the problem [112]. Here are four methods.

12.1   HERON OF ALEXANDRIA

It is in *Metrica*, in the course of finding the area of a certain triangle using what we now refer to as Heron's formula, that Heron of Alexandria (third century AD)[1] provides a method for approximating square roots. As Heath [69] puts it,

> 'Since', says Heron,[2] '720 has not its side rational, we can obtain its side within a very small difference as follows. Since the next succeeding square number is 729, which has 27 for its side, divide 720 by 27. This gives $26\frac{2}{3}$. Add 27 to this, making $53\frac{2}{3}$, and take half of this or $26\frac{1}{2}\frac{1}{3}$. The side of 720 will therefore be very nearly $26\frac{1}{2}\frac{1}{3}$. In fact, if we multiply $26\frac{1}{2}\frac{1}{3}$ by itself, the product is $720\frac{1}{36}$, so that the difference (in the square) is $\frac{1}{36}$.
>
> 'If we desire to make the difference still smaller than $\frac{1}{36}$, we shall take $720\frac{1}{36}$ instead of 729 [or rather we should take $26\frac{1}{2}\frac{1}{3}$ instead of 27], and by proceeding in the same way we shall find that the resulting difference is much less than $\frac{1}{36}$.'

(Note that by $26\frac{1}{2}\frac{1}{3}$ is meant 26 *and* $\frac{1}{2}$ *and* $\frac{1}{3}$, or $26\frac{5}{6}$.)

In other words, to extract or approximate $\sqrt{A}$,

1. Find a number a such that $a^2 > A$ is "close" to A. If $a^2 = A$, then $\sqrt{A} = a$; if $a^2 \neq A$, then $\sqrt{A} \approx a$ with remainder in the square $A - a^2$.

2. To improve the approximation, find the number

$$\alpha_1 = \frac{1}{2}\left(a + \frac{A}{a}\right).$$

   (The symbol α is the Greek lowercase letter *alpha*. Note that α_1 is the arithmetic mean or average of the numbers a and $\frac{A}{a}$.)

3. To improve the approximation further, find the number

$$\alpha_2 = \frac{1}{2}\left(\alpha_1 + \frac{A}{\alpha_1}\right),$$

and so on, that is

$$\alpha_3 = \frac{1}{2}\left(\alpha_2 + \frac{A}{\alpha_2}\right), \quad \alpha_4 = \frac{1}{2}\left(\alpha_3 + \frac{A}{\alpha_3}\right), \ldots.$$

Remark 12.2 It can be shown that the approximations $\alpha_1, \alpha_2, \ldots$ found in this way always exceed $\sqrt{A}$ and that $\alpha_1 > \alpha_2 > \cdots$. Thus, the approximations will gradually get closer and closer to the true value. Also, today we would obtain the method of Heron using calculus and a general method for finding roots of equations called *Newton's method* or the *Newton-Raphson method*.   □

---

[1]There is much uncertainty about the dates of Heron of Alexandria [69, pp. 298ff]. By inferring from a variety of references, made both by and about Heron, historians over the years have placed Heron anywhere from 150 BC to the fourth century AD. The reason for not placing Heron before 150 BC, Heath tells us, is the discovery by R. Schöne in 1896 of the Greek text *Metrica* in which Heron alludes to Archimedes (287–212 BC) and Apollonius (262–190 BC); and the reason for the lower limit of the fourth century AD is the dubious suggestion that Heron dedicated his work *Definitions* to "a certain Dionysius who was *praefectus urbi* at Rome in AD 301." In the end, however, Heath places Heron in the third century AD, "and perhaps little, if anything, earlier than Pappus" (early fourth century AD).

[2]*Metrica*, i. 8, pp. 18 22–20. 5.

Think About It 12.1 What is happening in Heron's example where he finds half of the quantity $720 \div 27 + 27$? □

Example 12.1 Find the square root of 720.

We follow Heron. With $A = 720$, we seek a number a such that $a^2 > A$ is "close" to 720. By trial we find that $26^2 = 676$ and $27^2 = 729$, so we choose $a = 27$. Then

$$\sqrt{720} \approx 27.$$

To improve the approximation, we find

$$\alpha_1 = \frac{1}{2}\left(a + \frac{A}{a}\right) = \frac{1}{2}\left(27 + \frac{720}{27}\right) = 26\frac{5}{6},$$

so that

$$\sqrt{720} \approx 26\frac{5}{6},$$

which is the result that Heron had gotten. We note that $26\frac{5}{6} = 26.8333\ldots$ agrees with a calculator result of $\sqrt{720} = 26.8328\ldots$ to two decimal places.

To improve the approximation, we find

$$\alpha_2 = \frac{1}{2}\left(\alpha_1 + \frac{A}{\alpha_1}\right) = \frac{1}{2}\left(26\frac{5}{6} + \frac{720}{26\frac{5}{6}}\right) = 26\frac{1609}{1932},$$

so that

$$\sqrt{720} \approx 26\frac{1609}{1932}.$$

We note that $26\frac{1609}{1932} = 26.832\,815\,734\ldots$ agrees with a calculator result of $\sqrt{720} = 26.832\,815\,729\ldots$ to seven decimal places after only two steps or iterations. That is remarkable! □

Example 12.2 Find the square root of 10.

With $A = 10$, we seek a number a such that $a^2 > A$ is "close" to 10. By trial we find that $3^2 = 9$ and $4^2 = 16$, so we choose $a = 4$. Then

$$\sqrt{10} \approx 4.$$

To improve the approximation, we find

$$\alpha_1 = \frac{1}{2}\left(a + \frac{A}{a}\right) = \frac{1}{2}\left(4 + \frac{10}{4}\right) = 3\frac{1}{4},$$

so that

$$\sqrt{10} \approx 3\frac{1}{4}.$$

We note that $3\frac{1}{4} = 3.25$ agrees with a calculator result of $\sqrt{10} = 3.1623\ldots$ to the one place.

To improve the approximation, we find

$$\alpha_2 = \frac{1}{2}\left(\alpha_1 + \frac{A}{\alpha_1}\right) = \frac{1}{2}\left(3\frac{1}{4} + \frac{10}{3\frac{1}{4}}\right) = 3\frac{17}{104},$$

so that

$$\sqrt{10} \approx 3\frac{17}{104}.$$

We note that $3\frac{17}{104} = 3.163\,46\ldots$ agrees with a calculator result of $3.162\,27\ldots$ to two decimal places. □

Now You Try 12.1

1. Repeat the preceding example that approximates $\sqrt{10}$, but use $a = 3\frac{1}{5}$ instead.

2. Use the method of Heron of Alexandria to extract or approximate the square root of the following. Do not go beyond two steps to α_2. Also, use a calculator to find or approximate each square root. To how many decimal places does the approximation by Heron's method agree with the calculator result?

   a) 60   b) 500

   □

12.2   THEON OF ALEXANDRIA

Theon of Alexandria (fl. AD 375) was one of the more important commentators of Greek mathematics. In fact, it appears that it was upon his edition of Euclid's *Elements* [53] that all later editions were based [112]. Theon was also the father of Hypatia (d. AD 415), the earliest known female mathematician.

Theon's method for finding square roots is an application of Proposition 4 in Book II of the *Elements* [53, p. 39], which is an assertion about the equality of certain areas.

> If a straight line be cut at random, the square on the whole is equal to the squares on the segments and twice the rectangle contained by the segments.

Let us see what *Elements* II.4 is saying.

- *If a straight line be cut at random,*

  Let AB be a straight line segment that is cut at random at a point C.

- *the square on the whole*

  This is the square on AB, that is, the square with edge AB.

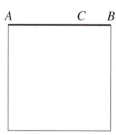

- *is equal to the squares on the segments*

  The squares on the segments are the squares with edge AC and with edge BC shown in yellow.

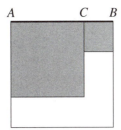

- *and twice the rectangle contained by the segments.*

    This is the rectangle with length AC and height BC; the two are shown in cyan.

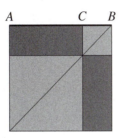

Thus, if AC has length a and BC has length b, we see that the area of

- *the square on the whole* equals $(a + b)^2$;

- *the squares on the segments* equals a^2 and b^2;

- *the rectangle contained by the segments* equals ab;

A	a	C b B
b	ab	b^2
a	a^2	ab

and, therefore, *Elements* II.4 asserts the algebraic identity

$$(a + b)^2 = a^2 + b^2 + 2 \cdot ab.$$

We illustrate how Theon of Alexandria, in his commentary on Ptolemy's *Syntaxis* [68], used *Elements* II.4 to extract or approximate square roots.

In general, to find $\sqrt{A}$,

1. Note that $\sqrt{A}$ is the side of a square with area A.

2. Find a number a, by trial, such that $a^2 \le A$. If $a^2 = A$, then $\sqrt{A} = a$; if $a^2 < A$, then $\sqrt{A} \approx a$ with remainder in the square $A - a^2$. The remainder is the gnomon or L-shaped region shaded in light gray below.

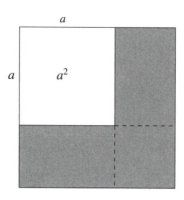

3. To improve the approximation, find the greatest number b, by trial, such that $b^2 + 2ab \leq A - a^2$, the remainder. (Where are b^2 and $2ab$ in the diagram above?) If $b^2 + 2ab = A - a^2$, then $\sqrt{A} = a + b$; if $b^2 + 2ab < A - a^2$, then $\sqrt{A} \approx a + b$ with a lesser remainder $A - a^2 - b^2 - 2ab$. The lesser remainder is the gnomon shaded in light gray below.

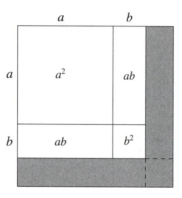

4. To improve the approximation further, find the greatest number c such that $c^2 + 2(a + b)c$ is less than or equal to the last remainder. Then either $\sqrt{A} = a + b + c$ or $\sqrt{A} \approx a + b + c$ with a yet lesser remainder, and so on.

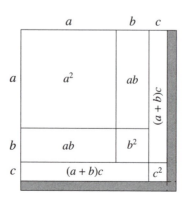

Example 12.3 Find the square root of 144.

Heath [68] tells us that this is an example of Theon of Alexandria in his commentary on the *Syntaxis*, and so we follow Heath's description of the solution.

To begin, we note that $A = 144$, so we seek a number a such that $a^2 \leq 144$. Now, since $10^2 = 100$ is less than 144, we let $a = 10$. Then

$$\sqrt{144} \approx 10$$

with remainder in the square

$$A - a^2 = 144 - 10^2 = 44.$$

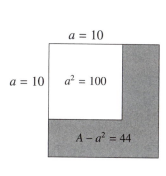

To improve the approximation, we seek a number b, by trial, such that $b^2 + 2ab = b^2 + 20b$ is less than or equal to the remainder, 44. We divide the remainder $A - a^2$ by $2a$ to obtain a guess for b; that is, we divide 44 by 20 to obtain a guess for b. Since $44 \div 20 = 2$ with a remainder, we try $b = 2$. We find that

$$b^2 + 2ab = 2^2 + 2(10)(2) = 44$$

exactly.

Therefore, we have succeeded in extracting the square root of 144:

$$\sqrt{144} = 10 + 2 = 12.$$

☐

Think About It 12.2 Why would $\frac{A-a^2}{2a}$ be a good approximation for b? (Hint: What does $A = (a + b)^2$ imply about b?)
☐

Example 12.4 Find the square root of 10.

Following the last example, we first note that $A = 10$, so we seek a number a such that $a^2 \leq 10$. If we choose $a = 2$, then

$$\sqrt{10} \approx 2$$

with remainder in the square

$$A - a^2 = 10 - 2^2 = 6.$$

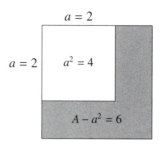

To improve the approximation, we seek a number b, by trial, such that $b^2 + 2ab = b^2 + 4b$ is less than or equal to the remainder, 6. We divide the remainder $A - a^2$ by $2a$ to obtain a guess for b, that is, we divide 6 by 4 to obtain a guess for b. Since $6 \div 4 = 1$ with a remainder, we try $b = 1$. We find that

$$b^2 + 2ab = 1^2 + 2(2)(1) = 5 < 6.$$

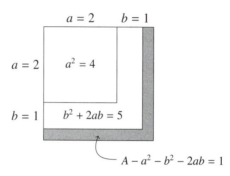

Hence, an improved approximation is

$$\sqrt{10} \approx 2 + 1 = 3.$$

To improve the approximation further, we seek a number c, by trial, such that $c^2 + 2(a + b)c = c^2 + 6c$ is less than or equal to the remainder, 1. We see that c must be a fraction. We divide the remainder $A - a^2 - b^2 - 2ab$ by $2(a + b)$ to obtain a guess for c, that is, we divide 1 by 6 to obtain a guess for c. Since $1 \div 6 = \frac{1}{6}$, we try $c = \frac{1}{6}$. However, we find that

$$\left(\frac{1}{6}\right)^2 + 2(2 + 1)\left(\frac{1}{6}\right) = 1\frac{1}{36} > 1,$$

so we have to choose a number c that is less than $\frac{1}{6}$. We choose $c = \frac{1}{9}$ because 9 will cancel conveniently with $a + b = 3$. We find that

$$\left(\frac{1}{9}\right)^2 + 2(2 + 1)\left(\frac{1}{9}\right) = \frac{55}{81} < 1.$$

Hence, a further improved approximation is

$$\sqrt{10} \approx 2 + 1 + \frac{1}{9} = 3\frac{1}{9}.$$

We note that $3\frac{1}{9} = 3.111\ldots$ agrees with a calculator result of $\sqrt{10} = 3.1623\ldots$ to one decimal place. We could improve the approximation further still, but we shall stop here instead.   □

Now You Try 12.2 Use the method of Theon of Alexandria to extract or approximate the square root of the following. Do not go beyond three steps. Also, use a calculator to approximate each square root. To how many decimal places does the approximation by Theon's method agree with the calculator result?

1. 60   2. 500

☐

Think About It 12.3 Look at the diagram for Theon's method. Describe why $b^2 + 2ab$ must be less than or equal to the first remainder, and why $c^2 + 2(a + b)c$ must be less than or equal to the second remainder. ☐

12.3   BAKHSHĀLĪ MANUSCRIPT

The Bakhshālī Manuscript is an Indian birch-bark manuscript that contains rules and examples for a variety of arithmetic techniques. The dating of the manuscript is still uncertain (see page 53). Following the rule-and-example style, rules (*sūtras*) and examples are given in verse, and the solutions explained in prose. In the following example, explanations and commentary are rendered in sans serif text.

Example 12.5 Among the examples in the Bakhshālī Manuscript is a method for approximating square roots. Here is an example from the Bakhshālī Manuscript that approximates the square root of 481 [95, pp. 439–440]. The example is rather cryptic, and we will present it again in modern notation following the example.

[2nd solution of Example 1 for Sūtra N18.]

[. . .] "And squared. When one has added [the square] to it" [Sūtra N18]: 481. The square root [obtained before] is $21\frac{20}{21}$. This is inaccurate. Therefore, [we recall Sūtra Q2 again].

[Sūtra Q2:] The divisor for the remainder from the diminution of the non-square number by the square of the first approximation, is twice [the first approximation]. Division of half the square of that (i.e., the quotient just obtained) by the second approximation [is made]. Subtraction [of the result from the second approximation gives the third approximation]. Less the square.

[The first half of this rule], "The divisor for the remainder [. . .] is twice," has been already applied, [the result being $21\frac{20}{21}$]. "The square of that,"

21				"the 2nd approx" 21	: divisor
$\frac{20}{21}$	$\frac{400}{441}$	"half"	$\frac{1}{2}$	$\frac{20}{21}$	

The remainder, [400], should be removed [and preserved separately for verification]. When one has applied [the rule], "a factor into a factor" [Sūtra Q1], [the result is] divided: $\frac{21}{461}$. "The lower should be multiplied by the lower, and the upper by the upper" [Sūtra Q3]. [The result is $\frac{400}{19362}$.] One should [then] subtract the square. There, reduction to the same [denominator] is made:

$$\frac{425042}{19362} \quad - \frac{400}{19362} \quad \text{Remainder:} \quad \frac{424642}{19362}$$

☐

The complaint in the example above is that the approximation $\sqrt{481} \approx 21\frac{20}{21}$ is not accurate, and a more accurate approximation is sought. The result is the improved approximation $\sqrt{481} \approx \frac{424\,642}{19\,362}$ or $21\frac{18\,040}{19\,362}$. We compare the two approximations using a calculator.

$$\text{First:} \qquad 21\tfrac{20}{21} = 21.952\,38\ldots$$

$$\text{Second:} \quad 21\tfrac{18\,040}{19\,362} = 21.931\,72\ldots$$

$$\text{Calculator:} \quad \sqrt{481} = 21.931\,71\ldots$$

So, we see that the first approximation agrees with a calculator result of $\sqrt{481} = 21.931\,71\ldots$ to one decimal place, while the second approximation agrees to *four* decimal places—a significant improvement.

If the above method was a bit cryptic, we will now present it in modern algebraic form. If a positive integer $A = a^2 + r$, where a and r are integers, then

$$\sqrt{A} \approx a + \frac{r}{2a} - \frac{(r/(2a))^2}{2(a + r/(2a))}.$$

In the example above, $A = 481$ and $a = 21$, the greatest square that is less than 481. Since $A = a^2 + r$, the remainder

$$r = A - a^2 = 481 - 21^2 = 40.$$

Thus, substituting $a = 21$ and $r = 40$ into the formula

$$\sqrt{A} \approx a + \frac{r}{2a} - \frac{(r/(2a))^2}{2(a + r/(2a))},$$

we obtain

$$\sqrt{481} \approx 21 + \frac{40}{2 \cdot 21} - \frac{(40/(2 \cdot 21))^2}{2(21 + 40/(2 \cdot 21))}$$

$$= 21 + \frac{40}{42} - \frac{1600/1764}{42 + 80/42}$$

$$= 21 + \frac{40}{42} - \frac{1600/1764}{922/21}$$

$$= 21 + \frac{40}{42} - \frac{33\,600}{1\,626\,408}$$

$$= \frac{34\,154\,568}{1\,626\,408} + \frac{1\,548\,960}{1\,626\,408} - \frac{33\,600}{1\,626\,408}$$

$$= \frac{35\,669\,928}{1\,626\,408}$$

$$= \frac{424\,642}{19\,362},$$

which agrees with the result in the Bakhshālī example (Remainder: $\frac{424642}{19362}$).

Example 12.6 Use the method of the Bakhshālī Manuscript to approximate the square root of 720.

Here $A = 720$. We seek a number a such that a^2 is the greatest square less than 720; we find that $a = 26$. Thus, the remainder is

$$r = A - a^2 = 720 - 26^2 = 44.$$

Thus, using the formula for $\sqrt{A}$,

$$\sqrt{A} \approx a + \frac{r}{2a} - \frac{(r/(2a))^2}{2(a + r/(2a))},$$

we obtain

$$\sqrt{720} \approx 26 + \frac{44}{2 \cdot 26} - \frac{(44/(2 \cdot 26))^2}{2(26 + 44/(2 \cdot 26))}$$

$$= 26 + \frac{44}{52} - \frac{1936/2704}{698/13}$$

$$= 26 + \frac{1\,597\,024}{1\,887\,392} - \frac{25\,168}{1\,887\,392}$$

$$= 26\frac{1\,571\,856}{1\,887\,392}.$$

We note that $26\frac{1\,571\,856}{1\,887\,392} = 26.832\,819\ldots$ agrees with a calculator result of $\sqrt{720} = 26.832\,815\ldots$ to five decimal places. □

Example 12.7 We use the method of the Bakhshālī Manuscript to approximate the square root of 720 again, but we use a different value of a.

Here, again, $A = 720$. In the last example, we chose $a = 26$ because $26^2 = 676$ is the greatest square that is less than 720; however, 676 is not the greatest square that is *closest* to 720: Note that $27^2 = 729$ is closer to 720 than is $26^2 = 676$. Thus, in this example, we choose $a = 27$. Then, $r = A - a^2 = -9$. (Yes, r is negative here.) Now, using the formula

$$\sqrt{A} \approx a + \frac{r}{2a} - \frac{(r/(2a))^2}{2(a + r/(2a))},$$

we obtain

$$\sqrt{720} \approx 27 + \frac{-9}{2 \cdot 27} - \frac{(-9/(2 \cdot 27))^2}{2(27 - 9/(2 \cdot 27))}$$

$$= 27 - \frac{9}{54} - \frac{81/2916}{2898/54}$$

$$= 27 - \frac{1408428}{8450568} - \frac{4374}{8450568}$$

$$= 27 - \frac{1412802}{8450568}$$

$$= 26\frac{7037766}{8450568}.$$

We note that $26\frac{7037766}{8450568} = 26.832\,815\,735\ldots$ agrees with a calculator result of $\sqrt{720} = 26.832\,815\,729\ldots$ to *seven* decimal places, which is even better than the approximation that we obtained in the last example. (Compare this to the result obtained using Heron's method in EXAMPLE 12.1 on page 123.) □

Now You Try 12.3 Use the method of the Bakhshālī Manuscript to extract or approximate the square root of the following. Also, use a calculator to approximate each square root. To how many decimal places does the approximation agree with the calculator result?

1. 60

2. 500

□

12.4 NICOLAS CHUQUET

Nicolas Chuquet (d. 1487) was a French physician who, in mathematics, is known for his treatise on arithmetic and algebra, *Triparty en la science des nombres* (*Science of Numbers in Three Parts*), that he wrote near the end of his life. However, it was never printed, and only a handwritten manuscript exists today [73, 112].

In *Triparty*, Chuquet, coined the terms *billion* for million million (bi-million: 1,000,000,000,000) and *trillion* for million million million (tri-million: 1,000,000,000,000,000,000), and so on, which, Suzuki [112] notes, "corresponds to the use of these terms in England and in Germany but not, oddly enough, in France, or in the United States"; in the United States one billion is one thousand million (1,000,000,000) and one trillion is one million million (1,000,000,000,000). In *Triparty* Chuquet also generalized al-Khwārizmī's methods for solving quadratic equations to solving equations of any degree that are of quadratic type [73]. (Al-Khwārizmī was a prominent Arabic mathematician whom we shall encounter later; see page 312.)

Think About It 12.4 Recall that the place value we commonly use in writing a numeral is base 10, but that the American word name is base 1000 (page 4). In what base is the British word name? □

It is in *Triparty* that Chuquet noted that "[to] find a number between two fractions, add numerator to numerator and denominator to denominator" [112]. In other words, if $0 < \frac{a}{b} < \frac{c}{d}$, then

$$\frac{a}{b} < \frac{a+c}{b+d} < \frac{c}{d}.$$

Chuquet used this observation to approximate square roots, as well as roots of equations like $x^2 + x = 39\frac{13}{81}$ [73, 112].

Example 12.8 Katz [73] relates how Chuquet applied his observation to find $\sqrt{6}$.

1. Bound $\sqrt{6}$ by two positive integers. Since $2^2 = 4$, $3^2 = 9$, and $4 < 6 < 9$,

$$2 < \sqrt{6} < 3.$$

2. Now, a convenient number to choose between 2 and 3 is $2\frac{1}{2}$. Since $(2\frac{1}{2})^2 = 6\frac{1}{4} > 6$,

$$2 < \sqrt{6} < 2\frac{1}{2}.$$

3. Now, a convenient number to choose between 2 and $2\frac{1}{2}$ is $2\frac{1}{3}$. Since $(2\frac{1}{3})^2 = 5\frac{4}{9} < 6$,

$$2\frac{1}{3} < \sqrt{6} < 2\frac{1}{2}.$$

4. Next, to find a number between $2\frac{1}{3}$ and $2\frac{1}{2}$, add the numerators and denominators, respectively, of $\frac{1}{3}$ and $\frac{1}{2}$ to obtain $\frac{2}{5}$; then $2\frac{1}{3} < 2\frac{2}{5} < 2\frac{1}{2}$. Since $(2\frac{2}{5})^2 = 5\frac{19}{25} < 6$,

$$2\frac{2}{5} < \sqrt{6} < 2\frac{1}{2}.$$

5. Next, to find a number between $2\frac{2}{5}$ and $2\frac{1}{2}$, add the numerators and denominators, respectively, of $\frac{2}{5}$ and $\frac{1}{2}$ to obtain $\frac{3}{7}$; then $2\frac{2}{5} < 2\frac{3}{7} < 2\frac{1}{2}$. Note that $2\frac{3}{7} = 2.4285\ldots$ agrees with a calculator result of $\sqrt{6} = 2.4495\ldots$ to one decimal place.

Continuing in this way, Chuquet obtained the further approximations $2\frac{4}{9}$, $2\frac{5}{11}$, and $2\frac{9}{20}$, the last approximating $\sqrt{6}$ with error less than 5×10^{-3} or 0.005, that is, $2\frac{9}{20}$ approximates $\sqrt{6}$ to two decimal places. □

Now You Try 12.4 Use the method of Nicolas Chuquet to extract or approximate the square root of the following. Do not go beyond five steps. Also, use a calculator to approximate each square root. To how many decimal places does the approximation by Chuquet's method agree with the calculator result?

1. 60   2. 500

☐

Think About It 12.5 Who would have used the various methods of calculation discussed above? Where would they have been taught? Why are they not taught in schools today? Are any of these methods related to methods used today?   ☐

Think About It 12.6 Given the slow evolution of mathematical numeration and calculation that we have looked at, would you expect mathematics to look the same in another one hundred or five hundred years from now? What is it about our modern society that would accelerate change in mathematical practices? What is it about our society that would help lock current notation and methods in place?   ☐

Chapter 13

Exercises

Ex. 13.1 — Use the grating or lattice method to find *fifty-three times eighteen* in the following bases.

1. In base 6 using the base-6 numerals $\alpha, \beta, \gamma, \delta$, and ϵ, and place holder $\cdot$. (See SECTION 1.1.)

2. In base 4 using the base-4 numerals symbols $|$, $\wedge$, and $\vee$, and place holder $\square$. (See SECTION 1.2.)

Ex. 13.2 — Every positive integer either is a power of 2 or can be written as a sum of powers of 2; for example, $1 = 2^0$ and $13 = 2^0 + 2^2 + 2^3$ or $1 + 4 + 8$. Write each of the following positive integers either as a power of 2 or as a sum of powers of 2.

1. 15   2. 37   3. 100   4. 1024

Ex. 13.3 — Simplify the following expressions, used in one interpretation of Plimpton 322.

1. $\dfrac{1}{\frac{n+1}{2}} + \dfrac{1}{\frac{n(n+1)}{2}}$

2. $\dfrac{1}{p \cdot \frac{p+q}{2}} + \dfrac{1}{q \cdot \frac{p+q}{2}}$

Ex. 13.4 — Simplify the following expression.

$$\frac{1}{2p} + \frac{1}{6p}$$

Ex. 13.5 — To explain the decomposition of fractions in the $2 \div N$ table (page 102), Calinger [35, p. 45] says that,

> A'hmosè generally utilizes the verbal equivalent of what is today called the splitting identity:
>
> $2/n = 1/[(n + 1)/2] + 1/[n(n + 1)/2]$.

1. Verify that $2/n = 1/[(n + 1)/2] + 1/[n(n + 1)/2]$.

2. Use this rule to find a decomposition of each of the following fractions, and then compare your results to those in the $2 \div N$ table. What do you conclude?

    a) $\frac{2}{5}$ b) $\frac{2}{7}$ c) $\frac{2}{23}$ d) $\frac{2}{55}$

Ex. 13.6 — Calinger goes on to say that,

Showing a preference for 2/3, for all fractions of the form $2/(3p)$, A'hmosè employs another general rule: $2/(3p) = 1/(2p) + 1/(6p)$.

1. Verify that $2/(3p) = 1/(2p) + 1/(6p)$.

2. Use this rule to find a decomposition of each of the following fractions, and then compare your results to those in the $2 \div N$ table. What do you conclude?

a) $\frac{2}{9}$ b) $\frac{2}{15}$ c) $\frac{2}{21}$ d) $\frac{2}{39}$

Ex. 13.7 — Use the distributive law to verify the algebraic identity,

$$(a + b)^2 = a^2 + b^2 + 2ab,$$

that is asserted by *Elements* II.4.

Ex. 13.8 — Verify that if $0 < \frac{a}{b} < \frac{c}{d}$, then

$$\frac{a}{b} < \frac{a + c}{b + d} < \frac{c}{d}.$$

Ex. 13.9 — In EXAMPLE 12.8, we saw how Chuquet approximated $\sqrt{6}$. However, the example was incomplete because it did not show how Chuquet obtained the last approximations, namely, $2\frac{4}{9}$, $2\frac{5}{11}$, and $2\frac{9}{20}$. Complete the example by obtaining these last approximations.

Ex. 13.10 — The *arithmetic mean* and the *geometric mean* of two numbers x and y are given by

$$\text{arithmetic mean} = \frac{x + y}{2},$$

$$\text{geometric mean} = \sqrt{xy}.$$

The two means always satisfy the inequality

$$\frac{x + y}{2} \le \sqrt{xy}$$

that is called the *arithmetic mean-geometric mean inequality (AGM)*. Use the AGM to show that the approximations $\alpha_1, \alpha_2, \dots$ found in the method of Heron of Alexandria to approximate $\sqrt{A}$ always exceed $\sqrt{A}$.

Ex. 13.11 — Show that the method for approximating the square root of a non-square number in the Bakhshālī Manuscript is the same as using two steps of the method of Heron.

Ex. 13.12 — In *Metrica*, Heron of Alexandria gave two formulas for finding the area Δ of a scalene triangle given the lengths of the sides [69]. The first method, based on *Elements* II.12 and 13, determines the height of the triangle and then applies the familiar formula

$$\Delta = \frac{1}{2} \times \text{base} \times \text{height}.$$

The second method is the following formula that we call *Heron's formula*:

$$\Delta = \sqrt{s(s - a)(s - b)(s - c)},$$

where a, b, and c are the lengths of the sides of the triangle, and

$$s = \frac{a + b + c}{2}$$

is the *semiperimeter*.

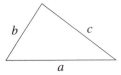

Use Heron's formula to find the areas of the following triangles with sides a, b, and c. If the square root needs to be approximated, use two steps of Heron's method (page 122).

1. $a = 7, b = 8, c = 9$ 2. $a = 10, b = 8, c = 12$

Ex. 13.13 — A *quadrilateral* is a closed figure with four straight sides. (For example, a rectangle is a particular quadrilateral; a quadrilateral may also be called a *quadrangle*.) Brahmagupta, in chapter 12 of *Brāhmasphuṭasiddhānta*, gives the following formulas for finding the area of a triangle and a quadrilateral given the lengths of the sides [95, p. 423]:

> 12.21. The approximate area is the product of the halves of the sums of the sides and opposite sides of a triangle and a quadrilateral. The accurate [area] is the square-root from the product of the halves of the sums of the sides diminished by [each] side of the quadrilateral.

Thus, for a triangle with sides a, b, and c, Brahmagupta tells us that the approximate area is $\frac{a+b}{2} \cdot \frac{c}{2}$; and for a quadrilateral with sides a, b, c, and d, he tells us that the approximate area is $\frac{a+b}{2} \cdot \frac{c+d}{2}$. (Note that the formula to approximate the area of a triangle can be obtained by setting $d = 0$ in the formula to approximate the area of a quadrilateral.) Our interest here, however, are the "accurate" formulas.

For the accurate formulas, Brahmagupta tells us that we need to use the "halves of the sums of the sides." These are the semiperimeters, which are $s_\triangle$ for a triangle and $s_\square$ for a quadrilateral, namely,

$$s_\triangle = \frac{a + b + c}{2} \quad \text{and} \quad s_\square = \frac{a + b + c + d}{2}.$$

Then, to be "diminished by [each] side" means to subtract the length of a side from the semiperimeter:

$$s - a, \quad s - b, \quad s - c, \quad s - d.$$

Therefore, the accurate formulas are $A_\triangle$ for the area of a triangle and $A_\square$ for the area of a quadrilateral[1] given by

$$A_\triangle = \sqrt{(s - a)(s - b)(s - c)s},$$
$$A_\square = \sqrt{(s - a)(s - b)(s - c)(s - d)}.$$

(Note that Brahmagupta's accurate formula for the area of a triangle is precisely Heron's formula. Also, just as with the approximate formulas, the accurate formula to find the area of a triangle can be obtained

---

[1] The accurate formula for the area of a quadrilateral is *truly accurate only for* "cyclic quadrilaterals." A *cyclic quadrilateral* is one that can be *circumscribed* by a circle, that is, a circle can be drawn to touch each of the *vertices* or corners of the quadrilateral. Indeed, if a quadrilateral is not cyclic, then $A_\square$ only approximates its area (with no general sense of how good or bad is the approximation).

cyclic quadrilateral

non-cyclic quadrilateral

by setting $d = 0$ in the formula to find the area of a quadrilateral.) Use Brahmagupta's accurate formula to find the areas of the following quadrilaterals with sides a, b, c, and d; if the square root needs to be approximated, use the method of the Bakhshālī Manuscript (page 130).

    1. $a = 7, b = 8, c = 9, d = 10$ 2. $a = 10, b = 8, c = 12, d = 14$

Ex. 13.14 — A trapezoid or trapezium is a particular quadrilateral in which one pair of opposite sides is parallel. Determine if the following trapezoid is a cyclic quadrilateral by computing its area in two ways: using the familiar formula for the area of a trapezoid,

$$A = \frac{1}{2}(B + b)h;$$

using Brahmagupta's formula (exercise 13.13).

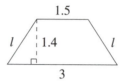

If Brahmagupta's formula gives the correct area, then the trapezoid is a cyclic quadrilateral; otherwise, the trapezoid is not a cyclic quadrilateral. Hint: You may use Pythagoras's theorem to find the lateral side length l. See page 216.

* **Ex. 13.15** — Research different ways the abacus was used in Europe to perform calculations.

* **Ex. 13.16** — Describe the similarities and the differences between the Chinese suan pan and the Japanese soroban.

* **Ex. 13.17** — Discover the type of abacus featured on the cover of this book. What makes it different from the Chinese and Japanese abacuses?

* **Ex. 13.18** — Research how the Chinese rod numerals have evolved and spread over time.

* **Ex. 13.19** — Trace the etymology of the word "gnomon."

* **Ex. 13.20** — What is *prosthaphæresis* as a method for finding the product ab?

* **Ex. 13.21** — Write about Euclid's *Elements* and its significance to modern mathematics.

* **Ex. 13.22** — Write about Hypatia, one of the earliest known women mathematicians.

* **Ex. 13.23** — Write about John Napier and describe "Napier's bones" mentioned on page 97.

* **Ex. 13.24** — In CHAPTER 11 (page 117 ff.), we presented a "simplified version" of casting out nines. What is the "proper version" of casting out nines? Demonstrate it with a few examples, and explain why the method of casting out nine works.

* **Ex. 13.25** — Explain how "casting out thirteens" works, and write about its history.

* **Ex. 13.26** — Explain why the method of division in *Liber Abaci* shown in EXAMPLE 10.5 on page 111 works.

* **Ex. 13.27** — How is the date of Easter calculated? Why would that require intricate mathematics?

* **Ex. 13.28** — Leonardo of Pisa is also known as Fibonacci. What does "Fibonacci" mean?

* **Ex. 13.29** — Explain how Leonardo of Pisa arrived at the Fibonacci sequence, and give examples in which the Fibonacci sequence appears in nature.

* **Ex. 13.30** — Research the city of Alexandria, Egypt, and its role in science and mathematics during the Hellenistic period.

Part III

Foundations

Chapter 14

Sets

It is usual to divide the world around us into groups according to some common attribute or attributes. For example, we may group cars by their make (Ford, Toyota, BMW) or by their color (red, green, blue) or by their number of doors (two door, four door), and so on. We also do this in mathematics, where we call such groups of objects "sets."

Surprisingly, the history of sets is very recent. Although the notion of sets in a mathematical setting seems fairly elementary and convenient, it was not developed slowly over time by many people as much of the history of mathematics and the sciences were. The invention of the theory of sets is primarily that of one man, Georg Cantor (1845–1918) [71].

Cantor was born in Denmark, spent his early childhood in St. Petersburg, Russia, and moved to Germany at the age of eleven. (See FIGURE 14.1.) After attending university in Zurich and Berlin, he spent the remainder of his life and career in Halle as a professor of mathematics. Even from the comparative remoteness of Halle, Cantor launched the Association of German Mathematicians in 1890 and was a leader in the organization of the first International Congress of Mathematicians in 1897. Cantor's life work was the study of the infinite, which we will say more about later. In order to talk about things finite and infinite, Cantor needed to develop a way of comparing sets of finite and infinite sizes. He realized that he needed a way to think of sets as entities in their own right—thus set theory was born. Cantor's work in set theory, and the infinite in particular, was revolutionary—so revolutionary, in fact, that it was met with stiff resistance from leading mathematicians of the time. Cantor suffered from bouts of mental instability, perhaps caused or exacerbated by the resistance his theory of the infinite met. He died while in a sanitarium in 1918.

We start this chapter with a naive, but adequate, definition of a set.

Definition 14.1 A *set* is a collection of objects. The objects that belong to a set are called the *elements* of the set. □

We indicate that objects belong to the same set by enclosing a list or a description of the objects within curly brackets, "{" and "}." We call this *set notation*.

If an object a belongs to a set A, we write $a \in A$, read "a is an element of (the set) A" or "a is in (the set) A"; otherwise, if a does not belong to A, we write $a \notin A$, read "a is not an element of A" or "a is not in A."

For example, you are most likely already familiar with these sets of numbers, even if you may not be familiar with their names.[1]

1. The set of *natural numbers* or *counting numbers*, denoted $\mathbb{N}$:

$$\mathbb{N} = \{1, 2, 3, \dots\}.$$

---

[1] You may also recall having heard of the sets of *rational numbers*, *irrational numbers*, and *real numbers*. We shall describe these sets in CHAPTER 15.

Figure 14.1: Antique map of Europe. (Source: © Thinkstock.)

2. The set of *whole numbers*, denoted $\mathbb{W}$:

$$\mathbb{W} = \{0,\ 1,\ 2,\ 3, \dots\}.$$

3. The set of *integers*, denoted $\mathbb{Z}$:[2]

$$\mathbb{Z} = \{\dots,\ -3,\ -2,\ -1,\ 0,\ 1,\ 2,\ 3, \dots\}.$$

Moreover, note that zero is an element of the set of whole numbers, but zero is not an element of the set of natural numbers: $0 \in \mathbb{W}$, but $0 \notin \mathbb{N}$.

Now You Try 14.1 Let $A = \{2,\ 4,\ 6,\ 8\}$ and $B = \{0,\ 2,\ 4,\ 6\}$. Fill in the blank with either $\in$ or $\notin$.

1. 0 _____ A 2. 0 _____ B 3. 8 _____ A 4. 8 _____ B

◻

Example 14.1 The set containing the natural numbers from 1 to 5 may be written

$$\{1,\ 2,\ 3,\ 4,\ 5\};$$

---

[2] We use $\mathbb{Z}$ to denote the set of integers because, in German, *die Zahl* means number.

however, listing all the elements of a set may be impractical, and so we may also describe the elements of the set. For instance, the set $\{1, 2, 3, 4, 5\}$ may also be written

$$\{n : n \in \mathbb{N}, 1 \le n \le 5\} \quad \text{or} \quad \{n \in \mathbb{N} \mid 1 \le n \le 5\}.$$

Note that the colon and the vertical bar may be read as "such that." Thus,

$$\{n : n \in \mathbb{N}, 1 \le n \le 5\}$$

may be read: "The set of objects n, such that n is a natural number that is greater than or equal to one and less than or equal to five"; and

$$\{n \in \mathbb{N} \mid 1 \le n \le 5\}$$

may be read: "The set of natural numbers n, such that n is greater than or equal to one and less than or equal to five." □

Example 14.2 The set

$$\left\{\frac{1}{1}, \frac{1}{2}, \frac{1}{3}, \frac{1}{4}, \frac{1}{5}\right\}$$

may also be written

$$\left\{\frac{1}{n} : n \in \mathbb{N}, 1 \le n \le 5\right\}.$$

As another example, the elements of the set

$$\{2n - 1 : n \in \mathbb{N}, 1 \le n \le 5\}$$

are 1, 3, 5, 7, 9. This is because, when

$$
\begin{aligned}
n = 1 : \quad & 2n - 1 = 2(1) - 1 = 1, \\
n = 2 : \quad & 2n - 1 = 2(2) - 1 = 3, \\
n = 3 : \quad & 2n - 1 = 2(3) - 1 = 5, \\
n = 4 : \quad & 2n - 1 = 2(4) - 1 = 7, \\
n = 5 : \quad & 2n - 1 = 2(5) - 1 = 9.
\end{aligned}
$$

□

Now You Try 14.2 List the elements of the set.

1. $\{2n : n \in \mathbb{N}, 1 \le n \le 10\}$

2. $\left\{\dfrac{1}{2n + 1} : n \in \mathbb{N}, 5 \le n < 10\right\}$

□

Much of the notation of set theory and logic was given to us by Giuseppe Peano (1858–1932). Peano was born in Italy and schooled in Turin. Peano worked in many fields of mathematics, but he is most well known for his work in early set theory and logic, and for the continuous "space-filling curve" named after him. In particular, following the work of Cantor, Peano's work put elementary mathematics on a firmer basis by defining the natural numbers in terms of sets. He did this by providing axiomatic definitions, thus doing for arithmetic what Euclid did for geometry.

14.1   SET RELATIONS

Set concepts can be illustrated using a *Venn diagram*. Now, diagrams had been used to delineate the logic of problems during the previous few centuries; however, much like with set theory, Venn diagrams are largely the work of one man. In an 1880 paper titled, "On the diagrammatic and mechanical representation of propositions and reasonings," Englishman John Venn (1834–1923) introduced the use of circular diagrams to aid in the visualization of propositional arguments. At the time, Venn referred to his diagrams as "Euler circles" after the famous eighteenth century mathematician Leonhard Euler. Although the name "Venn diagram" was not used until after the turn of the twentieth century, Venn still deserves the credit for Venn diagrams because he elaborated on and systematized their use.

Just as graphs help us to visualize some equations, so Venn diagrams help us to visualize some set relations and set operations. But unlike using a collection of points to represent an equation, called the graph of the equation, a Venn diagram uses a circle or other simple closed curve to represent a set. We illustrate this in the following examples.

Example 14.3 Consider the set $A = \{2, 4, 6, 8\}$. Then the following are Venn diagrams that represent A.

Note that a Venn diagram does not need to list the elements of the set.   □

Example 14.4 Consider the sets $A = \{2, 4, 6, 8\}$ and $B = \{0, 2, 4, 6\}$ shown in the following "two-circle" Venn diagram, with A the circle on the left and B on the right.

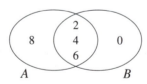

We see that, together, the two sets contain the elements 0, 2, 4, 6, and 8: we say that the *union* of A and B is the set $\{0, 2, 4, 6, 8\}$. We also see that the two sets have in common the elements 2, 4, and 6: we say that the *intersection* of A and B is the set $\{2, 4, 6\}$.   □

Definition 14.2 Let A and B be sets. The *union* of A and B, denoted $A \cup B$, is the set that contains elements of A or B:[3]

$$A \cup B = \{x \in A \quad \text{or} \quad x \in B\}.$$

---

[3]In mathematics the word "or" is understood generally to be inclusive, meaning also "or both." Hence, saying "A or B" is understood generally to mean "A or B, or both," so that $A \cup B$ is the set that contains the elements of A or B, or of both. Otherwise, we may say, "A or B, exclusively" to mean "A or B, but not both." An "exclusive or" is often denoted XOR in computer science, for example.

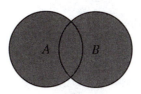

The *intersection* of A and B, denoted $A \cap B$, is the set that contains elements of A and B:

$$A \cap B = \{x \in A \quad \text{and} \quad x \in B\}.$$

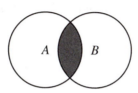

Put another way, the intersection of sets contains the elements that are common to the sets. So, in the previous example in which $A = \{2,\ 4,\ 6,\ 8\}$ and $B = \{0,\ 2,\ 4,\ 6\}$, we find that

$$A \cup B = \{0, 2, 4, 6, 8\} \quad \text{and} \quad A \cap B = \{2, 4, 6\}.$$

Now You Try 14.3

1. Consider the sets
   $$A = \{2, 5\}, \quad B = \{2, 5, 8\}, \quad \text{and} \quad C = \{3, 5, 7\}.$$
   Find each of the following and draw a two-circle Venn diagram for each.

   a) $A \cup B$　　　　　　c) $A \cup C$　　　　　　e) $B \cup C$

   b) $A \cap B$　　　　　　d) $A \cap C$　　　　　　f) $B \cap C$

2. Consider the sets
   $$A = \{2, 3, 5, 7\}, \quad B = \{3, 7, 11\}, \quad \text{and} \quad C = \{3, 5, 7\}.$$
   Find each of the following and draw a "three-circle" Venn diagram for each.

   a) $(A \cup B) \cup C$　　　b) $A \cup (B \cup C)$　　　c) $(A \cap B) \cap C$　　　d) $A \cap (B \cap C)$

3. Let $A = \{2,\ 3,\ 5,\ 7,\ 13,\ 17\}$ and $B = \{7,\ 13,\ 17,\ 19,\ 23,\ 29,\ 31\}$. Show that

   $$\text{card}(A \cup B) = \text{card}(A) + \text{card}(B) - \text{card}(A \cap B),$$

   where card($\cdot$) denotes the number of elements in the set. For example, card(A) = 6 here.

4. Everyone in the mathematics department drinks either coffee or grape milk at breakfast. (Grape milk is a delicious drink that is made by mixing grape juice and milk.) If eighteen persons drink grape milk, thirty-four persons drink coffee, and ten drink both grape milk and coffee, how many persons are in the mathematics department? Use a Venn diagram to support your answer.

5. Draw a Venn diagram for the following situation: A computer science class has 20 students, in which 10 of them already know Java, 8 already know C++, and 3 know both.

□

Think About It 14.1 Is the relation

$$\operatorname{card}(A \cup B) = \operatorname{card}(A) + \operatorname{card}(B) - \operatorname{card}(A \cap B)$$

true in general for any finite sets A and B? Why do we need to subtract the number of elements in the intersection?

□

Example 14.5 The following are two Venn diagrams: one for the set of natural numbers, $\mathbb{N}$, and one for the set of whole numbers, $\mathbb{W}$.

We may also draw a two-circle Venn diagram to depict both sets.

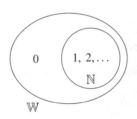

Thus, we see that

$$\mathbb{N} \cup \mathbb{W} = \mathbb{W} \quad \text{and} \quad \mathbb{N} \cap \mathbb{W} = \mathbb{N}.$$

This is because, as we see from the Venn diagram, every element of $\mathbb{N}$ is also an element of $\mathbb{W}$, but not every element of $\mathbb{W}$ is an element of $\mathbb{N}$ (specifically, $0 \in \mathbb{W}$, but $0 \notin \mathbb{N}$). We say that the set $\mathbb{N}$ is a *subset* of the set $\mathbb{W}$, but that $\mathbb{W}$ is *not a subset* of $\mathbb{N}$.

□

Definition 14.3 A set A is a *subset* of a set B if every element of A is an element of B. We write $A \subseteq B$, read "A is a subset of B" or "A is contained in B." In other words, $A \subseteq B$ if

$$x \in B \quad \text{whenever} \quad x \in A.$$

On the other hand, A is *not a subset* of B if there is an element of A that is not in B. We write $A \not\subseteq B$, read "A is not a subset of B" or "A is not contained in B." In other words, $A \not\subseteq B$ if

$$x \notin B \quad \text{for some} \quad x \in A.$$

□

Example 14.6 Consider the sets

$$A = \{-2,\ 1,\ 3\}, \quad B = \{-2,\ 3\}, \quad \text{and} \quad C = \{0,\ 1,\ 3\}.$$

1. $A \nsubseteq B$ because $1 \in A$, but $1 \notin B$.

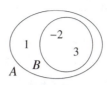

2. $B \subseteq A$ because every element of B is an element of A:

$$-2 \in B \quad \text{and} \quad -2 \in A;$$
$$3 \in B \quad \text{and} \quad 3 \in A.$$

3. $A \nsubseteq C$ and $C \nsubseteq A$:

$$-2 \in A, \quad \text{but} \quad -2 \notin C;$$
$$0 \in C, \quad \text{but} \quad 0 \notin A.$$

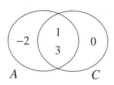

Now You Try 14.4

1. Consider the sets

$$A = \{2,\ 3,\ 5,\ 7\}, \quad B = \{3,\ 7,\ 11\}, \quad \text{and} \quad C = \{3,\ 5,\ 7\}.$$

Draw a two-circle Venn diagram for each of the following to justify your answer.

a) Is $B \subseteq A$?   b) Is $C \subseteq A$?   c) Is $C \subseteq B$?

2. Draw a three-circle Venn diagram to show the relation among the three sets

$$A = \{2,\ 5\}, \quad B = \{2,\ 5, 8\}, \quad \text{and} \quad C = \{2,\ 3,\ 5,\ 7\}.$$

a) Is $A \subseteq B$?   c) Is $B \subseteq C$?

b) Is $A \subseteq C$?   d) Is $C \subseteq C$?

Remark 14.1 Any set is a subset of itself: If A is a set, then $x \in A$ whenever $x \in A$, naturally. Hence, the definition of subset is satisfied, so that $A \subseteq A$. Therefore, it is not true that a subset is necessarily "smaller in size" than the set in which it is contained. In other words, that $A \subseteq B$ does not imply necessarily that A is "smaller in size" than B.

Definition 14.4 A set A is a *proper subset* of a set B if A is a subset of B, but B is not a subset of A, that is,

$$\text{if } A \subseteq B, \quad \text{but } B \nsubseteq A.$$

We write $A \subset B$, read "A is a proper subset of B" or "A is properly contained in B." □

Put another way, A is a proper subset of B if every element of A is also an element of B, but there is some element of B that is not an element of A.

Example 14.7

1. Let $A = \{2, 5\}$ and $B = \{2, 5, 8\}$. Then A is a subset of B because every element of A is in B, but B is not a subset of A because not every element of B is in A (specifically, $8 \in B$, but $8 \notin A$); hence, A is a proper subset of B.

2. Because $\mathbb{N} \subseteq \mathbb{W}$, but $\mathbb{W} \nsubseteq \mathbb{N}$ (because $0 \in \mathbb{W}$, but $0 \notin \mathbb{N}$), $\mathbb{N} \subset \mathbb{W}$.

□

Definition 14.5 The set that does not contain any elements is called the *empty set* or the *null set*, denoted $\{\,\}$ or $\varnothing$. □

We may think of the empty set as an empty box.

Example 14.8 Recall that a set A is a subset of a set B if *every element of A is an element of B*. Equivalently, we may say that A is a subset of B if *there is* no *element of A that is not also an element of B*.
Now, consider the sets

$$A = \{2, 3\}, \quad B = \{1, 2\}, \quad C = \varnothing, \quad \text{and} \quad D = \{0, 1, 2\}.$$

1. $A \nsubseteq D$ because there is an element of A that is not an element of D, to wit, $3 \in A$, but $3 \notin D$.

2. $B \subseteq D$ because there is *no* element of B that is not an element of D. In fact, $B \subset D$.

3. $C \subseteq D$ because there is *no* element of C that is not an element of D. (It does not matter that C is the empty set and, so, has no element at all. For C not to be a subset of D, there would need to be an element of C that is not an element of D.)

□

In the preceding example, that the empty set C is a subset of the nonempty set D may seem strange. Nevertheless, it is true when we apply the definition. (In this case, we say that it is *vacuously true*.)

> The empty set is a subset of *every* set.

Example 14.9 List all the subsets of the set $A = \{p, q, r\}$.

To begin, the subsets of A that contain one element are

$$\{p\}, \quad \{q\}, \quad \{r\}.$$

The subsets of A that contain two elements are

$$\{p, q\}, \quad \{p, r\}, \quad \{q, r\}.$$

Now, because every set is a subset of itself, the subset of A with three elements is

$$\{p, q, r\}.$$

And, because the empty set is a subset of every set, the subset of A with no element is

$$\{\ \}.$$

Therefore, all the subsets of $A = \{p, q, r\}$ are

$$\{\ \}, \quad \{p\}, \quad \{q\}, \quad \{r\}, \quad \{p, q\}, \quad \{p, r\}, \quad \{q, r\}, \quad \{p, q, r\},$$

being eight subsets in all. □

Example 14.10 List all the nonempty proper subsets of $S = \{a, b, c\}$.

All the subsets of S are

$$\{\ \}, \quad \{a\}, \quad \{b\}, \quad \{c\}, \quad \{a, b\}, \quad \{a, c\}, \quad \{b, c\}, \quad \{a, b, c\};$$

however, the *proper* subsets of S are

$$\{\ \}, \quad \{a\}, \quad \{b\}, \quad \{c\}, \quad \{a, b\}, \quad \{a, c\}, \quad \{b, c\},$$

and the *nonempty* proper subsets of S are

$$\{a\}, \quad \{b\}, \quad \{c\}, \quad \{a, b\}, \quad \{a, c\}, \quad \{b, c\}.$$

Therefore, we see that $S = \{a, b, c\}$ has eight subsets in all, but only seven proper subsets, and only six nonempty proper subsets. □

Now You Try 14.5 Let the set $S = \{2, 4, 6, 8\}$. List all of the subsets of S. Which are the proper subsets? Which are the nonempty proper subsets? □

Think About It 14.2 A set with 5 elements would have how many subsets? What about a set with 6 elements? With n elements? With an infinite number of elements? □

Now, we would naturally want to say that two sets A and B are equal if they have exactly the same elements, but we need to be precise about this, particularly when the sets may be infinite.

Definition 14.6 A set A *equals* a set B, denoted $A = B$, if both $A \subseteq B$ and $B \subseteq A$; otherwise, A *does not equal* B, denoted $A \neq B$. □

Think About It 14.3 Does the definition of $A = B$ capture what we would naturally want it to mean, namely, that the two sets have exactly the same elements? □

Example 14.11 Consider the sets

$$A = \{1, 2, 3\}, \quad B = \{2, 3, 1\}, \quad \text{and} \quad C = \{0, 1, 2\}.$$

We see that both $A \subseteq B$ and $B \subseteq A$, so that $A = B$. Of course, we can also see that $A = B$ because both sets plainly contain exactly the same elements, namely, the numbers 1, 2, and 3. (The ordering of the elements in a set does not matter.)

On the other hand, $A \not\subseteq C$ because $3 \in A$, but $3 \notin C$, so that $A \neq C$. (We see that $C \not\subseteq A$, as well, but it takes only one noncontainment for two sets not to be equal.) Of course, again, we can plainly see that A and C do not both have exactly the same elements. □

Now You Try 14.6 Explain why in each case.

1. Consider the sets

$$A = \{2, 4, 6, 8\}, \quad B = \{0, 2, 4, 6\}, \quad \text{and} \quad C = \{2, 6, 4, 8\}.$$

a) Does $A = B$?   b) Does $A = C$?   c) Does $B = C$?

2. Consider the sets $\mathbb{N}$, $\mathbb{W}$, and $\mathbb{Z}$.

a) Does $\mathbb{N} = \mathbb{W}$?   b) Does $\mathbb{N} = \mathbb{Z}$?   c) Does $\mathbb{W} = \mathbb{Z}$?

□

14.2   FINDING 2^n

The powers of two are used throughout modern mathematics, from computing to coding to counting, and will become important in our upcoming discussion of power sets. (See DEFINITION 15.4.) But were the powers of two important historically? Not particularly; however, they do show up in an unlikely context—the ancient Indian Vedics.

Prosody or metrics is one place in poetry where a little mathematics may be useful. Specifically, the number of different metrics or syllable patterns that can be formed in a line that contains n syllables, where each syllable can be either heavy (stressed) or light (unstressed), equals the number of subsets of a set that contains n elements. This was of use to Indian mathematicians because all of early Indian mathematics was written in verse or prose.

Example 14.12 Suppose that we have a line with four syllables that we denote (a, b, c, d) in that order, and associate to it the set $\{a,\ b,\ c,\ d\}$. If a subset of $\{a, b, c, d\}$ specifies the syllables that are to be heavy, then, for example, the subset $\{a,\ c,\ d\}$ specifies that the first, third, and fourth syllables are to be **heavy**, and the second is to be light:

a <u>b</u> c d.

So, all the ways of choosing which syllables in $(a,\ b,\ c,\ d)$ are to be heavy or light are given by all the subsets of $\{a,\ b,\ c,\ d\}$.

$$\{\}$$
$$\{a\},\quad \{b\},\quad \{c\},\quad \{d\}$$
$$\{a,\ b\},\quad \{a,\ c\},\quad \{a,\ d\},\quad \{b,\ c\},\quad \{b,\ d\},\quad \{c,\ d\}$$
$$\{a,\ b,\ c\},\quad \{a,\ b,\ d\},\quad \{a,\ c,\ d\},\quad \{b,\ c,\ d\}$$
$$\{a,\ b,\ c,\ d\}$$

Notice that there are sixteen subsets in all; $\{a,\ b,\ c,\ d\}$ has four elements, and $16 = 2^4$.   □

In essence, there are two choices (heavy or light) for each of the n elements, which leads to

$$\underbrace{2 \cdot 2 \cdot 2 \cdots 2}_{n \text{ factors}} = 2^n$$

choices altogether.

Rule 14.1 A set that contains n elements has 2^n subsets.   □

The Indian mathematician Piṅgala (dated prior to 200 BC), in his work *Chandahsūtra (Rules of Metrics)* found in the Vedic limb of metrics, gave the following prescription for computing 2^n in the context of prosody or metrics [95, p. 393]:

When halved, [record] two. When unity [is subtracted, record] zero. When zero, [multiply by] two; when halved, [it is] multiplied [by] so much [i.e., squared].

We illustrate this rather cryptic prescription in the following example.

Example 14.13 The number of different syllable patterns that can be formed in a line that contains 13 syllables is $2^{13} = 8192$. We follow Piṅgala's prescription to find 2^{13} as follows.

- 13 is odd, so subtract 1 and record 0

- 12 is even, so halve it and record 2

- 6 is even, so halve it and record 2

- 3 is odd, so subtract 1 and record 0

- 2 is even, so halve it and record 2

- 1 is odd, so subtract 1 and record 0

- 0 marks the end

Now we work backward to determine the number of different syllable patterns. Beginning with the number 1, for every 0 that was recorded, we double the present result, and for every 2 that was recorded, we square it.

$$1 \xrightarrow[\times 2]{0} 2 \xrightarrow[()^2]{2} 4 \xrightarrow[\times 2]{0} 8 \xrightarrow[()^2]{2} 64 \xrightarrow[()^2]{2} 4096 \xrightarrow[\times 2]{0} 8192$$

Thus, we conclude that we can form 8192 different syllable patterns in a line that contains 13 syllables, where each syllable can be either heavy or light. □

Note that we found 2^{13} in six steps instead of 12 (by repeated multiplication):

$$2^{13} = 2((2((1 \times 2)^2))^2)^2 = 8192.$$

Now You Try 14.7

1. Find the number of different syllable patterns that can be formed in a line that contains 3 syllables using the method Piṅgala prescribed. Also, write this number as a power of 2. Verify that this is the same as the number of subsets of the set $\{a, b, c\}$ that has three elements by listing all of its subsets.

2. Repeat the preceding exercise for a line that contains 8 syllables. (Feel free to list them all if you have nothing else to do.)

□

Think About It 14.4 Will Piṅgala's method always end with the last two steps being a 1 and a 0 respectively? Why?

□

14.3   ONE-TO-ONE CORRESPONDENCE AND CARDINALITY

Suppose you were a sheep herder in an ancient land. You send your sheep out in the morning to graze the day away, and at dusk you drive them back home. How would you know that your herd has come home intact if you do not know how to count? What could you do if you had a basket and lots of small stones?

We wish to be able to compare the "sizes" of two sets. One way we may do this would be to count the number of objects in each set. Another way—an apparently more primitive way that avoids counting altogether—would be to pair the objects in each set. For instance, suppose that you have two large bins, one filled with red balls and the other filled with blue balls (or a herd of sheep and a basket filled with stones). One way you may determine if you have the same number of red and blue balls would be to count the number of balls in each bin. Another way would be to reach into the bins and retrieve one red ball and one blue ball, set them aside, and then repeat doing this: one red ball and one blue ball, one red ball and one blue ball, one red ball and one blue ball, and so on. If both bins were to empty simultaneously, then you would know that you have the same number of red and blue balls, even though you may not know how many in number. If one bin were to empty before the other, then you would know that you do not have the same number of red balls and blue balls. Let us make this process of "one red ball and one blue ball" precise.

Definition 14.7 A *one-to-one correspondence* (or *1-1 correspondence*) between two sets A and B is a set of pairings (a, b), where $a \in A$ and $b \in B$, such that every element of A is paired with exactly one element of B, and every element of B is paired with exactly one element of A.   □

Example 14.14

1. Let $A = \{2,\ 4,\ 6,\ 8\}$ and $B = \{0,\ 2,\ 4,\ 6\}$. Then

$$
\begin{array}{ccc}
A & & B \\
2 & \longleftrightarrow & 0 \\
4 & \longleftrightarrow & 2 \\
6 & \longleftrightarrow & 4 \\
8 & \longleftrightarrow & 6
\end{array}
$$

   is a one-to-one correspondence between A and B because every element of A is paired with exactly one element of B, and every element of B is paired with exactly one element of A:

$$\{(2,0),(4,2),(6,4),(8,6)\}.$$

   Another one-to-one correspondence between A and B is

$$\{(2,2),(4,4),(6,6),(8,0)\}.$$

   In fact, there are 24 different one-to-one correspondences between the sets A and B here.

2. Let $A = \{-2,\ 3,\ 5\}$ and $B = \{1,\ 2\}$. Then

$$\{(-2,1),(3,2),(5,2)\}$$

   is *not* a one-to-one correspondence between A and B because $2 \in B$ is paired with more than one element in A, namely, with $3 \in A$ and with $5 \in A$. In fact, there is no one-to-one correspondence between A and B. (Why?)

3. Let $A = \{-2,\ 3\}$ and $B = \{1,\ 2,\ 3\}$. Then there is no one-to-one correspondence between A and B because one element of A must be paired with more than one element of B.

□

Think About It 14.5 The set of natural numbers, $\mathbb{N}$, is a proper subset of the set of whole numbers, $\mathbb{W}$. Nevertheless, is there a one-to-one correspondence between $\mathbb{N}$ and $\mathbb{W}$? Why or why not? □

Example 14.15 There is a one-to-one correspondence between the set $\mathbb{N}$ and the set $\mathbb{W}$, namely,

$$\{(1,0),\ (2,1),\ (3,2),\ (4,3),\ (5,4),\dots\}$$

or

$$\{(n, n-1) : n \in \mathbb{N}\}$$

or

$$
\begin{array}{ccc}
\mathbb{N} & & \mathbb{W} \\
1 & \longleftrightarrow & 0 \\
2 & \longleftrightarrow & 1 \\
3 & \longleftrightarrow & 2 \\
4 & \longleftrightarrow & 3 \\
5 & \longleftrightarrow & 4 \\
& \vdots &
\end{array}
$$

□

Think About It 14.6

1. We saw in the preceding example that there is a the one-to-one correspondence between $\mathbb{N}$ and $\mathbb{W}$, namely, $\{(n, n-1) : n \in \mathbb{N}\}$. Is this the only one-to-one correspondence between $\mathbb{N}$ and $\mathbb{W}$?

2. There is no one-to-one correspondence between the sets $A = \{-2,\ 3\}$ and $B = \{1,\ 2,\ 3\}$ because $A \subset B$, yet there is a one-to-one correspondence between $\mathbb{N}$ and $\mathbb{W}$ even though $\mathbb{N} \subset \mathbb{W}$. Why can that be?

□

Now You Try 14.8

1. Find a one-to-one correspondence between the set of vowel letters, $V = \{a,\ e,\ i,\ o,\ u\}$, and the set $S_5 = \{1,\ 2,\ 3,\ 4,\ 5\}$.

2. Find a one-to-one correspondence between the set of all the subsets of $A = \{p,\ q,\ r\}$, and the set $S_8 = \{1,\ 2,\ 3,\ 4,\ 5,\ 6,\ 7,\ 8\}$.

3. Explain why there is no one-to-one correspondence between the set $A = \{p,\ q,\ r,\ s,\ t\}$ and the set $S_3 = \{1,\ 2,\ 3\}$.

4. Find a one-to-one correspondence between the set of even natural numbers,

$$2\mathbb{N} = \{2,\ 4,\ 6,\dots\},$$

and the set of natural numbers, $\mathbb{N}$.

5. Find a one-to-one correspondence between the set of integers, $\mathbb{Z}$, and the set $\mathbb{N}$.

□

Think About It 14.7

1. What is a requirement for there to be a one-to-one correspondence between two finite sets?

2. Why are there 24 different one-to-one correspondences between the sets $A = \{2,4,6,8\}$ and $B = \{0,2,4,6\}$?

☐

For finite sets to be in a one-to-one correspondence, they must have the same number of elements, which is exactly what you would expect when you say that two sets have the same size; but what if the sets were infinite?

Definition 14.8 A set A is a *finite set* if one of the following holds:

- There is *no* one-to-one correspondence between A and any proper subset of A.

- There is a one-to-one correspondence between A and the set $S_n = \{1,2,3,\ldots,n\}$ for some natural number n. In this case, we say that A has n elements.

☐

What if a set is not finite?

Definition 14.9 A set A is an *infinite set* if there is a one-to-one correspondence between A and a proper subset of A.   ☐

Example 14.16

1. The set of vowels $V = \{a, e, i, o, u\}$ is finite because there is a one-to-one correspondence between V and the set $S_5 = \{1, 2, 3, 4, 5\}$, for instance,

$$\{(a, 1),\ (e, 2),\ (i, 3),\ (o, 4),\ (u, 5)\}.$$

In fact, V has 5 elements. Indeed, the correspondence between V and S_5 is a means of counting the number of elements in V.

Note also that there is no one-to-one correspondence between V and any proper subset of V because all the proper subsets of V contain fewer than 5 elements.

2. The set of natural numbers is infinite because it can be put in one-to-one correspondence with a proper subset of itself, for example, with the set of all even natural numbers $2\mathbb{N} \subset \mathbb{N}$: one way is simply to map each natural number n to twice n.

$$
\begin{array}{ccc}
\mathbb{N} & & 2\mathbb{N} \\
1 & \longleftrightarrow & 2 \\
2 & \longleftrightarrow & 4 \\
3 & \longleftrightarrow & 6 \\
4 & \longleftrightarrow & 8 \\
& \vdots &
\end{array}
$$

☐

Think About It 14.8 Which set is larger: the set of all multiples of 10 or the set of all multiples of 100?
☐

Now You Try 14.9

1. Show that the set of vowels is finite by showing that there is no one-to-one correspondence between the set and any of its proper subsets.

2. Show that the set C of consonants is finite by showing that there is a one-to-one correspondence between C and the set $S_n = \{1, 2, \ldots, n\}$ for some natural number n. How many elements are in the set?

$\square$

We are now ready to investigate a very interesting question: Are all infinite sets the same size? To answer this question, we first need to define formally what we mean by the "size" of a set, which we call the *cardinality* of a set.

Definition 14.10 Two sets A and B are said to have the same *cardinality* if there is a one-to-one correspondence between A and B. We write card(A) for the cardinality of A.

A finite set has *finite cardinality*. In particular, if there is a one-to-one correspondence between a finite set A and the set $S_n = \{1, 2, 3, \ldots, n\}$ for some natural number n, then we say that card(A) = n.

An infinite set has *infinite cardinality*. In particular, if there is a one-to-one correspondence between an infinite set A and the set of all natural numbers $\mathbb{N}$, then we say that the cardinality of A is *countably infinite* or that A is *countable*. Otherwise, we say that the cardinality of A is *uncountably infinite* or that A is *uncountable*.

$\square$

Think About It 14.9 Are there sets that are uncountable?

$\square$

The cardinality of a set, then, is the "size" of the set. Thus, because there is a one-to-one correspondence between $\mathbb{W}$ and $\mathbb{N}$, both sets have the same cardinality or the same "size" (both are countably infinite), even though $\mathbb{N}$ is a proper subset of $\mathbb{W}$. Similarly, the set of integers is no "larger" than the set of whole numbers or the set of natural numbers (all are countably infinite) even though the set of integers may appear to have twice as many elements.

$$
\begin{array}{ccc}
\mathbb{Z} & & \mathbb{N} \\
0 & \longleftrightarrow & 1 \\
1 & \longleftrightarrow & 2 \\
-1 & \longleftrightarrow & 3 \\
2 & \longleftrightarrow & 4 \\
-2 & \longleftrightarrow & 5 \\
& \vdots &
\end{array}
$$

So even though these sets appear to be of different sizes,

$$
\begin{array}{lllllllll}
\mathbb{Z}: & \ldots, & -3, & -2, & -1, & 0, & 1, & 2, & 3, \ldots \\
\mathbb{W}: & & & & & 0, & 1, & 2, & 3, \ldots \\
\mathbb{N}: & & & & & & 1, & 2, & 3, \ldots
\end{array}
$$

all three sets are the same "size": they are all countably infinite.

$$\mathbb{N} \subset \mathbb{W} \subset \mathbb{Z}, \quad \text{but} \quad \text{card}(\mathbb{Z}) = \text{card}(\mathbb{W}) = \text{card}(\mathbb{N})$$

This surprising result illustrates why the conception of "infinity" confounded mathematicians for a very long time.

Now You Try 14.10

1. Let V denote the set of vowels and C denote the set of consonants. Find card(V) and card(C).

2. Let $\mathbb{E}$ and $\mathbb{O}$ denote, respectively, the set of even integers and the set of odd integers, that is,

$$\mathbb{E} = \{\ldots, \ -4, \ -2, \ 0, \ 2, \ 4, \ldots\},$$
$$\mathbb{O} = \{\ldots, \ -3, \ -1, \ 1, \ 3, \ldots\}.$$

Show that both sets are countably infinite, so that card($\mathbb{E}$) = card($\mathbb{O}$).

3. Show that the set of all multiples of 10 and the set of all multiples of 100 are both countably infinite. Thus, neither set is "larger" than the other.

□

Although the notion of "infinite" is ancient, defining it and using it in mathematics was generally avoided until the work of Georg Cantor (1845–1918). The earliest concrete discussion of the infinite is found in the work of the Greek philosopher Zeno of Elea (ca. 490–420 BC). Zeno's famous paradoxes discuss the nature of the infinitely small, and show how difficult it is to work with the infinite. As we were taught in school, infinity is not a number, but an idea; however, starting with the ancient Greek mathematicians, most notably Archimedes, infinite processes were used to find formulas for the areas and volumes of geometric shapes and solids. These infinite processes found their ultimate application and justification in the development of the calculus in the seventeenth and eighteenth centuries.

With only a few exceptions, the infinite has been thought of as an unending process, such as counting. The idea of the natural numbers going on forever was termed "*potentially* infinite"; one rarely talked about "all" the natural numbers. However, with the work of Cantor, the infinite could now be viewed as "complete" so that, for instance, the (infinite) collection of natural numbers could be thought of as a *complete set* in itself. Cantor saw that infinite sets are mathematical objects that could be manipulated like numbers or other mathematical objects. His groundbreaking work changed the face of mathematics. Nevertheless, even today the notion of a "potential" versus a "completed" infinity still causes disputes.

Chapter 15

Rational, Irrational, and Real Numbers

The natural numbers have their roots in prehistoric times; however, to be more useful, there was a need for numbers to describe not just nature, but also human activities. Thus, there was a need to expand the conception of number beyond the set of natural numbers.

Definition 15.1 A *common fraction* or *vulgar fraction* is an expression of the form $\frac{a}{b}$, where a and b are integers with $b \neq 0$. The common fraction $\frac{a}{b}$ is said to be in *lowest terms* if b is positive and a and b have no common factor greater than one; the fraction $\frac{0}{b}$ in lowest terms is $\frac{0}{1}$. □

Remark 15.1 Since $a = a \div 1 = \frac{a}{1}$ for any number a, we see that every natural number a can be written as a common fraction, namely, $\frac{a}{1}$. In this way, the set of natural numbers (and, in the same way, also the set of whole numbers and the set of integers) is a subset of the set of common fractions. □

To the ancient Greeks, however, what we call a common fraction today (for instance, $\frac{1}{2}$ or $\frac{2}{3}$) really represented a *relationship* between two whole numbers (for instance, "the ratio of one to two" or "the ratio of two to three"), and it was only over a long period of time that this perspective evolved to where this relationship between two whole numbers was extended to represent not only a ratio, but also a non-whole number as well as a quotient. As we saw in part I, decimal fractions were not developed until very much later.

In this chapter, we discuss what are called the *rational numbers*, *irrational numbers*, and *real numbers*. To begin, it will be useful to discuss decimal fractions.

We know from experience that

$$\tfrac{1}{2} = 0.5 \quad \text{and} \quad \tfrac{2}{3} = 0.666\,666\ldots = 0.\overline{6},$$

where a bar is used to indicate that a block of digits repeats. By carrying out the long division, we see that every common fraction can be represented as either a decimal fraction that terminates ($\frac{1}{5} = 0.2$) or a decimal fraction that does not terminate ($\frac{1}{6} = 0.1\overline{6}$), but has a repeating block of digits. The latter is what we call a *repeating decimal fraction* because a block of digits repeats without interruption.

Think About It 15.1 Can a common fraction have two different decimal fraction representations? □

It turns out, however, that every terminating decimal fraction can also be expressed as a nonterminating decimal fraction. For example, the terminating decimal fraction 0.2 can be expressed as a nonterminating decimal fraction, namely,

$$0.200\,000\ldots = 0.2\overline{0} \quad \text{or} \quad 0.199\,999\ldots = 0.1\overline{9}.$$

It is evident that $0.2\overline{0} = 0.2$. To see that $0.1\overline{9} = 0.2$, let $r = 0.199\,999\ldots$. Then,

$$10r = 10(0.199\,999\ldots) = 1.999\,999\ldots$$

and

$$100r = 100(0.199\,999\ldots) = 19.999\,999\ldots$$

Next, subtracting $10r$ from $100r$, we find that

$$100r = 19.999\,999\ldots$$
$$10r = 1.999\,999\ldots$$
$$\text{so} \quad 90r = 18.000\,000\ldots$$

because the repeating block of digits cancels. Finally, solving for r, we see that $r = \frac{18}{90} = 0.2$. Therefore, we conclude that

$$0.199\,999\ldots = 0.2.$$

Now You Try 15.1 Show the following.

1. $6.\overline{9} = 6.999\,999\ldots = 7$. (Let $r = 6.\overline{9}$ and subtract $10r$ from $100r$.)

2. $0.124\overline{9} = 0.124\,999\ldots = 0.125$. (Let $r = 0.124\overline{9}$ and subtract $1000r$ from $10,000r$.)

$\square$

In order to build a succinct definition of *rational number*, we will make the following convenient choice, although the choice itself is arbitrary.

> In order to express a decimal fraction that terminates as a nonterminating decimal fraction, we will always choose the expression that has a tail end of 9's instead of a tail end of 0's.

With this choice, we may amend what we said above—that every common fraction can be represented as either a decimal fraction that terminates or a decimal fraction that repeats—to the following.

> Every common fraction can be represented by a unique nonterminating decimal fraction.

When considering the set of all decimal fractions, therefore, we may consider the set of all nonterminating decimal fractions only.

Now You Try 15.2 Express the following terminating decimal fractions as nonterminating decimal fractions.

1. 0.1   2. 0.5   3. 0.75   4. 14

$\square$

Think About It 15.2 Does every decimal fraction represent a common fraction? For example, does 0.2 1 22 1 222 1 2222 1…represent a common fraction? $\square$

Let $\mathbb{R}$ denote the set of all nonterminating decimal fractions:

$$\mathbb{R} = \{\text{nonterminating decimal fractions}\}.$$

Now, note that, because any nonterminating decimal fraction either has a repeating block of digits or it does not have a repeating block of digits—that is to say, either it is a repeating decimal fraction or it is a nonrepeating decimal fraction—the set $\mathbb{R}$ can be partitioned into two disjoint subsets,[1] namely,

$$\mathbb{Q} = \{x \in \mathbb{R} : x \text{ has a repeating block of digits}\},$$

$$\mathbb{I} = \{x \in \mathbb{R} : x \text{ does not have a repeating block of digits}\}.$$

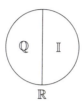

For example, the following repeating decimal fractions are in $\mathbb{Q}$.

$$7.\overline{6} = 7.666\,666\ldots$$

$$0.1\overline{9} = 0.199\,999\ldots$$

$$0.\overline{142\,857} = 0.142\,857\,142\,857\ldots$$

$$-0.0\overline{45} = -0.045\,454\ldots$$

On the other hand, the following nonrepeating decimal fractions are in $\mathbb{I}$.

$$0.2\,1\,22\,1\,222\,1\,2222\,1\ldots$$

$$1.414\,213\,562\,373\,1\ldots$$

$$-3.141\,592\,653\,589\,7\ldots$$

$$2.718\,281\,828\,459\,0\ldots$$

(Note that, although $0.2\,1\,22\,1\,222\,1\,2222\,1\ldots$ has a clear pattern, it does not repeat, that is, it has no repeating block of digits.)

We are now ready to define what is a rational number, an irrational number, and a real number.[2]

Definition 15.2 A nonterminating decimal fraction that has a repeating block of digits is called a *rational number*, and a nonterminating decimal fraction that does not have a repeating block of digits is called an *irrational number*. A *real number* is either a rational number or an irrational number. The set of rational numbers is denoted $\mathbb{Q}$, the set of irrational numbers is denoted $\mathbb{I}$, and the set of real numbers is denoted $\mathbb{R}$; hence, $\mathbb{R} = \mathbb{Q} \cup \mathbb{I}$, where the sets $\mathbb{Q}$ and $\mathbb{I}$ are disjoint. □

---

[1] The sets A and B are *disjoint sets* if $A \cap B = \varnothing$.

[2] There are other ways to define a rational number, an irrational number, and a real number that are equivalent to the ones used here.

The letter $\mathbb{Q}$ is used to denote the set of rational numbers because a rational number can also be thought of as a *quotient*: $\frac{1}{2}$ is the quotient 1 divided by 2. (We give an equivalent definition of a rational number on page 164 as a set of equivalent fractions.) One of the earliest, if not the first, use of the term "real" number was by René Descartes (1596–1650) in his work *la géométrie* (1637), in which he states, "Neither the true nor the false roots [of an equation] are always real; sometimes they are imaginary…" [46, p. 175].

15.1   COMMENSURABLE AND INCOMMENSURABLE MAGNITUDES

The above definitions of rational and irrational numbers are very modern, using decimal notation that did not appear until the Renaissance. But how do these definitions compare to ancient ideas of numbers? In particular, how did the ancient Greeks view these ideas? Recall that the Greeks did not have "fractions"; they had lengths represented as ratios of whole numbers. Did they have the notion of a length that could not be written as a ratio of two numbers? The best place to start is with Euclid's *Elements* [53].

Book X of Euclid's *Elements*, Heath acclaims [68, p. 402], "is perhaps the most remarkable, as it is the most perfect in form, of all the Books of the *Elements*." Book X studies "commensurable" (rational) and "incommensurable" (irrational) magnitudes. Citing a scholium to Euclid X.9 and a passage from a commentary on Book X by Pappus (ca. 290–350) that is preserved in the Arabic, Heath acknowledges that Euclid did not invent the theory of commensurability and incommensurability, himself; Theaetetus (ca. 415–369 BC) is credited with that. Nevertheless, Euclid did make notable contributions, regarding which Heath quotes from Pappus's commentary [68, p. 403]:

> 'As for Euclid, he set himself to give rigorous rules, which he established, relative to commensurability and incommensurability in general; he made precise the definitions and the distinctions between rational and irrational magnitudes, he set out a great number of orders of irrational magnitudes, and finally he made clear their whole extent.'

We will discuss rational numbers and irrational numbers as we understand them today in sections 15.2 and 15.3. The notions of "rational" and "irrational" that we find in Euclid's *Elements*, however, do not agree exactly with how we define them today. Let us get a taste for how Euclid treats these notions. We begin with the definitions at the beginning of Book X of the *Elements* [53, p. 237]:

1. Those magnitudes are said to be *commensurable* which are measured by the same measure, and those *incommensurable* which cannot have any common measure.

2. Straight lines are *commensurable in square* when the squares on them are measured by the same area, and *incommensurable in square* when the squares on them cannot possibly have any area as a common measure.

3. With these hypotheses, it is proved that there exist straight lines infinite in multitude which are commensurable and incommensurable respectively, some in length only, and others in square also, with an assigned straight line. Let then the assigned straight line be called *rational*, and those straight lines which are commensurable with it, whether in length and in square or in square only, *rational*, but those which are incommensurable with it *irrational*.

4. And let the square on the assigned straight line be called *rational* and those areas which are commensurable with it *rational*, but those which are incommensurable with it *irrational*, and the straight lines which produce them *irrational*, that is, in case the areas are squares, the sides themselves, but in case they are any other rectilineal figures, the straight lines on which are described squares equal to them.

Thus, the Greeks did realize that some measures could not be represented as ratios of numbers. In fact, their geometric proof of the irrationality of $\sqrt{2}$ relies on showing that the ratio of a diagonal of a square to a side of the square results in another smaller triangle; and that smaller triangle similarly results in yet another smaller triangle; and so on without end. That is to say, their proof demonstrates an infinite process that produces ever smaller triangles.

15.2 RATIONAL NUMBERS

We defined a rational number to be a nonterminating decimal fraction that has a repeating block of digits in DEFINITION 15.2 on page 161. We now see how the set of common fractions is related to the set of rational numbers.

Recall that we may plot numbers (as dots) on a number line. Here, for instance, are the numbers -2, $\frac{2}{3}$, and 3 plotted on a number line.

Now, recall further that the fractions $\frac{4}{6}$ and $\frac{20}{30}$, for example, would be plotted at the same position on a number line as the fraction $\frac{2}{3}$. This is because the fractions $\frac{4}{6}$, $\frac{20}{30}$, and $\frac{2}{3}$ are *equivalent fractions*:[3]

$$\frac{2}{3} = \begin{cases} \dfrac{2}{3} \cdot \dfrac{2}{2} = \dfrac{4}{6} \\[2mm] \dfrac{2}{3} \cdot \dfrac{10}{10} = \dfrac{20}{30} \end{cases}, \quad \frac{4}{6} = \begin{cases} \dfrac{\cancel{2} \cdot 2}{\cancel{2} \cdot 3} = \dfrac{2}{3} \\[2mm] \dfrac{4}{6} \cdot \dfrac{5}{5} = \dfrac{20}{30} \end{cases}, \quad \frac{20}{30} = \begin{cases} \dfrac{2 \cdot \cancel{10}}{3 \cdot \cancel{10}} = \dfrac{2}{3} \\[2mm] \dfrac{4 \cdot \cancel{5}}{\cancel{5} \cdot 6} = \dfrac{4}{6} \end{cases}.$$

In fact, all the fractions of the form $\frac{2x}{3x}$, where x is a nonzero integer, are equivalent and may be collected together into a *set of equivalent fractions*:

$$\left\{ \frac{2}{3}, \frac{4}{6}, \frac{10}{30}, \frac{12}{18}, \frac{-2}{-3}, \frac{-6}{-9}, \dots \right\}.$$

Note that this set contains a distinguished element, namely, $\frac{2}{3}$, the fraction that is in *lowest terms* (see DEFINITION 15.1 on page 159).

Now, because $\frac{2}{3}$ is the fraction in the set $\{\frac{2}{3}, \frac{4}{6}, \frac{10}{30}, \frac{12}{18}, \frac{-2}{-3}, \frac{-6}{-9}, \dots\}$ that is in lowest terms, it is reasonable to let $\frac{2}{3}$ represent all the fractions in this set. Indeed, we may go so far as to *abuse* the $=$ symbol and write

$$\frac{2}{3} = \left\{ \frac{2}{3}, \frac{4}{6}, \frac{10}{30}, \frac{12}{18}, \frac{-2}{-3}, \frac{-6}{-9}, \dots \right\}.$$

In this way, we find that we may partition the set of all common fractions (or vulgar fractions) $\frac{a}{b}$, where a and b are integers with $b \neq 0$, into disjoint subsets q_1, q_2, q_3, and so on, of equivalent fractions, for example, perhaps

$$q_1 = \frac{2}{3} = \left\{ \frac{2}{3}, \frac{4}{6}, \frac{10}{30}, \frac{12}{18}, \frac{-2}{-3}, \frac{-6}{-9}, \dots \right\},$$

$$q_2 = \frac{1}{5} = \left\{ \frac{1}{5}, \frac{2}{10}, \frac{10}{50}, \frac{100}{500}, \frac{-1}{-5}, \frac{-10}{-50}, \dots \right\},$$

$$q_3 = \frac{-1}{22} = \left\{ \frac{-1}{22}, \frac{-3}{66}, \frac{-6}{132}, \frac{-10}{220}, \frac{1}{-22}, \frac{6}{-132}, \dots \right\}.$$

---

[3] We define two fractions, $\frac{a}{b}$ and $\frac{c}{d}$, to be equivalent without reference to "building" or "reducing" either fraction as follows: The fractions $\frac{a}{b}$ and $\frac{c}{d}$ are equivalent if $ad = bc$. For example, $\frac{2}{3}$ is equivalent to $\frac{4}{6}$ because $2 \cdot 6 = 3 \cdot 4$, and $\frac{4}{6}$ is equivalent to $\frac{20}{30}$ because $4 \cdot 30 = 6 \cdot 20$.

and so on. Moreover, it should not be a surprise that *common fractions that are equivalent have the same decimal fraction representation*. For instance, upon doing the division ($\frac{a}{b} = a \div b$), we see that the equivalent fractions

$$\frac{2}{3} = 0.666\ldots = 0.\overline{6},$$

$$\frac{4}{6} = 0.666\ldots = 0.\overline{6},$$

$$\frac{20}{30} = 0.666\ldots = 0.\overline{6},$$

and so on. In this way, we may find the decimal fraction representations

$$q_1 = \frac{2}{3} = \underbrace{\left\{\frac{2}{3}, \frac{4}{6}, \frac{10}{30}, \frac{12}{18}, \frac{-2}{-3}, \frac{-6}{-9}, \ldots\right\}}_{0.666\ldots = 0.\overline{6}},$$

$$q_2 = \frac{1}{5} = \underbrace{\left\{\frac{1}{5}, \frac{2}{10}, \frac{10}{50}, \frac{100}{500}, \frac{-1}{-5}, \frac{-10}{-50}, \ldots\right\}}_{0.2 = 0.199\ldots = 0.1\overline{9}},$$

$$q_3 = \frac{-1}{22} = \underbrace{\left\{\frac{-1}{22}, \frac{-3}{66}, \frac{-6}{132}, \frac{-10}{220}, \frac{1}{-22}, \frac{6}{-132}, \ldots\right\}}_{-0.045\,454\ldots = -0.0\overline{45}},$$

and so on for every set of equivalent common fractions.

> Every common fraction $\frac{a}{b}$, where a and b are integers with $b \neq 0$, has a decimal fraction representation that has a repeating block of digits, and every decimal fraction with a repeating block of digits represents a common fraction.

In other words, every set of equivalent common fractions corresponds to a rational number. For example,

$$q_1 = \frac{2}{3} = \left\{\frac{2}{3}, \frac{4}{6}, \frac{10}{30}, \frac{12}{18}, \frac{-2}{-3}, \frac{-6}{-9}, \ldots\right\} \qquad \longleftrightarrow \qquad 0.\overline{6},$$

$$q_2 = \frac{1}{5} = \left\{\frac{1}{5}, \frac{2}{10}, \frac{10}{50}, \frac{100}{500}, \frac{-1}{-5}, \frac{-10}{-50}, \ldots\right\} \qquad \longleftrightarrow \qquad 0.1\overline{9},$$

$$q_3 = \frac{-1}{22} = \left\{\frac{-1}{22}, \frac{-3}{66}, \frac{-6}{132}, \frac{-10}{220}, \frac{1}{-22}, \frac{6}{-132}, \ldots\right\} \qquad \longleftrightarrow \qquad -0.0\overline{45},$$

and so on. In fact, here is another way to define a rational number that is equivalent to DEFINITION 15.2.

> A rational number is a *set* of *equivalent* common fractions.

Hence, because we also understand that $\frac{a}{b} = a : b$ (the ratio of a to b), we see that every *rational number* may be expressed as the *ratio* of two integers.

Example 15.1 Find a common fraction $\frac{a}{b}$ that the decimal fraction represents.

1. $0.\overline{9}$

   To find a common fraction $\frac{a}{b}$ that the decimal fraction represents, let $r = 0.\overline{9} = 0.999\,999\ldots$. Then,

   $$10r = 10(0.999\,999\ldots) = 9.999\,999\ldots.$$

   Next, subtracting r from $10r$, we find that

   $$10r = 9.999\,999\ldots$$
   $$r = 0.999\,999\ldots$$
   $$\text{so} \quad 9r = 9.000\,000\ldots$$

   because the repeating block of digits cancels. Finally, solving for r, we conclude that $r = \frac{9}{9}$, that is,

   $$r = 0.999\,999\ldots = \frac{9}{9}.$$

   Therefore, a common fraction that the decimal fraction $0.\overline{9}$ represents is $\frac{9}{9}$ or $\frac{1}{1}$, that is, $0.\overline{9} = \frac{1}{1}$. Of course, $\frac{1}{1} = 1$, so that we also conclude that $0.\overline{9} = 1$.

2. $0.\overline{12}$

   Let $r = 0.\overline{12} = 0.121\,212\ldots$. Then,

   $$100r = 100(0.121\,212\ldots) = 12.121\,212\ldots.$$

   Next, subtracting r from $100r$, we find that

   $$100r = 12.121\,212\ldots$$
   $$r = 0.121\,212\ldots$$
   $$\text{so} \quad 99r = 12.000\,000\ldots$$

   because the repeating block of digits cancels. Finally, solving for r, we conclude that $r = \frac{12}{99}$, that is,

   $$r = 0.121\,212\ldots = \frac{12}{99}.$$

   Therefore, a common fraction that the decimal fraction $0.\overline{12}$ represents is $\frac{12}{99}$ or $\frac{4}{33}$, that is, $0.\overline{12} = \frac{4}{33}$.

   $\square$

Now You Try 15.3 Find a common fraction $\frac{a}{b}$ that the decimal fraction represents.

1. $0.\overline{3}$
2. $0.0\overline{1}$
3. $0.\overline{45}$
4. $0.1\overline{29}$

$\square$

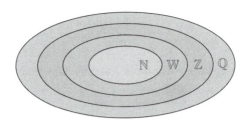

Figure 15.1: $\mathbb{N} \subset \mathbb{W} \subset \mathbb{Z} \subset \mathbb{Q}$

The sets of natural numbers, whole numbers, integers, and rational numbers are related by set inclusion: $\mathbb{N} \subset \mathbb{W} \subset \mathbb{Z}$. (See FIGURE 15.1.) Now, because $n = \frac{n}{1}$ for any integer n, we see that

$$\mathbb{N} \subset \mathbb{W} \subset \mathbb{Z} \subset \mathbb{Q}.$$

Think About It 15.3 We have seen that card($\mathbb{W}$) = card($\mathbb{N}$) and card($\mathbb{Z}$) = card($\mathbb{N}$), so what about card($\mathbb{Q}$)? Can you describe a systematic way to list all the rational numbers that would not leave any out so that they can be counted? □

We saw earlier that the set of natural numbers, the set of whole numbers, and the set of integers all have the same cardinality or the same "size" (they are all countably infinite sets) despite the fact that $\mathbb{N} \subset \mathbb{W} \subset \mathbb{Z}$. Surely, however, the set $\mathbb{Q}$ of rational numbers must be "larger" than $\mathbb{N}$ or $\mathbb{W}$ or $\mathbb{Z}$. After all, $\mathbb{Q}$ is *dense* (there is always a rational number that lies between any two given rational numbers on a number line), whereas $\mathbb{N}$ and $\mathbb{W}$ and $\mathbb{Z}$ are not (there may not be a natural number or a whole number or an integer that lies between, respectively, two given natural numbers or whole numbers or integers on a number line). For example, between the rational numbers $\frac{2}{1}$ and $\frac{3}{1}$ on a number line is the rational number $\frac{5}{2}$ (and many others), whereas there is no natural number between the natural numbers 2 and 3 on a number line.

Now You Try 15.4

1. To get a feel for what "dense" looks like, draw a number line and mark on it the whole numbers from zero to 5. Now start filling in rational numbers such as $\frac{1}{2}$, $\frac{5}{4}$, $\frac{11}{3}$, and so on, between zero and 5. Stop when you get tired.

2. Find a rational number that is between the two given rational numbers.

   a) 9.999 999 999 999 and 10

   b) $0.\overline{27}$ and $0.2\overline{9}$

   c) $\frac{17}{16}$ and $\frac{13}{12}$

   d) $-2\frac{5}{6}$ and $-2\frac{3}{4}$

                                                                                                                    □

So, it appears that the set of rational numbers *must* be much "larger" than the set natural numbers. The surprising fact is that the set of rational numbers is countably infinite—the set of rational numbers has the same cardinality as the set of natural numbers—that is, there are "just as many" rational numbers as natural numbers.

$$\boxed{\text{card}(\mathbb{Q}) = \text{card}(\mathbb{Z}) = \text{card}(\mathbb{W}) = \text{card}(\mathbb{N})}$$

To see this, we need to demonstrate a one-to-one correspondence between $\mathbb{Q}$ and $\mathbb{N}$.

We enumerate or list the rational numbers as common fractions systematically as follows. This enumeration or listing of the rational numbers to show that $\mathbb{Q}$ is countably infinite follows what is commonly called the "Cantor diagonal argument" that is due to Georg Cantor.

$$
\begin{array}{ccccccccc}
\frac{0}{1} & & & & & & & \\
\frac{1}{1} & \frac{-1}{1} & \frac{2}{1} & \frac{-2}{1} & \frac{3}{1} & \frac{-3}{1} & \frac{4}{1} & \cdots \\
\frac{1}{2} & \frac{-1}{2} & \frac{2}{2} & \frac{-2}{2} & \frac{3}{2} & \frac{-3}{2} & \frac{4}{2} & \cdots \\
\frac{1}{3} & \frac{-1}{3} & \frac{2}{3} & \frac{-2}{3} & \frac{3}{3} & \frac{-3}{3} & \frac{4}{3} & \cdots \\
\frac{1}{4} & \frac{-1}{4} & \frac{2}{4} & \frac{-2}{4} & \frac{3}{4} & \frac{-3}{4} & \frac{4}{4} & \cdots \\
\vdots & & \ddots & & & & &
\end{array}
$$

Now we need to put them in a one-to-one correspondence with the natural numbers, that is, in essence, we need to count the rational numbers: but how?

Think About It 15.4 Write down the next two rows of the list above. Will this scheme really capture all of the rational numbers? Would we want to start counting by going straight across the top row or down the first column? □

If we go across the top row or down the first column, we would never begin the next row or column, and, thus, we would not count all of the rational numbers. So, we need a way to proceed that *guarantees* that *every* rational number will be counted. Cantor's diagonal argument is to move diagonally back and forth: We define a one-to-one correspondence between $\mathbb{Q}$ and $\mathbb{N}$ by moving diagonally back and forth, and skipping over any equivalent fractions that have already been counted. We proceed as follows: Begin with $\frac{0}{1}$ and go down; then to the right; then down diagonally to the left to $\frac{1}{2}$; then down; then up diagonally to the right until $\frac{2}{1}$; then to the right; then down diagonally to the left until $\frac{1}{4}$; and so on. (See FIGURE 15.2.) Thus, the one-to-one correspondence we get in this way is

$$
\begin{array}{ccc}
\frac{0}{1} & \longleftrightarrow & 1 \\
\frac{1}{1} & \longleftrightarrow & 2 \\
\frac{1}{2} & \longleftrightarrow & 3 \\
\frac{-1}{1} & \longleftrightarrow & 4 \\
\frac{1}{3} & \longleftrightarrow & 5 \\
\frac{-1}{2} & \longleftrightarrow & 6 \\
\frac{2}{1} & \longleftrightarrow & 7 \\
\frac{1}{4} & \longleftrightarrow & 8 \\
\frac{-1}{3} & \longleftrightarrow & 9 \\
& \vdots &
\end{array}
$$

Think About It 15.5 Does every rational number indeed get counted using the diagonal argument above? □

Therefore, the set of rational numbers has the same cardinality or is the same "size" as the set of natural numbers and, hence, also as the set of whole numbers and the set of integers.

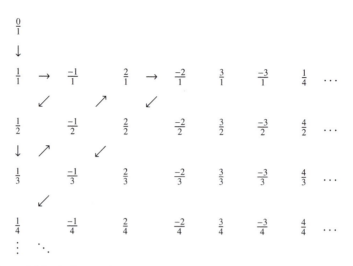

Figure 15.2: A Cantor diagonal argument to show the countability of the set of rational numbers.

15.3   IRRATIONAL NUMBERS

We defined an irrational number to be a nonterminating decimal fraction that does not have a repeating block of digits in DEFINITION 15.2 on page 161. Thus, an irrational number is any real number that is not a rational number. One example that we gave is the decimal fraction

$$0.2\,1\,22\,1\,222\,1\,2222\,1\ldots,$$

but perhaps more familiar examples come from square roots.

To be sure, some square roots are rational numbers. For example, $\sqrt{9} = 3$, and 3 is a rational number; also, $\sqrt{\frac{4}{121}} = \frac{2}{11}$, and $\frac{2}{11}$ is also a rational number. Many square roots, however, are not rational numbers.

Example 15.2 Suppose that we did not realize that

$$\sqrt{\frac{4}{121}} = \frac{2}{11} = 0.\overline{18}.$$

How would we find or approximate the square root? One way, certainly, is to use a calculator; however, perhaps a more instructive way is to use an *iterative method* like that of Heron of Alexandria that we introduced in SECTION 12.1. Here, we relax Heron's use of "the next succeeding square number," and choose *any convenient number a* whose square is "close" to the square of the desired root. In other words, we use Heron's method that is given on page 122 to extract or approximate $\sqrt{A}$, but, instead of finding a number a such that $a^2 > A$ is "close" to A, we find *any* convenient number a such that a^2 is "close" to A.

To use Heron's method to find or approximate $\sqrt{\frac{4}{121}}$, let[4]

$$A = \tfrac{4}{121} = 0.033057851, \quad \text{and} \quad a = \tfrac{1}{10} = 0.1$$

because $a^2 = 0.01$ is "close" to $A \approx 0.03$. The remainder or error is

$$A - a^2 = 0.033057851 - 0.01 = 0.023057851,$$

---

[4]The typewriter font used in "$\frac{4}{121} = 0.033057851$," for instance, indicates the result that is displayed by a calculator. In fact, $\frac{4}{121} = 0.0\overline{330\,578\,512\,396\,694\,214\,876\,0} > 0.033\,057\,851$.

which is less than 0.05, so we may say roughly that the approximation is accurate to one decimal place:

$$\sqrt{\tfrac{4}{121}} \approx 0.1.$$

For a better approximation, we find α_1:

$$\alpha_1 = \frac{1}{2}\left(a + \frac{A}{a}\right) = \frac{1}{2}\left(0.1 + \frac{0.033057851}{0.1}\right) = 0.215289255.$$

The error is

$$A - \alpha_1^2 = 0.033057851 - 0.046349463 = -0.013291612,$$

which is again less than 0.05 in absolute value, so we have not improved our confidence in the approximation beyond 1 decimal place:

$$\sqrt{\tfrac{4}{121}} \approx 0.2.$$

For a better approximation, we find α_2:

$$\alpha_2 = \frac{1}{2}\left(\alpha_1 + \frac{A}{\alpha_1}\right) = \frac{1}{2}\left(0.215289255 + \frac{0.033057851}{0.215289255}\right) = 0.184420059.$$

The error is now

$$A - \alpha_2^2 = 0.033057851 - 0.034010758 = -0.000952907,$$

which is less than 0.005 in absolute value, so we may say roughly that the approximation is accurate to 2 decimal places:

$$\sqrt{\tfrac{4}{121}} \approx 0.18.$$

For yet a better approximation, we find α_3:

$$\alpha_3 = \frac{1}{2}\left(\alpha_2 + \frac{A}{\alpha_2}\right) = \frac{1}{2}\left(0.184420059 + \frac{0.033057851}{0.184420059}\right) = 0.181836535.$$

The error is now

$$A - \alpha_3^2 = 0.033057851 - 0.033064525 = -0.000006674,$$

which is now less than 0.000 05 in absolute value, so we may say roughly that the approximation is accurate to 4 decimal places:

$$\sqrt{\tfrac{4}{121}} \approx 0.1818.$$

Let us continue with one more iteration and find α_4:

$$\alpha_4 = \frac{1}{2}\left(\alpha_3 + \frac{A}{\alpha_3}\right) = \frac{1}{2}\left(0.181836535 + \frac{0.033057851}{0.181836535}\right) = 0.181818182.$$

The error is now

$$A - \alpha_4^2 = 0.033057851 - 0.033057851 = 0.$$

Note that this does not mean that $\sqrt{\tfrac{4}{121}} = 0.181\,818\,182$ exactly, for we know that in fact

$$\sqrt{\tfrac{4}{121}} = \tfrac{2}{11} = 0.\overline{18} < 0.181\,818\,182\ldots.$$

That the computed error, $A - \alpha_4^2$, is zero only indicates that we have reached the limit of the calculator we are using; however, we may guess that the 2 at the end of the display is a result of rounding, and so we may *guess* that $\sqrt{\tfrac{4}{121}} = 0.\overline{18}$. To be sure, we would need to check.

To check if, indeed, $\sqrt{\tfrac{4}{121}} = 0.\overline{18}$, let $r = 0.\overline{18} = 0.181\,818\ldots.$ Then,

$$100r = 100(0.181\,818\ldots) = 18.181\,818\ldots.$$

Next, subtracting r from $100r$, we find that

$$100r = 18.181\,818\ldots$$

$$r = 0.181\,818\ldots$$

$$\text{so} \quad 99r = 18.000\,000\ldots$$

because the repeating block of digits cancels. Finally, solving for r, we conclude that $r = \frac{18}{99}$ or $\frac{2}{11}$, that is,

$$r = 0.181\,818\ldots = \tfrac{2}{11}.$$

Now, squaring, we find that

$$\left(\tfrac{2}{11}\right)^2 = \tfrac{4}{121},$$

so that *now* we may conclude that $\sqrt{\tfrac{4}{121}} = \tfrac{2}{11} = 0.\overline{18}.$ □

Now You Try 15.5 Use Heron's method to find or approximate the square root with the given initial guess a. If a pattern appears in the decimal fraction, guess an exact value of the square root, then check it.

1. $\sqrt{4}$; $a = 1$

2. $\sqrt{\tfrac{16}{289}}$; $a = \tfrac{1}{20}$

□

Square roots, of course, are related to squares:

$$\sqrt{a} = b \quad \text{if} \quad b^2 = a.$$

Geometrically, $b^2 = a$ if a is the area of a square with side b. For example, $2^2 = 4$, and a square with area 4 square units has side 2 units; in this case, $\sqrt{4} = 2$, and 2 is a rational number. As another example, $\left(\tfrac{4}{17}\right)^2 = \tfrac{16}{289}$, and a square with area $\tfrac{16}{289}$ square units has side $\tfrac{4}{17}$ unit; in this case, $\sqrt{\tfrac{16}{289}} = \tfrac{4}{17} = 0.235\,294\,117\,647\,058\,8$ is also a rational number. What about a square that has area 2 square units?

If a square has area 2 square units, then the square has side $\sqrt{2}$ units; but what is $\sqrt{2}$? In other words, what is a number ρ such that $\rho^2 = 2$? (The symbol ρ is the Greek lowercase letter *rho.*) We can use Heron's method to see if a pattern appears that would allow us to guess the number ρ.

To use Heron's method to find or approximate $\rho = \sqrt{2}$, let

$$A = 2 \quad \text{and} \quad a = 1.$$

The remainder or error is

$$A - a^2 = 2 - 1 = 1,$$

which is less than 5, so we may say roughly that the approximation is accurate to the ten place value:

$$\sqrt{2} \approx 0.$$

This is a very bad approximation.

For a better approximation, we find α_1:

$$\alpha_1 = \frac{1}{2}\left(a + \frac{A}{a}\right) = \frac{1}{2}\left(1 + \frac{2}{1}\right) = 1.5.$$

The error is

$$A - \alpha_1^2 = 2 - 2.25 = -0.25,$$

which is less than 0.5 in absolute value, so we may say roughly that the approximation is accurate to the one place:

$$\sqrt{2} \approx 2.$$

This is still a very bad approximation: it is not any better than $\sqrt{2} \approx 0$.

Still hoping to find a better approximation, we find α_2:

$$\alpha_2 = \frac{1}{2}\left(\alpha_1 + \frac{A}{\alpha_1}\right) = \frac{1}{2}\left(1.5 + \frac{2}{1.5}\right) = 1.416666667.$$

The error is

$$A - \alpha_2^2 = 2 - 2.006944445 = -0.00644445,$$

which is less than 0.05, so we may say roughly that the approximation is accurate to 1 decimal place:

$$\sqrt{2} \approx 1.4.$$

This is better, finally, but we can hope to do better still.

Hoping to find yet a better approximation, we find α_3:

$$\alpha_3 = \frac{1}{2}\left(\alpha_2 + \frac{A}{\alpha_2}\right) = \frac{1}{2}\left(1.416666667 + \frac{2}{1.416666667}\right) = 1.414215686.$$

The error is

$$A - \alpha_3^2 = 2 - 2.000006007 = -0.000006007,$$

which is less than 0.00005 in absolute value, so we may say roughly that the approximation is accurate to 4 decimal places:

$$\sqrt{2} \approx 1.4142.$$

This seems to be pretty good now, but let us try to improve the approximation once more.

Let us continue with one more iteration and find α_4:

$$\alpha_4 = \frac{1}{2}\left(\alpha_3 + \frac{A}{\alpha_3}\right) = \frac{1}{2}\left(1.414215686 + \frac{2}{1.414215686}\right) = 1.414213562.$$

The error is

$$A - \alpha_4^2 = 2 - 1.999999999 = 0.000000001,$$

which is less than 0.000000005, so we may say roughly that the approximation is accurate to 8 decimal places:

$$\sqrt{2} \approx 1.41421356.$$

Finally, this is a very good approximation; however, even to 8 decimal places, we do not yet notice a repeating block of digits in the decimal fraction approximation of $\sqrt{2}$. Thus, it is *possible* that $\sqrt{2}$ is an irrational number (or perhaps we do not have a good enough approximation yet to notice a repeating block of digits).

Think About It 15.6 It turns out that computers that have computed $\sqrt{2}$ to millions of decimal places have not shown that its decimal fraction representation has a repeating block of digits. Is this enough evidence to conclude that $\sqrt{2}$ is in fact an irrational number? ☐

The discovery of what we today call irrational numbers is credited to Pythagoras and the Pythagoreans of the sixth century BC. The Pythagoreans, a cult of sorts that had mystical as well as mathematical interests, believed that number—meaning the natural numbers or ratios of natural numbers—was the basis of all things. This was the bedrock of their beliefs. The discovery of irrational numbers—numbers that were neither the natural numbers nor ratios of natural numbers—shook this bedrock. Depending on what source you read, legend has it that Hippasus, the Pythagorean who discovered the irrationality of $\sqrt{2}$, was either exiled, drowned, or subjected to all manner of other punishments for this discovery.

Given that no written records were left by Pythagoras, and that all discoveries by Pythagoreans were attributed to the master himself, all we have is legend. As is true of all legends, they are picked up and added to over time. A lot has been written about Pythagoras and the Pythagoreans, but very little of it is substantiated. (See SUBSECTION 17.1.1 for more on Pythagoras.)

It has long been believed that the first irrational number to be discovered was $\sqrt{2}$ by looking at the ratio of a diagonal of a square to its side. Heath [68, pp. 90–91], for example, is quite sure of this and tells us so in no uncertain terms.

> But it is certain that the incommensurability of the diagonal of a square to its side, that is, the 'irrationality' of $\sqrt{2}$, was discovered in the school of Pythagoras.

Heath goes on to explain the Pythagoreans' proof, which is in essence identical to a proof of the irrationality of $\sqrt{2}$ that one would find in any modern book on number theory.

> The actual method by which the Pythagoreans proved the fact that $\sqrt{2}$ is incommensurable with 1 was doubtless that indicated by Aristotle, a *reductio ad absurdum* showing that, if the diagonal of a square is commensurable with its side, it will follow that the same number is both odd and even. This is evidently the proof interpolated in the texts of Euclid as X. 117, which is in substance as follows:
>
> Suppose AC, the diagonal of a square, to be commensurable with AB, its side; let $\alpha : \beta$ be their ratio expressed in the smallest possible numbers.
>
> Then $\alpha > \beta$, and therefore α is necessarily > 1.
>
> Now
> $$AC^2 : AB^2 = \alpha^2 : \beta^2;$$
> and, since
> $$AC^2 = 2AB^2, \quad \alpha^2 = 2\beta^2.$$
>
> Hence α^2, and therefore α, is even.
> Since $\alpha : \beta$ is in its lowest terms, it follows that β must be odd.
> Let $\alpha = 2\gamma$; therefore $4\gamma^2 = 2\beta^2$, or $2\gamma^2 = \beta^2$, so that β^2, and therefore β, is *even*.
> But β was also *odd*; which is impossible.
> Therefore the diagonal AC cannot be commensurable with the side AB.

Think About It 15.7 Since any decimal number you can think of is a real number, are there any real numbers that are neither rational nor irrational?   □

Heath's presentation is terse, so we elaborate upon it. It is a "proof by contradiction," and relies on the fact that $\sqrt{2}$ either is a rational number (so that it can be expressed as $\frac{a}{b}$ in lowest terms) or is an irrational number (it cannot be expressed as $\frac{a}{b}$ in lowest terms), and that it cannot be both, nor can it be neither. Hence, if we *assume* that $\sqrt{2}$ is a rational number, and logical deductions made from that assumption lead to a contradictory statement, then the assumption must be incorrect. Consequently, we must conclude that $\sqrt{2}$ must be an irrational number after all.

Let ρ be a positive real number whose square is 2: $\rho^2 = 2$. Then, either ρ is a rational number or it is not. Let us suppose that ρ is a rational number and write

$$\rho = \tfrac{a}{b},$$

where, without any loss of generality, we assume that $\frac{a}{b}$ is in lowest terms, that is, we assume that a and b have no common factor besides 1. Let us see where this leads.

If $\rho^2 = 2$, then
$$\left(\tfrac{a}{b}\right)^2 = 2, \quad \text{so that} \quad a^2 = 2b^2$$

after multiplying through the equation by b^2. This implies that a^2 is an even number, so that a must be an even number; hence, $a = 2m$ for some positive integer m, or $a^2 = (2m)^2 = 4m^2$. Replacing a^2 by $4m^2$, we see that

$$4m^2 = 2b^2 \quad \text{or} \quad 2m^2 = b^2$$

after dividing through the equation by 2. This, in turn, implies that b^2 is an even number, so that b must be an even number; hence, $b = 2n$ for some positive integer n. However, if

$$a = 2m \quad \text{and} \quad b = 2n,$$

then a and b have a common factor besides 1, namely, a and b have a common factor of 2, contrary to our assumption that $\frac{a}{b}$ is in lowest terms. Therefore, either we were wrong to assume that a and b have no common factor besides 1, or we were wrong to assume that ρ is a rational number. (These were the only two assumptions that we made.)

The assumption that a and b have no common factor besides 1 cannot be disputed because the set of common fractions that are equivalent to $\frac{a}{b}$ corresponds to the same rational number, *a fortiori* there is no harm in assuming that $\frac{a}{b}$ is in lowest terms, that is, that a and b have no common factor besides 1. Hence, it must be the assumption that ρ is a rational number that is false. Consequently, we conclude that ρ is not a rational number after all. In other words, the positive real number whose square is 2 is an irrational number that we call the square root of 2.

> The real number $\sqrt{2}$ is an irrational number.

Think About It 15.8 Explain why if a^2 is an even number, then a must be an even number. □

Along with Euclid's proof of the infinitude of prime numbers, this proof of the irrationality of $\sqrt{2}$ is generally considered to be one of the more elegant proofs in mathematics. More recently, some scholars have argued that the first irrational number to be discovered may have been $\sqrt{5}$, and not $\sqrt{2}$, by looking at the ratio of a diameter of a regular pentagon to its side [123].

Now You Try 15.6 Show that the real numbers $\sqrt{5}$ and $\sqrt{6}$ are irrational numbers. □

Remark 15.2 The preceding demonstration that $\sqrt{2}$ (the positive real number whose square is 2) is an irrational number illustrates a method of proof known as *proof by contradiction* or, in Latin, *reductio ad absurdum*. It is a useful method of proof, especially when a direct proof would be very difficult, if not impossible. For example, it would be impossible to demonstrate directly that the decimal fraction representation of $\sqrt{2}$ has no repeating block of digits, for we would never know if we had not detected a repeating block of digits only because we had not gone out far enough in the decimal expression. We shall encounter proofs by contradiction again. □

Think About It 15.9 The set of real numbers is the union of the set of rational numbers and the set of irrational numbers. We already know the set of rational numbers is countably infinite. What do you think are the cardinalities of the set of irrational numbers and the set of real numbers? If you think that the sets are countably infinite, can you find a systematic way to count them? □

We have been surprised to see that, although $\mathbb{N} \subset \mathbb{W} \subset \mathbb{Z} \subset \mathbb{Q}$, all of these sets have the same cardinality: they are all countably infinite, that is, each is in one-to-one correspondence with the set of natural numbers, $\mathbb{N}$. We may expect now that all infinite sets are in one-to-one correspondence with the set of natural numbers, so that the set of irrational numbers is also countably infinite. The surprising fact here is that *the set of irrational numbers is uncountably infinite*, that is, $\mathbb{I}$ has a cardinality that is different from the cardinality of $\mathbb{N}$ (or $\mathbb{W}$ or $\mathbb{Z}$ or $\mathbb{Q}$); in fact, $\mathbb{I}$ has greater cardinality or is "larger" than $\mathbb{N}$. How can we see this?

Think About It 15.10 If the cardinality of the set of irrational numbers is greater than that of the natural numbers, how "large" is the set of irrational numbers? □

We demonstrate first that $\mathbb{I}$ is uncountably infinite. After that, we demonstrate that $\mathbb{I}$ has greater cardinality than $\mathbb{N}$.

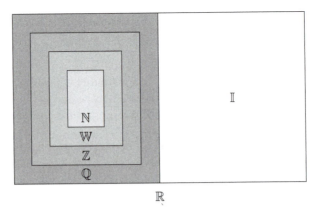

Figure 15.3: The set of real numbers.

15.4 $\mathbb{I}$ IS UNCOUNTABLY INFINITE

That the irrational numbers, $\mathbb{I}$, is uncountably infinite follows most easily from the following two facts: the union of two countably infinite sets is a countably infinite set, and the set of real numbers is uncountably infinite. (We will see why.) We show these two facts in turn.

1. *The union of two countably infinite sets is a countably infinite set.*

   Let A and B be countably infinite sets, say

   $$A = \{a_1,\ a_2,\ a_3, \ldots\} \quad \text{and} \quad B = \{b_1,\ b_2,\ b_3, \ldots\},$$

   so that

   $$A \cup B = \{a_1,\ a_2,\ a_3, \ldots,\ b_1,\ b_2,\ b_3, \ldots\}.$$

   Then, we define a one-to-one correspondence between $A \cup B$ and $\mathbb{N}$ by interlacing the elements of A and B:

   $$
   \begin{array}{ccc}
   a_1 & \longleftrightarrow & 1 \\
   b_1 & \longleftrightarrow & 2 \\
   a_2 & \longleftrightarrow & 3 \\
   b_2 & \longleftrightarrow & 4 \\
   a_3 & \longleftrightarrow & 5 \\
   b_3 & \longleftrightarrow & 6 \\
   & \vdots &
   \end{array}
   $$

   Therefore, $A \cup B$ is countably infinite, that is, the union of two countably infinite sets is a countably infinite set.

   Think About It 15.11 What is the cardinality of the union of three countably infinite sets? Of ten countably infinite sets? Of one hundred? Of a countably infinite union of countably infinite sets? □

2. *The set of real numbers is uncountably infinite.*

We will show this in two steps. We start by showing that the set of all real numbers that are strictly between zero and one is uncountably infinite. This set is called the *open unit interval*:

$$I^{(0,1)} = \{x \in \mathbb{R}: 0 < x < 1\}.$$

Second, we show that card($\mathbb{R}$) = card($I^{(0,1)}$).

We employ a proof by contradiction to show that $I^{(0,1)}$ is uncountably infinite. We note that either $I^{(0,1)}$ is countably infinite or it is uncountably infinite, that is, either there is a one-to-one correspondence between $I^{(0,1)}$ and $\mathbb{N}$ or there is not. Let us suppose that $I^{(0,1)}$ is countably infinite so that there is a one-to-one correspondence between $I^{(0,1)}$ and $\mathbb{N}$; in other words, we suppose that *every* decimal fraction in $I^{(0,1)}$ is paired with exactly one natural number, and vice versa. We can only *assume* that such a list can be made. With the rationals, we saw that Cantor was able to list them by arraying them in two dimensions. But unlike the rationals, there is no way to list all the decimal fractions in $I^{(0,1)}$ systematically. So, we suppose that we do have a listing of all of them, say,

$$
\begin{array}{ccc}
\alpha_1 & \longleftrightarrow & 1 \\
\alpha_2 & \longleftrightarrow & 2 \\
\alpha_3 & \longleftrightarrow & 3 \\
& \vdots &
\end{array}
$$

Here, the α_i ($i = 1, 2, 3, \ldots$) are presumed to be *all* the decimal fractions in $I^{(0,1)}$. Now we will come up with a decimal fraction

$$\beta = 0.\beta_1\beta_2\beta_3 \ldots$$

that is not any of the α_i.

Chose any two distinct digits between zero and nine. We will use 3 and 5, but it will work with any two distinct digits. We will build β one digit at a time by looking at the digit of each α_i "on the diagonal." For the first digit of β we look at the first digit of α_1: if the first digit of α_1 is 3, we let $\beta_1 = 5$; if the first digit is not 3, we let $\beta_1 = 3$. For the second digit of β, we look at the second digit of α_2: if the second digit of α_2 is 3, we let $\beta_2 = 5$; if the second digit is not 3, we let $\beta_2 = 3$; and so on. Here is an illustration.

$$
\begin{array}{llll}
\alpha_1 = 0.0\underline{4}5\,454\,545\ldots & \longleftrightarrow & 1 \qquad & \beta = 0.3\ldots & \ne \alpha_1 \\
\alpha_2 = 0.1\underline{4}1\,421\,356\ldots & \longleftrightarrow & 2 \qquad & \beta = 0.33\ldots & \ne \alpha_2 \\
\alpha_3 = 0.31\underline{4}\,159\,265\ldots & \longleftrightarrow & 3 \qquad & \beta = 0.33\underline{3}\ldots & \ne \alpha_3 \\
\alpha_4 = 0.333\,\underline{3}33\,333\ldots & \longleftrightarrow & 4 \qquad & \beta = 0.333\,\underline{5}\ldots & \ne \alpha_4 \\
& \vdots & & \vdots &
\end{array}
$$

Observe that β is different from every decimal fraction α_i because β differs from every α_i in at least one decimal place. Yet $\beta \in I^{(0,1)}$. Thus, we have arrived at a contradiction: We assumed that $I^{(0,1)}$ is countably infinite so that the α_i constitute *all* the decimal fractions in $I^{(0,1)}$, yet $\beta \in I^{(0,1)}$ is not one of the α_i. Hence, our assumption that $I^{(0,1)}$ is countably infinite must be false.

Therefore, we conclude that $I^{(0,1)}$ is *not* countably infinite after all. In other words, $I^{(0,1)}$ is uncountably infinite. The above proof is also due to Cantor, and is also known as a Cantor diagonal argument.

Think About It 15.12 We created β using 3 and 5. Can we create another number not on our list? How many numbers can we create that are not listed? □

Next, we show that card($\mathbb{R}$) = card($I^{(0,1)}$). To do this, we need to exhibit a one-to-one correspondence between the set of real numbers and the open unit interval. For our purposes we will appeal to a diagram, although this is not a formal proof.

To begin, we "roll up" the open unit interval $I^{(0,1)}$ into a circle as shown below. Note that the "north pole" of the circle is missing because 0 and 1 are not included in the open unit interval.

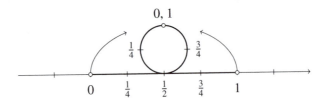

Then we draw lines from the north pole of the circle to the number line. Each of these lines intersects the circle once and the number line once and gives a one-to-one correspondence between the points on the circle and the points on the number line.

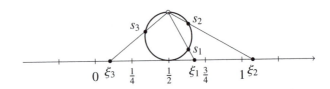

Think About It 15.13 Clearly every point on the circle is mapped to a real number on the number line, but is every real number on the number line mapped to a point on the circle? □

In this way, with the exception of the north pole, every point on the circle is paired with exactly one point on the number line, and vice versa:

$$(s_1, \xi_1), \ (s_2, \xi_2), \ (s_3, \xi_3), \dots .$$

(The symbol ξ is the Greek lowercase letter *xi*.) Since the circle *sans* the north pole is $I^{(0,1)}$, these pairs constitute a one-to-one correspondence between $I^{(0,1)}$ and $\mathbb{R}$.

Therefore, card($\mathbb{R}$) = card($I^{(0,1)}$), and so the set of real numbers is uncountably infinite.

Think About It 15.14 To what would the north pole be mapped if it were included in the circle? □

We are now in a position to show that the set of irrational numbers is uncountably infinite.

Toward this end, recall that the union of two countably infinite sets is a countably infinite set. Hence, knowing now that $\mathbb{R}$ is uncountably infinite, we see that at least one of $\mathbb{Q}$ or $\mathbb{I}$ must be uncountably infinite; for if both $\mathbb{Q}$ and $\mathbb{I}$ were countably infinite, then so must be $\mathbb{R} = \mathbb{Q} \cup \mathbb{I}$. However, we already know that $\mathbb{Q}$ is countably infinite (page 166); therefore, we are led to conclude that $\mathbb{I}$ is uncountably infinite.

Both $\mathbb{I}$ and $\mathbb{R}$ are uncountably infinite.

Now You Try 15.7 Recall the open unit interval

$$I^{(0,1)} = \{x \in \mathbb{R}: 0 < x < 1\}.$$

Now consider an *arbitrary* open interval

$$I^{(a,b)} = \{x \in \mathbb{R}: a < x < b\}.$$

Provide a visual argument to show there is a one-to-one correspondence between $I^{(0,1)}$ and $I^{(a,b)}$, so that $\text{card}(I^{(a,b)}) = \text{card}(I^{(0,1)})$. Consequently, because $\text{card}(I^{(0,1)}) = \text{card}(\mathbb{R})$, the cardinality of any open interval of the real number line is the same as the cardinality of the entire real number line, that is, any open interval of the real number line has the "same number" of points as the entire real number line! □

15.5   CARD(ℚ), CARD(𝕀), AND CARD(ℝ)

We have seen that the cardinality of the set of irrational numbers 𝕀 (nonterminating decimal fractions that do not have a repeating block of digits) is different from the cardinality of the set of rational numbers ℚ (nonterminating decimal fractions that have a repeating block of digits): the former is uncountably infinite and the latter is countably infinite. Put another way, there is not the same number of rational numbers as irrational numbers. This begs the question, Which one of the sets, 𝕀 or ℚ, is "larger" than the other?

Before we can answer that question, we need to decide what it means for one infinite set to be "larger" than another infinite set. After all, what could be larger than "infinite"?

Definition 15.3 Let A and B be sets. The cardinality of A is said to be *less than or equal to* the cardinality of B if there is a one-to-one correspondence between A and a subset of B. In this case, we write $\text{card}(A) \leq \text{card}(B)$.

The cardinality of A is said to be *(strictly) less than* the cardinality of B if both $\text{card}(A) \leq \text{card}(B)$ and there is no one-to-one correspondence between A and (all of) B. In this case, we write $\text{card}(A) < \text{card}(B)$. □

Remark 15.3

1. If $A \subseteq B$, then $\text{card}(A) \leq \text{card}(B)$ automatically.

2. The sets A and B have the same cardinality, $\text{card}(A) = \text{card}(B)$, if and only if both $\text{card}(A) \leq \text{card}(B)$ and $\text{card}(B) \leq \text{card}(A)$.

3. It is possible that $\text{card}(A) = \text{card}(B)$ even if A is a proper subset of B.

□

Example 15.3 Let $V = \{a, e, i, o, u\}$ and $S_5 = \{1, 2, 3, 4, 5\}$. We may define a one-to-one correspondence between V and a subset of S_5, here between V and $S_5 \subseteq S_5$.

$$
\begin{array}{cccccc}
V: & a & e & i & o & u \\
 & \updownarrow & \updownarrow & \updownarrow & \updownarrow & \updownarrow \\
S_5: & 1 & 2 & 3 & 4 & 5
\end{array}
$$

Hence, $\text{card}(V) \leq \text{card}(S_5)$.

We see, however, that we also have a one-to-one correspondence between S_5 and $V \subseteq V$; hence, we also have $\text{card}(S_5) \leq \text{card}(V)$. That both $\text{card}(V) \leq \text{card}(S_5)$ and $\text{card}(S_5) \leq \text{card}(V)$ is consistent with the fact that V and S_5 have the same cardinality, that is, $\text{card}(V) = \text{card}(S_5)$. In fact, both V and S_5 have cardinality 5. □

Example 15.4 Let V denote the set of vowels and C denote the set of consonants. We may define a one-to-one correspondence between V and a subset of C, here between V and $\{b, c, d, f, g\} \subset C$.

$$
\begin{array}{lcccccccccccccccccccccc}
V: & a & e & i & o & u \\
   & \updownarrow & \updownarrow & \updownarrow & \updownarrow & \updownarrow & \updownarrow & \updownarrow & \updownarrow & \updownarrow & \updownarrow & \updownarrow & \updownarrow & \updownarrow & \updownarrow & \updownarrow & \updownarrow & \updownarrow & \updownarrow & \updownarrow & \updownarrow & \updownarrow \\
C: & b & c & d & f & g & h & j & k & l & m & n & p & q & r & s & t & v & w & x & y & z
\end{array}
$$

Hence, $\mathrm{card}(V) \leq \mathrm{card}(C)$; however, we see that there is no one-to-one correspondence between V and (all of) C because we would run out of elements of V before being able to pair every element of C with exactly one element of V.

Hence, $\mathrm{card}(V) < \mathrm{card}(C)$. □

Remark 15.4 These two examples may seem a little contrived—since these sets are finite, we can just count the elements (which is putting the elements into a one-to-one correspondence with a subset of the natural numbers.) However, things get tricky when we deal with infinite sets. □

Now You Try 15.8 Consider the set of unit fractions

$$
U = \left\{\frac{1}{n} : n \in \mathbb{N}\right\} = \left\{\frac{1}{1}, \frac{1}{2}, \frac{1}{3}, \ldots\right\}.
$$

1. Show that $\mathrm{card}(U) \leq \mathrm{card}(\mathbb{Q})$.

2. Show that $\mathrm{card}(U) = \mathrm{card}(\mathbb{N})$. Does this contradict $\mathrm{card}(U) \leq \mathrm{card}(Q)$?

□

We are now in a position to compare the cardinalities of the set of rational numbers and the set of irrational numbers. First, we define a one-to-one correspondence between $\mathbb{Q}$ and a subset of $\mathbb{I}$ by inserting successively many zeros between the successive digits following the decimal point of a decimal fraction in $\mathbb{Q}$. This changes the rational number into an irrational number and, thus, maps every element of $\mathbb{Q}$ to an element of $\mathbb{I}$. For example,

$$
31.0\overline{45} = 31.045\,454\,545\ldots \quad \longleftrightarrow \quad 31.\underline{0}\,0\,\underline{4}\,00\,\underline{5}\,000\,\underline{4}\,0000\,\underline{5}\,00000\,\underline{4}\ldots,
$$
$$
0.\overline{3} = 0.333\,333\,333\ldots \quad \longleftrightarrow \quad 0.\underline{3}\,0\,\underline{3}\,00\,\underline{3}\,000\,\underline{3}\,0000\,\underline{3}\,00000\,\underline{3}\ldots.
$$

Hence, $\mathrm{card}(\mathbb{Q}) \leq \mathrm{card}(\mathbb{I})$.

Next, we note that the second Cantor diagonal argument demonstrates that there are irrational numbers that are not elements of the subset of $\mathbb{I}$ that is in a one-to-one correspondence with $\mathbb{N}$, and thus $\mathbb{Q}$. In fact, for any one-to-one correspondence we define between $\mathbb{Q}$ and a subset of $\mathbb{I}$, the Cantor diagonal argument demonstrates that there are irrational numbers that are not elements of any subset of $\mathbb{I}$ that is in a one-to-one correspondence with $\mathbb{Q}$. In other words, there is no one-to-one correspondence between $\mathbb{Q}$ and (all of) $\mathbb{I}$, so that $\mathrm{card}(\mathbb{Q}) < \mathrm{card}(\mathbb{I})$.

$$
\boxed{\mathrm{card}(\mathbb{Q}) < \mathrm{card}(\mathbb{I})}
$$

Thus, we have the astounding result that *one infinity may be "larger" than another* because the set of irrational numbers is "larger" than the set of rational numbers!

Think About It 15.15 In the Cantor diagonal argument method above, why will the new number created also be irrational? □

Georg Cantor's most lasting contribution to mathematics was his idea that there are different levels of infinity: that there are infinite sets of different "sizes." This caused a major paradigm shift. How can one infinite set be "larger" than another infinite set? Cantor's ingenious proofs were groundbreaking, and gained favor among many leading mathematicians; for example, David Hilbert (1862–1943) stated, "No one shall drive us from the paradise which Cantor has created for us." On the other hand, his results were also repudiated by many other famous mathematicians; for example, Leopold Kronecker (1823–1891), who was probably Cantor's harshest critic, in disparaging Cantor's work stated, "God made the natural numbers, all the rest is the work of man." Today Cantor's work is accepted by the vast majority of mathematicians.

15.6 TRANSFINITE NUMBERS

We have observed that any two countably infinite sets have the same cardinality, namely, $\text{card}(\mathbb{N})$. By establishing that $\text{card}(\mathbb{Q}) < \text{card}(\mathbb{I})$, we have shown that the cardinality of a countably infinite set is less than the cardinality of an uncountably infinite set. What about the cardinalities of two uncountably infinite sets: do two uncountably infinite sets have the same cardinality? We attempt to answer this question by considering $\mathbb{I}$ and $\mathbb{R}$, both of which are uncountably infinite sets.

To begin, note that $\text{card}(\mathbb{I}) \leq \text{card}(\mathbb{R})$ because $\mathbb{I} \subset \mathbb{R}$. If it were also true that $\text{card}(\mathbb{R}) \leq \text{card}(\mathbb{I})$, then it would follow from both inequalities that $\text{card}(\mathbb{I}) = \text{card}(\mathbb{R})$, and this would suggest that any two uncountably infinite sets have the same cardinality, just as any two countably infinite sets have the same cardinality.

We claim in fact that $\text{card}(\mathbb{R}) \leq \text{card}(\mathbb{I})$. To see this, we define a one-to-one correspondence between $\mathbb{R}$ and a subset of $\mathbb{I}$ by inserting the following sequence of zeros and ones,

$$0, \ 11, \ 000, \ 1111, \ 00000, \ 111111, \ldots$$

(one zero, two ones, three zeros, four ones, and so on) between successive digits following the decimal point for *every* real number. For example,

$$31.045\,454\,545\ldots \quad \longleftrightarrow \quad 31.0\underline{0}4\,11\,5\,000\,4\,1111\,5\,000004\ldots,$$
$$1.414\,213\,562\ldots \quad \longleftrightarrow \quad 1.4\underline{0}1\,11\,4\,000\,2\,1111\,1\,000003\ldots.$$

Clearly these numbers are irrational, but they do not account for all irrational numbers. Hence, $\text{card}(\mathbb{R}) \leq \text{card}(\mathbb{I})$.

Therefore, because both

$$\text{card}(\mathbb{I}) \leq \text{card}(\mathbb{R}) \quad \text{and} \quad \text{card}(\mathbb{R}) \leq \text{card}(\mathbb{I}),$$

we conclude that $\text{card}(\mathbb{I}) = \text{card}(\mathbb{R})$: the set of irrational numbers and the set of real numbers have the same cardinality.

$$\boxed{\text{card}(\mathbb{I}) = \text{card}(\mathbb{R})}$$

Think About It 15.16 Instead of inserting the sequence

$$0, \ 11, \ 000, \ 1111, \ 00000, \ 111111, \ldots$$

of zeros and ones between successive digits following the decimal point of a real number to ensure that we have an irrational number, could we just as well have used zeros only? For example,

$$31.045\,454\,545\ldots \quad \longleftrightarrow \quad 31.0\underline{0}4\,00\,5\,000\,4\,0000\,5\,000004\ldots,$$
$$1.414\,213\,562\ldots \quad \longleftrightarrow \quad 1.4\underline{0}1\,00\,4\,000\,2\,0000\,1\,000003\ldots,$$

where we have inserted the sequence

$$0, \ 00, \ 000, \ 0000, \ 00000, \ 000000, \ldots?$$

□

By definition, the cardinality of any countably infinite set is card($\mathbb{N}$). Is the cardinality of any uncountably infinite set is card($\mathbb{R}$)?

The official birth of Cantor's set theory is 1874 when he published his first paper on the subject. By 1879 he was well into his work on the cardinality or "size" of infinite sets. Cantor denoted the cardinality of the set of natural numbers, and so the cardinality of any countably infinite set, $\aleph_0$ (read "*aleph* naught"; $\aleph$ is the first letter of the Hebrew alphabet). The cardinality of the set of real numbers, and so the cardinality of the set of irrational numbers, is denoted c (because the set of real numbers is in one-to-one correspondence with the points on the real number line or the *continuum*).

$$\boxed{\quad \text{card}(\mathbb{N}) = \aleph_0 \quad \text{and} \quad \text{card}(\mathbb{R}) = c \quad}$$

The cardinalities $\aleph_0$ and c are called *transfinite numbers*.

After determining that different infinite sets could have different cardinalities, Cantor started to think about whole sequences of transfinite numbers. In particular, he looked at power sets, which we will examine below. This work led to the famous *continuum hypothesis* that conjectures that the cardinality of the "power set" of the set of natural numbers is c and, moreover, that there is no infinite set that has a cardinality that is in between $\aleph_0$ and c. The continuum hypothesis divided mathematicians for years until the groundbreaking work of Kurt Gödel (1906–1978) in the 1930s and Paul Cohen (1934–2007) in the 1960s put the matter to rest in a very surprising way. (You will be asked about this in the exercises.)

When we established that card($\mathbb{Q}$) < card($\mathbb{I}$), we determined that $\aleph_0 < c$. The continuum hypothesis asserts that there is no transfinite number between $\aleph_0$ and c. To see if there are transfinite numbers greater than c, we introduce the notion of a *power set*.

Definition 15.4 Let A be a set. The *power set* of A is the collection of all the subsets of A. We write $\mathcal{P}(A)$ for the power set of A. □

Example 15.5 Let $A = \{p, q, r\}$. Then, all the subsets of A are

$$\{\ \}, \quad \{p\}, \quad \{q\}, \quad \{r\}, \quad \{p, q\}, \quad \{p, r\}, \quad \{q, r\}, \quad \{p, q, r\},$$

being eight subsets in all. Thus, the power set of A is

$$\mathcal{P}(A) = \{\{\ \}, \{p\}, \{q\}, \{r\}, \{p, q\}, \{p, r\}, \{q, r\}, \{p, q, r\}\}.$$

Note that card(A) = 3 and card($\mathcal{P}(A)$) = 8, so that card(A) < card($\mathcal{P}(A)$). □

Now You Try 15.9 List the power sets of the following sets.

   1. $S = \{1, 2, 3\}$ 2. $T = \{w, x, y, z\}$

□

Think About It 15.17 For any finite set of cardinality n, what is the cardinality of its power set? □

The best way to think about the elements of the power set of a set is to list all the ways there are to make a choice about which elements from a set to include in a subset. For each element of a set, there are two choices: to include it in the subset or not to include it in the subset. In the example above, for each element of the set $A = \{p, q, r\}$ there are two choices: include that element in the subset or not. Thus, there are a total of $2 \times 2 \times 2 = 2^3$ or 8 choices, so that there are 8 subsets of A, that is, 8 elements in the power set of A.

> If $\text{card}(A) = n$, then $\text{card}(\mathcal{P}(A)) = 2^n$.

Now You Try 15.10 Calculate the cardinality of the power set of a set A with the given cardinality.

1. $\text{card}(A) = 5$
2. $\text{card}(A) = 10$

□

Now what would be the power set of a power set? If we let $A = \{p, q, r\}$ and let

$$a = \{\ \}, \qquad b = \{p\}, \qquad c = \{q\}, \qquad d = \{r\},$$
$$e = \{p, q\}, \quad f = \{p, r\}, \quad g = \{q, r\}, \quad h = \{p, q, r\},$$

then

$$B = \{a, b, c, d, e, f, g, h\} = \mathcal{P}(A),$$

so that $\mathcal{P}(B) = \mathcal{P}(\mathcal{P}(A))$. Moreover, the subsets of B are

$$\{\ \}, \quad \{a\}, \quad \{b\}, \quad \{c\}, \quad \{d\}, \quad \{e\}, \quad \{f\}, \quad \{g\}, \quad \{h\},$$

$$\{a, b\} \quad \{a, c\} \quad \{a, d\} \quad \{a, e\}, \quad \{a, f\}, \quad \{a, g\}, \quad \{a, h\},$$

$$\{b, c\}, \quad \{b, d\}, \quad \{b, e\}, \quad \{b, f\}, \quad \{b, g\}, \quad \{b, h\},$$

$$\{c, d\}, \quad \{c, e\}, \quad \{c, f\}, \quad \{c, g\}, \quad \{c, h\},$$

$$\{d, e\}, \quad \{d, f\}, \quad \{d, g\}, \quad \{d, h\},$$

$$\{e, f\}, \quad \{e, g\}, \quad \{e, h\}, \quad \{f, g\}, \quad \{f, h\}, \quad \{g, h\},$$

$$\{a, b, c\}, \quad \{a, b, d\}, \quad \{a, b, e\}, \quad \{a, b, f\}, \quad \{a, b, g\}, \quad \{a, b, h\},$$

$$\{a, c, d\}, \quad \{a, c, e\}, \quad \{a, c, f\}, \quad \{a, c, g\}, \quad \{a, c, h\},$$

$$\{a, d, e\}, \quad \{a, d, f\}, \quad \{a, d, g\}, \quad \{a, d, h\},$$

$$\{a, e, f\}, \quad \{a, e, g\}, \quad \{a, e, h\}, \quad \{a, f, g\}, \quad \{a, f, h\}, \quad \{a, g, h\},$$

$$\{b, c, d\}, \quad \{b, c, e\}, \quad \{b, c, f\}, \quad \{b, c, g\}, \quad \{b, c, h\},$$

$$\{b, d, e\}, \quad \{b, d, f\}, \quad \{b, d, g\}, \quad \{b, d, h\},$$

$$\vdots$$

$$\{a, b, c, d, e, f, g, h\},$$

being $2^8 = 256$ subsets in all. These are the elements of $\mathcal{P}(B)$ or $\mathcal{P}(\mathcal{P}(A))$. Hence, $\text{card}(\mathcal{P}(\mathcal{P}(A))) = \text{card}(\mathcal{P}(B)) = 256$.

Theorem 15.1 (CANTOR'S THEOREM) *For any set A,*

$$\text{card}(A) < \text{card}(\mathcal{P}(A)).$$

☐

Cantor's theorem applies to infinite sets as well as to finite sets, so that $\text{card}(\mathbb{R}) < \text{card}(\mathcal{P}(\mathbb{R}))$; thus, $\text{card}(\mathcal{P}(\mathbb{R}))$ is a transfinite number that is greater than c. And if EXAMPLE 15.5 is any indication, $\text{card}(\mathcal{P}(\mathcal{P}(\mathbb{R})))$ is greater than $\text{card}(\mathcal{P}(\mathbb{R}))$, and much greater than c. In fact, in this way we may obtain an increasing sequence of transfinite numbers, to wit,

$$c = \text{card}(\mathbb{R}) < \text{card}(\mathcal{P}(\mathbb{R})) < \text{card}(\mathcal{P}(\mathcal{P}(\mathbb{R}))) < \text{card}(\mathcal{P}(\mathcal{P}(\mathcal{P}(\mathbb{R})))) < \cdots,$$

each of which is an "ever greater infinity." If your mind is spinning, you are in good company. In the one hundred-plus years since Cantor introduced his levels of infinity, many prominent mathematicians have grappled with these ideas.

Chapter 16

Logic

A mother tells her daughter, "You may have dessert if you finish all the spinach on your dinner plate." A little while later, the daughter's father gives her dessert even though she did not finish all the spinach on her dinner plate. Did the father violate the mother's condition for dessert?

In this chapter, we introduce the basics of *classical propositional logic* or *sentential logic*. While propositional logic is not rich enough to encode mathematical language or to be applied to areas such as the natural sciences, social sciences, linguistics, philosophy, law, and ethics—for these we require predicate logic, formalized theories, and algorithms and recursive functions—propositional logic nevertheless provides a very good introduction to the formalism of logic [51, 59].

Logic is traditionally said to have started with Aristotle and the early Greek philosophers. When talking about the birth of mathematical logic, many start with Euclid, whose *Elements* shows the value and beauty of an axiomatic system. An axiomatic system starts with a list of common notions and axioms. These provide the basic building blocks of terms and ideas that are agreed upon as fundamental to the system. These are then used to prove propositions. Any theorem or postulate must be proved using earlier postulates or axioms. Euclid set the bar for rigor for the future of mathematics, which led to our very powerful system of deduction and proof.

Modern mathematical logic, or symbolic logic, began in the late 1800s as a reaction to the work of Georg Cantor. To answer the questions that Cantor's work on infinite sets raised, mathematicians realized that the natural number system, upon which mathematics had been built, had never been formally and systematically verified to make sure it did not lead to any contradictions. Thus, a rush to establish the rigor of the foundations of mathematics was launched. Famous mathematicians such as David Hilbert, Bertrand Russell, Augustus De Morgan (1806–1871), and George Boole devoted years to researching and

(a) Boole.

(b) Russell.

Figure 16.1: (a) A portrait of George Boole (ca. 1860) and (b) Bertrand Arthur William Russell at 52 years old in America (1924). (Source: © Thinkstock.)

developing the foundations of mathematics. (See FIGURE 16.1.)

As research into logic and foundations progressed, new areas of study emerged, and now include such specialties as set theory, recursion theory, and many areas of computer science.

Two of the leaders in the development of logic were George Boole and Augustus De Morgan. As a child in England, Boole was primarily interested in the classics and classical languages, learning Latin from a tutor, and then teaching himself Greek as a teenager. His interest in philosophy led him to teach himself mathematics, which in turn led to his marrying the ideas of philosophy, logic, and the foundations of mathematics. His famous work in this area is his 1854, *An Investigation of the Laws of Thought* (usually referred to as the *Laws of Thought*), which shows how, by using symbolism, propositions could be reduced to equations. Then they could be verified using syllogisms, or rules of logic.

Augustus De Morgan was born in India to a British Army officer. Like many English gentry, he studied to become a lawyer, even though his passion was mathematics; however, at the age of 21, he applied for and received a position teaching mathematics at the new University College London. In 1866 he cofounded the London Mathematical Society. His research was in algebra, logic, and arithmetic. Of importance to our story, like Boole, De Morgan showed that the symbolic nature of algebra was very powerful. Given the rules of algebra, the results of symbolic manipulations are valid, regardless of any real-world application. He also realized that there are different kinds of algebras, thus paving the way for the study of *abstract algebra*. De Morgan is probably most known for *De Morgan's laws* that relate the operations of negation, union, and intersection.

In classical propositional logic, every expression is assigned either a value of 1 (or T for "true") or 0 (or F for "false"), but not both. These are the so-called *law of the excluded middle* (either 1 or 0) and the *law of contradiction* (but not both). An expression may be a variable or the result of a "truth function" applied to expressions. If p is an expression, then

$$\text{either} \quad p = 1 \quad \text{or} \quad p = 0,$$

but not both; if p and q are expressions, then there are four different ways that p and q can be assigned values of 1 or 0, namely,

$$(p, q) = (1, 1), \quad (p, q) = (1, 0), \quad (p, q) = (0, 1), \quad (p, q) = (0, 0);$$

if p, q, and r are expressions, then there are eight different ways that p, q, and r can be assigned values of 1 or 0, namely,

$$(p, q, r) = (1, 1, 1)$$
$$(p, q, r) = (1, 1, 0) \quad (p, q, r) = (1, 0, 1) \quad (p, q, r) = (0, 1, 1)$$
$$(p, q, r) = (1, 0, 0) \quad (p, q, r) = (0, 1, 0) \quad (p, q, r) = (0, 0, 1) \quad (p, q, r) = (0, 0, 0);$$

and so on for four or more expressions.

Think About It 16.1 How do the different assignments of 1 or 0 to n expressions compare to the different subsets of a set that contains n elements?   □

Now You Try 16.1 Let p, q, r, and s be expressions. Write down all the ways that (p, q, r, s) can be assigned the values of 1 or 0. How many ways are there in all?   □

There are five truth functions: *non, et, vel, seq,* and *aeq*. These functions and their "functorial notations" are defined below. We define the truth functions using so-called *truth tables*. The use of truth tables will also provide us a way to evaluate expressions.

Definition 16.1 Let p and q be expressions. We define the *truth functions* as follows.

not p (non)	
p	$\sim p$ or $\neg p$
1	0
0	1

p and q (et)		
p	q	$p \wedge q$
1	1	1
1	0	0
0	1	0
0	0	0

p or q (vel)		
p	q	$p \vee q$
1	1	1
1	0	1
0	1	1
0	0	0

if p, then q (seq)		
p	q	$p \to q$
1	1	1
1	0	0
0	1	1
0	0	1

p if and only if q (aeq)		
p	q	$p \leftrightarrow q$
1	1	1
1	0	0
0	1	0
0	0	1

Think About It 16.2 Do the values of the truth functions defined above make sense? Explain.

Remark 16.1 The last two lines of the definition of $p \to q$ are said to be *vacuously true* because $p \to q$ has value 1 (true) when p has value 0 (false), regardless of whether q has value 1 or 0. (We may think that $p \to q$ is "truly" true—as opposed to being vacuously true—when p has value 1 and q has value 1.)

p	q	$p \to q$
0	1	1
0	0	1

This may seem to be counterintuitive. After all, how can the statement, "If p is true, then q is true," be itself true when the condition p is false? The fact is that anytime an argument is based on a false condition, then any conclusion may be reached. The singular case when $p \to q$ has value 0 is when p has value 1 and q has value 0; we call this a *counterexample* to $p \to q$. For example, consider the statement "If cows fly, then the sky is purple." Until we find a cow that flies *and* see that the sky is not purple, we cannot say that the statement is false; and if we cannot conclude that the statement is false, then—by the law of the excluded middle (page 184)—we must conclude that it is (vacuously) true.

Example 16.1 We evaluate the statement,

"You may have dessert if you finish all the spinach on your dinner plate."

To begin, let

$$p = \text{"you finish all the spinach on your dinner plate,"}$$
$$q = \text{"you may have dessert."}$$

Then, the statement, "You may have dessert if you finish all the spinach on your dinner plate," may be expressed as $p \rightarrow q$.

Now, in the opening example, the father gives his daughter dessert even though she did not finish all the spinach on her dinner plate, that is, $p = 0$ (false) and $q = 1$ (true); thus, by definition, $p \rightarrow q$ has value 1 (true or not false).

p	q	$p \rightarrow q$
0	1	1

Therefore, we conclude that the father did *not* violate the mother's condition for dessert.   □

Example 16.2 A mother tells her daughter, "You may have dessert *if and only if* you finish all the spinach on your dinner plate." A little while later, the father gives his daughter dessert even though she did not finish all the spinach on her dinner plate. Did the father violate the mother's condition for dessert?

Let

$$p = \text{"you may have dessert,"}$$
$$q = \text{"you finish all the spinach on your dinner plate."}$$

Then, the statement, "You may have dessert if and only if you finish all the spinach on your dinner plate," may be expressed as $p \leftrightarrow q$.

Now, the father gives his daughter dessert even though she did not finish all the spinach on her dinner plate, that is, $p = 0$ (false) and $q = 1$ (true); thus, by definition, $p \leftrightarrow q$ has value 0 (false or not true).

p	q	$p \leftrightarrow q$
0	1	0

Therefore, we conclude that the father *did* violate the mother's condition for dessert in this instance. □

Two special cases are when a statement has the value 1 only, and when a statement has the value 0 only.

Definition 16.2 A *tautology* is an expression that has only the value 1 (is always true). A *contradiction* is an expression that has only the value 0 (is always false).   □

Example 16.3 The simplest tautology is $p \vee (\sim p)$ (p is true or p is not true—of course), and the simplest contradiction is $p \wedge (\sim p)$ (p is true and p is not true—violates the law of contradiction).

p	$\sim p$	$p \vee (\sim p)$
1	0	1
0	1	1

p	$\sim p$	$p \wedge (\sim p)$
1	0	0
0	1	0

□

Example 16.4 Determine if each of the following is a tautology, a contradiction, or neither.

1. $(p \to q) \leftrightarrow (q \to p)$

We use a truth table. First, fill in the table with the different ways that p and q can be assigned values of 1 or 0.

p	q	$p \to q$	$q \to p$	$(p \to q) \leftrightarrow (q \to p)$
1	1			
1	0			
0	1			
0	0			

Second, fill in the rest of the table using the definitions of $\to$ and $\leftrightarrow$.

p	q	$p \to q$	$q \to p$	$(p \to q) \leftrightarrow (q \to p)$
1	1	1	1	1
1	0	0	1	0
0	1	1	0	0
0	0	1	1	1

Therefore, because the expression has value 1 in some instances and it has value 0 in other instances, $(p \to q) \leftrightarrow (q \to p)$ is neither a tautology nor a contradiction.

2. $(p \to q) \land (p \land (\sim q))$

We use a truth table. As in the last example, fill in the table with the different ways that p and q can be assigned values of 1 or 0, then fill in the rest of the table using the definitions of $\sim$, $\to$, and $\land$.

p	q	$\sim q$	$p \to q$	$p \land (\sim q)$	$(p \to q) \land (p \land (\sim q))$
1	1	0	1	0	0
1	0	1	0	1	0
0	1	0	1	0	0
0	0	1	1	0	0

Therefore, because the expression has value 0 in all instances, $(p \to q) \land (p \land (\sim q))$ is a contradiction, meaning it is never true, regardless of the truth or falsity of p or q.

3. $(p \lor (q \land r)) \leftrightarrow ((p \lor q) \land (p \lor r))$

Let $\sigma = (p \lor (q \land r)) \leftrightarrow ((p \lor q) \land (p \lor r))$, let $\sigma_1 = p \lor (q \land r)$, and let $\sigma_2 = (p \lor q) \land (p \lor r)$, so that $\sigma = \sigma_1 \leftrightarrow \sigma_2$.

Now fill in a truth table with the different ways that p, q, and r can be assigned values of 1 or 0, then fill in the rest of the table using the definitions of $\vee$, $\wedge$, and $\leftrightarrow$. Your final columns should look like the following:

p	q	r	σ_1	σ_2	σ
1	1	1	1	1	1
1	1	0	1	1	1
1	0	1	1	1	1
1	0	0	1	1	1
0	1	1	1	1	1
0	1	0	0	0	1
0	0	1	0	0	1
0	0	0	0	0	1

Therefore, because the expression has value 1 in all instances, $\sigma = (p \vee (q \wedge r)) \leftrightarrow ((p \vee q) \wedge (p \vee r))$ is a tautology, meaning it is always true, regardless of the truth or falsity of p, q, or r.

$\square$

Now You Try 16.2 Use truth tables to show that the following are tautologies.

1. Distributive laws

   a) $(p \wedge (q \vee r)) \leftrightarrow ((p \wedge q) \vee (p \wedge r))$ b) $(p \vee (q \wedge r)) \leftrightarrow ((p \vee q) \wedge (p \vee r))$

2. Negation

   a) $(\sim(\sim p)) \leftrightarrow p$
   b) $(\sim(p \rightarrow q)) \leftrightarrow (p \wedge (\sim q))$
   c) $(\sim(p \leftrightarrow q)) \leftrightarrow ((p \wedge (\sim q)) \vee ((\sim p) \wedge q))$
   d) $(\sim(p \wedge q)) \leftrightarrow ((\sim p) \vee (\sim q))$
   e) $(\sim(p \vee q)) \leftrightarrow ((\sim p) \wedge (\sim q))$

   (The last two are called *De Morgan's laws*.)

3. Contraposition: $(p \rightarrow q) \leftrightarrow ((\sim q) \rightarrow (\sim p))$

4. Biconditional: $(p \leftrightarrow q) \leftrightarrow ((p \rightarrow q) \wedge (q \rightarrow p))$

$\square$

Definition 16.3 Two expressions A and B are said to be *logically equivalent* if $A \leftrightarrow B$ is a tautology. We write $A \equiv B$.

$\square$

Hence, NOW YOU TRY 16.2 lists pairs of statements that are logically equivalent.

Now You Try 16.3

1. Use a truth table to:

a) Show that $((p \to q) \wedge (q \to p)) \equiv ((p \to q) \leftrightarrow (q \to p))$.

b) Show that $(p \wedge q) \not\equiv (p \leftrightarrow q)$, that is, the statements $p \wedge q$ and $p \leftrightarrow q$ are not logically equivalent.

2. Is $(p \wedge q) \leftrightarrow (p \leftrightarrow q)$ a contradiction?

□

Now, we turn our attention in particular to the statements $p \to q$ and $p \leftrightarrow q$, for many assertions in mathematics are expressed in one of these ways. For example, recall that, when we showed that $\sqrt{2}$ is an irrational number (page 173), we stated, "a^2 is an even number, so that a must be an even number." This is an example of a $p \to q$ statement:

If a^2 is an even number, then a is an even number.

Definition 16.4 The expression $A \to B$ is called a *conditional statement* or a *conditional* for short; A is called the *hypothesis* of the conditional and B is called the *conclusion*. Moreover, to every conditional statement there are three related statements, namely, its *contrapositive*, its *converse*, and its *inverse*.

Conditional	$A \to B$	Converse	$B \to A$
Contrapositive	$(\sim B) \to (\sim A)$	Inverse	$(\sim A) \to (\sim B)$

□

Note that we showed in NOW YOU TRY 16.2 that a conditional statement and its contrapositive are logically equivalent.

Now You Try 16.4

1. Each of $A \to B$, $B \to A$, $(\sim B) \to (\sim A)$, and $(\sim A) \to (\sim B)$ stands as a conditional statement on its own.

   a) For each of the above conditional statements, translate it into English. For example, $A \to B$ can be read, "if A, then B" or "A implies B."

   b) For each of the following conditional statements, write its related contrapositive, converse, and inverse statements.

       i. $B \to A$ ii. $(\sim B) \to (\sim A)$ iii. $(\sim A) \to (\sim B)$

2. Use a truth table to:

   a) Determine if the conditional statement $A \to B$ is logically equivalent to its converse, $B \to A$.

   b) Determine if the conditional statement $A \to B$ is logically equivalent to its contrapositive, $(\sim B) \to (\sim A)$.

   c) Determine if the conditional statement $A \to B$ is logically equivalent to its inverse, $(\sim A) \to (\sim B)$.

3. Are the converse and the inverse of a given conditional statement logically equivalent?

□

It turns out that we may use set relations to understand logical statements, and consequently we may use Venn diagrams to visualize logical statements. For example, let A and B be sets, and let p and q be the expressions

$$p = \text{``}x \in A\text{''} \quad \text{and} \quad q = \text{``}x \in B.\text{''}$$

Then $A \subseteq B$ means precisely that $p \to q$. Therefore, we use the same following Venn diagram

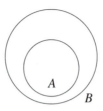

to represent either $A \subseteq B$ if A and B are sets, or $A \to B$ if A and B are expressions. In this way, we may use Venn diagrams to illustrate the fact that a conditional statement and its contrapositive are logically equivalent, and its converse and its inverse are logically equivalent; however, a conditional statement is neither logically equivalent to its converse nor to its inverse.

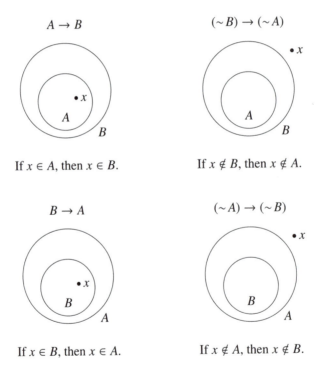

Example 16.5 Consider the conditional statement,

"If it is raining, then the road is wet."

State the hypothesis and the conclusion. Write its contrapositive, converse, and inverse. Write the negation of the conditional.

Let

$$R = \text{``it is raining''} \quad \text{and} \quad W = \text{``the road is wet.''}$$

Then, the hypothesis is R, "it is raining," and the conclusion is W, "the road is wet."

Conditional	$R \to W$:	If it is raining, then the road is wet.
Contrapositive	$(\sim W) \to (\sim R)$:	If the road is not wet, then it is not raining.
Converse	$W \to R$:	If the road is wet, then it is raining.
Inverse	$(\sim R) \to (\sim W)$:	If it is not raining, then the road is not wet.

Note that the conditional, "If it is raining, then the road is wet," is logically equivalent to its contrapositive, "If the road is not wet, then it is not raining," that is, both statements make the same assertion. Also, its converse, "If the road is wet, then it is raining," is logically equivalent to its inverse, "If it is not raining, then the road is not wet."

For how to negate a conditional, we refer to NOW YOU TRY 16.2 on page 188, where we see that

$$(\sim (p \to q)) \leftrightarrow (p \wedge (\sim q))$$

is a tautology, in other words,

$$(\sim (p \to q)) \equiv (p \wedge (\sim q)).$$

Here, this means that

$$\sim (R \to W) \quad \text{is logically equivalent to} \quad R \wedge (\sim W),$$

and this is precisely how we shall form the negation of the conditional, "If it is raining, then the road is wet," to wit,

It is raining and the road is not wet.

$\square$

Remark 16.2 The conditional statement, "If it is raining, then the road is wet," in the preceding example—as well as its contrapositive, converse, inverse, and negation—is not meant to reflect reality. It is possible, of course, for it to be raining and the road not to be wet (for example, if the road is covered); or for the road to be wet when it is not raining (for example, if a fire hydrant were spewing water onto the road). $\square$

Example 16.6 Consider the statement, "All Marines have courage." Write the statement as a conditional, then state the hypothesis and the conclusion. Write its contrapositive, converse, and inverse. Write the negation of the statement.

Let

$$M = \{\text{persons who are Marines}\},$$
$$C = \{\text{persons who have courage}\},$$

and consider the following two Venn diagrams.

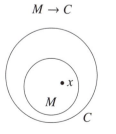

$M \to C$

If $x \in M$, then $x \in C$:
"All Marines have courage."

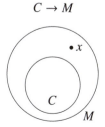

$C \to M$

$x \in M$, but $x \notin C$:
"All who have courage are Marines."

Thus, we see that that statement, "All Marines have courage," can be written as the conditional,

If a person is a Marine, then the person has courage.

The hypothesis is, "a person is a Marine," and the conclusion is, "the person has courage."

Conditional	$M \rightarrow C$:	If a person is a Marine, then the person has courage.
Contrapositive	$(\sim C) \rightarrow (\sim M)$:	If a person does not have courage, then the person is not a Marine.
Converse	$C \rightarrow M$:	If a person has courage, then the person is a Marine.
Inverse	$(\sim M) \rightarrow (\sim C)$:	If a person is not a Marine, then the person does not have courage.

Observe that the conditional and its contrapositive are both illustrated by the Venn diagram above left, and that its converse and inverse are both illustrated by the Venn diagram above right.

To negate the statement written as a conditional, we recall that

$$(\sim(p \rightarrow q)) \equiv (p \wedge (\sim q)),$$

so that

$\sim(M \rightarrow C)$ is logically equivalent to $M \wedge (\sim C)$.

Thus, the negation of the statement, "If a person is a Marine, then the person has courage," is the statement

A person is a Marine and the person does not have courage

or, more fluidly,

There is a Marine who does not have courage

(which is surely a false statement).   □

Remark 16.3 We say this only in passing: The statement, "All Marines have courage," is an example of a *universal statement* because of the *universal quantifier* "all"; and the statement, "There is a Marine who does not have courage," is an example of an *existential statement* because of the *existential quantifier* "there is a" (or "there *exists* a"). Note that a universal quantifier is negated by an existential quantifier. Moreover, to negate the statement, "All Marines...," we need to find only "*A* person..."; in other words, we need to find only *one* counterexample.   □

Think About It 16.3 Is the negation (or "opposite") of a conditional statement logically equivalent to the converse of the conditional?   □

Now You Try 16.5 For each of the following, write the statement as a conditional (if it is not already in this form), then state the hypothesis and the conclusion. Write its contrapositive, converse, and inverse. Write the negation of the statement.

1. If a^2 is an even number, then a is an even number.

2. All purple polka-dotted pigs can fly.

□

In general, a conditional and its converse are not logically equivalent; however, in some instances, a conditional and its converse are both true. For example, both the conditional, "If a^2 is an even number, then a is an even number," and its converse, "If a is an even number, then a^2 is an even number," are true. In this case, we may express both the conditional and its converse compactly as "A number a^2 is even *if and only if* a is even."

Definition 16.5 The expression $A \leftrightarrow B$ is called a *biconditional statement* or a *biconditional* for short, and is logically equivalent to $(A \rightarrow B) \wedge (B \rightarrow A)$. In the "forward direction," $A \rightarrow B$, A is the hypothesis and B is the conclusion; in the "backward direction," $B \rightarrow A$, B is the hypothesis and A is the conclusion.
□

Now You Try 16.6

1. Consider the biconditional statement, "A number a^2 is even if and only if a is even." What is the conditional statement in the forward direction, and what is the conditional statement in the backward direction? Identify the hypothesis and the conclusion for the statement in each direction.

2. Write the negation of the statement, "A number a^2 is even if and only if a is even."

□

There is a variety of ways to express conditional and biconditional statements in words, and it is useful to become familiar with some of them [89, pp. 26–27 and 29].

$A \rightarrow B$:

1. If A is true, then B is true.

2. If A holds, then B holds.

3. A implies B.

4. B is implied by A.

5. B follows from A.

6. A is a sufficient condition for B.

7. B is a necessary condition for A.

8. B is true if A is true.

9. A is true only if B is true.

10. If B is false, then A is false.

$A \leftrightarrow B$:

1. A is true if and only if B is true.

2. B is true if and only if A is true.

3. If A, then B, and conversely.

4. If B, then A, and conversely.

5. A implies B, and conversely.

6. B implies A, and conversely.

7. A is a necessary and sufficient condition for B.

8. B is a necessary and sufficient condition for A.

9. A and B are equivalent statements.

Example 16.7 The following are different ways of stating: "If a^2 is an even number, then a is an even number."

- That a^2 is an even number implies that a is an even number.

- That a is an even number follows from a^2 being an even number.

- It is sufficient that a^2 is an even number for a to be an even number.

- It is necessary that a is an even number for a^2 to be an even number.

- The number a^2 is even only if a is even.

- If a is not an even number, then a^2 is not an even number.

□

Now You Try 16.7 Provide at least five different ways of stating each of the following.

1. If m and n are even integers, then mn is an even integer.

2. A number a is divisible by 3 if and only if a^2 is divisible by 3.

□

We finish up with a word about definitions and about proofs. First, about definitions.

When a term is defined by a condition, the definition is always a biconditional, even if it may not be written that way. For example, consider the definition of a subset (DEFINITION 14.3) on page 148:

A set A is a *subset* of a set B if every element of A is an element of B.

If we let

$$p = \text{"every element of } A \text{ is an element of } B\text{,"}$$
$$q = \text{"}A \text{ is a subset of } B\text{,"}$$

then it appears that the definition of a subset may be expressed as

$$p \to q.$$

This, however, is just laziness on our part. The definition is actually, and should more properly be written,

A set A is a *subset* of a set B if *and only if* every element of A is an element of B,

that is to say,

$$p \leftrightarrow q.$$

It is common to state a definition as a conditional *with the understanding* that it is actually a biconditional.

Next, about proofs. As we mentioned earlier, many assertions in mathematics are stated as conditionals or biconditionals. When reading or writing proofs for such statements, it is useful to keep in mind some common constructions. Here are examples of how we may prove $A \to B$.

$A \rightarrow B$	
Direct proof	Assume A is true and deduce that B is true.
Indirect proof	Assume B is false and deduce that A is false.
	Assume A is true and B is false, and deduce a contradiction.

To prove that $A \leftrightarrow B$, we must prove *both*

$$A \rightarrow B \quad \text{and} \quad B \rightarrow A,$$

each of which may be proved in any of the ways described above, even perhaps directly in one direction and indirectly in the other.

Sometimes several statements are logically equivalent to one another. If the statements A, B, C, D, and E are conjectured all to be equivalent to one another (that is, they are all different ways of making the same assertion), then it is common to prove that they are in fact all equivalent to one another by showing that

$$A \rightarrow B \rightarrow C \rightarrow D \rightarrow E \rightarrow A.$$

Think About It 16.4 Why are the indirect proofs of $A \rightarrow B$ described above valid? Why is

$$A \rightarrow B \rightarrow C \rightarrow A \qquad \text{logically equivalent to}$$

□

Chapter 17

The Higher Arithmetic

In part I we explored various number systems and bases. In particular, we looked at the number systems that eventually matured into the decimal place-value system in use today, and witnessed how beneficial a place-value system is; and we will see in part IV how powerful that system is when combined with a symbolic algebraic system. In fact, our symbolic place-value system was essential to the development of such areas of mathematics as the calculus that have allowed us to build everything from modern skyscrapers to the Internet. In this part we will look at an area of mathematics that finds it roots in ancient times, but did not find any practical applications until modern times, namely, the area of mathematics that concerns numbers themselves that we today call *number theory*. As with many areas of mathematics, number theory remains largely unnoticed by the general public, yet its results touch our everyday lives.

There are two ways to study (natural) numbers: computation techniques and number relations. The ancient Greeks called the study of computation *logisti* or *logistica* and they called the study of number relations *arithmetic* or *arithmetica*. Today, we tend to refer to the study of computation techniques as arithmetic and to the study of number relations as the *higher* arithmetic or number theory, which can be broken down broadly into elementary number theory, algebraic number theory, analytic number theory, and probabilistic number theory.

The term "number theory" is modern, and usually refers to the modern field of study that explores the properties of the natural numbers, in particular the prime numbers. Research in number theory used to be considered completely pure, in that it had no real-world applications, until the advent of modern computers and the need to encrypt information.

Modern number theory aside, the broader idea of number theory goes back to the ancient Greeks, specifically to Pythagoras (ca. 585–500 BC). The history of Pythagoras and the Pythagorean brotherhood that he founded is shrouded in mystery and legend. The mysticism of the Pythagoreans manifests itself in many discoveries about the natural numbers, for example, figurate numbers and friendly or amicable numbers (SECTION 17.3 and SUBSECTION 17.6.4). And perhaps the most famous legend about the Pythagoreans revolves around the discovery that the square root of 2 is not a rational number (see page 171).

Another important modern offshoot of ancient Greek number theory is the theory of equations, such as equations of the form $nx^2 + 1 = y^2$ that are today called *Pell's equation*. The Greek mathematician Diophantus of Alexandria (fl. AD 250) presented his methods for solving certain types of problems in his seminal work, *Arithmetica* (see page 203). The *Arithmetica* is important because Diophantus uses for the first time what is called *syncopated* notation (a mixture of abbreviations, symbols, and words) to present relations between numbers. This is usually seen as the start of algebra. (Prior to Diophantus, and even for centuries after him, mathematics was completely *rhetorical*—presented only in words. Our modern algebraic notation is called *symbolic* and is comparatively recent.) Thus, Diophantus is often referred to as the father of algebra.

In this chapter we introduce some concepts of elementary number theory. By "elementary" we do not mean "simple." Instead, "elementary" here refers to the general methods employed in the subject. Indeed, some of the best mathematicians in history have worked on various problems in elementary number theory, and sometimes without success. Davenport gives us this description of the subject [45, Introduction]:

> The higher arithmetic, or the theory of numbers, is concerned with the properties of the natural numbers 1, 2, 3, ... These numbers must have exercised human curiosity from a very early period; and in all the records of ancient civilization there is evidence of some preoccupation with arithmetic over and above the needs of everyday life....
>
> A peculiarity of the higher arithmetic is the great difficulty which has often been experienced in proving simple general theorems which had been suggested quite naturally by numerical evidence. 'It is just this,' said Gauss, 'which gives the higher arithmetic that magical charm which has made it the favourite science of the greatest mathematicians, not to mention its inexhaustible wealth, wherein it so greatly surpasses other parts of mathematics.'

It is not uncommon to find that just to understand an assertion in a particular subject of mathematics—for example, in topology or in differential equations—one needs first to have a considerable background in that subject. What has attracted many people, amateur and professional mathematicians alike, to elementary number theory is that many of its assertions are easy for anyone to understand without much background in the subject, yet turn out to be extremely challenging to prove or disprove. Here are a few examples, ranging from the easily resolved to the still unresolved, and from ancient Greece to the nineteenth century [45].

- The series of primes never comes to an end. (Euclid's *Elements*, IX.20.)

- Every natural number is representable as the sum of four squares of integers. (Girard and Fermat; Lagrange.)

- The equation $x^n + y^n = z^n$ has no solution in natural numbers x, y, and z if n is an integer greater than 2. (Fermat's last theorem.)

- Every even number from 6 onward is representable as the sum of two primes other than 2. (Goldbach's Conjecture.)

Now You Try 17.1

1. *Every natural number is representable as the sum of four squares of integers.* For example,
$$14 = 3^2 + 2^2 + 1^2 + 0^2 \quad \text{and} \quad 37 = 5^2 + 2^2 + 2^2 + 2^2.$$

   Express the numbers 70 and 666 as the sum of four squares of integers. Make up two more examples.

2. *Every even number from 6 onward is representable as the sum of two primes other than 2.* For example,
$$20 = 3 + 17.$$

   (Also, 20 = 7 + 13, so we see that the sum may not be unique.) Express the numbers 6 and 50 as the sum of two primes other than 2. Make up two more examples.

□

Remark 17.1 In this chapter, "number" shall generally mean *natural number* unless otherwise stated. □

17.1   EARLY GREEK ELEMENTARY NUMBER THEORY

Four persons stand out in the history of early Greek number theory: Pythagoras (ca. 585–500 BC), Euclid (ca. 323–285 BC), Nicomachus of Gerasa (fl. ca. AD 100), and Diophantus of Alexandria (fl. AD 250). We introduce each of these persons here, drawing chiefly from Heath [68, 69] and Calinger [35].

17.1.1 Pythagoras

It is generally accepted that the first in the Greek tradition to study number relations was Pythagoras and his followers. (See FIGURE 17.1.) Pythagoras was the founder of the Pythagorean school, one of the two great Greek schools in southern Italy at the time; the other was the Eleatic that was founded by Parmenides (ca. 515–450 BC) [35, pp. 68ff.]. Pythagoras is perhaps best known today for the result that bears his name, *Pythagoras's theorem* or the *Pythagorean theorem*:

> The square on the hypotenuse of a right triangle is equal to the sum of the squares on the other two sides.

Anyone who has taken high school algebra or geometry will almost surely remember having encountered Pythagoras's theorem, perhaps as the phrase "*a* squared plus *b* squared equals *c* squared" ($a^2 + b^2 = c^2$).

Think About It 17.1 Why is Pythagoras's theorem stated as *the square* on *the hypotenuse* and *the squares* on *the other two sides*, rather than stated as *the square* of *the hypotenuse* and *the squares* of *the other two sides*? ☐

Figure 17.1: An artist's interpretation of Pythagoras. (Source: © Thinkstock.)

We have no firsthand records of Pythagoras or his immediate successors. This is due to the Pythagoreans' following an oral tradition and to an oath of secrecy among its initiates.[1] What we know, then, comes from later Pythagoreans, the Roman compiler Plutarch (AD 46–120), and commentators, such as Proclus (AD 410–485). We also find as sources for Pythagorean mathematics such persons as Aristotle (384–322 BC), Euclid, and Nicomachus.

Pythagoras was born on the Ionian island of Samos.[2] He journeyed to Egypt and Babylon, where he was probably greatly influenced by those societies. After returning to his home island of Samos around 532 BC, he was exiled and moved west toward Magna Graecia to settle in Croton, now Crotona in southern Italy, where he founded his school of Pythagoreans. The school was a religious society that distinguished itself from the many other mystical cults of the time by promoting the idea that one advances spiritually through an understanding of mathematics. Indeed, the Pythagoreans subscribed to the doctrine that "all things *are* numbers." As Heath explains it [68, pp. 67–69]:

---

[1]Heath [68] disputes the latter, saying: "Nor is the absence of any written record of Pythagorean doctrines down to the time of Philolaus [a later Pythagorean; fl. 430 BC] to be attributed to a pledge of secrecy binding the school; at all events, it did not apply to their mathematics or their physics; the supposed secrecy may even have been invented to explain the absence of documents. The fact appears to be that oral communication was the tradition of the school, while their doctrine would in the main be too abstruse to be understood by the generality of people outside."

[2]Given that all material on Pythagoras was written centuries after his death, there are no known hard and fast facts about his life. What we present here are some of the most prevalent stories of his life, but be warned that there is much written about him that has no historical basis.

Now any one who was in the habit of intently studying the heavens would naturally observe that each constellation has two characteristics, the number of the stars which compose it and the geometrical figure which they form. Here, as a recent writer has remarked,[3] we find, if not the origin, a striking illustration of the Pythagorean doctrine.

[...]

True, Aristotle seems to regard the theory as originally based on the analogy between the properties of numbers.

'They thought they found in numbers, more than in fire, earth, or water, many resemblances to things which are and become; thus such and such an attribute of numbers is justice, another is soul and mind, another is opportunity, and so on; and again they saw in numbers the attributes and ratios of the musical scales. Since, then, all other things seemed in their whole nature to be assimilated to numbers, while numbers seemed to be the first things in the whole of nature, they supposed the elements of numbers to be the elements of all things, and the whole heaven to be a musical scale and a number.'[4]

[...]

May we not infer from these scattered remarks of Aristotle about the Pythagorean doctrine that 'the number in the heaven' is the number of the visible stars, made up of units which are material points? And may this not be the origin of the theory that all things are numbers, a theory which of course would be confirmed when the further capital discovery was made that musical harmonies depend on numerical ratios, the octave representing the ratio 2 : 1 in length of string, the fifth 3 : 2 and the fourth 4 : 3?

Pythagoreans studied arithmetic (number theory), geometry, music, and astronomy, the four subjects that, in the Middle Ages, were called the *quadrivium*. (The quadrivium together with the *trivium*—grammar, logic, and rhetoric—make up the traditional liberal arts education [62, 63].) The number theory that Pythagoreans studied was shrouded in numerology. For instance, the number one stood for reason, two stood for man, three for woman, four for justice, five for marriage, and so on. The numbers one, two, three, and four also represented, respectively, the elements of earth, air, fire, and water. And, because $1 + 2 + 3 + 4 = 10$, the number ten was a particularly important number to Pythagoreans, which they called the *perfect number*. Indeed, the number ten was the essence of *tetractys*, one of two secret symbols of the Pythagoreans. The other secret symbol was the regular *pentagram* or five-pointed star. Pythagoreans would swear oaths by the tetractys, and they would recognize other members by the pentagram.

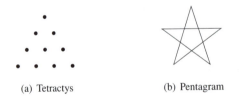

(a) Tetractys   (b) Pentagram

Figure 17.2: Two secret symbols of the Pythagoreans.

At its height, the school may have had as many as 300 members made up of young aristocrats. The Pythagoreans were disbanded in the fifth century BC, and their meeting houses destroyed, when the democratic revolution swept through Magna Graecia.

17.1.2 Euclid

Not much is known about Euclid's life: where he was born, when he was born, when he died, and so on. The approximate dates for Euclid, ca. 323–285 BC, are derived from Proclus's statement in his "Summary"[5] [68, p. 354]:

---

[3]L. Brunschvicg, *Les étapes de la philosophie mathématique*, 1912, p. 33.

[4]*Metaph.* A. 5, 985 b 27–986 a 2.

[5]The so-called "Summary" of Proclus refers to a few pages of his *Commentary on Euclid*, Book I, in which he gives an outline account of Greek geometry from the earliest times until Euclid, "with special reference to the evolution of the Elements" [68, p. 118].

'...This man [Euclid] lived in the time of the first Ptolemy [Ptolemy I Soter (ca. 367–283 BC)]. For Archimedes, who came immediately after the first (Ptolemy), makes mention of Euclid; and further they say that Ptolemy once asked him if there was in geometry any shorter way than that of the Elements, and he replied that there was no royal road to geometry. He is then younger than the pupils of Plato, but older than Eratosthenes and Archimedes, the latter having been contemporaries, as Eratosthenes somewhere says.'[6]

That places Euclid between Plato (427–347 BC) and Archimedes (287–212 BC). Moreover, because of the reference to Ptolemy I, it is believed that Euclid flourished around 300 BC, during which time he arrived in Alexandria and founded a school of mathematics at the Museum,[7] perhaps at the invitation of the governor Demetrius of Phalerum.

Euclid was interested in a variety of subjects, as evidenced by his books, some of which have been lost entirely, and some of which have survived only in translations: the *Phaenomena* on spherical geometry applied to astronomy; *Optica* on perspective; *Catoptrica* on mirrors; the *Elements of Music* on the theory of musical intervals; the *Pseudaria*, the *Porisms*, the *Conics*, and the *Surface Loci*, which are lost geometrical works; *On Divisions of Figures* on conic sections; and the *Data*, which is closely related to the first six books of Euclid's most influential work, the *Elements*.

The *Elements* [53] is a compilation of all the basic mathematics up to Euclid's time. Euclid wrote the *Elements* in thirteen books (each book being what we would consider to be a chapter today). What makes the *Elements* such an important contribution to mathematics is that it lays out mathematics in an axiomatic and deductive fashion that became the standard by which mathematical rigor was judged until the eighteenth century. The *Elements* has been used by countless generations as an introduction to both mathematics and deductive reasoning: Archimedes referred to the *Elements* as the standard textbook of basic mathematics, and Abraham Lincoln (1809–1865), at the age of forty, mastered the first six books to make himself a more exact reasoner.

It is a common misconception that the *Elements* is a work only on plane geometry, that is, on what we may consider to be high school geometry today. This may be because the Greeks made geometrical arguments in all their mathematics. In fact, the *Elements* covers not only plane geometry but also geometric algebra[8] and proportions, number theory (arithmetic), and solid geometry. On the other hand, the *Elements* omits such topics as "commercial mathematics," conic sections, and spherical geometry that were also known in that day. Calinger gives a brief description of the thirteen books of the *Elements* [35, pp. 133–134]:

Plane Geometry

    I. Preliminary principles (definitions, postulates, and axioms) along with congruence theorems and geometry of straight lines and rectilinear figures

    II. Transformation of areas

    III. Major propositions about circles

    IV. Construction of regular polygons of three, four, five, six, and fifteen sides

    V. Eudoxus' theory of proportions applied to commensurable and incommensurable magnitudes

    VI. Application of this general theory of proportions to similar figures

Theory of Numbers

    VII. Pythagorean theory of numbers

---

[6]Proclus on Eucl. I, p. 68. 6–20.

[7]There is an interesting anecdote regarding Euclid as a teacher. As Calinger tells it [35, p. 132]:

> In *Eclogues* II.31 Stobaeus relates that, after having understood only the first theorem, a beginning student asked [Euclid] what was to be got out of learning such things. Calling a slave, Euclid said, "Give [the student] three obols, since he must need make gain out of what he learns."

[8]As an example, Proposition II-4 states [53, p. 39], "If a straight line be cut at random, the square on the whole is equal to the squares on the segments and twice the rectangle contained by the segments," expresses the algebraic identity $(a+b)^2 = a^2 + b^2 + 2ab$.

VIII. Series of numbers in continued proportions

IX. Miscellany on the theory of numbers, including products and primes

Plane Geometry

X. The classification of certain incommensurable magnitudes (irrationals)

Solid Geometry

XI. The geometry of three dimensions, particularly parallelepipeds

XII. Areas and volumes found by the method of exhaustion

XIII. Inscription of the five regular solids in a sphere

Calinger goes on to say that,

> In later antiquity it was a frequent practice to attribute to famous authors books that they had not written. Older editions of the *Elements* added two books with further propositions on regular solids.
>
> [...]
>
> Seeking to avoid circular arguments and provide sound starting points for mathematical reasoning, Euclid begins the *Elements* with a set of three types of first principles that correspond to those of Aristotle and are concordant with Eudoxus' approach. His first principles consist of twenty-three explicitly stated (although sometimes vague) definitions together with five postulates and five common notions, or what Proclus would later call axioms. These definitions, postulates, and axioms are the foundational rules of the game. Building on them, Euclid derives almost exclusively by deductive reasoning an orderly progression of 465 propositions. His demonstrations of these propositions employ high standards of consistency, rejecting any deduction of contradictory theorems from the axioms and postulates. Proofs of each proposition conclude with the Greek formula for "which was to be proved," known to students by its Latin abbreviation Q.E.D. (*quod erat demonstrandum*); proofs for problems conclude with "which was to be done," the familiar Q.E.F. (*quod erat faciendum*).

Theon of Alexandria (fl. AD 375) was one of the more important commentators of Greek mathematics, and, until the nineteenth century, all but one of the surviving editions of the *Elements* were based on Theon's redaction, which may have been a written version of his lectures. There have been numerous translations, editions, and commentaries of the *Elements*, with over a thousand editions having appeared since its first printing in 1482 in Venice by Erhard Ratdolt. It has had more editions than any other book, except for the Bible.

17.1.3   *Nicomachus and Diophantus*

The expansion of the Roman empire saw an end to the "Golden Age of Greek Mathematics" that was dominated by Euclid, Archimedes, and Apollonius of Perga (262–190 BC) and centered about the Museum of Alexandria. As we mentioned in SECTION 3.2, to the Romans mathematics was a tool to an end, and spending time on theory was folly and even unpatriotic because it did not apparently contribute to the common good. Nevertheless, despite its decline, Greek "theoretical mathematics" was not lost altogether. One person who helped to preserve it to an extent was Nicomachus of Gerasa, a Neo-Pythagorean who flourished around AD 100.

Nicomachus is remembered for his *Introductio Arithmeticae* that was written in two books. Boyer [15, pp. 179–180] tells us that the *Introductio* "served as a model for later imitators and commentators," such as Theon of Smyrna (fl. ca. AD 125) and Beothius (d. AD 524), despite the fact that the *Introductio* contains little that is original (being chiefly a compilation of the work of early Pythagoreans) and, in fact, "makes serious mistakes in certain problems and has shortcomings in its method of exposition" [35, p. 207]; nevertheless, it was used as the standard textbook in arithmetic in Europe until the late sixteenth century. Indeed, Calinger [35, p. 208] laments that "Nicomachus even gained a reputation for being a leading mathematician. That indicates the limitations of arithmetical studies in the Latin West until the late Renaissance."

The state of Greek mathematics turned around during the so-called Silver Age of Greek mathematics, during which time there was a brief resurgence of Greek mathematical thought with advances by Diophantus of Alexandria and Pappus of Alexandria (early fourth century AD).

Calinger [35, p. 209] tells us that "almost nothing is known [about Diophantus's life] aside from his residence for some time in Alexandria. No biographical details have survived from personal records or comments by contemporaries. Presumably, Diophantus was a Greek or perhaps a Hellenized Babylonian." On the other hand, Heath [69, p. 448] asserts: "The date of DIOPHANTUS can now be fixed with fair certainty. He was later than Hypsicles, from whom he quotes a definition of a polygonal number, and earlier than Theon of Alexandria, who has a quotation from Diophantus's definitions. The possible limits of date are therefore, say, 150 B.C. to A.D. 350." There is a general agreement, however, that Diophantus flourished around AD 250 because a letter of Michael Psellus mentions that Anatolius, who became Bishop of Laodicea in the latter part of the third century AD, had dedicated a treatise on Egyptian computation to Diophantus. We do find a hint about Diophantus's age when he died in an epigram in the *Greek Anthology* [69, p. 448]:

> [H]is boyhood lasted $\frac{1}{6}$th of his life; his beard grew after $\frac{1}{12}$th more; he married after $\frac{1}{7}$th more, and his son was born 5 years later; the son lived to half his father's age, and the father died 4 years after his son.

Think About It 17.2 How old was Diophantus when he died? □

Diophantus is best known for his works the *Arithmetica*, written in thirteen books, but of which only the first six have survived, and *On Polygonal Numbers*; furthermore, Heath [69, p. 449] relates that the *Arithmetica* alludes to a third work, *Porisms*, for "in three propositions (3, 5, 16) of Book V [of the *Arithmetica*], Diophantus quotes as known certain propositions in the Theory of Numbers, prefixing to the statement of them the words 'We have it in the *Porisms* that....'"

Hypatia (d. AD 415), the daughter of Theon of Alexandria, is the first known commentator of the *Arithmetica*. It has been suggested that Hypatia commented on the first six books of the *Arithmetica*, but not on the remaining seven, and that may be a reason why only the first six books have survived. Even so, the *Arithmetica* has had a profound impact on modern mathematics. Countless mathematicians, over the years, have studied Diophantus's work, including Pierre de Fermat (1601–1665), the father of modern number theory. Indeed, it was in a copy of Bachet's edition of the *Arithmetica* that Fermat teased the world with what has come to be known as "Fermat's last theorem" (see page 339).

One of the distinguishing features of the *Arithmetica* is that it does not rely on the traditional Greek geometrical arguments. In fact, Diophantus may be the first person to have used symbols (instead of line segments, planes, and so on) to represent unknown quantities in his work (see page 340). Book I is chiefly on problems that lead to linear and determinate equations in two or more unknowns, and the remaining five books that have survived are chiefly on problems that lead to quadratic and indeterminate equations. For these problems, Diophantus sought solutions that are positive integers or common fractions. These sort of problems are now referred to as *Diophantine problems* or *Diophantine equations* and are included in any textbook on elementary number theory.

17.2 EVEN AND ODD NUMBERS

To Pythagoreans, what we call the number one was not a (natural) number at all, but rather a "generator" of numbers. Indeed, Euclid says as much in the *Elements*, Book VII, Definitions [53, p. 157]:[9]

1. An *unit* is that by virtue of which each of the things that exist is called one.

---

[9]A footnote in the translation of the *Elements* explains [53, p. 1]: "Euclid's definitions, postulates, and common notions—if Euclid is indeed the author—were not numbered, separated, or italicized until translators began to introduce that practice. The Greek text, however, as far back as the 1533 first printed edition, presented the definitions in a running narrative, more as a preface discussing how the terms would be used than as an axiomatic foundation for the propositions to come. We follow Heath's formatting here. —Ed."

2. A *number* is a multitude composed of units.

So, the unit is *named* "one," and is not itself a number, but all numbers are composed of units; for example, three is composed of 3 units, and twenty is composed of 20 units.

Now, evenness and oddness is the most elementary way of classifying numbers. We find definitions of *even* and *odd* in Euclid VII.Definitions:

6. An *even number* is that which is divisible into two equal parts.

7. An *odd number* is that which is not divisible into two equal parts, or that which differs by an unit from an even number.

Example 17.1 We apply the definitions of even number and odd number given in Euclid VII. Definitions to determine if the numbers 6 and 7 are even or odd.

Represent the number 6 by a collection of dots, $\{• • • • • •\}$. Then, we see that 6 can be divided into two parts as follows:

$$\{(•) • • • • •)\}, \quad \{(• •) (• • • •)\}, \quad \{(• • •) (• • •)\}.$$

Since 6 can be divided into two equal parts, $\{(• • •) (• • •)\}$, we conclude that 6 is an even number.

Next, consider the number 7, $\{• • • • • • •\}$. Then, we see that 7 can be divided into two parts as follows:

$$\{(•) (• • • • • •)\}, \quad \{(• •) (• • • • •)\}, \quad \{(• • •) (• • • •)\}.$$

Since 7 cannot be divided into two equal parts, we conclude that 7 is an odd number.

Note that we may also conclude that 7 is an odd number from the fact that "it differs by an unit from an even number":

$$\{• • • • • • \cancel{•} \} \quad \rightarrow \quad \{• • • • • •\} = \{(• • •) (• • •)\}$$
$$\text{delete one}$$

or

$$\{• • • • • • • ○\} \quad \rightarrow \quad \{• • • • • • • •\} = \{(• • • •) (• • • •)\}.$$
$$\text{add one}$$

□

Heath relates the following alternative definition of even and odd as given by Nicomachus [68, p. 70]:

'an *even* number is that which can be divided both into two equal parts and into two unequal parts (except the fundamental dyad which can only be divided into two equal parts), but, however it is divided, must have its two parts *of the same kind* without part in the other kind (i.e. the two parts are both odd or both even); while an *odd* number is that which, however divided, must in any case fall into two unequal parts, and those parts always belonging to the two *different* kinds respectively (i.e. one being odd and one being even).'[10]

Remark 17.2 Heath remarks that, in Nicomachus's definition, "we have a trace of the original conception of 2 (the dyad) as being, not a number at all, but the principle or beginning of the even, just as one was not a number but the principle or beginning of number; the definition implies that 2 was not originally regarded as an even number, the qualification made by Nicomachus with reference to the dyad being evidently a later addition to the original definition (Plato already speaks of two as even)."[11] □

---

[10]Nicom. i. 7. 4.
[11]Plato, *Parmenides*, 143 D.

We note that Nicomachus's definition of even and odd suggests that an even number is the sum of two even numbers or the sum of two odd numbers, and an odd number is the sum of an even number and an odd number. For example, 6 is an even number and

$$6 = 1 + 5 \quad (odd \text{ plus } odd),$$
$$6 = 2 + 4 \quad (even \text{ plus } even),$$
$$6 = 3 + 3 \quad (odd \text{ plus } odd),$$

whereas 7 is an odd number and

$$7 = 1 + 6 \quad (odd \text{ plus } even),$$
$$7 = 2 + 5 \quad (even \text{ plus } odd),$$
$$7 = 3 + 4 \quad (odd \text{ plus } even).$$

Now You Try 17.2 Use Euclid's or Nicomachus's definition to determine the *parity* (evenness or oddness) of 9 and 12. □

On the other hand, Euclid's definition of even and odd suggests that a number n is an even number if $n = k + k$ or $2k$ ("that which is divisible into two equal parts"), and n is an odd number if $n = (k + k) - 1 = 2k - 1$ or $n = (k + k) + 1 = 2k + 1$ ("that which differs by a unit from an even number"). For example, 6 is an even number and

$$6 = 3 + 3 = 2 \times 3,$$

whereas 7 is an odd number and

$$7 = (4 + 4) - 1 = 2 \times 4 - 1 \quad \text{or} \quad 7 = (3 + 3) + 1 = 2 \times 3 + 1.$$

In fact, this is how we define even and odd today.

Definition 17.1 A natural number n is said to be *even* if $n = 2k$ for some natural number k; otherwise, n is said to be *odd*. □

Now You Try 17.3 Twelve is an even number and 9 is an odd number. Find k, l, and m such that

$$12 = 2k, \quad \text{and}$$
$$9 = 2l - 1 \quad \text{or} \quad 9 = 2m + 1.$$

□

Think About It 17.3

1. Would the following be good definitions of even or odd? Explain.

   a) A natural number n is said to be *even* if $n = 4k$ for some natural number k.

   b) A natural number n is said to be *odd* if $n = 2k - 1$ for some natural number k.

   c) A natural number n is said to be *odd* if $n = 2k + 1$ for some natural number k.

2. Is it generally true that a number is even if and only if it is the sum of two even numbers or the sum of two odd numbers? Is it generally true that a number is odd if and only if it is the sum of an even number and an odd number?

□

More generally, we define even and odd for *integers*, instead of only for natural numbers or whole numbers.

Definition 17.2 An integer n is said to be *even* if $n = 2k$ for some integer k; otherwise, n is said to be *odd*.
□

Think About It 17.4 What does this definition say about the number 0? □

Now, if we assume that the set of integers is closed under addition and multiplication—that is, for any integers a and b, both $a + b$ and ab are also integers—and also assume that the associative, commutative, and distributive laws for addition and multiplication of integers hold—that is,

> 1. $(a + b) + c = a + (b + c)$ and $(ab)c = a(bc)$ (associative law)
> 2. $a + b = b + a$ and $ab = ba$ (commutative law)
> 3. $a(b + c) = ab + ac$ (distributive law)

—then we can easily show that generally the sum of two odd integers is even, and the product of two odd integers is odd, for example.

Example 17.2

1. The sum of two odd integers is even.

   Let m and n be two odd integers, so that $m = 2j - 1$ and $n = 2k - 1$ for some integers j and k. (Why?) Then

   $$
   \begin{aligned}
   m + n &= (2j - 1) + (2k - 1) \\
   &= 2j + (-1 + 2k) - 1 && \text{(associative law)} \\
   &= 2j + (2k - 1) - 1 && \text{(commutative law)} \\
   &= (2j + 2k) + (-1 - 1) && \text{(associative law)} \\
   &= 2j + 2k - 2 && \text{(simplify)} \\
   &= 2(j + k - 1) && \text{(distributive law)} \\
   &= 2l && (j + k - 1 = l)
   \end{aligned}
   $$

   Since l is another integer, because the set of integers is closed under addition, we see that $m + n = 2l$ for some integer l. Therefore, by definition, $m + n$ is even.

   Here are three numerical examples.

   a) $3 + 7 = 10$ b) $3 + (-7) = -4$ c) $-3 + (-7) = -10$

2. The product of two odd integers is odd.

   Let m and n be two odd integers, so that $m = 2j - 1$ and $n = 2k - 1$ for some integers j and k. Then,

   $$
   \begin{aligned}
   mn &= (2j - 1)(2k - 1) \\
   &= (2j)(2k - 1) + (-1)(2k - 1) && \text{(distributive law)} \\
   &= (2j)(2k) + (2j)(-1) \\
   &\quad + (-1)(2k) + (-1)(-1) && \text{(distributive law)}
   \end{aligned}
   $$

$$= 4jk - 2j - 2k + 1 \qquad \text{(simplify)}$$
$$= 2(2jk - j - k) + 1 \qquad \text{(distributive law)}$$
$$= 2l + 1 \qquad (2jk - j - k = l)$$

Since l is another integer, because the set of integers is closed under addition and multiplication, we see that $mn = 2l + 1$. Therefore, by definition, mn is odd.

Here are three numerical examples.

a) $(3)(7) = 21$ b) $(3)(-7) = -21$ c) $(-3)(-7) = 21$

□

Now You Try 17.4 Show the following in general.

1. The sum of two even integers is an even integer.

2. The product of two even integers is an even integer.

Provide three numerical examples for each of the statements. □

17.3 FIGURATE NUMBERS

The number 10 is an example of a triangular number because we may configure ten dots as a triangle (FIGURE 17.2 on page 200); and a triangular number is an example of a *figurate number* or *figured number*. Other figurate numbers include square numbers (those that, when represented as dots, may be configured as squares), pentagonal numbers (those that may be configured as pentagons), and hexagonal numbers (those that may be configured as hexagons).

According to Heath [68, p. 76], the theory of figurate numbers apparently goes back to Pythagoras himself:

> It seems clear that the oldest Pythagoreans were acquainted with the formulation of triangular and square numbers by means of pebbles or dots[12]; and we judge from the account of Speusippus's book, *On the Pythagorean Numbers*, which was based on works of Philolaus, that the latter dealt with linear numbers, polygonal numbers, and plane and solid numbers of all sorts, as well as with the five regular solid figures.[13] The varieties of plane numbers (triangular, square, oblong, pentagonal, hexagonal, and so on), solid numbers (cube, pyramidal, &c.) are all discussed, with the methods of their formation, by Nicomachus[14] and Theon of Smyrna.[15]

We explore various figurate numbers.

17.3.1 Triangular Numbers

The first four *triangular numbers* are shown in FIGURE 17.3. Observe that, from top to bottom, each row of a triangular number has one more point than the last. Thus, the triangular numbers are sums of consecutive natural numbers.

---

[12]Cf. Arist. *Metaph.* N. 5, 1092 b 12.
[13]*Theol. Ar.* (Ast), p. 61.
[14]Nicom. i. 7–11, 13–16, 17.
[15]Theon of Smyrna, pp. 26–42.

Figure 17.3: The first four triangular numbers.

Let $T(n)$ denote the nth triangular number.[16] Then $T(1)$ is the first triangular number, $T(2)$ is the second triangular number, and so on, so that

$$T(1) = 1, \quad T(2) = 3, \quad T(3) = 6, \quad \text{and} \quad T(4) = 10.$$

Moreover, from the figure above, we see that

$$T(1) = 1,$$
$$T(2) = 1 + 2,$$
$$T(3) = 1 + 2 + 3,$$
$$T(4) = 1 + 2 + 3 + 4.$$

In fact, we see that $T(n)$ in general is the sum of the first n natural numbers, that is,

$$T(n) = 1 + 2 + 3 + \cdots + n.$$

Now You Try 17.5 Express each of the following triangular numbers $T(n)$ as the sum of the first n natural numbers, find the sum, and draw the configuration of dots.

1. $T(5)$, the fifth triangular number.   2. $T(10)$, the tenth triangular number.

□

17.3.2 Square Numbers

A *square number* is a number that, when represented as dots, may be configured as a square. The first four square numbers are shown in FIGURE 17.4.

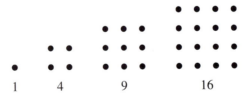

Figure 17.4: The first four square numbers.

Think About It 17.5 What patterns do you see in the first four square numbers? Come up with a formula for the nth square number.   □

---

[16]The notation $T(n)$ is function notation (read "T of n"), and is not the product of T and n. Recall that a function is a set of ordered pairs (x, y) for which every value x is paired with one and only one value y. Here, the ordered pairs $(x, y) = (x, T(x))$.

Let $S(n)$ denote the nth square number. Then

$$S(1) = 1, \quad S(2) = 4, \quad S(3) = 9, \quad \text{and} \quad S(4) = 16.$$

Moreover, from the figure above, we see that

$$S(1) = 1 \times 1,$$
$$S(2) = 2 \times 2,$$
$$S(3) = 3 \times 3,$$
$$S(4) = 4 \times 4.$$

In fact, we see that the nth square number is given in general by the formula

$$S(n) = n \times n \quad \text{or} \quad S(n) = n^2.$$

Now You Try 17.6 Find the following square numbers, and draw each configuration of dots.

1. The fifth square number, $S(5)$.

2. The tenth square number, $S(10)$.

☐

Now observe that another way to view square numbers is to break them into gnomons or L-shaped configurations of dots. FIGURE 17.5 shows the fourth square number broken into gnomons, for example. When viewed in this way, we see that the fourth square number is the sum of the first four *odd* natural numbers, that is,

$$S(4) = 1 + 3 + 5 + 7.$$

Now You Try 17.7 Express each of the following square numbers as the sum of the first n odd natural numbers, then draw the configuration of dots for each of the square numbers broken up into gnomons.

1. $S(1)$ 2. $S(2)$ 3. $S(3)$

☐

Figure 17.5: The fourth square number broken into gnomons.

In general, when a square number is broken into gnomons, we find that

$$S(n) = 1 + 3 + 5 + \cdots + (2n - 1).$$

Hence, from

$$S(n) = n^2 \quad \text{and} \quad S(n) = 1 + 3 + 5 + \cdots + (2n - 1),$$

we obtain the following remarkable formula for the sum of the first n odd natural numbers.

$$1 + 3 + 5 + \cdots + (2n - 1) = n^2$$

According to Heath [68, p. 77], "All this was known to Pythagoras. The odd numbers successively added were called *gnomons*; this is clear from Aristotle's allusion to gnomons placed around 1 which now produce different figures every time (oblong figures, each dissimilar to the preceding one), now preserve one and the same figure (squares)[17]; that latter is the case with the gnomons now in question."

Now You Try 17.8 Find the following sums.

1. $1 + 3 + 5 + 7 + 9$

2. $1 + 3 + 5 + 7 + 9 + 11 + 13 + 15 + 17 + 19$

3. $1 + 3 + 5 + \cdots + 99$

4. $21 + 23 + 25 + \cdots + 99$

$\square$

Another observation about square numbers is found in the theorem of Theon of Smyrna that any square number is composed of two successive triangular numbers.[18] For example, we see here that the fourth square number is composed of the third and fourth triangular numbers.

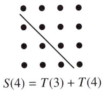

$$S(4) = T(3) + T(4)$$

Think About It 17.6 Try to come up with a formula for $S(n)$, the nth square number, in terms of two successive triangular numbers.

$\square$

17.3.3 Rectangular Numbers

A *rectangular number* or *oblong number* is a number that, when represented as dots, may be configered as a rectangle. According to Heath [68, p. 82], after Pythagoras, or the earliest Pythagoreans, had discovered the sum of the first n natural numbers in triangular numbers, and the sum of the first n odd natural numbers in square numbers, "it cannot be doubted that in the like manner they summed the series of even numbers …and discovered accordingly that the sum of any number of successive terms of the series beginning with 2 was an 'oblong' number (*eteromekes*), with 'sides' or factors differing by 1." The first four such rectangular numbers are shown in FIGURE 17.6. Additionally, Heath says later, "It is to be noted that the word *eteromekes* ('oblong') is in Theon of Smyrna and Nicomachus limited to numbers which are the product of two factors differing by unity, while they apply the term *promekes* ('prolate', as it were) to numbers which are the product of factors differing by two or more (Theon makes *promekes* include *eteromekes*). In Plato and Aristotle *eteromekes* has the wider sense of any non-square number with two unequal factors." We will consider here only rectangular numbers to be of the first kind, with factors that differ by one.

Heath claims that the Pythagoreans "would also see that the oblong number [with factors differing by one] is double of a triangular number." We illustrate this below, where each of the first four rectangular numbers is composed of two copies of the same triangular number (one using dots and an identical one using circles).

---

[17] Arist. *Phys*, iii. 4, 203 a 13–15.
[18] Heath [68, footnote on p. 83] cites Theon of Smyrna, p. 41. 3–8.

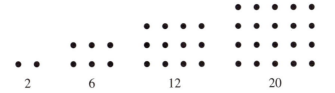

Figure 17.6: The first four rectangular or oblong numbers.

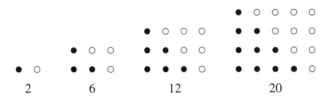

Think About It 17.7 What patterns do you see in the first four rectangular numbers? Try to come up with a formula for the nth rectangular number. □

Let $R(n)$ denote the nth rectangular number. Then

$$R(1) = 2, \quad R(2) = 6, \quad R(3) = 12, \quad \text{and} \quad R(4) = 20.$$

Moreover, from the figure above, we see that

$$R(1) = 1 \times 2,$$
$$R(2) = 2 \times 3,$$
$$R(3) = 3 \times 4,$$
$$R(4) = 4 \times 5.$$

In fact, we see that the nth rectangular number is given in general by the formula

$$R(n) = n(n + 1).$$

Now You Try 17.9 Find the following rectangular numbers and draw each configuration of dots.

1. The fifth rectangular number, $R(5)$.   2. The tenth rectangular number, $R(10)$.

□

Now observe that, just as with square numbers, another way to view rectangular numbers is to break them into gnomons. FIGURE 17.7 shows the fourth rectangular number broken into gnomons, for example. In this way, we see that the fourth rectangular number is the sum of the first four *even* natural numbers; that is,

$$R(4) = 2 + 4 + 6 + 8.$$

Now You Try 17.10 Express each of the following rectangular numbers as the sum of the first n even natural numbers, and draw the configuration of points for each of the rectangular numbers broken up into gnomons.

1. $R(1)$ 2. $R(2)$ 3. $R(3)$

Figure 17.7: The fourth rectangular number broken into gnomons.

In general, when a rectangular number is broken into gnomons, we find that

$$R(n) = 2 + 4 + 6 + \cdots + 2n.$$

Hence, from

$$R(n) = n(n + 1) \quad \text{and} \quad R(n) = 2 + 4 + 6 + \cdots + 2n,$$

we obtain the following remarkable formula for the sum of the first n even natural numbers.

$$2 + 4 + 6 + \cdots + 2n = n(n + 1)$$

Now You Try 17.11 Find the following sums.

1. $2 + 4 + 6 + 8 + 10$

2. $2 + 4 + 6 + 8 + 10 + 12 + 14 + 16 + 18 + 20$

3. $2 + 4 + 6 + \cdots + 100$

4. $22 + 24 + 26 + \cdots + 100$

Think About It 17.8 What relations among triangular, square, and rectangular numbers does the following figure suggest?

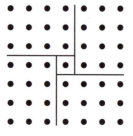

Now, recall that a rectangular number is obtained by adding a triangular number to itself; in other words,

$$R(1) = T(1) + T(1),$$
$$R(2) = T(2) + T(2),$$
$$R(3) = T(3) + T(3),$$
$$R(4) = T(4) + T(4),$$

and, in general,

$$R(n) = T(n) + T(n) \quad \text{or} \quad R(n) = 2T(n).$$

Solving the last equation for $T(n)$, we obtain a formula for the nth triangular number, namely,

$$T(n) = \frac{R(n)}{2} \quad \text{or} \quad T(n) = \frac{n(n+1)}{2}.$$

Now You Try 17.12 Verify the formula $T(n) = \frac{n(n+1)}{2}$ for $T(1)$, $T(2)$, $T(3)$, $T(4)$, $T(5)$, and $T(10)$. □

Hence, recalling further that $T(n)$ is the sum of the first n natural numbers, we obtain from

$$T(n) = \frac{n(n+1)}{2} \quad \text{and} \quad T(n) = 1 + 2 + 3 + \cdots + n$$

the following also remarkable formula for the sum of the first n natural numbers.

$$1 + 2 + 3 + \cdots + n = \frac{n(n+1)}{2}$$

Heath [68, pp. 76–77] tells us that Pythagoras probably discovered this formula precisely by considering triangular numbers. He relates that "[t]he particular triangle which has 4 for its side is mentioned in a story of Pythagoras by Lucian. Pythagoras told some one to count. He said 1, 2, 3, 4, whereon Pythagoras interrupted, 'Do you see? What you take for 4 is 10, a perfect triangle, and our oath'.[19] This connects the knowledge of triangular numbers with true Pythagorean ideas."

Now You Try 17.13 Find the sum of the first 100 natural numbers, that is, find

$$1 + 2 + 3 + \cdots + 100.$$

□

17.3.4 Other Figurate Numbers

We usually think of a gnomon to be an L-shaped figure. Heath [68, pp. 78–79] tells us, however, that Euclid extended the meaning in "[*Elements*] (II. Def. 2) to cover the figure similarly related to any parallelogram, instead of a square; it is defined as made up of 'any one whatever of the parallelograms about the diameter (diagonal) with the two complements'." See FIGURE 17.8. Heath goes on to say, "Later still ... Heron of Alexandria defines a *gnomon* in general as that which, when added to anything, number or figure, makes the whole similar to that to which it is added.[20]" In this way, we may form the pentagonal numbers, the hexagonal numbers, and so on. For example, FIGURE 17.9 shows the fourth triangular, square, and pentagonal numbers. The gnomons are shown as dashed lines.

[19] Lucian, Βίων πρᾶσις, 4.
[20] Heron, Def. 58 (Heron, vol. iv, Heib., p. 225).

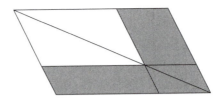

Figure 17.8: A general gnomon.

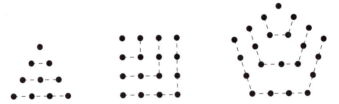

Figure 17.9: The fourth triangular, square, and pentagonal numbers are shown with their respective gnomons.

Think About It 17.9 What patterns do you see in the gnomons of the triangular, square, and pentagonal numbers shown in FIGURE 17.9? Draw the configuration of dots for the first four *hexagonal* numbers and show the gnomons. ☐

Let $P(n)$ denote the nth pentagonal number. Then, we see in FIGURE 17.9 that the first four pentagonal numbers are

$$P(1) = 1, \quad P(2) = 5, \quad P(3) = 12, \quad \text{and} \quad P(4) = 22.$$

Think About It 17.10 Try to come up with a formula for the nth pentagonal number. ☐

Now You Try 17.14

1. Find the following pentagonal numbers.

   a) The fifth pentagonal number, $P(5)$.

   b) The tenth pentagonal number, $P(10)$.

2. Let $H(n)$ denote the nth hexagonal number.

   a) Find $H(1)$, $H(2)$, $H(3)$, and $H(4)$, the first four hexagonal numbers.

   b) Find $H(5)$ and $H(10)$, the fifth and tenth hexagonal numbers, respectively.

☐

Think About It 17.11 Try to come up with a formula for the nth hexagonal number. ☐

17.4 PYTHAGOREAN TRIPLES

We mentioned at the beginning of this section that Pythagoras is perhaps best known today for the result that bears his name, Pythagoras's theorem—a result primarily about the lengths of the sides of right triangles or right-angle triangles—that "is arguably the most important elementary theorem in mathematics, since its consequences and generalizations have wide-ranging application" [75, p. 19]. Moreover, there are many proofs of Pythagoras's theorem; in fact, Elisha Loomis [78] has collected 255 different proofs of the theorem, and the count may have increased since then.[21]

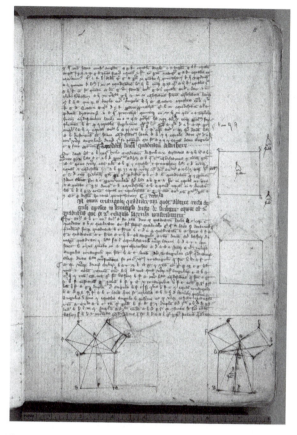

Figure 17.10: Pythagoras's theorem in Euclid's *Elements*, Book I Proposition 47. (Source: Rare Book and Manuscript Library, Columbia University.)

Euclid stated and proved Pythagoras's theorem in the last two propositions in Book I of the *Elements* [53, pp. 35–36]:

Proposition 47 *In right-angled triangles the square on the side subtending the right angle is equal to the squares on the sides containing the right angle.*

Proposition 48 *If in a triangle the square on one of the sides be equal to the squares on the remaining two sides of the triangle, the angle contained by the remaining two sides of the triangle is right.*

We first remark that, in line with the Greek mathematical mentality—namely, geometric—Euclid uses the phrase "square *on* the side" (our emphasis) instead of "square *of* the side" because squares were thought literally to be placed *on* the sides of the triangle. We see this in the figure that accompanies Euclid I.47 shown in FIGURE 17.10 on page 215. Today, Pythagoras's theorem is often stated as follows.

---

[21]According to Loomis [78, pp. 6–7], "During the middle ages this proposition [Pythagoras's theorem] ... won the honor-designation *Magister matheseos*, and the knowledge thereof was some decades ago still the proof of a solid mathematical training (or education). In examinations to obtain the master's degree this proposition was often given; there was indeed a time, as is maintained, when from every one who submitted himself to the test a master of mathematics a new (original) demonstration was required.

"This latter circumstance, or rather the great significance of the proposition under consideration was the reason why numerous demonstrations of it were thought out." Loomis classifies his collection of proofs into four kinds: algebraic proofs through linear relations; geometric proofs; quarternionic proofs; and dynamic proofs.

Theorem 17.1 (PYTHAGORAS'S THEOREM) *In a right triangle, the square of one side equals the sum of the squares of the other two sides, and conversely.* □

Notice the phrase "square *of* one side" (our emphasis) now instead of Euclid's "square *on* one side." This reflects a shift away from the Greek geometric thinking toward algebraic thinking. In fact, today it is common to remember Pythagoras's theorem simply as the algebraic relation "*a* squared plus *b* squared equals *c* squared" ($a^2 + b^2 = c^2$), where a, b, and c are the lengths of the sides of a right triangle. The longest side that is opposite the right angle is called the *hypotenuse* of the right triangle, and the two shorter sides that include the right angle are called the *legs*.

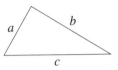

Definition 17.3 Let a, b, and c be natural numbers. The triple (a, b, c) is called a *Pythagorean triple* if $a^2 + b^2 = c^2$. Equivalently, the sides of any right triangle with sides of integral lengths forms a Pythagorean triple. □

For example, $(3, 4, 5)$ is a Pythagorean triple, since

$$3^2 + 4^2 = 5^2 \quad \text{or} \quad 9 + 16 = 25.$$

Moreover, $(6, 8, 10)$, where we have *scaled* or multiplied $(3, 4, 5)$ by a factor of 2, is also a Pythagorean triple,

$$6^2 + 8^2 = 10^2 \quad \text{or} \quad 36 + 64 = 100.$$

In fact, for any natural number k, the triple $(3k, 4k, 5k)$ is a Pythagorean triple, since

$$\begin{aligned}
(3k)^2 + (4k)^2 &= 9k^2 + 16k^2 \\
&= (9 + 16)k^2 \\
&= 25k^2 \\
&= (5k)^2.
\end{aligned}$$

See FIGURE 17.11.

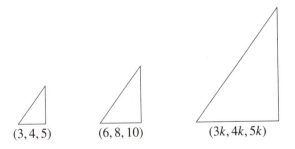

Figure 17.11: Right triangles based on the Pythagorean triple $(3, 4, 5)$.

Now You Try 17.15

1. Show that $(5, 12, 13)$ is a Pythagorean triple.

2. Show that $(15, 36, 39)$, where we have scaled $(5, 12, 13)$ by a factor of 3, is also a Pythagorean triple.

3. Show that, for any natural number k, the triple $(5k, 12k, 13k)$ is a Pythagorean triple.

□

Think About It 17.12 Let x, y, z, and k be natural numbers. Is it true that (xk, yk, zk) is a Pythagorean triple if (x, y, z) is a Pythagorean triple? Is it true that (xk, yk, zk) is a Pythagorean triple only if (x, y, z) is a Pythagorean triple?

□

The early Egyptians are generally credited with the beginnings of geometry. This stems not only from the construction of the great pyramids of Egypt, but also from the fact that the annual flooding of the Nile River wiped out farm boundaries and required the *harpendonáptai* or "rope stretchers" (surveyors) to re-mark the boundaries so that appropriate taxes could be collected. The geometry of the Egyptians was completely practical, however, and, unlike the Greeks, there is no evidence that the Egyptians generalized any of the results they used. For example, it is commonly thought that the rope stretchers used a 12-knotted rope to create a $(3, 4, 5)$ right triangle for surveying; however, there is no evidence that they knew Pythagoras's theorem at all.

On the other hand, there is evidence that the Babylonians knew Pythagoras's theorem at least empirically long before the time of Pythagoras. For example, the cuneiform tablet Plimpton 322 of the Plimpton collection at Columbia University shown in FIGURE 2.3 on page 14 contains a list of numbers from which one can easily produce Pythagorean triples. As another example, FIGURE 19.2 on page 279 shows the cuneiform tablet YBC 7289 from the Yale Babylonian Collection that depicts a square with one side marked 30 and one 1 24 51 10 ($= 1 + \frac{24}{60} + \frac{51}{60^2} + \frac{10}{60^3} = 1.41421 \approx \sqrt{2}$), and below that 42 25 35 ($= 42 + \frac{25}{60} + \frac{35}{60^2} = 42.42639 \approx \sqrt{1800}$). Noting that $\sqrt{2}$ is the length of the diagonal of a square that has side 1 unit, and that $\sqrt{1800}$ is the length of the diagonal of a square that has side 30 units, we conclude from YBC 7289 that the Babylonians were aware of Pythagoras's theorem.

We also find Pythagoras's theorem in the earliest Chinese texts, where it is called the *gou-gu theorem*; in reference to a right triangle, the *gou* is the base, the *gu* is the height, and the *xian* is the hypotenuse that joins the *gou* and *gu* [44, p. 215]. For example, we find the *gou-gu* theorem in the *Zhou bi suan jing* (translated variously as *Mathematical Classic of the Zhou Gnomon* [44] or *Arithmetical Classic of the Gnomon and the Circular Paths of Heaven* [75], for example), whose author is unknown but was most likely compiled no later than the first century BC. This work is commonly referred to as the *Zhou bi* for short. Interestingly, we hear from Dauben [44, p. 215] that

> As for the applications of the "Pythagorean theorem" in the *Zhou bi*, these led the historian of mathematics Mikami Yoshio to conjecture that Pythagoras may indeed have learned his famous theorem from the Chinese: "is he not said to have traveled in the east to Babylon and perhaps further on probably to India? Will it not be probable that he had seen the theorem that was destined to be connected with his name on his travel? Might he not have encountered it brought from China in some unknown way? The lapse of time between him and Chou-Kong (Zhou Gong) justly makes the matter an open question" [83, 7]. More blatantly, this idea served to provoke the same possibility in the title of a translation of chapter 9 of the *Jiu zhang suan shu* published in 1977: *Was Pythagoras Chinese? An Examination of Right Triangle Theory in Ancient China* [115]. Although the translators insist their title was meant only half-seriously, more to provoke readers' interest than to argue that Pythagoras had actually heard of earlier results in China, the fact remains that from a very early date Chinese mathematicians seem to have appreciated the special properties of right triangles. Additionally, the dialogue goes on to refer to gnomons, the circle and square, and methods of measuring heights and distances that cannot be measured by ordinary means.

(Swetz, in his article, "Similarity vs. the 'in-and-out complementary principle': a cultural faux pas" [114], corrects some of the assertions that were made in his earlier work with Kao, *Was Pythagoras Chinese?* [115].) We also find Pythagoras's theorem in early Indian and Arabic mathematics [22, 41].

Returning to Pythagorean triples, Heath [68, pp. 79–80] posits that Pythagoras was certainly aware that any triangle with sides of lengths $(3k, 4k, 5k)$ is a right triangle; furthermore, he says, "This fact could not but add strength to his conviction that all things were numbers because it established a connexion between numbers and the *angles* of geometrical figures." Not only that, but Heath also attributes to Pythagoras the following formula for generating Pythagorean triples (a, b, c).

$$a = m \text{ (an odd number)}, \quad b = \frac{m^2 - 1}{2}, \quad c = \frac{m^2 + 1}{2}.$$

Think About It 17.13 Would Pythagoras's formula produce a Pythagorean triple (a, b, c) if $a = m$ were an *even* natural number? □

Example 17.3 Let $a = m = 3$. Then,

$$b = \frac{m^2 - 1}{2} = \frac{3^2 - 1}{2} = 4 \quad \text{and} \quad c = \frac{m^2 + 1}{2} = \frac{3^2 + 1}{2} = 5.$$

Thus, we have generated the Pythagorean triple $(3, 4, 5)$.
    As another example, let $a = m = 5$. Then,

$$b = \frac{m^2 - 1}{2} = \frac{5^2 - 1}{2} = 12 \quad \text{and} \quad c = \frac{m^2 + 1}{2} = \frac{5^2 + 1}{2} = 13.$$

Thus, we have generated the Pythagoren triple $(5, 12, 13)$. □

Now You Try 17.16 Use the formula

$$a = m \text{ (an odd number)}, \quad b = \frac{m^2 - 1}{2}, \quad c = \frac{m^2 + 1}{2}.$$

1. Show that $a^2 + b^2 = c^2$, that is,

$$m^2 + \left(\frac{m^2 - 1}{2}\right)^2 = \left(\frac{m^2 + 1}{2}\right)^2.$$

2. Show that $c - b = 1$, so that it is always the case that two of the numbers in a Pythagorean triple differ by one if the Pythagorean triple is generated using Pythagoras's formula.

3. Generate Pythagorean triples (a, b, c) using the given value of a.

    a) $a = 5$ b) $a = 9$ c) $a = 11$

□

How might Pythagoras have discovered this formula for generating Pythagorean triples? He knew that the sum of the first n odd natural numbers is a square number, that is,

$$1 + 3 + 5 + \cdots + (2n - 1) = n^2.$$

Then the genius would be to choose the last odd number in the sum, $2n - 1$, such that it is itself a square number. In other words, we want a square number such that the gnomon that is to be added to produce the next square number is itself a square number. Note that the gnomon must be an odd number. See FIGURE 17.12.

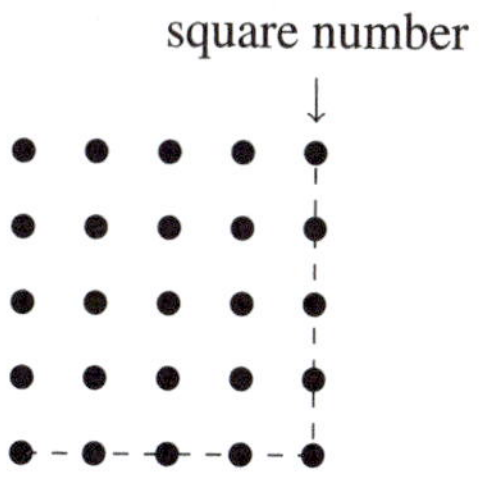

square number
↓

Figure 17.12: A gnomon of a square number that is itself a square number.

Think About It 17.14 How can we find odd square numbers? ☐

For example, $9 = 3^2$ is an odd square number, so we want the square number such that the gnomon that is to be added to produce the next square number is 9:

$$\underbrace{1 + 3 + 5 + 7 +}_{16} \overset{\text{gnomon}}{\underset{\downarrow}{9}} = 25 \quad \text{or} \quad 4^2 + 3^2 = 5^2,$$

which makes $(3, 4, 5)$ a Pythagorean triple.

As another example, $25 = 5^2$ is an odd square number and

$$\underbrace{1 + 3 + 5 + \cdots + \underset{\underset{2(12)-1}{\uparrow}}{23} + \overset{\overset{2(13)-1}{\downarrow}}{25}}_{12^2} = \overset{\overset{13^2}{\downarrow}}{169} \quad \text{or} \quad 12^2 + 5^2 = 13^2,$$

which makes $(5, 12, 13)$ a Pythagorean triple.

Now, to find odd square numbers, we need only square the odd numbers.

m	1	3	5	7	9	$\cdots$
m^2	1	9	25	49	81	$\cdots$

So, as a third example, $81 = 9^2$ is an odd square number and

$$\underbrace{1 + 3 + 5 + \cdots + \underset{\underset{2(40)-1}{\uparrow}}{79} + \overset{\overset{2(41)-1}{\downarrow}}{81}}_{40^2} = \overset{\overset{41^2}{\downarrow}}{1681} \quad \text{or} \quad 40^2 + 9^2 = 41^2,$$

which makes $(9, 40, 41)$ a Pythagorean triple.

In summary, if m is an odd natural number, then Pythagoras's formula for producing Pythagorean triples states that

$$\underbrace{1 + 2 + 3 + \cdots + (2k - 3)}_{\left(\frac{m^2-1}{2}\right)^2} + \overset{\overset{m^2}{\downarrow}}{(2k - 1)} = \overset{\overset{\left(\frac{m^2+1}{2}\right)^2}{\downarrow}}{n^2},$$

which makes

$$\left(m, \frac{m^2 - 1}{2}, \frac{m^2 + 1}{2}\right)$$

a Pythagorean triple, of which two of the numbers differ by one.

You will find in the exercises other formulas for producing Pythagorean triples.

17.5 DIVISORS, COMMON FACTORS, AND COMMON MULTIPLES

The *prime numbers* are the multiplicative building blocks of the natural numbers. Many results in number theory depend upon *prime factorization*. More practically, the encryption and encoding schemes used in today's electronic communication, for example, generally depend upon prime factorization in some way because prime factorization is assumed to be generally a very hard problem. Before we introduce prime numbers and prime factorizations, we introduce factorization in general.

17.5.1 Factors and Multiples

Definition 17.4 A natural number a is called a *factor* or *divisor* of a natural number b if $b = ac$ for some natural number c; in this case, we also call b a *multiple* of a. We write $a \mid b$, read "a divides b." If $a \mid b$ and $a < b$, we say that a is a *proper factor* or *proper divisor* of a.[22] □

Remark 17.3

1. One may confuse $a \mid b$ and a/b. Note that $a \mid b$—using a *vertical* bar—means that a is a factor of b (or that b is a multiple of a), whereas a/b—using a solidus or slash—means that a is divided by b ($a \div b$ or $\frac{a}{b}$).

2. If $a \mid b$, then there is a number c such that $b = ac$; but then $b = ca$ because multiplication is commutative, so that $c \mid b$ also.

3. More generally, we may define factor and multiple for integers in essentially the same way as for natural numbers, being careful not to allow zero to be a factor.

□

Think About It 17.15 Recall that, in general, when we divide a number a by a number b, we find that

$$a = bq + r, \quad 0 \le r < b,$$

where q is the quotient and r is the remainder. This is often called the *division algorithm*. How would you define "factor" or "multiple" in terms of the operation of division and the remainder upon division? □

Example 17.4 For any natural number (or, more generally, nonzero integer) n, we have both $1 \mid n$ and $n \mid n$. □

Example 17.5 Four is a factor of 12 because $12 = 4 \cdot 3$; we may write $4 \mid 12$. It follows also that $3 \mid 12$, and that 12 is a multiple of 4 and a multiple of 3.

On the other hand, 5 is not a factor or 12 (because there is no natural number c such that $12 = 5c$), and so 12 is not a multiple of 5; we write $5 \nmid 12$. □

Example 17.6 The set of factors of 36 is $\{1, 2, 3, 4, 6, 9, 12, 18, 36\}$, for

$$36 = 1 \cdot 36 = 2 \cdot 18 = 3 \cdot 12 = 4 \cdot 9 = 6 \cdot 6.$$

The set of *proper* factors of 36 is $\{1, 2, 3, 4, 6, 9, 12, 18\}$. □

Think About It 17.16 To find all the factors of a number, do we need to try dividing the number by every number that is less than it? □

The last example suggests the following useful fact.

---

[22]Some authors, for example, Davenport [45, p. 16], also exclude 1 as a proper factor.

> To find the factors of a natural number n, it is enough to find those factors that are less than or equal to $\sqrt{n}$.

For observe that if $n = ab$, then b gets smaller as a gets larger, or vice versa. Thus, either a or b is less than $\sqrt{n}$. Now, if $a > \sqrt{n}$ and $b > \sqrt{n}$, then

$$n = ab > \sqrt{n} \cdot \sqrt{n} = n,$$

which is absurd. Consequently, we need to attempt to divide into n only those numbers between 1 and $\sqrt{n}$. For example, if $n = 1,000,001$, then $\sqrt{n}$ is between 1000 and 1001, so that there are potentially only 999 numbers (from 2 to 1000) to attempt to divide into n, which—although it is not a small number—is certainly a much smaller number of potential factors than is 999,999.

Example 17.7 Find all the factors of 70.

To begin, we note that $8^2 = 64$, $9^2 = 81$, and $64 < 70 < 81$, so that $8 < \sqrt{70} < 9$. Thus, to find all the factors of 70, it will be enough to divide 70 by the natural numbers less than 9. Those divisors and quotients that result in a remainder of zero are factors of 70.

$$70 \div 1 = 70 \quad \text{R } 0, \quad 70 \div 2 = 35 \quad \text{R } 0, \quad 70 \div 3 = 23 \quad \text{R } 1, \quad 70 \div 4 = 17 \quad \text{R } 2,$$
$$70 \div 5 = 14 \quad \text{R } 0, \quad 70 \div 6 = 11 \quad \text{R } 4, \quad 70 \div 7 = 10 \quad \text{R } 0, \quad 70 \div 8 = 8 \quad \text{R } 6$$

Therefore, we see that the set of factors of 70 is $\{1, 2, 5, 7, 10, 14, 35, 70\}$. ☐

Now You Try 17.17 Find the set of factors of each of the following numbers.

1. 28
2. 41
3. 91
4. 191

☐

Before we continue, we note some facts about factors. Let a, b, c, k, and l, be natural numbers.

1. If $a \mid b$, then $a \le b$.

2. If $a \mid b$ and $b \mid c$, then $a \mid c$.

3. If $a \mid b$ and $a \mid c$, then $a \mid (kb + lc)$ and $a \mid (kb - lc)$; in particular, $a \mid (b + c)$ and $a \mid (b - c)$ (if $c < b$).

The third item above means that if a number divides two numbers, it divides what is called a *linear combination* of those two numbers.

Now You Try 17.18

1. If $a \mid 10$ and $10 \mid 30$, does $a \mid 300$? Why? What might a be?

2. If $a \mid 10$ and $a \mid 15$, does $a \mid 145$? Why? What might a be?

☐

What we have given above are modern definitions of factor and multiple. We turn now to the definitions of factor and multiple in Euclid's *Elements* to see how these compare to the modern definitions. We find these listed among the definitions at the beginning of Book VII:

3. A number is *a part* of a number, the less of the greater, when it measures the greater;

4. but *parts* when it does not measure it.

5. The greater number is a *multiple* of the less when it is measured by the less.

At first glance, these do not seem to define "factor" or "multiple," but they do. For example, we see here that 4 is "a part" of 12 because 4 "measures" 12 ($4 + 4 + 4 = 12$); also, 12 is a "multiple" of 4 because 12 "is measured by" 4.

In the same way, 6 is "a part" of 12 and 12 is a "multiple" of 6.

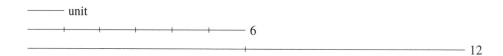

Now, we see that "4 measures 12" means that 12 can be divided by 4 evenly (indeed, $12 = 4 \cdot 3$), so that $4 \mid 12$; thus, "4 is a part of 12" means that 4 is a factor of 12. Similarly, "6 measures 12" means that 12 can be divided by 6 evenly (indeed, $12 = 6 \cdot 2$), so that $6 \mid 12$; thus, "6 is a part of 12" means that 6 is a factor of 12.

On the other hand, we see here that 5 is "parts" of 12 because 5 does not measure 12.

It can be easy to confuse "a part" and "parts" as they are defined in Euclid's *Elements*. Perhaps a way to remember what is "parts" is this: when a given number is divided by a number that is not a factor of the given number, there are "parts" left over that we call the remainder.

Now You Try 17.19 Use a diagram like those above to

1. Show that 1, 2, and 3 are each a part of 12. Note that this means $1 \mid 12$, $2 \mid 12$, and $3 \mid 12$.

2. Find every number that is a part of 28.

3. Find all the parts of 31.

4. Show that 4 is not a part of 15.

□

Therefore, Euclid VII.Def.3 indeed defines what we call a "factor" of a number, and VII.Def.5 defines what we call a "multiple" of a number.

Think About It 17.17 As with the "square *on* the hypotenuse" in Pythagoras's theorem, why would the ancient Greeks talk about one number *measuring* another number? An important historical concept to notice is that the ancient Greeks thought of numbers as lengths. This influenced their terminology, such as "measures." With this in mind, what then would the terms *commensurate* and *incommensurate* mean? □

17.5.2 Euclid's Algorithm

Finding the *greatest common factor (GCF)* or *highest common factor* or *greatest common divisor* of two numbers is used in many areas of computing and is seen in various historical settings.[23] *Euclid's algorithm* or the *Euclidean algorithm* for finding the GCF of two numbers (integers) is found in book VII of Euclid's *Elements*, Propositions VII-1 and VII-2. It was most likely extended to find the common measure of two magnitudes by Theaetetus in the fourth century BC. Known as the Chinese remainder problem, the method of repeated subtractions as used in Euclid's algorithm was also used by the Chinese in the third century and the Indians in the fourth century to solve systems of linear congruences for use in astronomy. Today such systems would be written using modular arithmetic, but they would have been written in words or a syncopated style in antiquity. The Indian method was first mentioned in the work of Aryabhata and later clarified by Brahmagupta, who aptly named the method the *pulverizer* for how it reduced the numbers down to their common divisors or measures.

If A denotes the set of factors of 12 and B denotes the set of factors of 18, then $A \cap B$ is the set of *common factors* of 12 and 18:

$$A = \{1, 2, 3, 4, 6, 12\}, \quad B = \{1, 2, 3, 6, 9, 18\},$$
$$A \cap B = \{1, 2, 3, 6\}.$$

It should be clear that the *least* common factor of any set of numbers is 1, but in general the *greatest* common factor will depend on the particular numbers in the sets. In this instance, we see that the GCF of 12 and 18 is 6, or $\gcf(12, 18) = 6$.

Example 17.8 The factors of

$$
\begin{array}{lll}
12 & \text{are} & 1, 2, 3, 4, 6, 12, \\
15 & \text{are} & 1, 3, 5, 15, \\
18 & \text{are} & 1, 2, 3, 6, 9, 18, \\
35 & \text{are} & 1, 5, 7, 35.
\end{array}
$$

Hence, we see that

$$\gcf(12, 18) = 6, \quad \gcf(15, 18) = 3, \quad \gcf(18, 35) = 1, \quad \gcf(12, 15, 18) = 3.$$

$\square$

Now You Try 17.20 Find the GCF of the following sets of numbers.

1. 24 and 36
2. 220 and 284
3. 24, 36, and 57

$\square$

In Proposition 2 of Book VII of the *Elements*, Euclid gives a method for finding the GCF, what he calls the *greatest common measure*, of two numbers without having first to find all the factors of the numbers. Proposition 2, however, refers to Proposition 1 that, in turn, refers to Definition 12. So, we begin with Euclid VII.Def.12 [53, p. 157]:

Numbers *prime to one another* are those which are measured by an unit alone.

For example, the numbers 6 and 9 are *not* prime to one another because they can be measured by 3; that is, 3 is a common factor of 6 and 9.

---

[23]The terms "greatest common factor," "highest common factor," and "greatest common divisor" are all commonly used.

On the other hand, 8 and 9 *are* prime to one another because they can be "measured by an unit alone," that is, one is the only common factor of 8 and 9.

In modern terminology, numbers that are "prime to one another" are said to be *relatively prime*.

Definition 17.5 The natural numbers a, b, c, ... are said to be *relatively prime* if $\gcf(a, b, c, \ldots) = 1$; the numbers are said to be *relatively prime in pairs* if no two of them have a common factor greater than 1. ☐

Think About It 17.18 If three natural numbers are relatively prime, are they then relatively prime in pairs? In other words, if $\gcf(a, b, c) = 1$, would it be true that $\gcf(a, b) = 1$, $\gcf(a, c) = 1$, and $\gcf(b, c) = 1$? What about conversely? ☐

Now that we have defined "numbers prime to one another," we give Euclid VII.1 [53, p. 158]:

> *Two unequal numbers being set out, and the less being continually subtracted in turn from the greater, if the number which is left never measures the one before it until an unit is left, the original numbers will be prime to one another.*

We omit the proof here, but, instead, only illustrate the proposition with the following examples.

Example 17.9

- "Two unequal numbers being set out": Consider the numbers 9 and 20.

- "and the less being continually subtracted in turn from the greater":

    - Subtract 9 from 20 as many times as possible:

$$20 - 9 = 11$$
$$11 - 9 = 2$$

```
──── unit

────────────────── 9

──────────────────────────────── 20
```

  Since $2 < 9$, we may not subtract 9 from 2.

  - Subtract 2 from 9 as many times as possible:

$$9 - 2 = 7$$
$$7 - 2 = 5$$
$$5 - 2 = 3$$
$$3 - 2 = 1$$

- "if the number which is left never measures the one before it until an unit is left, the original numbers will be prime to one another":

Since the last difference is a unit, $3 - 2 = 1$, we conclude that 9 and 20 are prime to one another. We verify this by listing all the factors of 9 and 20.

| factors of 9: | 1, 3, 9 |
| factors of 20: | 1, 2, 4, 5, 10, 20 |

Therefore, we see that the $\gcf(9, 20) = 1$, so that 9 and 20 are relatively prime.

☐

Example 17.10

- "Two unequal numbers being set out": Consider the numbers 27 and 60.

- "and the less being continually subtracted in turn from the greater":

  – Subtract 27 from 60 as many times as possible:

$$60 - 27 = 33$$
$$33 - 27 = 6$$

  Since $6 < 27$, we may not subtract 27 from 6.

  – Subtract 6 from 27 as many times as possible:

$$27 - 6 = 21$$
$$21 - 6 = 15$$
$$15 - 6 = 9$$
$$9 - 6 = 3$$

  Since $3 < 6$, we may not subtract 3 from 6.

  – Subtract 3 from 6 as many times as possible:

$$6 - 3 = 3$$
$$3 - 3 = 0$$

- "if the number which is left never measures the one before it until an unit is left, the original numbers will be prime to one another":

Since the last difference is zero, $3 - 3 = 0$, we see that 3 "measures the one before it," so we conclude that 27 and 60 are *not* prime to one another. We verify this by listing all the factors of 27 and 60.

| factors of 27: | 1, 3, 9, 27 |
| factors of 60: | 1, 2, 3, 4, 5, 6, 10, 12, 15, 20, 30, 60 |

Therefore, we see that the gcf$(27, 60) = 3$, so that 27 and 60 are *not* relatively prime.

□

Example 17.11 We show that the numbers 29,309 and 37,647 are prime to one another.

- Subtract 29,309 from 37,647 as many times as possible:

$$37,647 - 29,309 = 8338$$

Since $8338 < 29,309$, we may not subtract 29,309 from 8338.

- Subtract 8338 from 29,309 as many times as possible:

$$29,309 - 8338 = 20,971$$
$$20,971 - 8338 = 12,633$$
$$12,633 - 8338 = 4295$$

Since $4295 < 8338$, we may not subtract 8338 from 4295.

- Subtract 4295 from 8338 as many times as possible:

$$8338 - 4295 = 4043$$

Since $4043 < 4295$, we may not subtract 4295 from 4043.

- Subtract 4043 from 4295 as many times as possible:

$$4295 - 4043 = 252$$

Since $252 < 4043$, we may not subtract 4043 from 252.

- Subtract 252 from 4043 as many times as possible:

$$4043 - 252 = 3791$$
$$3791 - 252 = 3539$$
$$3539 - 252 = 3287$$

$$\vdots$$

$$515 - 252 = 263$$
$$263 - 252 = 11$$

Since $11 < 252$, we may not subtract 252 from 11.

- Subtract 11 from 252 as many times as possible:

$$252 - 11 = 241$$
$$241 - 11 = 230$$

$$\vdots$$

$$21 - 11 = 10$$

Since $10 < 11$, we may not subtract 11 from 10.

- Subtract 10 from 11 as many times as possible:

$$11 - 10 = 1$$

Since the difference is a unit, $11 - 10 = 1$, we conclude that 29,309 and 37,647 are prime to one another, in other words, that the gcf(29,309, 37,647) = 1 or 29,309 and 37,647 are relatively prime.

□

Now You Try 17.21

1. Show that 29,309 and 37,647 are relatively prime by listing all the factors of each number.

2. Show that 497 and 527 are prime to one another using the method in Euclid VII.1. Verify that gcf(497, 527) = 1 by listing all the factors of each number.

3. Use Euclid's method to show that 36 and 99 are not prime to one another.

□

Think About It 17.19 Using the fact that repeated subtraction corresponds to the operation of division, how can Euclid VII.1 be restated using division instead of repeated subtraction? Check this against the examples that follow the statement of Euclid VII.1. □

We now give *Euclid's algorithm* for finding the greatest common measure (GCF) of two numbers without having first to find all the factors of the numbers. Please note that given the geometric nature of Greek methods, the following is fairly long, so just take your time. We note that the Roman numerals here, (I), (IIa), (IIb), do not appear in the text of the *Elements*; we placed the numerals there so that we may refer to those sections of the proposition in the examples later.

The method is Euclid VII.2 [53, pp. 158–159]:

Given two numbers not prime to one another, to find their greatest common measure.

Let AB, CD be the two given numbers not prime to one another.

Thus it is required to find the greatest common measure of AB, CD.

(I) If now CD measures AB—and it also measures itself—CD is a common measure of CD, AB.

And it is manifest that it is also the greatest; for no greater number than CD will measure CD.

(IIa) But, if CD does not measure AB, then, the less of the numbers AB, CD being continually subtracted from the greater, some number will be left which will measure the one before it. [See FIGURE 17.13.]

For an unit will not be left; otherwise AB, CD will be prime to one another, [VII. 1] which is contrary to the hypothesis.

Therefore some number will be left which will measure the one before it.

Now let CD, measuring BE, leave EA less than itself,
let EA, measuring DF, leave FC less than itself,
and let CF measure AE.

Since, then, CF measures AE, and AE measures DF,

therefore CF will also measure DF.

But it also measures itself;

therefore it will also measure the whole CD.

But CD measures BE;

therefore CF also measures BE.

But it also measures EA;

therefore it will also measure the whole BA.

But it also measures CD;

therefore CF measures AB, CD.

Therefore CF is a common measure of AB, CD.

(IIb) I say next that it is also the greatest.

For, if CF is not the greatest common measure of AB, CD, some number which is greater than CF will measure the numbers AB, CD.

Let such a number measure them, and let it be G.

Now, since G measures CD, while CD measures BE,

G also measures BE.

But it also measures the whole BA;

therefore it will also measure the remainder AE.

But AE measures DF;

therefore G will also measure DF.

But it also measures the whole DC;
therefore it will also measure the remainder CF, that is, the greater will measure the less: which is impossible.

Therefore no number which is greater than CF will measure the numbers AB, CD;

therefore CF is the greatest common measure of AB, CD.

Q.E.D.

PORISM. From this it is manifest that, if a number measure two numbers, it will also measure their greatest common measure.

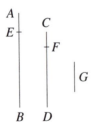

Figure 17.13: Euclid VII.2: Euclid's algorithm.

If nothing else, the above should give you an appreciation of the relative simplicity and efficency of our modern algebraic notation.

Recall that numbers for the Greeks were represented geometrically by lines, planes, and so on. In Euclid VII.2, all the numbers are represented by line segments. Thus, AB (which acts like a variable, say, x) is the number represented by the segment with endpoints A and B, CD (which acts like a variable, say, y) is the number represented by the segment with endpoints C and D, and so on; and, near the end, G is the number—assumed to be a common factor of AB and CD—represented by a line segment without identified endpoints. Note that in this convention, for example, AB and BA represent the same quantity, so that if $AB = x$, then $BA = x$ also.

Example 17.12 We illustrate Euclid VII.2. First, we let $AB = 60$ and $CD = 15$, where we note that $15 \,|\, 60$. Next, we let $AB = 60$ and $CD = 27$, where we note that $27 \nmid 60$. The Roman numerals here, (I), (IIa), (IIb), refer to those marked sections in Euclid's algorithm (page 227).

(I) Let $AB = 60$ and $CD = 15$.

Fifteen measures 60, for $60 = 15 \cdot 4$, and 15 measures itself, for $15 = 15 \cdot 1$. Hence, 15 is a common measure of 15 and 60. Moreover, 15 is the greatest common measure (GCF), for no greater number than 15 will measure 15.

(IIa) Let $AB = 60$ and $CD = 27$.

Note that 27 does not measure 60 (CD does not measure AB). Since $27 < 60$ ($CD < AB$), subtract 27 from 60 as many times as possible (subtract CD from AB as many times as possible):

$$60 - 27 = 33$$
$$33 - 27 = 6$$

Since $6 < 27$, we may not subtract 27 from 6; thus, $EA = 6$, so that $BE = 60 - 6 = 54$ and $54 = 27 \cdot 2$ (CD measures BE). (See FIGURE 17.13.)

Now, subtract 6 from 27 as many times as possible (subtract AE from CD as many times as possible):

$$27 - 6 = 21$$
$$21 - 6 = 15$$
$$15 - 6 = 9$$
$$9 - 6 = 3$$

Since $3 < 6$, we may not subtract 6 from 3; thus, $FC = 3$, so that $DF = 27 - 3 = 24$ and $24 = 6 \cdot 4$ (AE measures DF).

Note that 3 measures 6 (CF measures AE), for $6 = 3 \cdot 2$.

Since 3 measures 6 (CF measures AE), and

$$\begin{aligned}
24 &= 6 \cdot 4 & (AE = 6 \text{ measures } DF = 24) \\
&= (3 \cdot 2) \cdot 4 & (CF = 3 \text{ measures } AE = 6) \\
&= 3 \cdot (2 \cdot 4) & \text{(associative law)} \\
&= 3 \cdot 8 & \text{(simplify)}
\end{aligned}$$

therefore 3 measures 24 (CF will also measure DF).

But 3 also measures itself, for $3 = 3 \cdot 1$; therefore 3 measures 27 (CF will also measure the whole CD):

$$\begin{aligned}
27 &= 24 + 3 & (CD = DF + FC \\
&= (3 \cdot 8) + (3 \cdot 1) & (CF \text{ measures } DF, \\
& & \quad CF \text{ measures itself}) \\
&= 3(8 + 1) & \text{(distributive law)} \\
&= 3 \cdot 9 & \text{(simplify)}
\end{aligned}$$

But 27 measures 54 (CD measures BE); therefore 3 measures 54 (CF also measures BE).

But 3 also measures 6 (CF measures EA); therefore 3 measures 60 (CF will also measure the whole BA).

But 3 also measures 27 (CF also measures CD); therefore 3 measures 60 and 27 (CF measures AB, CD).

Therefore 3 is a common measure of 60 and 27 (CF is a common measure of AB, CD).

(IIb) We show that 3 is the greatest common measure of 60 and 27. We employ a proof by contradiction.

Suppose the 3 is not the greatest common measure of 60 and 27. Then there is a number $G > 3$ that is a common measure of 60 and 27, that is, $60 = G \cdot m$ (G measures AB) and $27 = G \cdot n$ (G measures CD) for some numbers m and n.

Now, G measures 27 (G measures CD), while

$$
\begin{aligned}
54 &= 27 \cdot 2 && (CD \text{ measures } BE) \\
&= (G \cdot n) \cdot 2 && (G \text{ measures } CD) \\
&= G \cdot (n \cdot 2) && (\text{associative law}) \\
&= G \cdot 2n && (\text{simplify})
\end{aligned}
$$

G measures 54 (G measures BE).

But G also measures 60 (G also measures the whole BA); therefore G measures 6 (G measures the remainder AE):

$$
\begin{aligned}
6 &= 60 - 54 && (AE = AB - BE) \\
&= (G \cdot m) - (G \cdot 2n) && (G \text{ measures } AB, \\
& && \quad G \text{ measures } BE) \\
&= G \cdot (m - 2n) && (\text{distributive law}) \\
&= G \cdot p && (p = m - 2n)
\end{aligned}
$$

But 6 measures 24 (AE measures DF), for $24 = 6 \cdot 4$; therefore G measures 24 (G will also measure DF):

$$
\begin{aligned}
24 &= 6 \cdot 4 && (AE \text{ measures } DF) \\
&= (G \cdot p) \cdot 4 && (G \text{ measures } AE) \\
&= G \cdot (p \cdot 4) && (\text{associative law}) \\
&= G \cdot q && (q = p \cdot 4)
\end{aligned}
$$

But G also measures 27 (G also measures the whole DC); therefore G will also measure 3 (G will also measure CF):

$$
\begin{aligned}
3 &= 27 - 24 && (CF = CD - DF) \\
&= (G \cdot n) - (G \cdot q) && (G \text{ measures } CD, \\
& && \quad G \text{ measures } DF) \\
&= G \cdot (n - q) && (\text{distributive law}) \\
&= G \cdot r && (r = n - q)
\end{aligned}
$$

that is, the greater will measure the less ($G > 3$ will measure 3). This is impossible.

Therefore, there is no $G > 3$ ($G > CF$) that is a common measure of 60 and 27 (AB, CD); that is, 3 is the greatest common measure of 60 and 27.

                                                                   □

You may have found Euclid's algorithm as it is stated in the *Elements* not so easy to follow. For this reason, we restate the algorithm using modern notation and replace the repeated subtraction with division.

Theorem 17.2 (EUCLID'S ALGORITHM) *Let a and b be natural numbers, and suppose that a > b. Perform the following sequence of divisions:*

$$a \div b = q_1 \quad R \; r_1,$$
$$b \div r_1 = q_2 \quad R \; r_2,$$
$$r_1 \div r_2 = q_3 \quad R \; r_3,$$
$$r_2 \div r_3 = q_4 \quad R \; r_4,$$
$$\vdots$$
$$r_{k-2} \div r_{k-1} = q_k \quad R \; r_k$$

until the last remainder $r_k = 1$ or 0.

If the last remainder $r_k = 1$, then the $\gcf(a, b) = 1$, so that a and b are relatively prime (Euclid VII.1: a and b are prime to one another).

If the last remainder $r_k = 0$, then the $\gcf(a, b) = r_{k-1}$, the last divisor (Euclid VII.2: r_{k-1} is the greatest common measure of a and b). □

In this formulation, the French mathematician Gabriel Lamé (1795–1870) showed that the number of divisions required in Euclid's algorithm is no more than five times the number of digits in the smaller number [92, p. 43].

Example 17.13 We repeat the example that illustrates Euclid VII.2.

(I) Let $a = 60$ and $b = 15$. Then

$$60 \div 15 = 4 \quad R \; 0.$$

Therefore, the $\gcf(60, 15) = 15$, the last divisor (in this case, also the first divisor because $15 \mid 60$).

(II) Let $a = 60$ and $b = 27$. Then,

$$60 \div 27 = 2 \quad R \; 6,$$
$$27 \div 6 = 4 \quad R \; 3,$$
$$6 \div 3 = 2 \quad R \; 0.$$

Therefore, the $\gcf(60, 27) = 3$, the last divisor.

□

Example 17.14 Find the GCF of 29,309 and 37,647.

We divide:

$$37{,}647 \div 29{,}309 = 1 \quad R \; 8338,$$
$$29{,}309 \div 8338 = 3 \quad R \; 4295,$$
$$8338 \div 4295 = 1 \quad R \; 4043,$$
$$4295 \div 4043 = 1 \quad R \; 252,$$
$$4043 \div 252 = 16 \quad R \; 11,$$
$$252 \div 11 = 22 \quad R \; 10,$$
$$11 \div 10 = 1 \quad R \; 1.$$

Therefore, the $\gcf(29{,}309, \; 37{,}647) = 1$. □

Now You Try 17.22 Use Euclid's algorithm (THEOREM 17.2) to find the GCF of the following pairs of numbers.

1. 497 and 527   2. 220 and 284   3. 863,835 and 802,725

□

We observe that Euclid's algorithm is "reversible" in the following sense. For example, take the numbers 60 and 27.

$$60 \div 27 = 2 \quad \text{R } 6 \quad \text{means} \quad 60 = 27 \cdot 2 + 6, \quad \text{so that} \quad 6 = 60 - 27 \cdot 2,$$
$$27 \div 6 = 4 \quad \text{R } 3 \quad \text{means} \quad 27 = 6 \cdot 4 + 3, \quad \text{so that} \quad 3 = 27 - 6 \cdot 4.$$

Hence,

$$
\begin{aligned}
3 &= 27 - 6 \cdot 4 \\
&= 27 - (60 - 27 \cdot 2) \cdot 4 && (6 = 60 - 27 \cdot 2) \\
&= 27 - 60 \cdot 4 + (27 \cdot 2) \cdot 4 && \text{(distributive law)} \\
&= 27 - 60 \cdot 4 + 27 \cdot 8 && \text{(simplify)} \\
&= 27 \cdot 9 - 60 \cdot 4 && \text{(combine like terms)}
\end{aligned}
$$

or

$$3 = 60(-4) + 27(9).$$

In other words, the GCF of 60 and 27 is a linear combination of 60 and 27, namely,

$$\gcf(60, 27) = 60(-4) + 27(9).$$

For a good challenge, we return to our previous example in which we found that the $\gcf(29{,}309, \, 37{,}647) = 1$. Let us express the GCF of 29,309 and 37,647 as a linear combination of these two numbers. To begin, we reverse the steps in Euclid's algorithm.

$$
\begin{array}{llll}
11 \div 10 = 1 & \text{R } 1, & \text{so that} & 1 = 11 - 10 \cdot 1, \\
252 \div 11 = 22 & \text{R } 10, & \text{so that} & 10 = 252 - 11 \cdot 22, \\
4043 \div 252 = 16 & \text{R } 11, & \text{so that} & 11 = 4043 - 252 \cdot 16, \\
4295 \div 4043 = 1 & \text{R } 252, & \text{so that} & 252 = 4295 - 4043 \cdot 1, \\
8338 \div 4295 = 1 & \text{R } 4043, & \text{so that} & 4043 = 8338 - 4295 \cdot 1, \\
29{,}309 \div 8338 = 3 & \text{R } 4295, & \text{so that} & 4295 = 29{,}309 - 8338 \cdot 3, \\
37{,}647 \div 29{,}309 = 1 & \text{R } 8338, & \text{so that} & 8338 = 37{,}647 - 29{,}309 \cdot 1.
\end{array}
$$

Hence,

$$
\begin{aligned}
1 &= 11 - 10 \cdot 1 \\
&= 11 - (252 - 11 \cdot 22) \cdot 1 && (10 = 252 - 11 \cdot 22) \\
&= 11 - 252 + 11 \cdot 22 && \text{(distributive law and simplify)} \\
&= 11 \cdot 23 - 252 && \text{(combine like terms)} \\
&= (4043 - 252 \cdot 16) \cdot 23 - 252 && (11 = 4043 - 252 \cdot 16) \\
&= 4043 \cdot 23 - 252 \cdot 368 - 252 && \text{(distribute and simplify)} \\
&= 4043 \cdot 23 - 252 \cdot 369 && \text{(combine like terms)} \\
&= 4043 \cdot 23 - (4295 - 4043 \cdot 1) \cdot 369 && (252 = 4295 - 4043 \cdot 1) \\
&= 4043 \cdot 23 - 4295 \cdot 369 + 4043 \cdot 369 && \text{(distributive law and simplify)} \\
&= 4043 \cdot 392 - 4295 \cdot 369 && \text{(combine like terms)}
\end{aligned}
$$

$$= (8338 - 4295 \cdot 1) \cdot 392 - 4295 \cdot 369 \qquad (4043 = 8338 - 4295 \cdot 1)$$
$$= 8338 \cdot 392 - 4295 \cdot 392 - 4295 \cdot 369 \qquad \text{(distributive law and simplify)}$$
$$= 8338 \cdot 392 - 4295 \cdot 761 \qquad \text{(combine like terms)}$$
$$= 8338 \cdot 392 - (29{,}309 - 8338 \cdot 3) \cdot 761 \qquad (4295 = 29{,}309 - 8338 \cdot 3)$$
$$= 8338 \cdot 392 - 29{,}309 \cdot 761 + 8338 \cdot 2283 \qquad \text{(distribute and simplify)}$$
$$= 8338 \cdot 2675 - 29{,}309 \cdot 761 \qquad \text{(combine like terms)}$$
$$= (37{,}647 - 29{,}309 \cdot 1) \cdot 2675 - 29{,}309 \cdot 761 \qquad (8338 = 37{,}647 - 29{,}309 \cdot 1)$$
$$= 37{,}647 \cdot 2675 - 29{,}309 \cdot 2675 - 29{,}309 \cdot 761 \qquad \text{(distributive law and simplify)}$$
$$= 37{,}647 \cdot 2675 - 29{,}309 \cdot 3436 \qquad \text{(combine like terms)}$$

Therefore,

$$\gcf(37{,}647, \ 29{,}309) = 1 = 37{,}647(2675) + 29{,}309(-3436).$$

The last two examples suggest the following corollary to THEOREM 17.2.

Corollary 17.1 (TO EUCLID'S ALGORITHM) *Let a and b be natural numbers, and suppose that the* $\gcf(a, b) = g$. *Then there are integers m and n for which*

$$g = am + bn.$$

☐

Now You Try 17.23 For each of the following pairs of numbers a and b, find integers m and n such that $\gcf(a, b) = am + bn$.

1. $a = 497$ and $b = 527$

2. $a = 220$ and $b = 284$

3. $a = 863{,}835$ and $b = 802{,}725$

☐

We end this subsection by looking briefly at multiples of numbers.

17.5.3 *Multiples*

Recall that a natural number b is called a *multiple* of a natural number a if $b = ac$ for some natural number c (DEFINITION 17.4). For example, the multiples of 12 and 18 are

×	1	2	3	4	5	6	7	8	9	10	…
12	12	24	36	48	60	72	84	96	108	120	…
18	18	36	54	72	90	108	126	144	162	180	…

Thus, if A denotes the set of multiples of 12 and B denotes the set of multiples of 18, then $A \cap B$ is the set of *common multiples* of 12 and 18:

$$A = \{12, \ 24, \ 36, \ 48, \ 60, \ 72, \ 84, \ 96, \ 108, \ 120, \dots\},$$
$$B = \{18, \ 36, \ 54, \ 72, \ 90, \ 108, \ 126, \ 144, \ 162, \ 180, \dots\},$$
$$A \cap B = \{36, \ 72, \ 108, \dots\}.$$

It should be clear that there is no *greatest* common multiple of any set of numbers, but there is always a *least* (or *lowest*) *common multiple* or *LCM*. In this instance, we see that the LCM of 12 and 18 is 36, or $\lcm(12, 18) = 36$.

Think About It 17.20 Given two numbers a and b, what is the largest that the LCM can be and when would this happen? What is the smallest that the LCM can be and when would this happen? □

Example 17.15 Find the lcm(8, 10).

The multiples of

$$8 \quad \text{are} \quad 8,\ 16,\ 24,\ 32,\ 40,\ 48,\ 56,\dots,$$
$$10 \quad \text{are} \quad 10,\ 20,\ 30,\ 40,\ 50,\ 60,\dots .$$

So, the lcm(8, 10) = 40. □

We note that the largest that the lcm(a, b) could be is ab. This happens when a and b are relatively prime.

Now You Try 17.24 Find the LCM of the following pairs of numbers by listing several multiples of the numbers.

1. 15 and 18   2. 9 and 20   3. 12, 15, and 18

□

It turns out that we may find the LCM of two numbers without having to list the multiples of the numbers. The method can be found in Euclid's *Elements*, Book VII, Propositions 33 and 34. We note that while Euclid VII.33 is required in the proof of VII.34, Proposition 34 actually tells us how to find the least common multiple of two numbers.

To begin, Euclid VII.33 states [53, pp. 178–179]:

Given as many numbers as we please, to find the least of those which have the same ratio with them.

Let A, B, C be the given numbers, as many as we please; thus it is required to find the least of those which have the same ratio with A, B, C. ⟋

A, B, C are either prime to one another or not.

Now, if A, B, C are prime to one another, they are the least of those which have the same ratio with them.
[VII. 21][24]

But, if not, let D the greatest common measure of A, B, C be taken   [VII. 3][25]
and, as many times as D measures the numbers A, B, C respectively, so many units let there be in the numbers E, F, G respectively.

Therefore the numbers E, F, G measure the numbers A, B, C respectively according to the units in D.
[VII. 16][26]

Therefore E, F, G measure A, B, C the same number of times;
therefore E, F, G are in the same ratio with A, B, C.   [VII. Def. 20][27]

I say next that they are the least that are in that ratio.

[...]

Q.E.D.

---

[24][52, p. 83]: Numbers prime to one another are the least of those which have the same ratio with them.

[25][52, p. 77]: Given three numbers not prime to one another, to find their greatest measure.

[26][52, p. 81]: If two numbers by multiplying one another make certain numbers, the numbers so produced will be equal to one another.

[27][52, p. 76]: Numbers are *proportional* when the first is the same multiple, or the same part, or the same parts, of the second that the third is of the fourth.

Before we state Proposition 34, we explain Proposition 33.

Euclid VII.33 begins by considering the numbers A, B, and C, and then proceeds to find the least numbers x, y, and z such that

$$\frac{A}{x} = \frac{B}{y}, \quad \frac{A}{x} = \frac{C}{z}, \quad \frac{B}{y} = \frac{C}{z}.$$

Euclid proceeds by acknowledging that either the $\gcf(A, B, C) = 1$ or the $\gcf(A, B, C) \neq 1$ ("A, B, C are either prime to one another or not"). First, suppose that the $\gcf(A, B, C) = 1$. Then Euclid claims that the proportions

$$\frac{A}{x} = \frac{B}{y}, \quad \frac{A}{x} = \frac{C}{z}, \quad \frac{B}{y} = \frac{C}{z},$$

imply that $x = A$, $y = B$, and $z = C$ ("if A, B, C are prime to one another, they are the least of those which have the same ratio with them"), and cites Book VII, Proposition 21.

Next, suppose that the $\gcf(A, B, C) \neq 1$, say, the $\gcf(A, B, C) = D$. If $A = E \times D$, $B = F \times D$, and $C = G \times D$, then Euclid claims that E, F, and G are the least numbers such that

$$\frac{A}{E} = \frac{B}{F}, \quad \frac{A}{E} = \frac{C}{G}, \quad \frac{B}{F} = \frac{C}{G}.$$

Euclid goes on to prove this; however, we omit the proof.

That is the essence of Proposition 33. We now go on to Propostion 34.

Next, Euclid VII.34 states [53, pp. 180–181]:

Given two numbers, to find the least number which they measure.

Let A, B be the two given numbers; thus it is required to find the least number which they measure.

Now A, B are either prime to one another or not.

First, let A, B be prime to one another,
and let A by multiplying B make C;
therefore also B by multiplying A has made C.

Therefore A, B measure C.

I say next that it is also the least number they measure.

[...]

Next, let A, B not be prime to one another,
and let F, E, the least numbers of those which have the same ratio with A, B, be taken; [VII. 33]
therefore, the product of A, E is equal to the product of B, F. [VII. 19][28]

And let A by multiplying E make C;
therefore also B by multiplying F has made C;
therefore A, B measure C.

I say next that it is also the least number that they measure.

[...]

Q.E.D.

Recall that while Euclid VII.33 was required in the proof of VII.34, it is Proposition 34 that actually tells us how to find the least common multiple of two numbers.

---

[28][52, p. 82]: If four numbers be proportional, the number produced from the first and fourth will be equal to the number produced from the second and third; and, if the number produced from the first and the fourth be equal to that produced from the second and third, the four numbers will be proportional.

Euclid VII.34 begins by considering the numbers A and B and, as in VII.33, acknowledges that either the $\gcf(A, B) = 1$ or not ("A, B are either prime to one another or not"). First, Euclid claims that

$$\gcf(A, B) = 1 \quad \text{implies} \quad \operatorname{lcm}(A, B) = A \times B$$

("let A by multiplying by B make C.... Therefore A, B measure C. I say next that it [C] is also the least number they measure"). Euclid goes on to prove this; however, we omit the proof.

Next, suppose that the $\gcf(A, B) \neq 1$ and let F, E be the least numbers such that

$$\frac{A}{F} = \frac{B}{E}.$$

(Proposition 33 tells us how to find F and E.) Then $A \times E = B \times F$ (by cross multiplying), say

$$A \times E = C, \quad \text{and so} \quad B \times F = C.$$

Euclid claims that, in this case, the

$$\operatorname{lcm}(A, B) = C.$$

Euclid goes on to prove this; however, we again omit the proof.

Note that to find C, the LCM of A and B, we first have to find F and E, the least numbers such that $\frac{A}{F} = \frac{B}{E}$; and the key to finding F and E is given in Proposition 33, namely, the $\gcf(A, B)$. Here is an example.

Example 17.16 Find the LCM of 12 and 18.

We follow Euclid VII.34. Let $A = 12$ and $B = 18$. First, we find the least numbers F and E such that $\frac{A}{F} = \frac{B}{E}$, that is, such that

$$\frac{12}{F} = \frac{18}{E}.$$

We note that (say, using Euclid's algorithm) the $\gcf(12, 18) = 6$. Thus, if we let $D = 6$, then, by Euclid VII.33,[29]

$$A = F \times D \implies 12 = F \times 6 \quad \text{and} \quad B = E \times D \implies 18 = E \times 6$$

implies that

$$F = 2 \quad \text{and} \quad E = 3.$$

Hence,

$$\frac{12}{2} = \frac{18}{3}.$$

Therefore, because (cross multiplying)

$$12 \times 3 = 36 \quad \text{and} \quad 18 \times 2 = 36,$$

by Euclid VII.34, the

$$\operatorname{lcm}(12, 18) = 36.$$

$\square$

FIGURE 17.14 summarizes the steps to finding the least common multiple of two numbers.

Example 17.17 Find the LCM of 60 and 27.

Let $A = 60$ and $B = 27$.

1. Find the GCF of A and B: Using Euclid's algorithm, we find that the $\gcf(60, 27) = 3$. Let $D = 3$.

2. Find the least numbers F and E such that $\frac{A}{F} = \frac{B}{E}$:

---

[29]Note that Euclid VII.33 refers to the ratio $\frac{A}{E} = \frac{B}{F}$, but Euclid VII.34 refers to the same ratio as $\frac{A}{F} = \frac{B}{E}$. Hence, we exchange the roles of E and F here.

> To find the LCM of A and B,
>
> 1. Find the GCF of A and B: Let $\gcf(A, B) = D$.
>
> 2. Find the least numbers F and E such that $\frac{A}{F} = \frac{B}{E}$:
> $$F = \frac{A}{D} \quad \text{and} \quad E = \frac{B}{D}.$$
>
> 3. Compute the LCM of A and B:
> $$\text{lcm}(A, B) = A \times E \quad \text{or} \quad B \times F.$$

Figure 17.14: Finding the least common multiple of two numbers.

$$F = \frac{A}{D} = \frac{60}{3} = 20 \quad \text{and} \quad E = \frac{B}{D} = \frac{27}{3} = 9.$$

3. Compute the LCM of A and B:

$$\text{lcm}(60, 27) = A \times E = 60 \times 9 = 540 \quad \text{or}$$
$$\text{lcm}(60, 27) = B \times F = 27 \times 20 = 540.$$

Therefore, the $\text{lcm}(60, 27) = 540$. □

Now You Try 17.25 Use Euclid VII.33 and 34, as summarized above, to find the LCM of the following pairs of numbers.

1. 497 and 527
2. 220 and 284
3. 863,835 and 802,725

□

Finally, observe that
$$F = \frac{A}{D} \quad \text{and} \quad E = \frac{B}{D}$$

imply that
$$A \times E = \frac{A \times B}{D} \quad \text{and} \quad B \times F = \frac{B \times A}{D}.$$

Hence, we may restate Euclid VII.33 and 34 as follows.

Theorem 17.3 *Let a and b be natural numbers. Then, the*

$$\text{lcm}(a, b) = \frac{ab}{\gcf(a, b)}.$$

□

Example 17.18

1. The $\gcf(12, 18) = 6$. Thus, the

$$\text{lcm}(12, 18) = \frac{12 \cdot 18}{6} = 36.$$

2. The gcf$(60, 27) = 3$. Thus, the

$$\text{lcm}(60, 27) = \frac{60 \cdot 27}{3} = 540.$$

3. The gcf$(37{,}647,\ 29{,}309) = 1$. Thus, the

$$\text{lcm}(37{,}647,\ 29{,}309) = \frac{37{,}647 \cdot 29{,}309}{1} = 1{,}103{,}395{,}923.$$

$\square$

Now You Try 17.26 Use THEOREM 17.3 to find the LCM of the following pairs of numbers.

1. 497 and 527   2. 220 and 284   3. 863,835 and 802,725

$\square$

17.6   PRIME NUMBERS

Prime numbers are the basic building blocks of the natural numbers, and, thus, they play a very important role in the higher arithmetic. This has been known since antiquity.

Definition 17.6 A natural number is called a *prime number* or a *prime* (or is said *to be prime*) if it has exactly two distinct factors, namely, one and itself. A natural number is called a *composite number* or a *composite* (or is said *to be composite*) if it has more than two distinct factors. The number 1 is neither prime nor composite. $\square$

For example, the first several primes are

$$2,\ 3,\ 5,\ 7,\ 11,\ 13,\ 17,\ 19,\ \text{and}\ 23,$$

and the first several composites are

$$4,\ 6,\ 8,\ 9,\ 10,\ 12,\ 14,\ 15,\ 16,\ 18,\ 20,\ 21,\ \text{and}\ 22.$$

As we saw earlier, the Greeks considered 1 to be the generator of all the numbers. Let us take a look at how Euclid defined primes and composites in the *Elements*, as well as at a few of the propositions in the *Elements* concerning primes.

To begin, we find the following definitions of prime and composite numbers in Euclid VII.Definitions:

11. A *prime number* is that which is measured by an unit alone.

13. A *composite number* is that which is measured by some number.

Example 17.19 We see that the number 9 can be "measured by some number," namely, by 3, as well as by a unit.

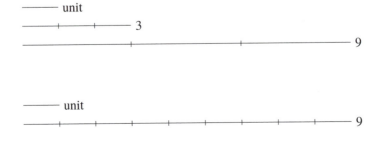

Therefore, we conclude that 9 is a composite number.

On the other hand, we see that the number 7 can be "measured by an unit alone."

Therefore, we conclude that 7 is a prime number.   □

Think About It 17.21 Do the definitions of prime and composite in Euclid's *Elements* agree with DEFI-
NITION 17.6?   □

Now, the most rudimentary way to determine if a given number n is prime is to attempt to divide
n by numbers that are less than n: If we find that n has a factor between 1 and n, then n is composite.
Thankfully, just as with finding all the factors of a number, we need to divide into n only numbers that are
less than $\sqrt{n}$. (See page 221.)

Example 17.20 Determine if n is a prime number.

1. $n = 101$

   First, we note that $\sqrt{101}$ is between 10 and 11 because $10^2 = 100$ and $11^2 = 121$. Thus, we attempt
   to divide 101 by the numbers 2, 3, 4, ..., 10, in turn, to see if any is a factor of 101.

   $$101 \div 2 = 50 \ \ \text{R } 1, \qquad 101 \div 3 = 33 \ \ \text{R } 2, \qquad 101 \div 4 = 25 \ \ \text{R } 1,$$
   $$101 \div 5 = 20 \ \ \text{R } 1, \qquad 101 \div 6 = 16 \ \ \text{R } 5, \qquad 101 \div 7 = 14 \ \ \text{R } 3,$$
   $$101 \div 8 = 12 \ \ \text{R } 5, \qquad 101 \div 9 = 11 \ \ \text{R } 2, \qquad 101 \div 10 = 10 \ \ \text{R } 1$$

   Since none of 2, 3, 4, ..., 10 divides into 101, the number 101 has exactly two distinct factors
   (namely, 1 and 101). Therefore, we conclude that 101 is a prime number.

2. $n = 1001$

   To begin, we note that $\sqrt{1001}$ is between 31 and 32 because $31^2 = 961$ and $32^2 = 1024$. Thus, we
   attempt to divide 1001 by the numbers 2, 3, 4, ..., 31, in turn, to see if any is a factor of 1001.

   $$1001 \div 2 = 500 \ \ \text{R } 1, \qquad 1001 \div 3 = 333 \ \ \text{R } 2, \qquad 1001 \div 4 = 250 \ \ \text{R } 1,$$
   $$1001 \div 5 = 200 \ \ \text{R } 1, \qquad 1001 \div 6 = 166 \ \ \text{R } 5, \qquad 1001 \div 7 = 143 \ \ \text{R } 0$$

   Thus, $1001 = 7 \cdot 143$, and we do not need to continue dividing.

   Since 7 (and also 143) is a factor of 1001, the number 1001 has more than two distinct factors
   (namely, 1, 7, 143, 1001, and potentially others as well). Therefore, we conclude that 1001 is *not* a
   prime number.

3. $n = 143$

   Since $\sqrt{143}$ is between 11 and 12 ($11^2 = 121$ and $12^2 = 144$), we divide 143 by the numbers 2, 3, 4,
   ..., 11, in turn, to see if any is a factor of 143.

   $$143 \div 2 = 71 \ \ \text{R } 1, \qquad 143 \div 3 = 47 \ \ \text{R } 3, \qquad 143 \div 4 = 35 \ \ \text{R } 3,$$
   $$143 \div 5 = 28 \ \ \text{R } 3, \qquad 143 \div 6 = 26 \ \ \text{R } 5, \qquad 143 \div 7 = 20 \ \ \text{R } 3,$$
   $$143 \div 8 = 17 \ \ \text{R } 7, \qquad 143 \div 9 = 15 \ \ \text{R } 9, \qquad 143 \div 10 = 14 \ \ \text{R } 3,$$

$$143 \div 11 = 13 \quad R \ 0$$

Thus, $143 = 11 \cdot 13$.

Since 11 (and also 13) is a factor of 143, the number 143 has more than two distinct factors. Therefore, we conclude that 143 is not prime.

$\square$

Think About It 17.22 When checking by division if a number n is prime, do we need to divide n by *every* number beginning with 2 until we find a factor or exceed $\sqrt{n}$, or may we skip some numbers? Why? $\square$

17.6.1   The Sieve of Eratosthenes

There is a simple method for finding all the prime numbers that are less than a given number. The method is called the *sieve of Eratosthenes*.

Eratosthenes (276–195 BC) was born in Cyrene (now Libya). Around 246 BC, at the invitation of King Ptolemy III, Eratosthenes moved to Alexandria to tutor the king's son, the crown prince of Egypt, Philopator. While in Alexandria, Eratosthenes was appointed chief librarian of the Museum. The appointment not only accorded Eratosthenes great prestige, but it also gave him access to a vast collection of knowledge.

Eratosthenes studied and became proficient in a wide variety of subjects—astronomy, geometry, grammar, history, philosophy—for which he earned two nicknames: Pentathis, a champion of five athletic events, and Beta, the second letter of the Greek alphabet and also the Greek numeral two. It is easy to understand why the nickname Pentathis would have been given by admirers of Eratosthenes, but what about the nickname Beta? Burton [22, p. 183] tells us that Eratosthenes was given the nickname Beta by his detractors, who were "insinuating that while Eratosthenes stood at least second in all fields, he was first in none"; but then Burton goes on to say, "Perhaps a kinder explanation . . . is that certain lecture halls in the Museum were marked with letters, and Eratosthenes was given the name of the room in which he taught." However, Calinger [35, p. 171] posits an even kinder explanation:

> The nickname *Beta* may suggest that he was second only to Archimedes. . . . But perhaps it meant only that a poll of scholars would accord him the "vote of Themistocles" in every branch of knowledge. The Greek defeat of the Persians in the sea battle of Salamis depended greatly upon the stratagems, valor, and fortitude of the Athenian leader Themistocles. After the victory when the generals voted to decide who was the bravest in battle "each voted for himself as the most valorous [*alpha*] and Themistocles as the second [*beta*]."

Eratosthenes was a younger contemporary of Archimedes, who is perhaps the most significat mathematician of all time. The two met, became friends, and corresponded on mathematics problems. Two significant letters that Archimedes wrote to Eratosthenes are the tract *On the Method* in which Archimedes explains how he arrived at some of his results, but does not provide any demonstrations, and *Cattle Problem*. Eratosthenes eventually became almost blind and committed suicide by starving himself to death (a "philosopher's death").

Today, Eratosthenes is best remembered for two great achievements. The first was his calculation of the circumference of the earth that appeared in his three-book collection, *Geography* [35, pp. 173–174]. By observing the length of the shadow of a gnomon at noon on summer solstice in Alexandria and in Syene (modern Aswan), due south of Alexandria, and knowing the distance between the two cities, Eratosthenes calculated the circumference of the earth to be 250,000 stades, that he later modified to be 252,000 stades. Now, distances measured in stades were not uniform: commonly $7\frac{1}{2}$ to 10 stades equaled one Roman mile. Thus, using $7\frac{1}{2}$ stades to one Roman mile, Eratosthenes's measurement of 252,000 stades converts to 33,600 Roman miles; at the other end, using the Egyptian equivalent of just over 10 stades to one Roman mile (about 10.2 stades to one Roman mile), Eratosthenes's measurement converts to 24,706 Roman miles. Today, the circumference of the earth is known to be about 24,860 miles. Since one Roman mile is taken to be 5000 feet, and a U.S. mile is 5280 feet, we see that the error in Eratosthenes's measurement was between about 28% over and 1% under modern measurement.

Eratosthenes's second great achievement was his "sieve" for finding all the prime numbers that are less than a given number. We learn about Eratosthenes's sieve from Nicomachus [68, p. 100], and it is his sieve that interests us here. Let us illustrate how one may use the sieve.

Example 17.21 Find all the prime numbers that are less than 40 using Eratosthenes's sieve.

First, note that 2 is prime, but multiples of 2, that is, the succeeding even numbers, are not. (This is because every $n > 2$ that is a multiple of 2 will have 2 as a factor, so that n has at least three distinct factors, namely, 1, 2, and n.) Next, list all the odd numbers that are less than 40.

$$3 \quad 5 \quad 7 \quad 9 \quad 11 \quad 13 \quad 15 \quad 17 \quad 19 \quad 21$$
$$23 \quad 25 \quad 27 \quad 29 \quad 31 \quad 33 \quad 35 \quad 37 \quad 39$$

The first number in the list, 3, is prime, but multiples of 3 are not (for the same reason that the multiples of 2 are not prime), so we cross off all the odd multiples of 3, namely, 9, 15, 21, and so on.

$$\boxed{3} \quad 5 \quad 7 \quad \cancel{9} \quad 11 \quad 13 \quad \cancel{15} \quad 17 \quad 19 \quad \cancel{21}$$
$$23 \quad 25 \quad \cancel{27} \quad 29 \quad 31 \quad \cancel{33} \quad 35 \quad 37 \quad \cancel{39}$$

The next number that has not been crossed off, 5, is prime, but multiples of 5 are not, so we cross off all the odd multiples of 5, namely, 15, 25, and 35.

$$\boxed{3} \quad \boxed{5} \quad 7 \quad \cancel{9} \quad 11 \quad 13 \quad \cancel{15} \quad 17 \quad 19 \quad \cancel{21}$$
$$23 \quad \cancel{25} \quad \cancel{27} \quad 29 \quad 31 \quad \cancel{33} \quad \cancel{35} \quad 37 \quad \cancel{39}$$

At this point, we may stop because the next number that has not been crossed off is 7, and $7^2 > 40$. All the numbers that have not been crossed off are prime.

$$\boxed{3} \quad \boxed{5} \quad \boxed{7} \quad \cancel{9} \quad \boxed{11} \quad \boxed{13} \quad \cancel{15} \quad \boxed{17} \quad \boxed{19} \quad \cancel{21}$$
$$\boxed{23} \quad \cancel{25} \quad \cancel{27} \quad \boxed{29} \quad \boxed{31} \quad \cancel{33} \quad \cancel{35} \quad \boxed{37} \quad \cancel{39}$$

Hence, the prime numbers that are less than 40 are

$$2, \ 3, \ 5, \ 7, \ 11, \ 13, \ 17, \ 19, \ 23, \ 29, \ 31, \ \text{and } 37.$$

Note that if we had wanted to find all the primes that are less than a number $n > 49$, then we would have repeated the process by crossing off all the odd multiples of 7 next, and so on, until the list begins with a number (a prime number) whose square is greater than n. □

Now You Try 17.27 Find all the prime numbers that are less than 100 using Eratosthenes's sieve. □

17.6.2   The Fundamental Theorem of Arithmetic

Consider the number 150. There are several ways to factor 150, namely,

$$2 \cdot 75, \quad 3 \cdot 50, \quad 5 \cdot 30,$$
$$2 \cdot 3 \cdot 25, \quad 2 \cdot 5 \cdot 15, \quad 3 \cdot 5 \cdot 10,$$
$$2 \cdot 3 \cdot 5 \cdot 5.$$

Among the different factorizations of 150, we note that the last factorization, $2 \cdot 3 \cdot 5 \cdot 5$, is distinguished because its factors are primes only. We call $2 \cdot 3 \cdot 5 \cdot 5$ a *prime factorization* of 150.

Think About It 17.23 Is there more than one prime factorization of a number?   □

An important result is that every natural number greater than one has a prime factorization and that this factorization is essentially unique. This result is known as the *fundamental theorem of arithmetic* or the *unique factorization theorem*.

Theorem 17.4 (FUNDAMENTAL THEOREM OF ARITHMETIC) *Every natural number greater than 1 either is a prime or can be factored into a product of primes in essentially one way. In other words, if $n > 1$ is a natural number, then either n is prime or*

$$n = p_1 p_2 p_3 \cdots p_k,$$

where the p are primes; moreover, if

$$n = p_1 p_2 p_3 \cdots p_k = q_1 q_2 q_3 \cdots q_l,$$

where the q are also primes, then $l = k$ and the primes q can be rearranged so that

$$q_1 = p_1, \quad q_2 = p_2, \quad q_3 = p_3, \ldots, \quad q_k = p_k.$$

□

Example 17.22 Find the prime factorization of each of the following numbers.

1. 150

    We use a "factor tree" to find the prime factorization of 150. In a factor tree, we express the number at each node as a product of any two numbers, and then branch until we arrive at only prime factors. Here are two possible factor trees for 150.

    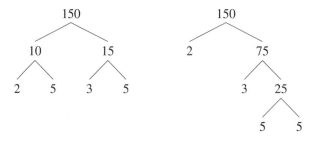

    We see that both factor trees result in essentially the same prime factorization of 150, for

    $$2 \cdot 5 \cdot 3 \cdot 5 = 2 \cdot 3 \cdot 5 \cdot 5.$$

    This is precisely what the fundamental theorem of arithmetic asserts.

    Therefore, the prime factorization of 150 is

    $$2 \cdot 3 \cdot 5 \cdot 5 \quad \text{or} \quad 2 \cdot 3 \cdot 5^2.$$

2. 525

We use a factor tree again.

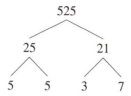

Therefore, the prime factorization of 525 is

$$3 \cdot 5 \cdot 5 \cdot 7 \quad \text{or} \quad 3 \cdot 5^2 \cdot 7.$$

□

Now You Try 17.28

1. Find the prime factorization of 525. Begin the factor tree as follows.

2. Find the prime factorization of the following.

a) 51   c) 101   e) 1001

b) 64   d) 500

□

Davenport [45, pp. 18–19] tells us that the history of the fundamental theorem of arithmetic "is strangely obscure. It does not figure in Euclid's *Elements*, though some of the arithmetical propositions in Book VII of the *Elements* are almost equivalent to it. Nor is it stated explicitly even in Legendre's *Essai sur la théorie des nombres* of 1798. The first clear statement and proof seem to have been given by Gauss in his famous *Disquisitiones Arithmeticae* of 1801."

The propositions in Euclid's *Elements* that lead to a proof of the fundamental theorem of arithmetic are in Book VII, Propositions 31, 32, and 30, and in Book IX, Proposition 14 [75, p. 79]. We state these propositions here without any demonstration.

VII.31 [52, p. 85] Any composite number is measured by some prime number.

VII.32 [52, p. 86] Any number either is prime or is measured by some prime number.

VII.30 [52, p. 85] If two numbers by multiplying one another make some number, and any prime number measure the product, it will also measure one of the original numbers.

IX.14 [52, p. 105] If a number be the least that is measured by prime numbers, it will not be measured by any other prime number except those originally measuring it.

We may obtain all the factors of a number from its prime factorization. For example, consider

$$150 = 2 \cdot 3 \cdot 5^2.$$

Then any factor of 150 will have to have *at most* one factor of 2, one factor of 3, and two factors of 5.

Factor of 150	as a product of 2, 3, and 5
1	$2^0 \cdot 3^0 \cdot 5^0$
2	$2^1 \cdot 3^0 \cdot 5^0$
3	$2^0 \cdot 3^1 \cdot 5^0$
5	$2^0 \cdot 3^0 \cdot 5^1$
6	$2^1 \cdot 3^1 \cdot 5^0$
10	$2^1 \cdot 3^0 \cdot 5^1$
15	$2^0 \cdot 3^1 \cdot 5^1$
25	$2^0 \cdot 3^0 \cdot 5^2$
30	$2^1 \cdot 3^1 \cdot 5^1$
50	$2^1 \cdot 3^0 \cdot 5^2$
75	$2^0 \cdot 3^1 \cdot 5^2$
150	$2^1 \cdot 3^1 \cdot 5^2$

As another example, let

$$m = p^2 q r^3,$$

where p, q, and r are primes. Then any factor of m will have at most two factors of p, one factor of q, and three factors of r.

Factor of m	as a product of p, q, and r
1	$p^0 q^0 r^0$
p, q, r	$p^1 q^0 r^0, \; p^0 q^1 r^0, \; p^0 q^0 r^1$
p^2, pq, qr, pr, r^2	$p^2 q^0 r^0, \; p^1 q^1 r^0, \; p^0 q^1 r^1, \; p^1 q^0 r^1, \; p^0 q^0 r^2$
$p^2 q, p^2 r, pqr, pr^2, qr^2, r^3$	$p^2 q^1 r^0, \; p^2 q^0 r^1, \; p^1 q^1 r^1, \; p^1 q^0 r^2, \; p^0 q^1 r^2, \; p^0 q^0 r^3$
$p^2 q r^2, pqr^3, p^2 r^3$	$p^2 q^1 r^2, \; q^1 q^1 r^3, \; p^2 q^0 r^3,$
$p^2 q r^3$	$p^2 q^1 r^3$

Now You Try 17.29

1. Show that any factor of $525 = 3 \cdot 5^2 \cdot 7$ will have at most one factor of 3, two factors of 5, and one factor of 7.

2. Show that any factor of $n = p^3 r$, where p and r are primes, will have at most three factors of p and one factor of r.

□

Now, because

- any factor of $150 = 2 \cdot 3 \cdot 5^2$ will have *at most* one factor of 2, one factor of 3, and two factors of 5, and

- any factor of $525 = 3 \cdot 5^2 \cdot 7$ will have *at most* one factor of 3, two factors of 5, and one factor of 7,

we see that any *common* factor of 150 and 525, which is the intersection of these two lists, will have *at most* one factor of 3 and two factors of 5. Hence, the *greatest* common factor of 150 and 525 will have *exactly* one factor of 3 *and* two factors of 5, that is, the

$$\gcf(150, 525) = 3 \cdot 5^2 = 75.$$

Number	Prime factors			
150	2	3	5^2	
525		3	5^2	7
GCF		3	5^2	

On the other hand, because

- any multiple of $150 = 2 \cdot 3 \cdot 5^2$ will have *at least* one factor of 2, one factor of 3, and two factors of 5, and

- any multiple of $525 = 3 \cdot 5^2 \cdot 7$ will have *at least* one factor of 3, two factors of 5, and one factor of 7,

we see that any *common* multiple of 150 and 525, which is the union of these two lists, will have *at least* one factor of 2, one factor of 3, two factors of 5, and one factor of 7. Hence, the *least* common multiple of 150 and 525 will have *exactly* one factor of 2, one factor of 3, two factors of 5, and one factor of 7, that is, the

$$\lcm(150, 525) = 2 \cdot 3 \cdot 5^2 \cdot 7 = 1050.$$

Number	Prime factors			
150	2	3	5^2	
525		3	5^2	7
LCM	2	3	5^2	7

As another example, if p, q, and r are primes, because

- any factor of $m = p^2qr^3$ will have at most two factors of p, one factor of q, and three factors of r, and

- any factor of $n = p^3r$, where p and r are primes, will have at most three factors of p and one factor of r,

we see that any common factor of m and n will have at most two factors of p and one factor of r. Hence, the GCF of m and n will have exactly two factors of p and one factor of r, that is, the

$$\gcf(m, n) = p^2r.$$

Number	Prime factors		
m	p^2	q	r^3
n	p^3		r
GCF	p^2		r

On the other hand, if p, q, and r are primes, because

- any multiple of $m = p^2qr^3$ will have at least two factors of p, one factor of q, and three factors of r, and

- any multiple of $n = p^3 r$ will have at least three factors of p and one factor of r,

we see that any common multiple of m and n will have at least three factors of p, one factor of q, and three factors of r. Hence, the LCM of m and n will have exactly three factors of p, one factor of q, and three factors of r, that is, the

$$\mathrm{lcm}(m, n) = p^3 q r^3.$$

Number	Prime factors		
m	p^2	q	r^3
n	p^3		r
LCM	p^3	q	r^3

Think About It 17.24 Let the $\gcf(a, b) = g$ and the $\mathrm{lcm}(a, b) = l$. How are the prime factorizations of a, b, g, and l related? □

Example 17.23 Find the GCF and the LCM.

1. 60 and 27

   We find the prime factorizations of 60 and 27.

Number	Prime factors		
60	2^2	3	5
27		3^3	
GCF		3	
LCM	2^2	3^3	5

   Therefore, the

   $$\gcf(60, 27) = 3,$$
   $$\mathrm{lcm}(60, 27) = 2^2 \cdot 3^3 \cdot 5 = 540.$$

   Observe that this is consistent with THEOREM 17.3 on page 237:

   $$\mathrm{lcm}(60, 27) = \frac{60 \cdot 27}{\gcf(60, 27)} = \frac{(2^2 \cdot 3 \cdot 5)(3^3)}{3} = 2^2 \cdot 3^3 \cdot 5.$$

2. 225, 60, and 27

   We find the prime factorizations of 225, 60, and 27.

Number	Prime factors		
225		3^2	5^2
60	2^2	3	5
27		3^3	
GCF		3	
LCM	2^2	3^3	5^2

Therefore, the

$$\gcf(60, 45, 27) = 3,$$
$$\lcm(60, 45, 27) = 2^2 \cdot 3^3 \cdot 5^2 = 2700.$$

□

Now You Try 17.30 Use prime factorization to find the GCF and the LCM of the following numbers.

1. 497 and 527   2. 220 and 284   3. 220, 284, 497, and 527

□

Think About It 17.25 Is there a largest prime? If not, how could you use the ideas presented in this section to prove that there is no largest prime? □

We close this subsection with Proposition 20 in Book IX of Euclid's *Elements* that asserts that the series of primes never comes to an end. We present both the proposition and proof in Euclid's *Elements*, as well as a modern proof for comparison. Euclid's proof of IX.20 is a classic example of a proof by contradiction or *reductio ad absurdum*.

Euclid IX.20 [53, p. 227]

Prime numbers are more than any assigned multitude of prime numbers.

Let A, B, C be the assigned prime numbers;
I say that there are more prime numbers
than A, B, C. [See FIGURE 17.15.]

For let the least number measured by A, B, C
be taken,   [VII. 36][30]
and let it be DE;
let the unit DF be added to DE.

Then EF is either prime or not.
First, let it be prime;
then the prime numbers A, B, C, EF have been found which are more than A, B, C.

Next, let EF not be prime;
                            therefore it is measured by some prime number: [VII. 31]

Let it be measured by the prime number G.

I say that G is not the same with any of the numbers A, B, C.

For, if possible, let it be so.

Now A, B, C measure DE;
                                                            therefore G also will measure DE.

But it also measures EF.

Therefore G, being a number, will measure the remainder, the unit DF: which is absurd.

Therefore G is not the same with any one of the numbers A, B, C.

And by hypothesis it is prime.

Therefore the prime numbers A, B, C, G have been found which are more than the assigned multitude of A, B, C.

Q.E.D.

Figure 17.15: Euclid IX.20: The infinitude of primes.

Euclid's proof of IX.20 is truly quite impressive, and a modern proof of the proposition remains essentially the same. For example, here is Davenport's proof that *the series of primes never comes to an end* [45, p. 17–18].

> Let 2, 3, 5, ..., P be the series of primes up to a particular prime P. Consider the number obtained by multiplying all these primes together, and then adding 1, that is
>
> $$N = 2 \times 3 \times 5 \times \ldots \times P + 1.$$
>
> This number cannot be divisible by 2, for then both the numbers N and $2 \times 3 \times 5 \times \ldots \times P$ would be divisible by 2, and therefore their difference would be divisible by 2. This difference is 1, and is not divisible by 2. In the same way, we see that N cannot be divisible by 3 or by 5 or by any of the primes up to and including P. On the other hand, N is divisible by *some* prime (namely N itself if N is a prime, or any prime factor of N if N is composite). Hence there exists a prime which is different from any of the primes 2, 3, 5, ..., P, and so is greater than P. Consequently the series of primes never comes to an end.

Another reason that these proofs by contradiction are remarkable is that we have at once that there are infinitely many prime numbers without at all saying how one may generate these primes.

Think About It 17.26 Explain how Euclid's proof and Davenport's proof that there are infinitely many prime numbers are essentially the same. □

Think About It 17.27 Can the formula $N = 2 \times 3 \times 5 \times \cdots \times P + 1$, where 2, 3, 5, ..., P are primes, be used to generate prime numbers? □

17.6.3 Perfect Numbers

We noted earlier that the Pythagoreans considered 10 the "perfect number" because $10 = 1 + 2 + 3 + 4$, and that 10 is the essence of the tetractys (see FIGURE 17.2). Euclid, in the *Elements*, Book VII, Definitions, gives a different definition of perfect numbers:[31]

> 22. A *perfect number* is that which is equal to its own parts.

For example, 6 and 28 are perfect numbers in the sense of Euclid VII.Def.22.[32] Recall that Euclid VII.Def.3 (page 222) defines a part of a number to be a *proper* factor of the number; the proper factors of a number are also called its *aliquot parts*. Hence, the aliquot parts of 6 are 1, 2, and 3, and

$$1 + 2 + 3 = 6.$$

———— unit

———————————————————— 6

---

[30][52, p. 87]: Given three numbers, to find the least number which they measure.

[31]Heath [68, p. 74] tells us that this definition of a perfect number, Euclid VII.Def.22, has not been found in any works prior to the *Elements*.

[32]Ore [92, p. 91] relates that "perfect numbers are essential elements in all numerological speculations. God created the world in six days, a perfect number. The moon circles the earth in 28 days, again a symbol of perfection in the best of all possible worlds."

Similarly, the aliquot parts of 28 are 1, 2, 4, 7, and 14, and

$$1 + 2 + 4 + 7 + 14 = 28.$$

— unit

28

Now You Try 17.31

1. Show that 496 is a perfect number, in fact, the third perfect number after 6 and 28.

2. Is 10 a perfect number according to Euclid's definiton?

☐

Perfect numbers have fascinated many since antiquity. It turns out that all even perfect numbers have the same formulation given by Euclid in the *Elements*, Book XI, Proposition 36 [53, p. 234]:

If as many numbers as we please beginning from an unit be set out continuously in double proportion, until the sum of all becomes prime, and if the sum multiplied into the last make some number, the product will be perfect.

Euclid proves the proposition, but we omit the proof here; instead, we only explain the proposition's assertion.

To let numbers "beginning from an unit be set out continuously in double proportion" means to start with 1 and continually double, forming the sequence

$$1, 2, 4, 8, 16, 32, \ldots \quad \text{or} \quad 1, 2, 2^2, 2^3, 2^4, 2^5, \ldots .$$

Thus, "beginning from an unit be set out continuously in double proportion, until the sum of all becomes a prime" means to consider the sums and note

$$1 + 2 = 3 \qquad \text{is prime,}$$
$$1 + 2 + 2^2 = 7 \qquad \text{is prime,}$$
$$1 + 2 + 2^2 + 2^3 = 15 \qquad \text{is not prime,}$$
$$1 + 2 + 2^2 + 2^3 + 2^4 = 31 \qquad \text{is prime,}$$
$$1 + 2 + 2^2 + 2^3 + 2^4 + 2^5 = 63 \qquad \text{is not prime,}$$

and so on. Finally, "until the sum of all becomes prime, and if the sum multiplied into the last make some number, the product will be perfect" means that, for example, because $1 + 2 = 3$ is prime, then

$$(1 + 2) \cdot 2 = 6 \quad \text{is perfect,}$$

which we know is true. Similarly, because $1 + 2 + 2^2 = 7$ is prime, then

$$(1 + 2 + 2^2) \cdot 2^2 = 28 \quad \text{is perfect,}$$

which, again, we know is true. On the other hand, $1 + 2 + 2^2 + 2^3 = 15$ is not prime and

$$(1 + 2 + 2^2 + 2^3) \cdot 2^3 = 120 \quad \text{is not perfect.}$$

In fact, the parts of 120 are 1, 2, 3, 5, 6, 8, 10, 12, 15, 20, 24, 40, and 60, and

$$1 + 2 + 3 + 5 + 6 + 8 + 10 + 12 + 15 + 20 + 24 + 40 + 60 = 206 > 120.$$

In modern notation, then, Euclid IX.36 may be stated: If the sum

$$1 + 2 + 2^2 + 2^3 + \cdots + 2^n$$

is a prime number, then the product

$$(1 + 2 + 2^2 + 2^3 + \cdots + 2^n)2^n$$

is a perfect number.

Now You Try 17.32 Express the perfect number 496 in the form $(1 + 2 + 2^2 + 2^3 + \cdots + 2^n)2^n$. ☐

We may write the formulation for even perfect numbers, $(1 + 2 + 2^2 + 2^3 + \cdots + 2^n)2^n$, more compactly. Toward this end, let us find a more compact way of writing the sequence $1 + 2 + 2^2 + \cdots + 2^n$. Let

$$S_1 = 1 + 2^0, \quad S_2 = 1 + 2^1, \quad S_3 = 1 + 2^1 + 2^2, \quad S_4 = 1 + 2^1 + 2^2 + 2^3,$$

and, in general,

$$S_n = 1 + 2 + 2^2 + 2^3 + \cdots + 2^{n-1}.$$

Think About It 17.28 Find the sums S_1, S_2, S_3, and so, and try to find a formula (a short cut) for computing any sum S_n without actually adding up all the addends. ☐

It turns out that $S_n = 2^n - 1$. Let us deduce this in a systematic way. We begin with a concrete example using S_4.

Observe that the sum S_4 is 15, and that $15 = 16 - 1 = 2^4 - 1$, that is, observe that $S_4 = 2^4 - 1$. Now, since $S_4 = 1 + 2 + 2^2 + 2^3$,

$$2S_4 = 2(1 + 2 + 2^2 + 2^3) = (2 + 2^2 + 2^3 + 2^4).$$

Then,

$$\begin{aligned}
S_4 &= 2S_4 - S_4 \\
&= (2 + 2^2 + 2^3 + 2^4) - (1 + 2 + 2^2 + 2^3) \\
&= 2^4 - 1.
\end{aligned}$$

Therefore, we conclude that

$$S_4 = 1 + 2 + 2^2 + 2^3 = 2^4 - 1.$$

The general case proceeds in the same way. Since $S_n = 1 + 2 + 2^2 + 2^3 + \cdots + 2^{n-1}$,

$$2S_n = 2(1 + 2 + 2^2 + 2^3 + \cdots + 2^{n-2} + 2^{n-1}) = 2 + 2^2 + 2^3 + 2^4 + \cdots + 2^{n-1} + 2^n.$$

Then,

$$\begin{aligned}
S_n &= 2S_n - S_n \\
&= (2 + 2^2 + 2^3 + 2^4 + \cdots + 2^{n-1} + 2^n) - (1 + 2 + 2^2 + 2^3 + \cdots + 2^{n-1}) \\
&= 2^n - 1.
\end{aligned}$$

Therefore, we conclude that

$$S_n = 1 + 2 + 2^2 + 2^3 + \cdots + 2^{n-1} = 2^n - 1.$$

With this formula for S_n in hand, we may state Euclid IX.36 thus:

> If $2^n - 1$ is a prime number, then $(2^n - 1)2^{n-1}$ is a perfect number.

Now You Try 17.33 Express the perfect number 496 in the form $(2^n - 1)2^{n-1}$. □

Remark 17.4 A number that can be expressed in the form $2^n - 1$, where n is a natural number, is called a *Mersenne number* after the French mathematician Marin Mersenne (1588–1648). A Mersenne number that is prime is called a *Mersenne prime*. According to Ore [92, p. 71], "The historical justification for this nomenclature seems rather weak, since several perfect numbers and their corresponding primes have been known since antiquity and occur in almost every medieval numerological speculation. Mersenne did, however, discuss the primes named after him in a couple of places in his work *Cogita physico-mathematica* (Paris, 1644) and expressed various conjectures in regard to their occurrence." □

Perfect numbers are "rare." Indeed, Nicomachus (ca. AD 60–120) knew only four perfect numbers, namely, 6, 28, 496, and 8128. Based on this, Nicomachus asserted that there must be one perfect number among the ones (namely, 6), one perfect number among the tens (namely, 28), one among the hundreds (namely, 496), and one among the thousands (namely, 8128); moreover, he asserted that perfect numbers terminate alternately in 6 and 8. Later, Iamblichus (ca. AD 245–325) suggested that there is one perfect number in the first myriads (less than $10,000^2$), one in the second myriads (less than $10,000^3$), and so on.

Nicomachus was correct about 6, 28, 496, and 8128 being the first four perfect numbers; he was also correct that perfect numbers end in either 6 or 8, but not necessarily alternately. Iamblichus, however, was off the mark. Here are the first several perfect numbers in order:

$$2(2^2 - 1) = 6,$$
$$2^2(2^3 - 1) = 28,$$
$$2^4(2^5 - 1) = 496,$$
$$2^6(2^7 - 1) = 8128,$$
$$2^{12}(2^{13} - 1) = 33,550,336,$$
$$2^{16}(2^{17} - 1) = 8,589,869,056,$$
$$2^{18}(2^{19} - 1) = 137,438,691,328,$$
$$2^{30}(2^{31} - 1) = 2,305,843,008,139,952,128,$$
$$2^{60}(2^{61} - 1) = 2,658,455,991,569,831,744,654,692,615,953,842,176.$$

In general, then, the sum of the parts of a number is either greater than or less than the number itself. If the sum of the aliquot parts of a number is *greater than* the number itself, then the number is said to be *over-perfect* or *abundant*. On the other hand, if the sum of the aliquot parts of a number is *less than* the number itself, then the number is said to be *defective* or *deficient*. These definitions are due to Nicomachus and Theon of Smyrna (ca. AD 70–135).

Now You Try 17.34 Decide which numbers from 4 to 30 are over-perfect and which are defective. □

Finally, we note that it is not known whether there are infinitely many perfect numbers. Also, no odd perfect number has been found to this day.

17.6.4   Friendly Numbers

Two numbers are said to be *friendly numbers* or *amicable numbers* if each is the sum of the aliquot parts (proper factors) of the other. For example, the numbers 220 and 284 are friendly numbers, for the aliquot parts of 220 are 1, 2, 4, 5, 10, 11, 20, 22, 44, 55, and 110, and

$$1 + 2 + 4 + 5 + 10 + 11 + 20 + 22 + 44 + 55 + 110 = 284;$$

and the aliquot parts of 284 are 1, 2, 4, 71, and 142, and

$$1 + 2 + 4 + 71 + 142 = 220.$$

Historian Oystein Ore surmises that amicable numbers were probably discovered after perfect numbers. According to Heath [68, p. 75], "Iamblichus attributes the discovery of such numbers to Pythagoras himself, who, being asked 'what is a friend?' said '*Alter ego*', and on this analogy applied the term 'friendly'....[33]" Ore, however, claims that this tradition has "little credit with the historians of the mathematical sciences" [92, p. 97]. Nevertheless, friendly numbers figured in numerology, finding their way into astrology, the casting of horoscopes, and the making of talismans, for example. Friendly numbers play a significant roll in Arabic writings on astrology and mysticism, which were in turn passed on to European scholars of the Renaissance. Thābit ibn Qurra (836–901) wrote a treatise on friendly numbers, consisting of 10 propositions together with their proofs. According to Berggren [10, p. 560], "the mathematical context of his treatise is the deductive approach found in the number theoretical books (VII–IX) of Euclid's *Elements*." Here are the first several pairs of friendly numbers:

$$\{220, 284\}, \quad \{1184, 1210\}, \quad \{2620, 2924\}, \quad \{5020, 5564\},$$

$$\{6263, 6368\}, \quad \{10{,}744, 10{,}856\}, \quad \{12{,}285, 14{,}595\}, \quad \{16{,}296, 18{,}416\}.$$

Now You Try 17.35 Show that {1184, 1210} is a pair of friendly numbers.   ☐

---

[33] Iambl. *in Nicom.*, p. 35. 1–7. The subject of 'friendly' numbers was taken up by Euler, who discovered no less than sixty-one pairs of such numbers. Descartes and van Schooten had previously found three pairs but no more.

Chapter 18

Exercises

Ex. 18.1 — Let $A = \{1, 2, 5, 7\}$, $B = \{6, 4, 5, 7\}$, $C = \{2, 5, 1, 7\}$, and $D = \{6, 8, 4, 2\}$. Answer each of the following questions. If any of the questions do not make sense, explain why.

1. Is $2 \in A$?
2. Is $\{2\} \in A$?
3. Is $5 \subset B$?
4. Is $\{5\} \subset B$?
5. Does $A = C$?
6. Does $A = D$?

7. Is $B \subset A$?
8. Is $C \subset A$?
9. Find $(A \cap B) \cap D$.
10. Find $A \cap (B \cap D)$.
11. Find $(A \cup B) \cup C$.
12. Find $A \cup (B \cup C)$.

Ex. 18.2 — Let $A = \{1, 2, 5, 7\}$, $B = \{6, 4, 5, 7\}$, $C = \{2, 5, 1, 7\}$.

1. For the given sets A and B, verify the general formula

$$\text{card}(A \cup B) = \text{card}(A) + \text{card}(B) - \text{card}(A \cap B)$$

   that holds for any finite sets A and B. (See DEFINITION 14.10.)

2. For the given sets A, B, and C, find $\text{card}(A \cup B \cup C)$. What is a general formula for the cardinality of the union of any three finite sets A, B, and C?

Ex. 18.3 — Answer each of the following *true* if it is always true, or *false* if it is ever false. If it is false, then also give an example that demonstrates that.

1. The empty set is a subset of every set.
2. The set containing the empty set is empty, that is, $\{\varnothing\} = \varnothing$.
3. The intersection of two nonempty sets is a nonempty set.
4. The union of two nonempty sets is a nonempty set.
5. If $A \cap B \neq \varnothing$ and $B \cap C \neq \varnothing$, then $A \cap B \cap C \neq \varnothing$.
6. Let A, B, and C be nonempty sets. If the intersection of any two of the sets is empty, then $A \cup B \cup C$ is empty.
7. If $A \subseteq B$ and $B \subseteq A$, then $A = B$.
8. If $A \subset B$ and $B \subset C$, then $A \subset C$.

9. Let A, B, and C be subsets of $\mathbb{W}$ (the set of whole numbers). If $1 \in A \cap C$ and $B \subset C$, then $1 \in B$.

10. A set with 10 elements cannot be a subset of a set with seven elements.

11. If A has nine elements and B has six elements, then $A \cup B$ has 15 elements.

12. Let A and B be finite sets. If $A \neq B$, then A and B do not have the same cardinality.

Ex. 18.4 — Find the number of possible metrical patterns for a line with eleven syllables in two ways: (i) by using the method Piṅgala prescribed in his work *Chandahsūtra*; (ii) by finding 2^n for an appropriate integer n.

Ex. 18.5 — A simple electrical circuit has one battery, one light bulb, and five switches placed in series. The only way to switch on the light bulb is to switch on all five switches.

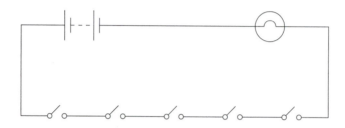

How many ways are there *not* to switch on the light bulb? Hint: Relate this problem to the problem of finding the number of subsets of the set $\{a, b, c, d, e\}$.

Ex. 18.6 — There are 42 persons in the mathematics department, and everyone of them drinks either coffee or grape milk at breakfast. If eighteen persons drink grape milk and ten persons drink both grape milk and coffee, how many persons drink grape milk only, and how many drink coffee only?

Ex. 18.7 — There are 25 persons in the mathematics department. If, at breakfast, seven persons drink grape milk, eighteen persons drink coffee, and four persons drink neither grape milk nor coffee, how many persons drink both grape milk and coffee at breakfast?

Ex. 18.8 — A one-to-one correspondence between the set $\mathbb{W}$ and the set $\mathbb{N}$ (the set of natural numbers or counting numbers) is

$$\{(0, 1),\ (1, 2),\ (2, 3),\ (3, 4),\ (4, 5), \dots \}.$$

Find another one-to-one correspondence between the two sets.

* **Ex. 18.9** — What is the International Congress of Mathematicians?

* **Ex. 18.10** — Write about Zeno's paradoxes.

* **Ex. 18.11** — What are "Euler circles"? Give at least three examples.

* **Ex. 18.12** — What is the Peano space-filling curve? What other space-filling curves are there?

Ex. 18.13 — Let a and b be rational numbers. Show that $\frac{a+b}{2}$ is a rational number that is between a and b.

Ex. 18.14 — A set S is said to be *closed* under a binary operation $*$ if $a * b \in S$ for every a, $b \in S$. (In other words, a set is *closed* if the result of performing a mathematical operation on the elements of the set does not "escape" to be outside the set.) Show that the set of rational numbers is closed under the operations of addition, subtraction, multiplication, and division (except for division by zero).

Ex. 18.15 — Show that the set of irrational numbers is not closed under the operation of addition or multiplication. (Find two irrational numbers whose sum is not irrational; similarly, find two irrational numbers whose product is not irrational.)

Ex. 18.16 — Let $S = \{0, 1\}$. Determine if S is closed under the operation of addition, and if S is closed under the operation of multiplication.

Ex. 18.17 — Show that the real number $\sqrt{3}$ is an irrational number. Hint: First show that a number a is divisible by 3 if and only if a^2 is divisible by 3. To do this, expand $(3n)^2$, $(3n + 1)^2$, and $(3n + 2)^2$. Then follow the proof that $\sqrt{2}$ is an irrational number (page 173).

* **Ex. 18.18** — What are some of the reasons why some scholars believe that $\sqrt{2}$ was the first incommensurable to be discovered, while other scholars believe that $\sqrt{5}$ was the first?

* **Ex. 18.19** — In what way does the Greek use of the terms "commensurate" and "incommensurate" correspond to "rational" and "irrational"?

Ex. 18.20 — Consider the sets

$$\Sigma_0 = \{\ \}, \quad \Sigma_1 = \{a\}, \quad \Sigma_2 = \{a, b\}, \quad \Sigma_3 = \{a, b, c\}, \quad \Sigma_4 = \{a, b, c, d\}.$$

(Σ is the Greek uppercase letter *sigma*.)

1. Find $\mathcal{P}(\Sigma_0)$, the power set of Σ_0. What is card($\mathcal{P}(\Sigma_0)$), the cardinality of $\mathcal{P}(\Sigma_0)$?
2. Find $\mathcal{P}(\Sigma_1)$. What is card($\mathcal{P}(\Sigma_1)$)?
3. Find $\mathcal{P}(\Sigma_2)$. What is card($\mathcal{P}(\Sigma_2)$)?
4. Find $\mathcal{P}(\Sigma_3)$. What is card($\mathcal{P}(\Sigma_3)$)?
5. Find $\mathcal{P}(\Sigma_4)$. What is card($\mathcal{P}(\Sigma_4)$)?
6. Let $\Sigma_5 = \{a, b, c, d, e\}$. Without listing all the subsets of Σ_5, determine how many subsets are there.
7. Let Σ_{10} be a set that contains 10 elements. Determine how many subsets of Σ_{10} are there.
8. Let Σ_n be a set that contains n elements, where $n \in \mathbb{N}$. Determine how many subsets of Σ_n are there.

Ex. 18.21 — Let $A = \{a\}$. Find

1. card(A)
2. card($\mathcal{P}(A)$)
3. card($\mathcal{P}(\mathcal{P}(A))$)
4. card($\mathcal{P}(\mathcal{P}(\mathcal{P}(A)))$)
5. card($\mathcal{P}(\mathcal{P}(\mathcal{P}(\mathcal{P}(A))))$)
6. card($\mathcal{P}(\mathcal{P}(\mathcal{P}(\mathcal{P}(\mathcal{P}(A)))))$)

Ex. 18.22 — List five finite and five infinite elements of $\mathcal{P}(\mathbb{N})$. Explain how the power set of the natural numbers can generate all positive real numbers.

* **Ex. 18.23** — How did Archimedes use infinity to find formulas for the area of various plane figures and the volumes of various solids?

* **Ex. 18.24** — Look up various histories of Pythagoras and the Pythagorean brotherhood. How different are the various accounts? Do any sound hard to believe?

* **Ex. 18.25** — What is a pentastar and a pentagram? What are some of their mystical properties?

* **Ex. 18.26** — Research the continuum hypothesis that takes Cantor's work to a higher level.

* **Ex. 18.27** — Describe briefly the work of Kurt Gödel and Paul Cohen with regard to the consistency and completeness of our number system. What is so profound about their results?

* **Ex. 18.28** — Why is card($\mathcal{P}(\mathbb{N})$) = c, the cardinality of the set of real numbers? Hint: Recall that $I^{(0,1)}$ can be put into a one-to-one correspondence with $\mathbb{R}$. Think of a subset of $\mathbb{N}$ as the digits of a decimal number in $I^{(0,1)}$.

* **Ex. 18.29** — What are "magic squares" and "Latin squares"? Write a brief history of them.

Ex. 18.30 — For each of the following, write the statement as a conditional—if A, then B—if it is not already in this form, then state the hypothesis and the conclusion. Write the contrapositive, converse, and inverse. Write the negation of the statement. (It is not important that you understand the statement to carry out the exercise.)

1. All hawks have good eyesight.

2. No penguin swims like a frog.

3. If the figure is a rectangle or a parallelogram, then its area is given by the formula, "base times height."

4. For any invertible element x in a monoid and any integer n, $(x^{-1})^n = (x^n)^{-1}$.

5. Let X^* be the dual space of a Banach space X. The closed unit ball in X^* is weak-$*$ compact.

Ex. 18.31 — Recall that $(\sim (A \to B)) \equiv (A \land (\sim B))$. Write the negation of the biconditional $A \leftrightarrow B$; you will need to use both $\land$ and $\lor$.

* **Ex. 18.32** — What are De Morgan's Laws? Write them in English. Give an example of how they can be used.

* **Ex. 18.33** — Write about Augustus De Morgan and his mathematical career.

* **Ex. 18.34** — What is a "syllogism"? Give at least three examples.

* **Ex. 18.35** — What is an axiomatic system? What are the axioms of the real number system?

* **Ex. 18.36** — There is an additional logical operation call xor. What is xor and how is it different from or? What are some of its uses?

* **Ex. 18.37** — From where do the logical terms "non," "et," "vel," "seq," and "aeq" come?

Ex. 18.38 — Would these be good definitions of even or odd? Explain.

1. An integer n is said to be *even* if $n = 4k$ for some integer k.

2. An integer n is said to be *even* if $n = -2k$ for some integer k.

3. A whole number n is said to be *odd* if $n = 2k - 1$ for some whole number k.

4. A whole number n is said to be *odd* if $n = 2k + 1$ for some whole number k.

5. An integer n is said to be *odd* if $n = 2k + 1$ for some integer k.

Ex. 18.39 — Determine if the following integers are even or odd using one of the defintions presented.

    1. 9   2. 12   3. -10 4. 0

Ex. 18.40 — Show that each of the following statements is true in general, and provide three numerical examples for each of the statements.

1. The sum of two odd integers is an even integer. Use $m = 2j + 1$ and $n = 2k + 1$.

2. The product of two odd integers is an odd integer. Use $m = 2j + 1$ and $n = 2k + 1$.

3. The sum of an even integer and an odd integer is an odd integer.

4. The product of an even integer and an odd integer is an even integer.

Ex. 18.41 — The triangular numbers lead to a formula for the sum of the first n natural numbers; the rectangular numbers lead to a formula for the sum of the first n even natural numbers; and the square numbers lead to a formula for the sum of the first n odd natural numbers. (See SECTION 17.3.) To what formulas, if any, do the pentagonal numbers and the hexagonal numbers lead?

Ex. 18.42 — The following figure suggests that

$$S(n) = T(n - 1) + T(n),$$

where $S(n)$ is the nth square number and $T(n)$ is the nth triangular number.

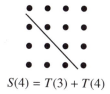

$$S(4) = T(3) + T(4)$$

Show in two ways that

$$\frac{1}{2}n(n - 1) + \frac{1}{2}n(n + 1) = n^2.$$

1. Expand the left-hand side and collect like terms.

2. Use $S(n) = T(n - 1) + T(n)$, where

$$S(n) = n^2 \quad \text{and} \quad T(n) = \frac{n(n + 1)}{2}.$$

Ex. 18.43 — According to Heath [68, p. 84], "that 8 times any triangular number $+ 1$ makes a square may easily go back to the early Pythagoreans. It is quoted by Plutarch[1] and used by Diophantus,[2] and is equivalent to the formula

$$8 \cdot \frac{1}{2}n(n + 1) + 1 = 4n(n + 1) + 1 = (2n + 1)^2.$$

It may easily have been proved by means of a figure made up of dots in the usual way"; and he displays the following figure that shows 7^2.

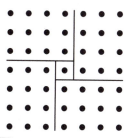

---

[1]Plutarch, *Plat. Quaest.* v. 2. 4, 1003 F.
[2]Dioph. IV. 38.

1. Use the figure to "prove" the formula.

2. Prove the formula algebraically by expanding, collecting like terms, and factoring.

Ex. 18.44 — Katz [73, p. 654] relates to us the following timeless story about Carl Friedrich Gauss (1777–1855), who is arguably the greatest of all German mathematicians [21, pp. 628–631].

> There are many stories told about Gauss's early-developing genius, one of which comes from his mathematics class when he was 9. At the beginning of the year, to keep his 100 pupils occupied, the teacher, J. G. Büttner, assigned them the task of summing the first 100 integers. He had barely finished explaining the assignment when Gauss wrote the single number 5050 on his slate and deposited it on the teacher's desk. Gauss had noticed that the sum in question was simply 50 times the sum 101 of the various pairs 1 and 100, 2 and 99, 3 and 98, ... and had performed the required multiplication in his head. Impressed by his young student, Büttner arranged for Gauss to have special textbooks, to have tutoring by his assistant Martin Bartels (1769–1836), who himself later became a professor of mathematics in Russia, and to be admitted to a secondary school where he mastered the classical curriculum.

Explain Gauss's reasoning, and explain why his method is equivalent to the formula

$$1 + 2 + 3 + \cdots + n = \frac{n(n+1)}{2}.$$

Ex. 18.45 — Loomis [78, pp. 19–2] provides several rules or formulas for generating Pythagorean triples. One rule is the *Rule of Pythagoras*:

Let n be odd; then n, $\frac{n^2-1}{2}$, and $\frac{n^2+1}{2}$ are three such numbers.

We recognize the rule of Pythagoras to be the formula given on page 218. Show that this formula leads to the algebraic identity

$$(2n)^2 + (n^2 - 1)^2 = (n^2 + 1)^2.$$

Ex. 18.46 — Another rule that Loomis provides is *Plato's Rule*:

Let m be any *even* number divisible by 4; then m, $\frac{m^2}{4} - 1$, and $\frac{m^2}{4} + 1$ are three such numbers.

1. Generate Pythagorean triples using Plato's rule and the given values of m.

   a) $m = 4$ b) $m = 8$ c) $m = 20$

2. Show that

$$m^2 + \left(\frac{m^2}{4} - 1\right)^2 = \left(\frac{m^2}{4} + 1\right)^2$$

   by expanding and collecting like terms.

3. Show that Plato's rule leads to the algebraic identity

$$(2m)^2 + (m^2 - 4)^2 = (m^2 + 4)^2.$$

Ex. 18.47 — Another rule that Loomis provides is *Euclid's Rule*:

Let x and y be any two even or odd numbers, such that x and y contain no common factor greater than 2, and xy is a square. Then $\sqrt{xy}$, $\frac{x-y}{2}$ and $\frac{x+y}{2}$ are three such numbers.

1. Use Euclid's rule to find two Pythagorean triples, one using even numbers and the other using odd numbers.

2. Show that

$$(\sqrt{xy})^2 + \left(\frac{x-y}{2}\right)^2 = \left(\frac{x+y}{2}\right)^2$$

by expanding and collecting like terms.

3. Show that Eulcid's rule leads to the algebraic identity

$$(2xy)^2 + (x-y)^2 = (x+y)^2.$$

Ex. 18.48 — Another rule that Loomis provides is *Rule of Maseres* (1721–1824):

Let m and n be any two even or odd, $m > n$, and $\frac{m^2+n^2}{2n}$ an integer. Then m^2, $\frac{m^2-n^2}{2n}$ and $\frac{m^2+n^2}{2n}$ are three such integers.

1. Use Maseres's rule to find two Pythagorean triples.

2. Show that

$$(m^2)^2 + \left(\frac{m^2 - n^2}{2n}\right)^2 = \left(\frac{m^2 + n^2}{2n}\right)^2.$$

3. Show that Maseres's rule leads to the algebraic identity

$$(2mn)^2 + (m^2 - n^2)^2 = (m^2 + n^2)^2.$$

Ex. 18.49 — Another rule that Loomis provides is *Dickson's Rule*:

Let m and n be any two prime integers, one even and the other odd, $m > n$ and $2mn$ a square. Then $m + \sqrt{2mn}$, $n + \sqrt{2mn}$ and $m + n + \sqrt{2mn}$ are three such numbers.

1. Use Dickson's rule to find two Pythagorean triples.

2. Show that

$$(m + \sqrt{2mn})^2 + (n + \sqrt{2mn})^2 = (m + n + \sqrt{2mn})^2.$$

Ex. 18.50 — Loomis [78, pp. 17–18] tells us that the "master Mathematical Analyst, Dr. Artemas Martin, of Washington D.C.…declares that no expression for two square numbers whose sum is a square can be found which are not deducible *from* this, or reducible *to* this formula,—that $(2pq)^2 + (p^2 - q^2)^2$ is always equal to $(p^2 + q^2)^2$," where p and q are any two natural numbers; in other words, that every Pythagorean triple is of the form

$$(2pq, p^2 - q^2, p^2 + q^2).$$

1. Show that

$$(2pq)^2 + (p^2 - q^2)^2 = (p^2 + q^2)^2.$$

2. Show that $p = n$ and $q = 1$ leads to the rule of Pythagoras.

3. Show that $p = m$ and $q = 2$ leads to Plato's rule.

4. Show that $p = x$ and $q = y$ leads to Euclid's rule.

5. Show that $p = m$ and $q = n$ leads to Maseres's rule.

6. Loomis provides

$$p = \sqrt{\frac{m + n + 2\sqrt{2mn} + \sqrt{m-n}}{2}},$$

$$q = \sqrt{\frac{m + n + 2\sqrt{2mn} - \sqrt{m-n}}{2}}$$

to deduce Dickson's rule, but they are unclear. Find the correct p and q.

Loomis [78, p. 21] tells us that all of the formulas above may be reduced to Dickson's rule, the advantage of which being that Dickson's rule gives every possible Pythagorean triple "in their lowest terms, and gives this set [triple] but once."

Ex. 18.51 — Suppose that you have 980 red balls, 735 green balls, and 1225 blue balls that you would like to place into buckets so that each bucket contains the same numbers of red, green, and blue balls. What is the most number of buckets that you can fill? How many red, green, and blue balls would be in each bucket?

Ex. 18.52 — Imagine that you are laying bricks of lengths 6 inches and 8 inches in two rows. At what distances will the two rows of bricks line up evenly? When does this happen the first time?

Ex. 18.53 — Mercury takes 88 (Earth) days to make one revolution about the sun, and Venus takes 225 days to make one revolution about the sun. If today Mercury, Venus, and the sun are lined up, after how many days will they be lined up again the first time? How often will Mercury, Venus, and the sun be lined up?

Ex. 18.54 — Suppose that you have 18 pencils and 15 erasers. What is the most number of packages that can be made so that each package contains the same numbers of pencils and erasers? How many pencils and erasers would there be in each package?

Ex. 18.55 — Show that 8128 is a perfect number, in fact, the fourth perfect number after 6, 28, and 496, by showing that 8128 is the sum of its parts.

Ex. 18.56 — Find all the prime numbers that are less than 500 using Eratosthenes's sieve.

Ex. 18.57 — The first several primes are 2, 3, 5, 7, 11, and 13. Observe that

$$2 + 1 = 3 \quad \text{is prime,}$$
$$2 \times 3 + 1 = 7 \quad \text{is prime,}$$
$$2 \times 3 \times 5 + 1 = 31 \quad \text{is prime.}$$

1. Determine if the following numbers are prime.

    a) $2 \times 3 \times 5 \times 7 + 1$ c) $2 \times 3 \times 5 \times 7 \times 11 \times 13 + 1$

    b) $2 \times 3 \times 5 \times 7 \times 11 + 1$

2. Let P be a prime number and let N be one more than the product of *all* of the primes not greater than P, that is,
$$N = 2 \times 3 \times 5 \times \cdots \times P + 1.$$

   Is N necessarily prime?

3. Let P be a prime number and let M be one more than the product of *some*, but *not all*, of the primes not greater than P. Is M necessarily prime?

Ex. 18.58 — Would the following be a good definition of a perfect number? Explain.

   A natural number n is called a *perfect number* if the sum of all the factors of n equals $2n$.

Ex. 18.59 — Let $M_n = 2^n - 1$, a Mersenne number. Find M_n for $n = 2, 3, 4, \ldots, 12$. Determine which M_n are prime numbers, Mersenne *primes*, and conjecture for which n, in general, M_n is a Mersenne prime.

Ex. 18.60 — Given that an even perfect number has the form $2^{n-1}(2^n - 1)$, show that an even perfect number ends in 6 or 8.

Ex. 18.61 — The numbers 3 and 5 are called *twin primes*, as are 5 and 7, as well as 11 and 13; the numbers 2 and 3, however, are not called twin primes, and also 7 and 11 are not. Characterize twin primes. How many more twin primes can you find? Do you think there is a largest pair?

* **Ex. 18.62** — The *trivium* and the *quadrivium* comprise the traditional "liberal arts" education. Write about the liberal arts and the place of mathematics in them, both historically and today. See Hardy Grant's *College Mathematics Journal* articles [62, 63].

* **Ex. 18.63** — Write about early elementary number theory in the following places.

1. China
2. Japan
3. India
4. the Middle East
5. Medieval Europe

* **Ex. 18.64** — Number theory was long thought to be the purest of mathematics because no one had conceived of any direct application of the subject to the everyday problems of the world. Today, however, number theory is being applied in several practical areas. For example, the March 2010 *Notices of the AMS* describes the uses of cryptography today.[3]

> Cryptography is a key technology widely deployed by private, commercial, and governmental users to ensure privacy and authenticity in secure electronic data communication. Its research directions are often driven by practical demands. For example, recently, various issues of privacy and electronic voting entered the world of cryptography. While most of us would agree that these activities are not the most pure and beautiful in our lives, cryptography has its own irresistible attraction and intrinsic motivation for further developments.
>
> [...]
>
> Cryptography is the best-known area of applications of number theory, but it is not the only one. The others include computer science, dynamical systems, physics, and even molecular chemistry. There are also emerging applications of number theory to quantum computing and financial mathematics.

Write about the practical application of number theory today. You may focus on one or two areas.

* **Ex. 18.65** — How are prime numbers used to encrypt information?

* **Ex. 18.66** — How are primes used in keeping data safe, like when you use your credit card online? (This is called *public key cryptography*.)

* **Ex. 18.67** — What is modular arithmetic? Give a few examples of numbers that are congruent mod 5. What modular arithmetic do we use every day?

* **Ex. 18.68** — What is syncopated algebra, and how does it compare with rhetorical algebra and with symbolic algebra?

* **Ex. 18.69** — Euclid's *Elements* is written in 13 "books." How were books constructed in ancient times? Would they be separate books if written now?

* **Ex. 18.70** — As we saw, the number 10 held significant meaning and properties for the Pythagoreans. Look up properties of other numbers. Did other cultures also ascribe mystical meanings to numbers?

* **Ex. 18.71** — How do people still ascribe special meaning to numbers today (*numerology*)? Give some examples of modern numerology.

---

[3]Igor E. Shparlinski, "Numbers at work and play," *Notices of the AMS*, Vol. 57, No. 3, (March 2010), pp. 334–342.

* Ex. 18.72 — Write about the mathematician Andrew Wiles (b. 1953) and some of his mathematical achievements.

* Ex. 18.73 — Find the comment that Fermat wrote in his copy of Diophantus's *Arithmetica* that sparked a search for a proof of what has come to be called "Fermat's last theorem" (FLT). What was the comment and why did it spark a search for FLT?

* Ex. 18.74 — Give an account of Fermat's last theorem up to its proof by Andrew Wiles in 1995.

* Ex. 18.75 — What is the currently known greatest prime number? Greatest perfect number? Greatest pair of amicable numbers? To whom is each attributed and how was it found?

* Ex. 18.76 — Proof Thirty-Five in Loomis [78, p. 49] depends on the following two figures.

1. In this figure, triangles AHB, BGF, FED, and DCA are congruent right triangles, and $ABFD$ is a square. Let $BH = a$, $AH = b$, and $AB = c$, and deduce Pythagoras's theorem by comparing the areas of the four right triangles and the squares $ABFD$ and $HGEC$.

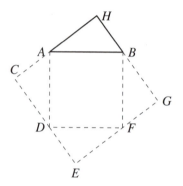

Loomis notes regarding the proof: "See Math. Mo., 1809, dem. 9, and there, p. 159, Vol. I, credited to Rev. A. D. Brunswick, Me.; also see Fourrey, p. 80, fig's a and b; also see 'Der Pythagoreisch Lehrsatz' (1930), by Dr. W. Leitzmann."[4]

2. In this figure, trianges ABH, ACB, FGH, and HEF are congruent right triangles, and $GHAK$, $EDBH$, and $FDCK$ are squares. Deduce Pythagoras's theorem, here $a^2 + b^2 = h^2$.

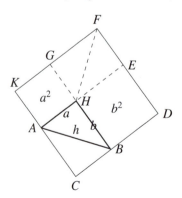

Loomis notes regarding the proof: "This proof was devised by Maurice Laisnez, a high school boy, in the Junior-Senior High School of South Bend, Ind., and sent to me, May 16, 1939, by his class teacher, Wilson Thornton."

---

[4]Math. Mo. = *Mathematical Monthly*.

* **Ex. 18.77** — Proof Two Hundred Thirty-One in Loomis [78, p. 231] depends on the following figure. In this figure, triangles AHB and BDC are congruent right triangles, where $AH = BD$ and $HDCA$ is a trapezoid. Deduce Pythagoras's theorem.

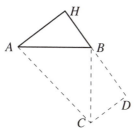

Loomis notes regarding this proof: This is the [President James A.] "Garfield Demonstration,"—hit upon by the General in a mathematical discussion with other M.C.'s about 1876." President Garfield was the twentieth president of the Unites States of America.

* **Ex. 18.78** — Write about Marin Mersenne and some of his mathematical achievements.

* **Ex. 18.79** — Write about GIMPS, the Great Internet Mersenne Prime Search.

* **Ex. 18.80** — Oystein Ore [92, p. 53] writes, "The simplest way to obtain the factorization of a number that is not too large is through the use of a *factor table*." What is a "factor table" and how is one used? What is the history of factor tables?

* **Ex. 18.81** — Explain how Eratosthenes measured the circumference of the earth. Include a diagram.

* **Ex. 18.82** — Observe the following factorizations:

$$150 = 2^1 \cdot 3^1 \cdot 5^2 \quad \text{has 12 factors,}$$
$$525 = 3^1 \cdot 5^2 \cdot 7^1 \quad \text{has 12 factors,}$$
$$m = p^2 q^1 r^3 \quad \text{has 24 factors,}$$
$$n = p^3 r \quad \text{has 8 factors.}$$

1. How are the number of factors of a given number related to the exponents in the prime factorization of the given number?

2. Without finding all the factors, determine the number of factors of each of the following numbers.

   a) $490{,}050 = 2 \cdot 3^4 \cdot 5^2 \cdot 11^2$

   b) $n = p^3 qr^2 s^5 t$, where p, q, r, s, and t are primes.

* **Ex. 18.83** — Research the usage of the terms "greatest common factor," "highest common factor," and "greatest common divisor." To compare their relative usage, look up Google Books "Ngram Viewer" <http://books.google.com/ngrams>.

Part IV

Solving Equations

Chapter 19

Linear Problems

The simplest problems to solve are the linear type, and so we begin with linear problems. Before we do so, however, we provide an overview of solving equations in general.

At the mention of "algebra," perhaps equations like

$$3(x - 1) = x + 7 \quad \text{and} \quad y = ax + b,$$

come to mind. Perhaps you encountered such equations when you solved word problems or graphed lines in a mathematics class. It may surprise you to learn, however, that while people may have been solving mathematical problems for over 4000 years, the writing of equations like the two above is a fairly recent development that dates back only a few hundred years—the culmination of the contributions of many people over a long period of time.

Take the two equations shown above: The use of letters to represent unknowns and constants is credited to François Viète (1540–1603); the use of brackets as grouping symbols is attributed to mathematicians of the sixteenth century, including Niccolò Tartaglia (1500–1557), Girolamo Cardano (1501–1576), and Rafael Bombelli (1526–1572); the first appearance in print of the symbols + and − for addition and subtraction, respectively, is found in Johannes Widman's (ca. 1460) *Behende vnnd hübsche Rechnūg auff allen Kauffmanschafften* (*Complete Business Calculation*); and the symbol = for equality was introduced by Robert Recorde (1510–1558) in his algebra book *The Whetstone of Witte* (FIGURE 19.1). Even then, it took some time for these symbols that we take for granted today to become widely accepted and used. So, what symbols were used to write equations before the ones we use today became standard notation?

For a very long time, everything was described in words (*rhetorical algebra*). Then, over time, individual mathematicians introduced symbols for particular objects to help themselves in their own work, and different contemporary mathematicians may have used different symbols for the same objects. At first the symbols were merely a type of shorthand (*syncopated algebra*). In this way, symbols came to be used very sparingly, if at all, and there were still no equations as we know them today until the seventeenth century (*symbolic algebra*). Cajori [31] provides a comprehensive and authoritative discussion of the many symbols used in arithmetic and algebra, and other branches of mathematics, from the earliest times and around the world.[1] TABLE 19.1 shows a few of the symbols that we commonly use today, when they first appeared, and the persons credited for the introduction of the symbols.

Remark 19.1 It is interesting to note Suzuki's [112, p. 309] remark that "the rules for the order of operations were not agreed upon until the twentieth century: $24 \div 4 \times 6$ might be 36 or 1, depending on the author, and $3 \times 4 - 2$ might mean 6 or 10." □

Finally, we point out that good mathematical notation is more than mere shorthand and more than a way to express mathematical statements or ideas compactly; rather, good mathematical notation makes the

---

[1] Another source is Jeff Miller's Web page, "Earliest Uses of Various Mathematical Symbols," <http://jeff560.tripod.com/mathsym.html>. And Berlinghoff and Gouvêa [11, Ch. 8] provide a good survey.

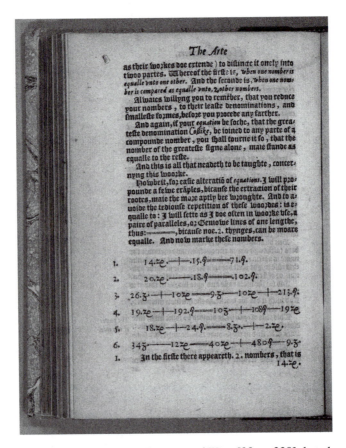

Figure 19.1: A page from *The Whetstone of Witte* [98, p. 238] that shows the
first appearance of the now-familiar = sign in print as a pair of long parallel
lines: "And to avoide the tediouse repetition of these woordes: is equalle to: I
will sette as I doe often in woorke use, a paire of paralleles, or Gemowe lines
of one lengthe, thus: ＝＝, bicause noe .2. thynges, can be moare equalle."
Notice also the elongated appearance of the + and − signs, which are also
printed using long lines. (Source: Rare Book and Manuscript Library, Columbia
University.)

mathematics at hand clearer and easier, and even leads to new avenues of mathematical pursuit, whereas
poor notation can, in fact, confuse. On exactly this point, Cajori [31, pp. 328–329] quotes J. W. L. Glaisher,
who marvels at how John Napier (1550–1617), during whose time "there was practically no notation,"
was able to discover or invent logarithms[2] "by mind alone without any aid from symbols":

> *J. W. L. Glaisher*[3] emphasizes the important rôle which notations have played in the development
> of mathematics: "Nothing in the history of mathematics is to me so surprising or impressive as the
> power it has gained by its notation or language. No one could have imagined that such 'trumpery tricks
> of abbreviation' as writing + and − for 'added to' and 'diminished by,' or x^2, x^3,... for xx, xxx,..., etc.,
> could have led to the creation of a language so powerful that it has actually itself become an instrument
> of research which can point the way to future progress. Without suitable notation it would have been
> impossible to express differential equations, or even to conceive of them if complicated, much less to

---

[2] See CHAPTER 24.

[3] J. W. L. Glaisher, "Logarithms and Computation" from the *Napier Tercentenary Memorial Volume* (ed. Cargill Gilston Knott;
London, 1915), p. 75.

+	plus (addition)	1489	Johannes Widmann
−	minus (subtraction)		
$\sqrt{}$	square root (no vinculum)	1525	Christoff Rudolff
=	equals (originally ====)	1557	Robert Recorde
<	less than	1631	Thomas Harriot
>	greater than		
×	times (multiplication)	1631	William Oughtred
÷	divided by (division)	1659	Johann Heinrich Rahn
a^n	exponent $(aa \cdots a)$	1637	René Descartes
$\sqrt{}$	square root (with vinculum)		
$\sqrt[n]{}$	nth root	1690	Michel Rolle
·	times	1698	Gottfried Wilhelm Leibniz

Table 19.1: Some commonly used operation symbols

deal with them; and even comparatively simple algebraic quantities could not be treated in combination. Mathematics as it has advanced has constructed its own language to meet its need, and the ability of a mathematician in finding the true means of representing his results as in the discovery of the results themselves.

"When mathematical notation has reached a point where the product of n x'es was replaced by x^n, and the extension of the law $x^m x^n = x^{m+n}$ had suggested $x^{\frac{1}{2}} \cdot x^{\frac{1}{2}} = x$ so that $x^{\frac{1}{2}}$ could be taken to denote $\sqrt{x}$, then fractional exponents would follow as a matter of course, and the tabulation of x in the equation $10^x = y$ for integral values of y might naturally suggest itself as a means of performing multiplication by addition. But in Napier's time, when there was practically no notation, his discovery or invention [of logarithms] was accomplished by mind alone without any aid from symbols."

Let us now turn to the topic of linear problems.

A *linear problem* is one that can be described by equations of the form

$$ax + by + cz + \cdots + dw = e,$$

where $a, b, c, \ldots, d$, and e are constants. An equation like this is called a *linear equation* in the variables $x, y, z, \ldots, w$. Here are a few examples of linear equations.

$$3(x - 1) = x + 7$$
$$4x - 3y + z = 10$$
$$p + 3q - 5 = 0$$

Any equation that is not a linear equation is called a *nonlinear equation*. Here are a few examples of nonlinear equations.

$$x^2 + x - 1 = 0$$
$$2xy = 10$$
$$D = \frac{q^2}{4} + \frac{p^3}{27}$$

Think About It 19.1 Why would the term "linear" be applied to the top group of equations? Why are the equations in the second group not "linear"? □

In broad terms, a linear equation—and, correspondingly, a linear problem—is typically "easy" to solve, whereas a nonlinear equation—and, correspondingly, a nonlinear problem—is typically "hard" to solve. Indeed, a general strategy for solving any type of nonlinear equation is to approximate the equation with a suitable linear equation. This is called *linearization* and, for example, it is the crux of differential calculus.[4]

In this chapter, we first review some modern methods of solving linear equations in one variable and graphing linear equations in two variables. After that, we take a look at a tried-and-true method for solving linear problems without having to write any equations at all, namely, the method of *false position*. This method seems to have been universal, with evidence of its use among the Babylonians, Egyptians, Chinese, Indians, Arabs, and Europeans.

19.1   REVIEW OF LINEAR EQUATIONS

A number is a *solution of an equation* or *root of an equation* if the substitution of the number for the variable in the equation produces a true statement. If an equation has more than one variable, then an *ordered set of numbers* is a solution of the equation if the substitution of the numbers for the respective variables in the equation produces a true statement. For example, 3 is a solution or root of the equation $x^2 = x + 6$, but 5 is not, for we see that

$$x = 3: \qquad 3^2 = 3 + 6 \qquad \text{(simplify)}$$
$$9 = 9 \qquad \text{(TRUE)},$$

but

$$x = 5: \qquad 5^2 = 5 + 6 \qquad \text{(simplify)}$$
$$25 = 11 \qquad \text{(FALSE)}.$$

As a second example, $(3, -2)$ is a solution of the equation $x - 3y = 9$, but $(5, -2)$ is not, for we see that

$$(x, y) = (3, -2): \qquad 3 - 3(-2) = 9 \qquad \text{(simplify)}$$
$$9 = 9 \qquad \text{(TRUE)},$$

but

$$(x, y) = (5, -2): \qquad 5 - 3(-2) = 9 \qquad \text{(simplify)}$$
$$11 = 9 \qquad \text{(FALSE)}.$$

You may recall having solved linear equations in one variable or having graphed linear equations in two variables in an algebra class. To solve a linear equation in the variable x, say, we may reduce the equation to the form $ax = b$, from which it follows that $x = \frac{b}{a}$ provided a is not zero. To graph a linear equation in the variables x and y, say, we may plot two or more solutions of the equation, or perhaps reduce the equation to the form $y = ax + b$ and use the slope, a, and the y intercept, $(0, b)$, to graph the equation.

Example 19.1 Solve the equation

$$5x + 7 - x = 2x + 23.$$

---

[4]Calculus was discovered independently by the well-known British mathematician and physicist Isaac Newton (1643–1727; of the "falling apple" fame) and the great German mathematician and philosopher Gottfried Wilhelm Leibniz (1646–1716) in the late seventeenth century. The term *differential calculus* is due to Leibniz. Newton called his discovery *fluxional calculus*. A tiff over the priority of the discovery led to one of the most storied feuds in mathematics history; see Eves's lecture, "Moving pictures versus still pictures" [55, Lecture 22] or, for a longer study, see Hall's account, *Philosophers at War* [64].

Here is a solution in which we first separate the variable terms from the constant terms. (There are other ways to proceed.)

$$5x + 7 - x = 2x + 23 \qquad \text{(add } -7 \text{ and } -2x \text{ to both sides)}$$

$$5x - x - 2x = 23 - 7 \qquad \text{(simplify each side by combining like terms)}$$

$$2x = 16 \qquad \text{(divide both sides by 2 or multiply both sides by } \tfrac{1}{2})$$

$$x = 8$$

We check that 8 is a solution:

$$x = 8: \qquad 5(8) + 7 - 8 = 2(8) + 23 \qquad \text{(simplify)}$$
$$39 = 39 \qquad \text{(TRUE)}.$$

Therefore, 8 is a solution. $\square$

Now You Try 19.1 Solve the following equations.

1. $x - 5 + 3x = 2x + 5$

2. $2(x + 2) - 1 = 11$

3. $5x + 4 = 8 - (2x + 6)$

4. $x + \tfrac{1}{4}x = 15$

$\square$

Example 19.2 Graph the equation $2x + y = 4$.

We note that, because this is a linear equation in x and y, the graph of the equation is a line in the xy plane. One way to graph the equation is to find three solutions of the equation, that is to say, to find three pairs of numbers (x, y) that satisfy the equation; plot the solutions as points in the xy plane; then draw the graph, a line, through the three points.

- To find three solutions, we choose three arbitrary (but reasonable) numbers for either x or y, then solve for the corresponding ordered pairs. We choose $x = 0$, $y = 0$, and $x = 1$, then we solve for the corresponding ordered pairs.

$$x = 0: \qquad 2(0) + y = 4$$
$$0 + y = 4$$
$$y = 4 \quad \Longrightarrow \quad (x, y) = (0, 4)$$
$$y = 0: \qquad 2x + 0 = 4$$
$$2x = 4$$
$$x = 2 \quad \Longrightarrow \quad (x, y) = (2, 0)$$
$$x = 1: \qquad 2(1) + y = 4$$
$$2 + y = 4$$
$$y = 2 \quad \Longrightarrow \quad (x, y) = (1, 2)$$

- Plot the solutions as points in the xy plane, then draw a line through the three points. The line is the graph of the equation.

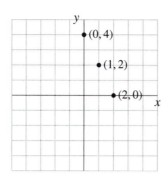 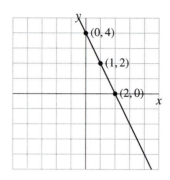

Therefore, we have graphed the equation $2x + y = 4$.   □

Recall that every point of the graph of an equation provides a solution of the equation. For example, we see above that the point $(4, -4)$ is on the graph of the equation $2x + y = 4$. We leave it to you to check that $(4, -4)$ is a solution of the equation.

Think About It 19.2 Only two points are needed to specify a unique line. What, then, may be a reason for finding *three* solutions instead of only two to graph a linear equation?   □

Now You Try 19.2 Graph the following equations.

1. $x - y = 3$ 3. $x = 2y + 4$

2. $y = 2x$ 4. $3y = 9x - 6$

    □

Recall that the *slope* of a line is a number that describes the inclination of the line. Given any two points on a line, we may compute the change in y and the change in x for the two points by computing the respective differences between the y coordinates and the x coordinates of the two points. Then the slope of the line is defined to be the ratio of the change in y to the change in x of the two points.

Think About It 19.3 How can we apply similar triangles to show that we may compute the slope of a line using *any* two points on the line?   □

Definition 19.1 Let (a, b) and (c, d) be two points on a given line in the xy plane. Let Δy denote the change in y of the two points, and let Δx denote the change in x; Δ is the upper-case Greek letter *delta*. If we compute the change in y as
$$\Delta y = d - b,$$
then the corresponding change in x is
$$\Delta x = c - a.$$
Equivalently, if we compute the change in y as
$$\Delta y = b - d,$$
then the corresponding change in x is
$$\Delta x = a - c.$$
Then the *slope* of the line is
$$\frac{\Delta y}{\Delta x} = \frac{d - b}{c - a} \quad \text{or} \quad \frac{b - d}{a - c}.$$

    □

Think About It 19.4 Are these two versions of the formula for slope equivalent? Why or why not? □

Example 19.3 We have seen that the graph of the equation $2x + y = 4$ is a line that passes through the points $(0, 4)$, $(2, 0)$, and $(1, 2)$. We compute the slope of the line using these points in a few ways.

- We use $(0, 4)$ and $(2, 0)$. If we compute the change in y as

$$\Delta y = 0 - 4 = -4,$$

then the corresponding change in x is

$$\Delta x = 2 - 0 = 2.$$

Thus, the slope of the line is

$$\frac{\Delta y}{\Delta x} = \frac{-4}{2} \quad \text{or} \quad -2.$$

- We use $(0, 4)$ and $(2, 0)$. If we compute the change in y as

$$\Delta y = 4 - 0 = 4,$$

then the corresponding change in x is

$$\Delta x = 0 - 2 = -2.$$

Thus, the slope of the line is

$$\frac{\Delta y}{\Delta x} = \frac{4}{-2} \quad \text{or} \quad -2.$$

- We use $(0, 4)$ and $(1, 2)$. If we compute the change in y as

$$\Delta y = 2 - 4 = -2,$$

then the corresponding change in x is

$$\Delta x = 1 - 0 = 1.$$

Thus, the slope of the line is

$$\frac{\Delta y}{\Delta x} = \frac{-2}{1} \quad \text{or} \quad -2.$$

Therefore, we see that the slope of the line is the ratio $\frac{-2}{1}$ or -2. □

Now You Try 19.3

1. Find the slope of the line that passes through the given points.

   a) $(1, 3)$ and $(2, 5)$ c) $(-2, -1)$ and $(0, 4)$

   b) $(0, 0)$ and $(2, 5)$ d) $(2, 2)$ and $(5, -3)$

2. Graph the four lines through the pairs of points given above in the same xy plane. What relation do you notice between the slope of a line and the inclination of the line?

□

19.2   FALSE POSITION

The method of *false position* for solving a linear problem is called such because the first step of the method is to posit or make a guess for the solution that, in all likelihood, will be incorrect: a false position; however, the next steps of the method use the incorrect guess (the false position) to obtain the correct solution. How does the method work? Let us try to discover it.[5]

---

[5] We are not suggesting that this is how the method was first discovered. In fact, we do not know how the method was first discovered.

Suppose that we seek a number such that, when one-third of the number is added to it, the result is 100, that is, we seek a number N such that

$$N + \frac{1}{3} \times N = 100.$$

Let us begin by positing or guessing that the number is 3 because it is easy to take one-third of 3: one-third of 3 is 1, so that

$$\text{Guess 3}: \quad 3 + \frac{1}{3} \times 3 = 4 \neq 100,$$

Since our guess of 3 led to a result of 4 instead of 100, we make another guess, say, 6 because it is easy to take one-third of 6: one-third of 6 is 2, so that

$$\text{Second guess 6}: \quad 6 + \frac{1}{3} \times 6 = 8 \neq 100.$$

But $8 \neq 100$ either, so let us make yet another guess, say, 30 because, again, it is easy to take one-third of 30: one-third of 30 is 10; hence, we see that

$$\text{Third guess 30}: \quad 30 + \frac{1}{3} \times 30 = 40.$$

But $40 \neq 100$ still.

Now, we could continue to guess in the hope of stumbling upon the solution—or we could pause and think.

We summarize our three guesses in the following table.

	Guess	Result	Observation
First	3	4	
Second	6	8	$6 = 2 \times 3$ and $8 = 2 \times 4$
Third	30	40	$30 = 10 \times 3$ and $40 = 10 \times 4$

From the table we see that if we guess $c \times 3$, then the result is $c \times 4$; thus, it is reasonable to conjecture that if we can find a number c such that c times 4 is the desired result, 100, then $c \times 3$ should be the solution.

By division, we find that $c = 100 \div 4 = 25$. So, we guess that the solution is 25×3 or 75. Let us check. Since one-third of 75 is 25,

$$\text{Guess 75}: \quad 75 + \frac{1}{3} \times 75 = 100.$$

So we have it: the number sought is 75.

Rule 19.1 (METHOD OF FALSE POSITION) Guess a solution, say, g. If g leads to a result h that is incorrect, but $c \times h$ is the desired result, then the solution is $c \times g$.   □

Example 19.4 Use the method of false position to solve the equation

$$x - \frac{1}{2}x + \frac{1}{3}x = 10.$$

Guess a number that makes the calculation easy. Since 6 is divisible by 2 and 3, we choose as our guess the number 6. Substituting 6 into the expression on the left-hand side of the equation, we find that

$$6 - \frac{1}{2}(6) + \frac{1}{3}(6) = 6 - 3 + 2 = 5.$$

But the desired result is 10, and not 5. We divide 10 by 5: $10 \div 5 = 2$. Thus, we conclude that the solution is 2×6 or 12. We check:

$$12 - \frac{1}{2}(12) + \frac{1}{3}(12) = 12 - 6 + 4 = 10.$$

Therefore, the solution is 12. ☐

Think About It 19.5 Would the method of false position work on a problem such as $\frac{1}{2}x + x^2 = 10$? Why or why not? What kinds of problems will it work on? ☐

Now You Try 19.4 Solve the following equations using the method of false position.

1. $x + \frac{1}{8}x = 12$

2. $x + \frac{1}{4}x + \frac{1}{6}x = 12$

☐

The method of false position is also called the method of *single* false position or *simple* false position to distinguish it from the method of "double false position" that we discuss in the next section. The method of double false position today would be understood as *linear interpolation*, although Chabert et al. [39] posit that, perhaps, the early Chinese may also have had this modern understanding. Some of the earliest examples of false position are found in Babylonian cuneiform tablets. The method is also known to have been used by the ancient Egyptians and, later, on the other side of the world, by Indian and Chinese mathematicians. In Europe, Leonardo of Pisa introduced the method from the Arabs in the twelfth century. Interestingly, even after the conception of symbolic algebra, the method of false position continued to be used up until the twentieth century [11, 39].

The following two examples are from the Rhind Mathematical Papyrus and the Moscow Mathematical Papyrus. The examples belong to the small group of "aha" problems ("aha" in Egyptian refers to the unknown quantity) that are purely mathematical. We continue to use the convention that a bar over an integer denotes a unit fraction, for instance, $\overline{5}$ is $\frac{1}{5}$ and $\overline{30}$ is $\frac{1}{30}$; as a special case, $\overline{3}$ is $\frac{2}{3}$. (See page 101.) Bold text here shows where the original text was marked in red ink. Red ink was used to mark the beginning of a problem and the result [70, p. 26]. (See TABLE 9.1 on page 102.) For multiplication by doubling, see examples 9.6 and 9.7 beginning on page 100, and for division, see examples 10.1 and 10.2.

Example 19.5 Rhind problem 26 [70, p. 26]

A quantity, its $\overline{4}$ (is added) to it so that 15 results

The problem is to find a number ("a quantity") such that, when one-fourth of the number is added to it, the total is 15 ("so that 15 results").

To find the number, A'hmosè first *guesses* that it is 4 ("Calculate with 4"; see A'hmosè's work below) because it is easy to take one-fourth of 4: one-fourth of 4 is 1 ("calculate its $\overline{4}$ as 1").[6] He then adds this result to the guess, 4, and obtains 5 ("Total 5"). But the desired total is 15 and not 5. A'hmosè then notes that $15 = 3 \times 5$ ("Divide 15 by 5 ... 3 shall result"), and concludes that the number sought is 3×4, three times the guessed quantity, or 12 ("Multiply 3 times 4 ... 12 shall result"). Finally, A'hmosè checks the result:

$$\frac{1}{4} \times 12 = 3, \quad 12 + 3 = 15$$

("its $\overline{4}$ 3, total 15").

---

[6]Imhausen [70, p. 27] remarks, "The instructions begin with 'Calculate with 4.' Since 4 is the inverse of the first datum ($\overline{4}$), there must have been one step in the calculation that has not been noted in the source text, namely the calculation of the inverse of $\overline{4}$." Imhausen inserts this extra step, $1 \div \overline{4} = 4$, in her analysis of the problem.

Guess:	**4**
Work:	$\frac{1}{4}$ of **4** added to **4** is $\underline{5}$, but the desired result is 15
Divide:	$15 \div \underline{5} = \underline{\underline{3}}$
Multiply:	$\underline{\underline{3}} \times 4 = 12$
Answer:	12
Check:	$\frac{1}{4}$ of 12 added to 12 is 15

Therefore, the number is 12. Here is A'hmosè's work.

A quantity, its $\overline{4}$ (is added) to it so that 15 results

Calculate with 4.
You shall calculate its $\overline{4}$ as 1. Total 5.
Divide 15 by 5.
$\quad$ \. $\quad$ 5
$\quad$ \ 2 $\quad$ 10
3 shall result.
Multiply 3 times 4.
$\quad\quad$. $\quad$ 3
$\quad\quad$ 2 $\quad$ 6
$\quad$ \ 4 $\quad$ 12
12 shall result.
$\quad\quad$. $\quad$ 12
$\quad\quad$ $\overline{4}$ $\quad$ 3 Total 15

The quantity 12
$\quad$ its $\overline{4}$ $\quad$ 3, **total 15.**

$\square$

Example 19.6 Moscow problem 25 [70, p 29]

Method of calculating a quantity calculated times 2
together with (it, i.e., the quantity), it has come to 9.
Which is the quantity that was asked for?
You shall calculate the sum of this quantity and this 2.
3 shall result.
You shall divide 9 by this 3.
3 times shall result.
Look, 3 is that which was asked for.
What has been found by you is correct.

The problem is to find a number ("a quantity") such that two times the number and the number itself ("calculated times 2 together with it") yields nine ("it has come to 9").

To find the number, the scribe first *guesses* that it is 1; then he adds twice his guess to the guess itself, $1 + 2 \times 1$, and obtains 3 ("calculate the sum of this quantity and this 2. 3 shall result"). But the desired total is 9 and not 3. The scribe then notes that $9 = 3 \times 3$ ("divide 9 by this 3. 3 times shall result"), and concludes that the number sought is 3×1, three times the guess quantity, or 3 ("3 is that which was asked for").

Guess:	1
Work:	2 times **1** added to **1** is $\underline{3}$, but the desired result is 9
Divide:	$9 \div \underline{3} = \underline{\underline{3}}$
Multiply:	$\underline{\underline{3}} \times 1 = 3$
Answer:	3

Therefore, the number is 3. ☐

Notice the difference in the styles of the presentations between problem 26 of the Rhind Papyrus and problem 25 of the Moscow Papyrus. While both state the problem, give the procedure for solving the problem, and state the solution, the former also shows the calculations and checks the solution, whereas the latter does not.

Now You Try 19.5 Solve the following problems using the method shown in examples 19.5 and 19.6.

1. (Rhind problem 27 [70, p. 28]) **A quantity, its $\bar{5}$ (is added) to it** so that 21 results.

2. A quantity, five times is added to it so that 2 results.

3. A quantity, its fifth is removed so that 12 results.

4. Two children are several months old. The age of one child is one-sixth of the other. Their ages together is 21 months.

☐

The following example is from the Old Babylonian period (1800–1600 BC). The problem is from the text VAT 8389.[7] Katz [74] presents the problem using base-10 numerals instead of base-60 numerals. It is an interesting example in that today we would likely solve it using a system of two linear equations; however, the Babylonian scribe cleverly uses false position.

Example 19.7 *VAT 8389* [74, p. 21]

One of two fields yields $\frac{2}{3}$ *sila* per *sar*, the second yields $\frac{1}{2}$ *sila* per *sar*. [*Sila* and *sar* are measures for capacity and area, respectively.] The yield of the first field was 500 *sila* more than that of the second; the areas of the two fields were together 1800 *sar*. How large is each field?

Katz tells us that the scribe solves the problem by first *guessing* that each field has an area of 900 *sar*, so that the two fields together would have a total area of 1800 *sar*. Then, the yield of the first field is $\frac{2}{3} \times 900$ *sila* or 600 *sila*, and the yield of the second field is $\frac{1}{2} \times 900$ *sila* or 450 *sila*. Hence, the yield of the first field is 150 *sila* more than that of the second; however, the problem states that the "yield of the first field was 500 *sila* more than that of the second," which is 350 more than 150.

Now, Katz tells us, "To adjust the answers, the scribe presumably realized that every unit increase in the value of x [the area of the first field] and a consequent unit decrease in the value of y [the area of the

---

[7] *Vorderasiatische Abteilung, Tontafeln* collection of the Berlin Museum.

second field] gave an increase in the 'function' $\frac{2}{3}x - \frac{1}{2}y$ of $\frac{2}{3} + \frac{1}{2} = \frac{7}{6}$." So, because $350 \div \frac{7}{6} = 300$, the scribe adjusts his guess of the two areas by adding and subtracting 300:

$$\text{first field :} \qquad 900 + 300 = 1200,$$
$$\text{second field :} \quad 900 - 300 = 600.$$

Therefore, the first field has an area of 1200 *sar* and the second an area of 600 *sar*.

Now the yield of the first field is $\frac{2}{3} \times 1200$ *sila* or 800 *sila*, and the yield of the second field is $\frac{1}{2} \times 600$ *sila* or 300 *sila*; hence, the yield of the first field is 500 more than that of the second. Moreover, the areas of the two fields together is $(1200 + 600)$ *sar* or 1800 *sar*.

In summary, the scribe finds the areas of the two fields in this way:

Guess:	first field **900**, second field **900**
Work:	$\frac{2}{3} \times \mathbf{900} = 600,$ $\frac{1}{2} \times \mathbf{900} = 450$
Difference:	$600 - 450 = \underline{150}$, but the desired difference is 500
Subtract:	$500 - \underline{150} = \underline{\underline{350}}$
Epiphany:	every unit increase in the area of the first field and consequent unit decrease in the area of the second field gives a net increase of $\frac{7}{6}$: $\frac{2}{3}(1) - \frac{1}{2}(-1) = \frac{2}{3} + \frac{1}{2} = \frac{7}{6}$
Divide:	$\underline{\underline{350}} \div \frac{7}{6} = 300$
Adjust:	first field **900** $+ 300 = 1200$, second field **900** $- 300 = 600$
Answer:	first field is 1200, second field is 600

Therefore, the first field has an area of 1200 *sar* and the second an area of 600 *sar*.   □

The next example is also from the Old Babylonian period. The problem, from a cuneiform tablet from Susa,[8] is to find the dimensions of a rectangle given the length of its diagonal and a linear relation between the sides. The translation uses our usual numerals instead of cuneiform script to express numerals in base 60 and a semicolon as the separatrix (to separate the whole number part of a number from the fraction part). Note that we would write 1;33 45, leaving a space between place values (see page 77), where the translation here writes 1;33;45.

Example 19.8 Tablet from Susa [39, p. 87]

Let the breadth (of the rectangle) measure a quarter less compared with the length.
40 (is the dimension of the diagonal). What are the length and breadth?
Thou, take 1 as a length, and put 1 for the length (of the rectangle).
15 the quarter, subtract from 1, you will find 45.
Put 1 as the length, put 45 as the breadth, square 1 the length, 1 you find.
Square 45, the breadth: 33;45 you find. From 1 and 33;45 (make) the sum: 1;33;45 you find.
What is the square root? 1;15 you find.
Expecting 40, the diagonal that was indicated to you, unravel the inverse of 1;15 the diagonal.
48 (you find). Take 48 to 40 [48 × 40] the diagonal which you were told, 32 you find.
Take 32 to 1 [32 × 1] the length that you put: 32 you find, 32 (that is) the length. Take 32 to 45
[32 × 45] the breadth that you put: 24 you find. 24 (that is) the breadth.

[8] *Textes mathématiques de Suse*, ed. and translated into French by E. M. Bruins & M. Rutten, Paris: Geuthner, 1961 (p. 102), problem xix, c.

Figure 19.2: YBC 7289 from the Yale Babylonian Collection. It pictures a square with one side marked 30 and one diagonal marked 1 24 51 10 and, below that, 42 25 35. This shows a knowledge of the Pythagoras's theorem. (Source: Courtesy of the Yale Babylonian Collection.)

The problem is to find the length and the breadth (width) of a rectangle given that its diagonal is 40 and its breadth is three-fourths its length ("the breadth ... measure a quarter less compared with the length": $b = l - \frac{1}{4}l$ or $\frac{3}{4}l$, where b denotes the breadth of the rectangle and l denotes the length). In the solution, the scribe uses Pythagoras's theorem that relates the sides of the right triangle formed by two sides and the diagonal of the rectangle.[9]

To solve the problem, the scribe first *guesses* that the length of the rectangle is 1 ("take 1 as the length, and put 1 for the length"). In this case, the breadth would be $\frac{3}{4}$ ("15 the quarter, subtract from 1, you will find 45": $1 - \frac{15}{60} = \frac{45}{60}$ or $1 - \frac{1}{4} = \frac{3}{4}$). Now the scribe uses Pythagoras's theorem.

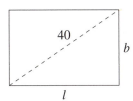
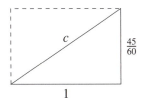

Applying Pythagoras's theorem, we find that the square of the hypotenuse is

$$c^2 = 1^2 + \left(\frac{3}{4}\right)^2 = 1\frac{9}{16}$$

("Square 45, the breadth: 33;45": $\left(\frac{45}{60}\right)^2 = \frac{33}{60} + \frac{45}{3600}$ or $\frac{9}{16}$; "From 1 and 33;45 ... sum: 1;33;45": $1 + \frac{33}{60} + \frac{45}{3600}$ or $1\frac{9}{16}$). Thus, the hypotenuse is

$$c = \sqrt{1\frac{9}{16}} = 1\frac{1}{4}$$

("...square root? 1;15: $\sqrt{1 + \frac{33}{60} + \frac{45}{3600}} = 1\frac{15}{60}$ or $1\frac{1}{4}$). But the hypotenuse of the right triangle is the diagonal of the rectangle, and so should be 40 and not $1\frac{1}{4}$ ("Expecting 40, the diagonal that was indicated to you").

---

[9]Although the Babylonians surely would not have called the result "Pythagoras's theorem," it appears that they knew it nonetheless. For example, Suzuki [112, p. 31] tells us that "[a] tablet in the possession of Yale University shows a square, whose diagonal is marked with the cuneiform number equivalent to 1;24,51,10. If this is interpreted as being the length of the diagonal (and there is no reason to suspect otherwise), then it shows the Babylonians were ... aware of the Pythagorean theorem...." (See FIGURE 19.2.) As another example, the numbers listed in the cuneiform tablet Plimpton 322 may be construed as arising from Pythagorean triples (see page 14).

The next step, then, is to divide 40 by $1\frac{1}{4}$ to get 32 ("unravel the inverse of 1;15 the diagonal. 48 ...[48 × 40] ...32 you find": the reciprocal of $1\frac{1}{4}$ is $\frac{4}{5}$ or $\frac{48}{60}$, so $40 \div 1\frac{1}{4} = 40 \times \frac{48}{60} = 32$). This means that $32 \times 1\frac{1}{4} = 40$, so 32 times the guess for the length, 1, is the correct length; that is, the correct length is $32 \times 1 = 32$ ("Take ...[32 × 1] the length ...32 you find"). Finally, the breadth is three-fourths of the length: $\frac{3}{4} \times 32 = 24$ ("Take ...[32 × 45] the breadth ...24 you find": $32 \times \frac{45}{60} = 24$).

Therefore, the length is 32 and the breadth is 24.

In summary, we proceeded as follows:

Guess length:	**1**
Breadth:	$\frac{3}{4} \times \mathbf{1} = \frac{3}{4}$
Hypotenuse:	$\sqrt{1^2 + \left(\frac{3}{4}\right)^2} = \frac{5}{4}$ or $1\frac{1}{4}$, but the desired result is 40
Divide:	$40 \div \frac{5}{4} = \underline{\underline{32}}$
Multiply:	length $\underline{\underline{32}} \times \mathbf{1} = 32$, breadth $\underline{\underline{32}} \times \frac{3}{4} = 24$
Answer:	length is 32, breadth is 24

Now You Try 19.6 Solve the following problems using the method shown in EXAMPLE 19.8.

1. Find the dimensions of a rectangle whose breadth is three-quarters its length and its diagonal is 100.

2. Two boats leave a floating platform at the same time, one traveling east and the other traveling north. Each boat travels at its own constant rate in miles per hour. If the northbound boat is traveling at five-twelfths the rate of the eastbound boat, how far from the platform will be each boat when they are 10 miles apart?

Think About It 19.6 For what types of linear problems will the method of false position work? Will it work for nonlinear problems?

We summarize the method of false position for solving linear problems. (See RULE 19.1.)

The goal is to find a number ξ that, after applying some sequence of steps, leads to a number b. (The symbol ξ is the lowercase Greek letter *xi*.) To begin, we guess that the number is g. If our guess g leads to the desired result b, then we were clever (or lucky) and have solved the problem. Otherwise, our guess g leads to a number h that is not the desired result; however, if $b = ch$ for some number c, then the correct number that leads to the desired result b is c times our guess g; that is, $\xi = cg$ will lead to the number b.

Guess:	g
Work:	leads to a number h, but the desired result is b
Divide:	$b \div h = c$
Multiply:	$c \times g = \xi$
Answer:	ξ

Why does this work?

Let us take a look at the method of false position for solving linear problems from a modern point of view.

The method of false position can be used when the linear equation that describes the problem reduces to $ax = b$ *without transposition*, that is, without our having to add terms to the equation. For example, let us consider these two linear equations:

1. $x + \dfrac{1}{6}x = 21$,

2. $x + \dfrac{1}{6}x + 7 = 28$.

Both equations are equivalent to the equation $\dfrac{7}{6}x = 21$, which is of the form $ax = b$.

$$x + \dfrac{1}{6}x = 21$$

$$\dfrac{7}{6}x = 21$$

$$x + \dfrac{1}{6}x + 7 = 28$$

$$\dfrac{7}{6}x + 7 = 28$$

$$\dfrac{7}{6}x + 7 - 7 = 28 - 7$$

$$\dfrac{7}{6}x = 21$$

However, note that the first equation (on the left above) reduces to $\dfrac{7}{6}x = 21$ immediately, whereas we had to add -7 to both sides of the second equation to reduce it to $\dfrac{7}{6}x = 21$.

It is easy to check that 18 solves both equations. Nevertheless, let us try to solve each equation using the method of false position.

1. $x + \dfrac{1}{6}x = 21$

We guess that the solution is 6:

Guess:	6
Work:	$6 + \frac{1}{6} \times 6 = \underline{7}$, but the desired result is 21
Divide:	$21 \div \underline{7} = \underline{\underline{3}}$
Multiply:	$\underline{\underline{3}} \times 6 = 18$
Answer:	18
Check:	$18 + \frac{1}{6} \times 18 = 21$, the desired result

Therefore, the solution is 18.

2. $x + \dfrac{1}{6}x + 7 = 28$

Again, we guess that the solution is 6:

Guess:	6
Work:	$6 + \frac{1}{6} \times 6 + 7 = \underline{14}$, but the desired result is 28
Divide:	$28 \div \underline{14} = \underline{\underline{2}}$
Multiply:	$\underline{\underline{2}} \times 6 = 12$
Answer:	12
Check:	$12 + \frac{1}{6} \times 12 + 7 = 21$, not the desired result 28

Therefore, the solution is *not* 12.

So, why did the method of false position work to solve the first equation but not the second when both equations are equivalent to the same equation, namely, $\frac{7}{6}x = 21$? The key difference between the two equations can be seen graphically in FIGURE 19.3. The solution of the equation $x + \frac{1}{6}x = 21$ is given by the point of intersection, $(\xi, 21)$, of the graphs of $y = x + \frac{1}{6}x$ and $y = 21$, and the solution of the equation $x + \frac{1}{6}x + 7 = 28$ is given by the point of intersection, $(\xi, 28)$, of the graphs of $y = x + \frac{1}{6}x + 7$ and $y = 28$. In both cases, $\xi = 18$, but observe that the the graph of $y = x + \frac{1}{6}x$ is a line that passes through the origin $(0, 0)$, whereas the graph of $y = x + \frac{1}{6}x + 7$ is a line that does *not* pass through the origin. This is precisely the reason why false position works for the first equation, but not for the second.

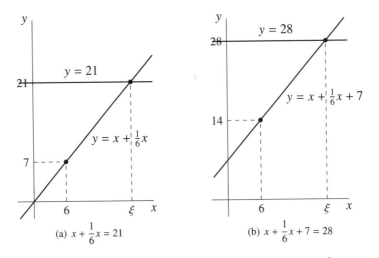

Figure 19.3: Even though both equations, $x + \frac{1}{6}x = 21$ and $x + \frac{1}{6}x + 7 = 28$, are equivalent to $\frac{7}{6}x = 21$—and so both have the same solution, $\xi = 18$—we see from the graphs that the two equations are fundamentally different.

Think About It 19.7 Explain why the method of false position does not work in the second case. $\square$

Let us take a more general look at solving the equation $ax = b$ by false position. FIGURE 19.4 shows the graphs of $y = ax$, $y = b$, and the numbers ξ, b, g, and h. We remark that g may be to the right of ξ; that is, our guess, g, may be greater than the number ξ we seek. We note that the line passes through the points $P_1 = (0,0)$, $P_2 = (g, h)$, and $P_3 = (\xi, b)$, where P_1 and P_2 are points known to be on the line and P_3 is to be found. Our strategy is to compute the slope of the line in two ways.

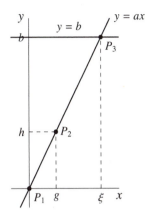

Figure 19.4: The method of false position.

On the one hand, using the known points P_1 and P_2, the slope of the line is computed to be

$$\frac{\Delta y}{\Delta x} = \frac{h - 0}{g - 0} = \frac{h}{g};$$

on the other hand, using the point P_1 and and the unknown point P_3, the slope of the line is computed to be

$$\frac{\Delta y}{\Delta x} = \frac{b - 0}{\xi - 0} = \frac{b}{\xi}.$$

Since the line has only one slope, this implies that we have the proportion

$$\frac{h}{g} = \frac{b}{\xi}.$$

Solving this proportion for ξ (by cross multiplying), we find that the number that will lead to the desired result b is

$$\xi = \frac{gb}{h}.$$

Now, suppose that $ch = b$. Then, multiplying $\frac{c}{c}$ times $\frac{bg}{h}$, we find that

$$\xi = \frac{bg}{h} = \frac{c}{c} \cdot \frac{gb}{h} = \frac{(cg)b}{ch} = \frac{(cg)b}{b} = cg.$$

Thus, we see that the number for which we are looking is $\xi = cg$; that is, cg will lead to the number b.

Think About It 19.8 Can you explain the method of false position using similar triangles instead? □

Recall EXAMPLE 19.5. The problem is to find a number such that, when one-fourth of the number is added to it, the total is 15. In other words, the goal is to find a number ξ that, after applying some steps ($\frac{1}{4}\xi$ added to ξ), leads to $b = 15$:

$$\xi + \frac{1}{4}\xi = 15.$$

If we let x denote a number, then the problem is described by the equation

$$x + \frac{1}{4}x = 15.$$

Simplifying the left-hand side, the equation reduces to

$$\frac{5}{4}x = 15,$$

which we see is of the form $ax = b$ without transposition, so that we may apply the method of false position.

Remark 19.2 Today, we would solve the equation $x + \frac{1}{4}x = 15$ by manipulating the symbols instead of using the method of false position:

$$x + \frac{1}{4}x = 15$$

$$\frac{5}{4}x = 15$$

$$x = \frac{15}{\frac{5}{4}} = 15\left(\frac{4}{5}\right) = 12.$$

This illustrates the power of symbolic algebra over rhetorical algebra: An equation can be manipulated without much thought from one step to the next in symbolic algebra, whereas much attention must be paid at each step in rhetorical algebra.   □

19.3   DOUBLE FALSE POSITION

We noted in the last section that, even though both equations

$$x + \frac{1}{6}x = 21 \quad \text{and} \quad x + \frac{1}{6}x + 7 = 28$$

are equivalent to the equation $\frac{7}{6}x = 21$, we may use the method of (single) false position to solve the first equation, but not the second. This is because the graph of $y = x + \frac{1}{6}x$ is a line that passes through the origin, so that we have *two* points that are known to be on the line (the point provided by our guess and the origin) that allow us to find the slope of the line; on the other hand, the graph of $y = x + \frac{1}{6}x + 7$ is a line that does *not* pass through the origin, so that we have *only one* point that is known to be on the line (the point that is provided by our guess), and this is insufficient to find the slope of the line. How can we obtain a second point on the graph of $y = x + \frac{1}{6}x + 7$? The answer is: Make a second guess. This leads us to the method of *double false position*.

We illustrate the method in the following example from *Daboll's Schoolmaster's Assistant* [43], an American arithmetic textbook of the early 1800s. Note that the layout used in this method is what was used in Europe and in America. We will also look at ways it was done in other areas.

Example 19.9 *Daboll's Schoolmaster's Assistant* [43, p. 202]

Daboll states the rule for double false position before giving an example.

DOUBLE POSITION,

TEACHES to resolve questions by making two suppositions of false numbers.

RULE.

1. Take any two convenient numbers, and proceed with each according to the conditions of the question.

2. Find how much the results are different from the results in the question.

3. Multiply the first position by the last error, and the last position by the first error.

4. If the errors are alike, divide the difference of the products by the difference of the errors, and the quotient will be the answer.

5. If the errors are unlike, divide the sum of the products by the sum of the errors, and the quotient will be the answer.

NOTE.—The errors are said to be alike when they are both too great, or both too small; and unlike, when one is too great, and the other too small.

EXAMPLES.

1. A purse of 100 dollars is to be divided among 4 men A, B, C, and D, so that B may have four dollars more than A, and C 8 dollars more than B, and D twice as many as C; what is each one's share of the money?

1st.	Suppose	A	6		2d.	Suppose	A	8
		B	10				B	12
		C	18				C	20
		D	36				D	40
			70					80
			100					100
	1st. error		30			2d. error		20

The errors being alike, are both too small, therefore,

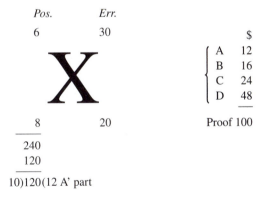

Pos.	Err.
6	30
8	20

$$
\begin{array}{ll}
 & \$ \\
A & 12 \\
B & 16 \\
C & 24 \\
D & 48 \\
\hline
\text{Proof} & 100
\end{array}
$$

240
120

10)120(12 A' part

We explain Daboll's example.

To solve this problem, Daboll first guesses that A receives $6, so that B has $10 ("four dollars more than A"), C has $18 ("8 dollars more than B"), and D has $36 ("twice as many as C"). Adding up these amounts, he finds a total of $70, which is $30 less than the $100 the four men are to share. So, another guess is made.

This time Daboll guesses that A gets $8, so that B gets $12, C gets $20, and D gets $40. Now, adding up these amounts, he finds a total of $80, which is $20 less than the $100 the four men are to share.

But, instead of continuing to guess, as Berlinghoff and Gouvêa [11, p. 102] say of this example, "Now comes the magic." Daboll displays the guesses and the corresponding error (the difference between the result and the required amount) as shown below, with the error across from the guess.

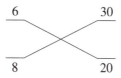

Now he computes:

$$\frac{8 \cdot 30 - 6 \cdot 20}{30 - 20} = \frac{120}{10} = 12.$$

Daboll, thus, concludes that A should get $12. Then B gets $16, C gets $24, and D gets $48. To check he adds up these amounts,

$$12 + 16 + 24 + 48 = 100,$$

to find a total of $100, which is exactly the amount that the four men are to share. That does seem magical!

We note that in this example the errors are "alike" (both are negative), so Daboll divides the difference between the (absolute values of the) products by the difference between the (absolute values of the) errors at the end. If the errors were "unlike" (one positive and the other negative), then Daboll would divide the *sum* of the (absolutes values of the) products by the *sum* of the (absolute values of the) errors, instead.

For example, suppose that Daboll had first guessed $6 for A as he did, but then had next guessed $14 for A.

Guess	A	B	C	D	Total	Error
1st	6	10	18	36	70	30
2nd	14	18	26	52	110	10

Now the errors are unlike, the first negative and the second positive. In this case, Daboll would compute

$$\frac{14 \cdot 30 + 6 \cdot 10}{30 + 10} = \frac{480}{40} = 12.$$

Daboll would then conclude, as he did, that A should get $12, and so on. □

Now You Try 19.7 Solve the following examples from *Daboll's Schoolmaster's Assistant* [43, p. 203–204] in the way that Daboll does. Note that there are 20 shillings in 1 pound: 20*s.* = 1*l.* or 20*s.* = £1.

1. 2. A, B, and C, built a house which cost 500 dollars, of which A paid a certain sum; B paid 10 dollars more than A, and C paid as much as A and B both; how much did each man pay?

Ans. A paid $120, B $130, and C $250.

2. 4. A labourer was hired for 60 days upon this condition that for every day he wrought he should receive 4s. and for every day he was idle should forfeit 2s; at the expiration of the time he received 7*l*. 10*s*.; how many days did he work, and how many was he idle?

Ans. He wrought 45 days, and was idle 15 days.

3. 7. Two men, A and B, lay out equal sums of money in trade; A gains 126*l*. and B loses 87*l*. and A's money is now double B's; what did each lay out?

Ans. £300.

4. 8. A farmer having driven his cattle to market, received for them all 130*l*. being paid for every ox 7*l*. for every cow 5*l*. and for every calf 1*l*. 10*s*. there were twice as many cows as oxen, and three times as many calves as cows; how many were there of each sort?

Ans. 5 oxen, 10 cows, and 30 calves.

□

We now state the method of double false position without regard to whether the errors are, as Daboll puts it, alike or unlike. We simply divide by the difference of the errors (accounting for their signs, positive or negative), disregarding whether they are both positive, both negative, or one is positive and the other is negative.

Rule 19.2 (METHOD OF DOUBLE FALSE POSITION)) Guess a solution, say, g_1. If the result should be b, but g_1 leads to a result h_1 that is incorrect, let $e_1 = b - h_1$ (where e_1 is the first error). Then make a second guess, say, g_2. If g_2 leads to a result h_2 that is also incorrect, let $e_2 = b - h_2$. Now arrange the numbers g_1, e_1, g_2, and e_2 as follows:

The solution is found by dividing the difference between the cross products of the guesses and the errors by the difference between the errors:

$$\text{solution} = \frac{g_1 e_2 - g_2 e_1}{e_2 - e_1} \quad \text{or} \quad \frac{g_2 e_1 - g_1 e_2}{e_1 - e_2}.$$

□

Think About It 19.9 Daboll has two versions of his rule for double false position based on the types of errors (alike or unlike), but the above general rule does not. Are these actually the same rule? □

Example 19.10 Solve the equation $x + \frac{1}{6}x + 7 = 28$ using the method of double false position.

We guess that the solution is 6:

Guess:	$g_1 = 6$
Work:	$6 + \frac{1}{6} \times 6 + 7 = \underline{14}$, but the desired result is 28

| Error: | $e_1 = 28 - \underline{14} = 14$ |

So we make a second guess, this time 24:

Guess:	$g_2 = \mathbf{24}$
Work:	$\mathbf{24} + \frac{1}{6} \times \mathbf{24} + 7 = \underline{35}$, but the desired result is 28
Error:	$e_2 = 28 - \underline{35} = -7$

Now we arrange the numbers 6, 14, 24, and −7 as follows:

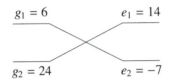

Then the solution is

$$\frac{g_1 e_2 - g_2 e_1}{e_2 - e_1} = \frac{6 \cdot (-7) - 24 \cdot 14}{-7 - 14} = 18.$$

Check:

$$18 + \frac{1}{6} \times 18 + 7 = 28.$$

☐

If you are still considering why our modern version works for both cases of the errors being alike or unlike, recalculate the above solution using the alternative version of the equation above and observe the −7 in the denominator.

Now You Try 19.8 Solve the following problems using the method of double false position (RULE 19.2).

1. Three children, A, B, and C, are several months old. The age of B is one-sixth A's, and C is four months older than B. Their ages together is 28 months; how old is each child?

2. (Rhind problem 26; see EXAMPLE 19.5) **A quantity, its $\overline{4}$ (is added) to it** so that 15 results.

3. Solve the equation $x - \frac{1}{2}x + \frac{1}{3}x + 5 = 15$. (See EXAMPLE 19.4.)

☐

Think About It 19.10 As with single false position, why is double false position exact only with linear problems? (Hint: Think about what happens to the graph of the equation as you make different guesses.)
☐

For our next examples, we turn to *Liber Abaci*, in which Leonardo discusses double false position in chapter 13, "On the Method of Elchataym and How with It Nearly All Problems of Mathematics Are Solved" [90].

Indeed the Arabic elchataym by which the solutions to nearly all problems are found is translated as the method of double false position.…

To introduce the method, Leonardo poses the problem:

> Indeed the value of one hundredweight, namely 100 rolls, is 13 pounds, and it is sought how much 1 roll is worth.

(There are 12 denari in one soldo, and 20 soldi in one pound or lira.) He then goes on to solve this problem in two ways, using two methods of elchataym. We are interested here in Leonardo's second method of elchataym[10] about which he says [90, p.449]:

> There is indeed another elchataym method which is called the augmented and diminished method in which the errors are put below their positions; the first error is multiplied by the second position, and the second error is multiplied by the first position. And if both the errors are minus, or both are plus, the lesser product is subtracted from the aforesaid greater product, and the difference is divided by the difference of the errors, and thus the solution of the problem is found; and if one of the errors is plus, and the other is minus, then their products are added together, and the sum is divided by the sum of their errors.

Leonardo then applies these instructions to his opening example: "Indeed the value of one hundredweight, namely 100 rolls, is 13 pounds, and it is sought how much 1 roll is worth." He does this three times to illustrate the three different cases of his method: The first time his guesses lead to two diminished values; the second time his guesses lead to two augmented values; and the third time his guesses lead to one augmented value and one diminished value. Leonardo then justifies the method. We present the first and the third, and leave the second to you as an exercise.

The first time

Leonardo's two guesses for the value of one roll are 1 soldo and 2 soldi. He says:

> For example, we put above the proportion of one roll to be 1 soldo with which we erred by minus 8 pounds; therefore you put the 8 below the 1, and you will note minus above the 8, as it is minus; next because we put 2 soldi in the second position for the price of the same roll, and we erred then by minus 3 pounds, you put the 2 soldi before[11] the first position and below this you put the error, namely the 3 pounds, above which you will note again minus as it is again deficient, and you will multiply the two soldi by the number of the first error; there will be 16 soldi, and you multiply the 1 soldo by the second error; there will be 3 soldi.
>
> And because both errors were minus you subtract the lesser product from the greater, namely the 3 from the 16 leaving 13 soldi which is divided by the error difference, namely by the 5, yielding $\frac{3}{5}$ 2 soldi, as we found before.

Explanation: First Leonardo writes down each guess and its error, in this case noting that the errors are "minus" (diminished or deficient).

soldi	soldi
1	2
minus	minus
8	3

---

[10]The first method is shown on page 450.

[11]Note that "before the first position" refers to being put *to the right* of the first numbers.

Then he cross multiplies each guess by the opposing error and notes the products above the corresponding guesses.

3	16
soldi	*soldi*
1	2
minus	*minus*
8	3

Now, because *both errors are minus*, Leonardo *subtracts* the cross products and also the errors.

13	
3	16
soldi	*soldi*
1	2
minus	*minus*
8	3
error	
diff. 5	

Finally, to find the sought result, Leonardo divides the differences:

$$\frac{16 - 3}{8 - 3} = \frac{13}{5} = 2\tfrac{3}{5}.$$

Thus, he concludes that if the value of 100 rolls is 13 pounds, then one roll is worth $\tfrac{3}{5}$ 2 soldi or 2 soldi and $\tfrac{1}{5}$ 7 denari (because $\tfrac{3}{5} \times 12 = 7\tfrac{1}{5}$).

The third time

This time Leonardo guesses 2 soldi and 3 soldi: the first leads to an error minus 3 and the second to an error plus 2. Then, as before, he cross multiplies each guess by the opposing error and notes the products above the corresponding guesses.

4	9
soldi	*soldi*
2	3
minus	*plus*
3	2

Now, because *one error is minus and the other is plus*, Leonardo *adds* the cross products and also the errors.

	13	
4		9
soldi		*soldi*
2		3
minus		*plus*
3		2
	error	
	sum. 5	

Finally, to find the sought result, Leonardo divides the sums and, again, concludes that if the value of 100 rolls is 13 pounds, then one roll is worth $\frac{3}{5}$ 2 soldi or 2 soldi and $\frac{1}{5}$ 7 denari.

Remark 19.3 In RULE 19.2 for the method of double false position (page 287), we said that if the result should be b, but the guess leads to a result h that is incorrect, then the error is $e = b - h$. Notice that Leonardo computes the error oppositely; that is, Leonardo computes $e = h - b$. The method will work regardless of whether the error is computed as

$$e = b - h \quad \text{or} \quad e = h - b$$

as long as both the first and the second errors are computed in the same way.   □

Think About It 19.11 How does Leonardo's method compare with the one in *Daboll's Schoolmaster's Assistant*?   □

Now You Try 19.9

1. The second time Leonardo solves the problem he guesses for the price of one roll 4 soldi and 3 soldi: both guesses lead to errors that are "plus" (augmented or excessive). He then goes through the same steps as in his first solution. And, because *both errors are plus*, Leonardo again *subtracts* the cross products and also the errors. Finally, to find the sought result, Leonardo again divides the differences and concludes that if the value of 100 rolls is 13 pounds, then one roll is worth $\frac{3}{5}$ 2 soldi or 2 soldi and $\frac{1}{5}$ 7 denari.

   Work through the details of Leonardo's second solution.

2. (*Liber Abaci* chapter 13 problem "Notable Problem on a Worker" [90, p. 453]) A certain worker received 7 bezants per month if he worked, and if he did not work he had to pay 4 bezants per month to the foreman; for whatever he worked or did not work he received at the end of the month 1 bezant from the foreman; it is sought how many days of the month he worked. Assume that there are 30 days in one month.

   a) Solve the problem using Leonardo's second elchataym method. First, guess that he worked 20 days; second, guess that he worked 15 days.

   b) Solve the problem using an algebraic equation. Let x denote the number of days he worked.

   □

Our last examples of false position are taken from the *Nine Chapters* (*Jiu zhang suan shu*; see page 37). The *Nine Chapters* contains "246 applied problems of a sort useful in teaching how to handle arithmetic and elementary algebra and how to apply them in commercial and administrative work" [41]. Yet, Shen et al. [108, p. viii] remark that in the *Nine Chapters* and Liu's comments "one learns a surprising amount about life in China at the beginning of the Christian era." The nine chapters of the *Nine Chapters* are as follows (the descriptions are taken from Liu's comments) [44]:

Chapter 1:   *Fang tian* (Field Measurement)   For determining the boundary and area of a field.

Chapter 2:   *Su mi* (Millet and Rice)   For barter trade and exchange of goods.

Chapter 3:   *Cui fen* (Proportional Distribution)   For distribution of grain and taxation.

Chapter 4:   *Shao guang* (Short Width)   For determining areas (volumes) of squares (cubes) and circles (spheres) and their relations.

Chapter 5:   *Shang gong* (Construction Consultations)   For calculating the number of laborers, the volume of earthworks, and the capacity of warehouses.

Chapter 6:   *Jun shu* (Fair Taxes)   To regulate the different expenses for distance and service, etc. in transportation.

Chapter 7:   *Ying bu zu* (Excess and Deficit)   For the treatment of intricate and implicit [problems].

Chapter 8:   *Fang cheng* (Rectangular Arrays)   Treating mixed positive and negative [numbers].

Chapter 9:   *Gou-gu* (Base-Height) For the treatment of altitude, depth, length, and width.

Each chapter presents a collection of specific problems on the topic at hand, solutions to the problems, and explanations of the methods used to obtain the solutions. The explanations are generally very terse, and so the comments provided by Liu Hui and others are invaluable additions to the work. No general methods or proofs are provided in the sense that we would expect, but clearly the reader is meant to take the specific problems as examples of general methods for solving problems.[12]

---

[12]See Joseph [72, pp. 130–139] for an overview of Chinese history and developments in mathematics from the Xia period to the Ming and the beginnings of contacts with the West. See Shen et al. [108, chapter 0] for a more extensive history of mathematics in China, including a synopsis of and comparison with the mathematics of ancient Egypt, Babylon, Greece, India, the Arab world, Europe, and Japan.

Chapter 7 of the *Nine Chapters*, titled "*Ying bu zu*" ("Excess and Deficit"), also treats the method of double false position. (Excess or deficit refers to whether a guess leads to a result that is greater than or less than the desired result.) Chabert et al. [39] remark that, if the editions of the *Nine Chapters* that have come down to us are faithful reproductions of the original, then the treatment of double false position that we find here "is certainly richer than that which appears in other ancient and mediaeval arithmetics." The following two examples from the *Nine Chapters* appear in [44]. The first example treats the case when there is an excess and a deficit; the second example treats the case when there are two deficits (the case of two excesses would be similar).

Example 19.11 *Nine Chapters* Chapter 7: "Excess and Deficit" [44, p. 270]

2. Now chickens are purchased jointly; everyone contributes 9, the excess is 11; everyone contributes 6, the deficit is 16. Tell: the number of people, the chicken price, what is each? Answer: 9 people, chicken 70.

The Excess and Deficit Rule: Display the contribution rates; lay down the [corresponding] excess and deficit below. Cross-multiply by the contribution rates; combine them as dividend; combine the excess and deficit as divisor. Divide the dividend by the divisor. [If] there are fractions, reduce them. To relate the excess and the deficit for the articles jointly purchased: lay down the contribution rates. Subtract the smaller from the greater, take the remainder to reduce the divisor and the dividend. The [reduced] dividend is the price of an item. The [reduced] divisor is the number of people.

We follow the "Excess and Deficit Rule":

- "Display the contributed rates"

  Dauben [44] tells us that the guesses and the excess and deficit would be displayed on a counting board thus:[13]

  $$\begin{array}{cc} 6 & 9 \\ 16 & 11 \end{array} \quad \begin{array}{l} \leftarrow \text{ contribution rates} \\ \leftarrow \text{ deficit and excess} \end{array}$$

- "Cross-multiply the contribution rates"

  $$\begin{array}{cc} 66 & 144 \\ 16 & 11 \end{array} \quad \leftarrow \quad 6 \cdot 11 = 66, \ 16 \cdot 9 = 144$$

- "combine them as dividend; combine the excess and deficit as divisor"

  $$\begin{array}{ll} 210 & \leftarrow \quad \text{dividend} : 66 + 144 = 210 \\ 27 & \leftarrow \quad \text{divisor} : 16 + 11 = 27 \end{array}$$

- "lay down the contribution rates. Subtract the smaller from the greater"

  $$\text{remainder} = 9 - 6 = 3$$

- "take the remainder and reduce the divisor and the dividend"

  $$\begin{array}{ll} 70 & \leftarrow \quad \text{reduced dividend} : 210 \div 3 = 70 \\ 9 & \leftarrow \quad \text{reduced divisor} : 27 \div 3 = 9 \end{array}$$

- "The [reduced] dividend is the price of an item. The [reduced] divisor is the number of people"

  $$\text{price of an item} = 70, \quad \text{number of people} = 9$$

---

[13]We would also obtain the correct result if we were to arrange the numbers $\begin{smallmatrix} 9 & 6 \\ 11 & 16 \end{smallmatrix}$.

Therefore, we conclude that there are 9 people contributing equally to buy a number of chickens at a price of 70 (monetary units) each.   □

Think About It 19.12 How does the method given in the *Nine Chapters* compare with the method given in *Daboll's Schoolmaster's Assistant* (page 284)? With the method given in *Liber Abaci* (page 290)? □

We explain the problem in EXAMPLE 19.11 a little further. To begin, we note that it is assumed that the price per chicken is the same for each chicken. Next, we define the variables n and p by

$$n \text{ persons} = \text{the number of persons who are buying the chickens,}$$

$$p \, \frac{\text{coins}}{\text{chicken}} = \text{price per chicken,}$$

where we are using "coin" as a monetary unit for convenience. Then, "everyone contributes 9, the excess is 11" may be expressed as

$$(n \text{ persons})\left(9 \, \frac{\text{coins per chicken}}{\text{person}}\right) = p \, \frac{\text{coins}}{\text{chicken}} + 11 \, \frac{\text{coins}}{\text{chicken}},$$

and "everyone contributes 6, the deficit is 16" may be expressed as

$$(n \text{ persons})\left(6 \, \frac{\text{coins per chicken}}{\text{person}}\right) = p \, \frac{\text{coins}}{\text{chicken}} + \left(-16 \, \frac{\text{coins}}{\text{chicken}}\right).$$

Hence, we see that the problem in the example may be expressed as the system of linear equations

$$9n = p + 11,$$
$$6n = p - 16,$$

which is straightforward to solve, and which yields the solution $n = 9$ and $p = 70$. This solution, of course, agrees with the solution that was found using the "excess and deficit" rule in EXAMPLE 19.11; however, this is not how the problem was solved in the example.

In the example, the author sought the rate

$$\xi \, \frac{\text{coins per chicken}}{\text{person}},$$

that is, the rate ξ at which each person contributs toward the price per chicken. To find ξ, the author made two guesses, namely,

$$g_1 \, \frac{\text{coins per chicken}}{\text{person}} = 9 \, \frac{\text{coins per chicken}}{\text{person}}$$

and

$$g_2 \, \frac{\text{coins per chicken}}{\text{person}} = 6 \, \frac{\text{coins per chicken}}{\text{person}}.$$

The author then finds that the first guess $g_1 = 9$ leads to an *excess* of 11, that is to say, if b is the desired result (the actual price per chicken), g_1 leads to a result h_1 that is 11 more than b:

$$g_1 = 9 \quad \text{leads to} \quad h_1 = b + 11, \quad \text{so that} \quad e_1 = h_1 - b = 11.$$

Likewise, the second guess $g_2 = 6$ leads to a *deficit* of 16, that is to say, g_2 leads to a result h_2 that is 16 less than b:

$$g_2 = 6 \quad \text{leads to} \quad h_2 = b - 16, \quad \text{so that} \quad e_2 = h_2 - b = -16.$$

We now lay out the numbers g_1, e_1, g_2, and e_2 as shown in EXAMPLE 19.9 on page 284, using the absolute values of the errors.

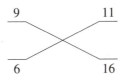

Daboll's rule tells us to "divide the sum of the products by the sum of the errors, and the quotient will be the answer" because the errors are unlike (one is an excess and the other a deficit):

$$\xi = \frac{9 \cdot 16 + 6 \cdot 11}{11 + 16} = \frac{210}{27} \quad \text{or} \quad \frac{70}{9} \quad \text{(reduced)}.$$

Note that this is equivalent to applying the method given in RULE 19.2 on page 287, namely,

$$\xi = \frac{g_1 e_2 - g_2 e_1}{e_2 - e_1} = \frac{9 \cdot (-16) - 6 \cdot 11}{-16 - 11} = \frac{-210}{-27} = \frac{70}{9}.$$

Therefore, we conclude that

$$\xi \frac{\text{coins per chicken}}{\text{person}} = \frac{70 \text{ coins per chicken}}{9 \text{ persons}},$$

that is, there are 9 people contributing equally to buy a number of chickens at a price of 70 (monetary units) each.

Think About It 19.13

1. Referring to the problem in EXAMPLE 19.11, determine how many chickens were bought altogether, or explain why we may not determine this number.

2. In the Chinese method, the deficit is not recorded as negative, as is done in the method given in *Liber Abaci* (page 290), yet the solution is correct. Compare the Chinese and European methods to determine why this is so.

☐

Example 19.12 *Nine Chapters* Chapter 7: "Excess and Deficit" [44, p. 271]

6. Now sheep are purchased jointly; everyone contributes 5, the deficit is 45 everyone contributes 7, the deficit is 3. Tell: the number of people, the sheep price, what is each? Answer: 21 people, sheep price 150.

The Double Excess and Double Deficit Rule: Lay down the contribution rates, with the corresponding excesses [or] deficits below. Cross-multiply by the contribution rates. Subtract the smaller from the greater. The surplus is the dividend. [Take] the two excesses [or] the two deficits, subtract the smaller from the greater. The surplus is the divisor. Divide the dividend by the divisor. If there are fractions, uniformize [the denominators]. In a joint purchase there may appear two excesses and two deficits. Lay down the contribution rates. Subtract the smaller from the greater, [take] the surplus to reduce the divisor and the dividend. The [reduced] dividend is the item price, the [reduced] divisor is the number of people.

We follow the 'Double Excess and Double Deficit Rule":

• "Lay down the contribution rates, with the corresponding excesses [or] deficits"

7	5	←	contribution rates
3	45	←	deficits

- "Cross-multiply by the contribution rates"

$$315 \quad 15 \quad \leftarrow \quad 7 \cdot 45 = 315, \ 3 \cdot 5 = 15$$
$$3 \quad 45$$

- "Subtract the smaller from the greater. The surplus is the dividend. [Take] the two excesses [or] the two deficits, subtract the smaller from the greater. The surplus is the divisor"

$$300 \quad \leftarrow \quad \text{dividend} : 315 - 15 = 300$$
$$42 \quad \leftarrow \quad \text{divisor} : 45 - 3 = 42$$

- "Lay down the contribution rates. Subtract the smaller from the greater"

$$\text{surplus} = 7 - 5 = 2$$

- "[take] the surplus to reduce the divisor and the dividend"

$$150 \quad \leftarrow \quad \text{reduced dividend} : 300 \div 2 = 150$$
$$21 \quad \leftarrow \quad \text{reduced divisor} : 42 \div 2 = 21$$

- "The [reduced] dividend is the item price, the [reduced] divisor is the number of people"

$$\text{item price} = 150, \quad \text{number of people} = 21$$

Therefore, we conclude that there are 21 people contributing equally to buy a number of sheep at a price of 150 (monetary units) each. □

Think About It 19.14 Find the step in the problem above that is different from the "excess and deficit" rule. How does this adjust for the new situation in this problem? □

Now You Try 19.10

1. Solve the problem in EXAMPLE 19.12 using the method in *Daboll's* shown in EXAMPLE 19.9.

2. Solve the following problems using the method in EXAMPLE 19.11 or 19.12.

   a) Some coworkers carpool to work and share equally in the cost of gasoline every week. If each person contributes $3.00 toward the price of one gallon of gasoline, the deficit is $0.60. If each person contributes $4.00, the excess is $2.40. Find the number of coworkers in the carpool and the price of one gallon of gasoline.

   b) A group of friends go out to dinner and share equally in the total cost. If each one were to contribute $30, they would have altogether $45 over the total, and if each one were to contribute $25 they would have altogether $20 over the total. Find the number of friends and the total cost of the dinner.

□

From a modern point of view, the method of double false position can be used when the linear equation that describes the problem reduces to $ax + c = b$ *without transposition*. To see this, we show in the FIGURE 19.5 the graphs of $y = ax + c$, $y = b$, and the numbers ξ, b, g_1, h_1, g_2, and h_2. (Recall that ξ is the solution sought that leads to the desired result b; g_1 and g_2 are guesses for ξ; but that the guess g_1 leads to an incorrect result h_1, and g_2 leads to another incorrect result h_2. The errors, then, are $e_1 = b - h_1$ and $e_2 = b - h_2$. See page 287.)

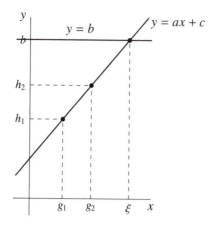

Figure 19.5: The method of double false position.

We remark that g_1 and g_2 could be on either side of the number we seek, ξ. We note that the line passes through the points $P_1 = (g_1, h_1)$, $P_2 = (g_2, h_2)$, and $P_3 = (\xi, b)$, where P_1 and P_2 are points known to be on the line and P_3 is to be found. As with the method of false position, our strategy is to compute the slope of the line in two ways. (See page 283.)

Using the known point P_1 and the unknown point P_3, the slope of the line is computed to be

$$\frac{\Delta y}{\Delta x} = \frac{b - h_1}{\xi - g_1};$$

on the other hand, using the known point P_2 and the point P_3, the slope of the line is computed to be

$$\frac{\Delta y}{\Delta x} = \frac{b - h_2}{\xi - g_2}.$$

Since the line has only one slope, this implies that we have the proportion

$$\frac{b - h_1}{\xi - g_1} = \frac{b - h_2}{\xi - g_2}$$

or, if we let $e_1 = b - h_1$ and $e_2 = b - h_2$,

$$\frac{e_1}{\xi - g_1} = \frac{e_2}{\xi - g_2}.$$

We leave the solving of this proportion for ξ to you as an exercise. You should find that the solution is

$$\xi = \frac{g_1 e_2 - g_2 e_1}{e_2 - e_1} \quad \text{or} \quad \frac{g_2 e_1 - g_1 e_2}{e_1 - e_2}.$$

Referring to FIGURE 19.5, we apply this general understanding to EXAMPLE 19.9 from *Daboll's Schoolmaster's Assistant*. In this case,

$$x \text{ dollars} = \text{the amount that A should receive,}$$
$$y \text{ dollars} = \text{the amount of the purse.}$$

The solution $x = \xi$, then, gives the amount that A should receive from the $100 purse ($b = 100$). Applied to EXAMPLE 19.11 from *Nine Chapters*,

$$x \frac{\text{coins per chicken}}{\text{person}} = \text{rate at which each person contributes toward the price per chicken,}$$

$$y \frac{\text{coins}}{\text{chicken}} = \text{price per chicken.}$$

The solution, $x = \xi$, then, gives both the price per chicken (the numerator) and the number of persons who are buying the chickens (the denominator).

Arabic mathematicians were also fond of using false position, principally double false position, to solve certain problems.[14] Chabert et al. [39, pp. 98–101] present a geometric justification of the method of double false position by the Christian Arabic mathematician Quṣtā ibn Lūqā of the late ninth century that considers areas of rectangles and gnomons (in the Euclidean tradition); but, because Arabic mathematicians as a rule did not consider negative numbers, ibn Lūqā's geometric justification treats separately the cases of excess and deficit, excess and excess, and deficit and deficit (as in the *Nine Chapters*, for example).

Finally, we mention one more example from chapter 7 of the *Nine Chapters*, namely, problem 19 [44, p. 273]. The purpose here is to provide a concrete example of how a method for solving a linear problem (double false position in this case) can be used to approximate the solution of a nonlinear problem. In this example, the nonlinear problem is described by a quadratic equation. Do not worry if you have difficulty following the solution of the quadratic equation below; we will discuss quadratic equations in detail in CHAPTER 20. For now, the point we want to make is illustrated in FIGURE 19.6.

> 19. Now a good horse and an inferior horse set out from Chang'an to Qi. Qi is 3000 *li* from Chang'an. The good horse travels 193 *li* on the first day and daily increases by 13 *li*, the inferior horse travels 97 *li* on the first day and daily decreases by 3 *li*. The good horse reaches Qi first [and] turns back to meet the inferior horse. Tell: how many days [till they] meet and how far has each traveled? Answer: 15 135/191 days [till they] meet, the good horse traveled 4534 46/191 *li*, the inferior horse traveled 1465 145/191 *li*.
>
> Method: Assume 15 days, deficit 337 1/2 *li*. Assume 16 days, excess 140 *li*. Cross-multiply the excess and the deficit by the assumed numbers [and] add to the dividend. Excess plus deficit as divisor. Divide the dividend by the divisor to obtain the number of days. Simplify the remainder by the *dengshu* [divide by the greatest common divisor] and express as a fraction.

This is followed by a long commentary by Liu Hui.

The method described above appears deceptively simple, but, in fact, before even applying the method of double false position, one would have to find the sums of two arithmetic progressions[15] to find the distances the horses travel. What is really interesting with this example, however, is that the problem actually is a *nonlinear* problem: If ξ denotes the number of days the horses travel, then ξ satisfies the quadratic equation

$$\frac{25}{4}x^2 + \frac{1135}{4}x - 6000 = 0,$$

and

$$\text{the distance the good horse travels} = [193 + \tfrac{13}{2}(\xi - 1)]\xi \; li,$$

$$\text{the distance the poor horse travels} = [97 - \tfrac{1}{4}(\xi - 1)]\xi \; li.$$

So, strictly speaking, false position should not be used to solve this problem. Nevertheless, the method of double false position, which is designed for linear problems, was used to obtain a good approximation of the solution of this nonlinear problem: The positive root of the quadratic equation above is[16]

$$\xi = \frac{-1135 + \sqrt{3{,}688{,}225}}{50} \approx 15.710,$$

and the solution (number of days the horses travel) determined in the problem is

$$15\tfrac{135}{191} \approx 15.707.$$

This brings us back to what we said earlier in the section (page 270): A general strategy for solving any type of nonlinear equation is to approximate the equation with a suitable linear equation. In this case, by

---

[14] Although the Indians had a method of false position, they did not have a method of double false position.

[15] See the footnote on arithmetic progressions on page 462.

[16] See page 330 for the quadratic formula.

using the values (guesses) 15 and 16 that are "close" to the root of the nonlinear equation, the root could be approximated by the linear method of double false position. A look at the graph makes this evident.

The graph shown on the left in FIGURE 19.6 is the graph of $y = \frac{25}{4}x^2 + \frac{1135}{4}x - 6000$; note that the graph is not a line because the equation is not linear. Nevertheless, the positive root of $\frac{25}{4}x^2 + \frac{1135}{4}x - 6000 = 0$ lies inside the small box in the upper right-hand quadrant—the vertices of the box are $(14, -400)$, $(14, 200)$, $(17, 200)$, and $(17, -400)$—and inside that box the graph of $y = \frac{25}{4}x^2 + \frac{1135}{4}x - 6000$ is very nearly a straight line; a blow up of the box is shown on the right below. Because the guesses were close to the solution, the linear method of double false position provides a good approximation of the solution of the nonlinear problem.

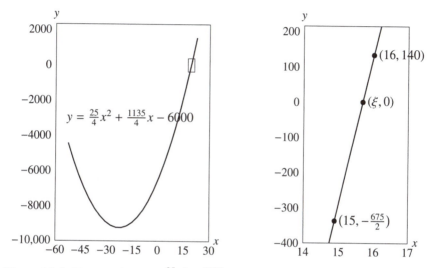

Figure 19.6: The graph of $y = \frac{25}{4}x^2 + \frac{1135}{4}x - 6000$ (left) and a blow up of a small section of the graph, which we see is practically a straight line.

Think About It 19.15 Looking at the graphs in FIGURE 19.6 for reference, when is double false position a good approximation and when is it not? □

Now You Try 19.11 Below is the graph of the equation $y = x^2 - x - 1$.

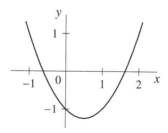

1. Use the method of double false position to approximate the positive root ϕ of the equation

$$x^2 - x - 1 = 0.$$

Chapter 20

Quadratic Problems

Many systems in nature are very complex and thus are not modeled using linear equations. As we mentioned at the start of the last chapter, nonlinear equations—and, correspondingly, nonlinear problems— are typically "hard" to solve. An example of a nonlinear system is the weather, and, for this reason, weather prediction is a very tricky business.

The simplest class—yet, a very important class—of nonlinear problems is that of *quadratic problems* and *quadratic equations*. Here are some examples of quadratic equations; in these examples, $a, b, c, \ldots, j$ are constants.

- In the variable x:

$$ax^2 + bx + c = 0, \quad a \neq 0$$

  Example: $x^2 + 10x = 39$ or $x^2 + 10x - 39 = 0$

- In the variables x and y:

$$ax^2 + by^2 + cxy + dx + ey + f = 0, \quad a, b, \text{ or } c \neq 0$$

  Example: $x^2 - 2x + y - 3 = 0$ or $(x - 1)^2 + y = 4$

- In the variables x, y, and z:

$$ax^2 + by^2 + cz^2 + dxy + exz + fyz + gx + hy + iz + j = 0,$$
$$a, b, c, d, e, \text{ or } f \neq 0$$

  Example: $x^2 + y^2 - 3z^2 + 2xy + 5yz - y - 3z + 10 = 0$

Quadratic equations are important because it turns out that they can be used to model many phenomena. For instance, if a stone is dropped from a bridge that is 700 feet high, then the height, h feet, of the stone at t seconds from the time it is dropped is given by the quadratic equation

$$h = -16t^2 + 700.$$

As another example, if an automobile slows at a constant deceleration[1] of a feet per second per second, then a simple model says that the stopping distance, s feet, to bring the automobile to a stop from a speed of v feet per second is given by

$$s = \frac{v^2}{2a}.$$

---

[1] Note that we use a in the equation because we designated our motion as a *deceleration*. If we want to use the term acceleration, but consider the case where an object is slowing down, then we would use $-a$ in the equation.

A third example comes from astronomy. Johannes Kepler (1571–1630) discovered that the orbit of a planet about the sun traces out an *ellipse*. In the xy plane, an ellipse centered at the origin has the equation

$$ax^2 + 2bxy + cy^2 = 1,$$

where a, b, and c are constants with $b^2 - ac \leq 0$.

Quadratic models even appear in fertilizer recommendations!

Now You Try 20.1

1. A stone is dropped from a bridge that is 700 feet high, so that the height, h feet, of the stone at t seconds from the time it is dropped is given by

$$h = -16t^2 + 700.$$

   a) Find the height of the stone at 1 second. At 2 seconds. At 3 seconds.

   b) Does the stone fall the same distance after each second?

   c) Between which two integral seconds will the stone reach the bottom (0 feet)?

2. If an automobile slows at a constant deceleration of a feet per second per second, then a simple model says that the stopping distance, s feet, to bring the automobile to a stop from a speed of v feet per second is given by

$$s = \frac{v^2}{2a}.$$

   a) Suppose that you are driving your automobile at 30 miles per hour when you apply the brakes. If you are slowing at a constant rate of 54,000 miles per hour per hour, what is your stopping distance in feet? Hint: You will need to convert miles per hour to feet per second and miles per hour per hour to feet per second per second. For this, note that there are 5280 feet in one mile and 3600 seconds in one hour.

   b) Suppose that you are driving your automobile at 60 miles per hour when you apply the brakes. If you are slowing at a constant rate of 54,000 miles per hour per hour, what is your stopping distance in feet?

   c) How does doubling your driving speed affect your stopping distance? Show this in general; that is, compare your stopping distance when traveling at v_0 feet per second and at $2v_0$ feet per second if, in both cases, you are slowing at a constant rate of a_0 feet per second per second.

□

The solution of quadratic problems dates back to the time of the ancient Babylonians and is seen in the mathematics of many cultures. In this chapter, we explore a few of the methods used at various times and by various people to solve problems that lead to quadratic equations when expressed in modern notation.

20.1   SOLVING QUADRATIC EQUATIONS BY COMPLETING THE SQUARE

Completing the square is a geometrical method for solving quadratic equations that falls under the broad category of *geometric algebra*. The method of completing the square has been used by many different cultures. In some instances—for example, with the Arabs—it is clear that the problems were being interpreted and solved geometrically by completing the square, whereas in other instances—for example, with the Babylonians—we may only interpret the solution presented geometrically as completing the square even though there is no explicit reference to the method. We explain the heart of "completing the square" in the following example.

Example 20.1 The shaded L-shaped region in the following figure is called a *gnomon*. Given that the shaded gnomon has area 45 square inches, find the length x inches in the figure.

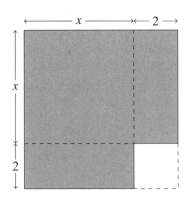

To find the length x inches, we note that the missing square corner (unshaded) has area 2×2 square inches or 4 square inches. By adding 4 to the area of the shaded region, we restore or complete the area of the larger square (we "complete the square"), which we find is $(45 + 4)$ square inches or 49 square inches. Since the completed square has area 49 square inches, and $7^2 = 49$, a side of the completed square has length 7 inches. Thus, we see that

$$(x + 2) \text{ inches} = 7 \text{ inches},$$

so that the length x inches $= 5$ inches. □

Now You Try 20.2 Find the length x inches in the following figure given that the shaded gnomon has area 160 square units.

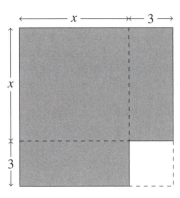

□

The idea behind the method of completing the square, then, is to cast a given problem as finding part of one side of a rectangle with known area; transform the rectangle into a gnomon that has the same area and is similar to the ones shown in the figures above; complete the square; and deduce the missing part.

20.1.1   Babylonian

Our first examples are from Mesopotamian mathematics, specifically, from the old Babylonian tablets YBC 6967 and BM 13901. The problem in YBC 6967 deals with "reciprocals," two numbers whose product is 60 (see page 18). The tablet BM 13901 contains twenty-four quadratic problems and their solutions. As Robson [102, p. 102] describes it, the set of problems in BM 13901 "progresses from very simple scenarios about single squares to complex situations involving two squares or more."

In these examples, we use our usual numerals instead of cuneiform script to express numerals in base 60, and a semicolon as the separatrix (to separate the whole number part of a number from the fraction part); see page 77.

Example 20.2 YBC 6967 [102, p. 102]

> A reciprocal exceeds its reciprocal by 7. What are the reciprocal and its reciprocal? You: break in two the 7 by which the reciprocal exceeds its reciprocal so that 3;30 (will come up). Combine 3;30 and 3;30 so that 12;15 (will come up). Add 1 00, the area, to the 12;15 which came up for you so that 1 12;15 (will come up). What squares 1 12;15? 8;30. Draw 8;30 and 8;30, its counterpart, and then take away 3;30, the holding square, from one; add to one. One is 12, the other is 5. The reciprocal is 12, its reciprocal is 5.

The problem is to find a number and its reciprocal given that the reciprocal exceeds the number by 7. We summarize the steps of the solution.

Babylonian	Step in YBC 6967	Indo-Arabic
7	A reciprocal exceeds its reciprocal by 7	7
$7 \div 2 = 3;30$	break in two the 7 ... so that 3;30 (will come up)	$7 \div 2 = 3\frac{1}{2}$
$3;30 \times 3;30 = 12;15$	Combine 3;30 and 3;30 so that 12;15 (will come up)	$3\frac{1}{2} \times 3\frac{1}{2} = 12\frac{1}{4}$
$12;15 + 1\ 00 = 1\ 12;15$	Add 1 00 ...to the 12;15 ...so that 1 12;15 (will come up)	$12\frac{1}{4} + (1 \times 60 + 0) = 72\frac{1}{4}$
$\sqrt{1\ 12;15} = 8;30$	What squares 1 12;15?	$\sqrt{72\frac{1}{4}} = 8\frac{1}{2}$
$8;30 - 3;30 = 5$	Draw 8;30 ...take away 3;30	$8\frac{1}{2} - 3\frac{1}{2} = 5$
$8;30 + 3;30 = 12$	add to one	$8\frac{1}{2} + 3\frac{1}{2} = 12$

And the scribe concludes that the number is 5 and its reciprocal is 12, that is to say, $\frac{12}{60}$ or $\frac{1}{5}$.

Now, you may say that, while 12 exceeds 5 by 7, certainly $\frac{12}{60}$ or $\frac{1}{5}$ does not exceed 5 by 7. This is true, and it brings home the point that to understand a mathematical problem we need to understand more than only the "rules of mathematics"; we also need to understand the way ideas were communicated by the culture at that time in history. In this case, by "a number and its reciprocal" was meant two numbers whose product is 60 (see BM 106444 on page 78)—which is reflected in the instruction "Add 1 00 ...to the 12;15" or, in other words, add 60 to $12\frac{1}{4}$—rather than two numbers whose product is 1 as we would assume today.[2] Today, we may pose the same problem in YBC 6967 in this way:

A number exceeds another by 7, and the product of the two numbers is 60. Find the numbers.

☐

Example 20.3 BM 13901 problem (i) [102, p. 104]

I summed the area and my square-side so that it was 0;45. You put down 1, the projection. You break off half of 1. You combine 0;30 and 0;30. You add 0;15 to 0;45. 1 squares 1. You take away 0;30 which you combined from inside 1 so that the square side is 0;30.

The statement of the problem is, "I summed the area and my square-side so that it was 0;45," in other words, it is given that the

$$\text{area of a square} + \text{length of one side of the square} = \frac{45}{60} \text{ or } \frac{3}{4}.$$

Implicitly, we are to find the length of one side of the square.

Now, you may object immediately, noting that we may not really add an area and a length because they have different units (and so are not like terms). But the scribe cleverly resolves this problem by issuing the instruction, "You put down 1, the projection," which means to multiply the length of one side of the square by 1, thereby converting a length into an area with the *same numerical value*: If the length of one side of the square is x units, then

$$x \text{ units} \cdot 1 \text{ unit} = x \text{ square units.}$$

Now the problem may be stated algebraically thus:

$$x^2 \text{ square units} + x \text{ square units} = \frac{45}{60} \text{ or } \frac{3}{4} \text{ of a square unit.}$$

Note that all the units in the equation are consistent. In this way, we have maintained "dimensional homogeneity" in the equation.

We summarize the steps of the solution.

Babylonian	Step in BM 13901 (i)	Indo-Arabic
0;45	given total area	$\frac{45}{60} = \frac{3}{4}$
1	put down 1	1
0;30	halve	$\frac{1}{2}$
0;15	square	$\frac{1}{4}$
1	add given total area	$\frac{3}{4} + \frac{1}{4} = 1$
1	square root	1

---

[2] Assuming that the product of a number and its reciprocal is 1 (instead of 60 as in YBC 6967) would have led to the two numbers

$$\frac{\sqrt{53}+7}{2} \quad \text{and} \quad \frac{\sqrt{53}-7}{2}.$$

We leave it to you to check that, indeed, one number exceeds the other by 7 and that the product of the two numbers is 1.

0;30	subtract halved amount	$\frac{1}{2}$

And the scribe concludes that the length of one side of the given square is 0;30 or $\frac{1}{2}$.

We check that the scribe's solution to the problem, $\frac{1}{2}$, satisfies the quadratic equation $x^2 + x = \frac{3}{4}$:

$$\left(\frac{1}{2}\right)^2 + \frac{1}{2} = \frac{1}{4} + \frac{1}{2} = \frac{3}{4}. \quad \checkmark$$

However, interestingly, we also note that $-\frac{3}{2}$ solves the equation:

$$\left(-\frac{3}{2}\right)^2 + \left(-\frac{3}{2}\right) = \frac{9}{4} - \frac{6}{4} = \frac{3}{4}. \quad \checkmark$$

But a Babylonian scribe would not expect a negative number, presumably because it would not make any sense geometrically. Moreover, Katz [73, p. 38] remarks that Babylonians also ignored those quadratic problems that produce *two positive* solutions, stating that "[t]he scribes evidently did not consider that a single equation could have two different values for the same unknown. Therefore, they used all sorts of ingenious devices to prevent this possibility."   □

Here is another example from BM 13901.

Example 20.4 BM 13901 problem (v) [102, p. 104]

> [I summed the area and my square-side and a third] of my square-side [so that it was 0;55]. You put down [1, the projection]. You add a third of [1, the projection to 1]: 1;20. [You combine] its half, 0;40, [and 0;40]. You add 0;26 40 to 0;55 and [1;21 40 squares 1;10. Take away 0;40 that you] combined from the middle of 1;10 so that the square side is [0;30].

The statement of the problem is, "[I summed the area and my square-side and a third] of my square-side [so that it was 0;55]," in other words, it is given that the

$$\text{area of a square + one and one-third the length of one side of the square} = \frac{55}{60} \text{ or } \frac{11}{12}.$$

Implicitly, we are to find the length of one side of the square.

We summarize the steps of the solution. Compare the steps here to those in EXAMPLE 20.3.

Babylonian	Step in BM 13901 (v)	Indo-Arabic
0;55	given total area	$\frac{55}{60} = \frac{11}{12}$
1	put down 1	1
1;20	add one-third	$1\frac{1}{3}$
0;40	halve	$\frac{2}{3}$
0;26 40	square ("[combine] its half")	$\frac{4}{9}$
1;21 40	add given total area ($\frac{11}{12} + \frac{4}{9}$)	$1\frac{13}{36}$ or $\frac{49}{36}$
1;10	square root	$1\frac{1}{6}$ or $\frac{7}{6}$
0;30	subtract halved amount (from $\frac{7}{6}$)	$\frac{1}{2}$

And the scribe concludes that the length of one side of the given square is 0;30 or $\frac{1}{2}$.   □

Now You Try 20.3

1. Verify the arithmetic in EXAMPLE 20.4.

2. Follow examples 20.2, 20.3, and 20.4 to solve the following. You may use Indo-Arabic numerals. Write your solutions in words as in the examples.

    a) A number exceeds another by 12, and the product of the two numbers is 540. Find the numbers.

    b) Solve the equation $x^2 + x = 6$.

    c) Solve the equation $x^2 + 4x = 165$.

    d) (BM 13901 problem (vi)) [I summed the area and two-thirds] of my square-side [so that it was 0;35]. (Find the length of one side of the original square.)

                                                          □

At this point you may be wondering, "How on Earth does this produce the correct answer, and how did the ancient Babylonians think of it?" If we interpret the solutions in examples 20.2, 20.3, and 20.4 geometrically, we find that they each essentially use the same procedure: the method of completing the square. We illustrate this below. Keep in mind, however, that no figures accompanied the solutions of the problems in the Babylonian tablets.

First, we take a look at EXAMPLE 20.2.

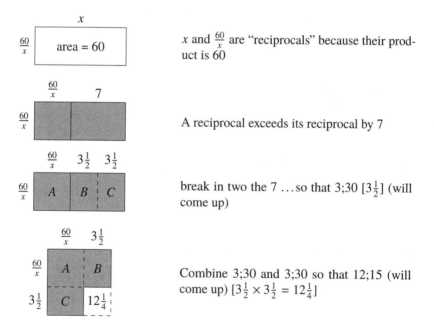

x and $\frac{60}{x}$ are "reciprocals" because their product is 60

A reciprocal exceeds its reciprocal by 7

break in two the 7 ... so that 3;30 [$3\frac{1}{2}$] (will come up)

Combine 3;30 and 3;30 so that 12;15 (will come up) [$3\frac{1}{2} \times 3\frac{1}{2} = 12\frac{1}{4}$]

So, we see that the completed square has area

$$\begin{array}{ccccc} 60 & + & 12\frac{1}{4} & = & 72\frac{1}{4}. \\ \text{(shaded gnomon)} & & \text{(added square)} & & \text{(completed square)} \end{array}$$

Thus, one side of the completed square, $\frac{60}{x} + 3\frac{1}{2}$, has length

$$\sqrt{72\frac{1}{4}} = 8\frac{1}{2},$$

that is,

$$\frac{60}{x} + 3\tfrac{1}{2} = 8\tfrac{1}{2}.$$

And, therefore, if we reconstruct the first rectangle $\boxed{A \mid B \mid C}$ from the gnomon, we see that the length of the first rectangle is

$$x = \left(\frac{60}{x} + 3\tfrac{1}{2}\right) + 3\tfrac{1}{2} = 8\tfrac{1}{2} + 3\tfrac{1}{2} = 12,$$

and the height is

$$\frac{60}{x} = \left(\frac{60}{x} + 3\tfrac{1}{2}\right) - 3\tfrac{1}{2} = 8\tfrac{1}{2} - 3\tfrac{1}{2} = 5.$$

Second, we take a look at EXAMPLE 20.3.

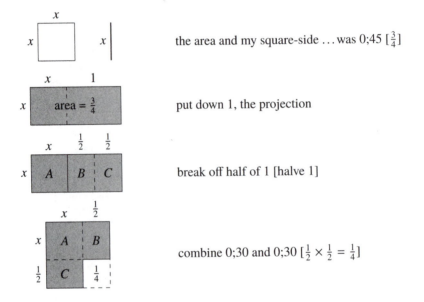

the area and my square-side ... was 0;45 $[\tfrac{3}{4}]$

put down 1, the projection

break off half of 1 [halve 1]

combine 0;30 and 0;30 $[\tfrac{1}{2} \times \tfrac{1}{2} = \tfrac{1}{4}]$

So, we see that the completed square has area

$$\begin{array}{ccccc} \dfrac{3}{4} & + & \dfrac{1}{4} & = & 1. \\[4pt] \text{\small(shaded gnomon)} & & \text{\small(added square)} & & \text{\small(completed square)} \end{array}$$

Thus, one side of the completed square, $x + \tfrac{1}{2}$, has length

$$\sqrt{1} = 1,$$

that is,

$$x + \frac{1}{2} = 1.$$

And, therefore, we see that the length of one side of the original square $\boxed{A}$ is the length of the larger square minus the length of the rectangle C or B, that is

$$x = \left(x + \frac{1}{2}\right) - \frac{1}{2} = 1 - \frac{1}{2} = \frac{1}{2}.$$

Hence, we see that we may interpret the solutions in both examples 20.2 and 20.3 geometrically as the method of completing the square.

Now You Try 20.4 Draw figures to show we may interpret the solution in EXAMPLE 20.4 as the method of completing the square. □

Here is one more example from BM 13901. This example is very similar to EXAMPLE 20.3; however, notice the difference in the last step of this example compared to the solutions in the previous two examples.

Example 20.5 BM 13901 problem (ii) [102, p. 104]

> I took away my square-side from inside the area so that it was 14 30. You put down 1, the projection. You break off half of 1. You combine 0;30 and 0;30. You add 0;15 to 14 30. 14 30;15 squares 29;30. You add 0;30 which you combined to 29;30 so that the square-side is 30.

The statement of the problem is, "I took away my square-side from inside the area so that it was 14 30," in other words, it is given that the

$$\text{area of a square} - \text{length of one side of the square} = 870$$

$(14\ 30 = 14 \times 60 + 30 = 870)$. Implicitly, we are to find the length of one side of the square.
We summarize the steps of the solution. Compare the steps here with those in EXAMPLE 20.3.

Babylonian	Step in BM 13901 (ii)	Indo-Arabic
14 30	given total area	870
1	put down 1	1
0;30	halve	$\frac{1}{2}$
0;15	square	$\frac{1}{4}$
14 30;15	add given total area	$870\frac{1}{4}$
29;30	square root	$29\frac{1}{2}$
30	add halved amount	30

And the scribe concludes that the length of one side of the given square is 30.
    Notice that the last step of the solution in this example differs from the last step of the solution in EXAMPLE 20.3: Here, the last step is to add, but in EXAMPLE 20.3, the last step is to subtract. This is because we subtract the square-side from the area in this example, whereas we add the square-side to the area in EXAMPLE 20.3. To elucidate this difference, we compare the geometrical interpretation of the solution of this example, given here, with the geometrical interpretation of the solution of EXAMPLE 20.3 on page 308. Notice that we ultimately use the method of completing the square to solve this problem.

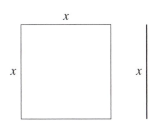

Begin with a square and a square-side

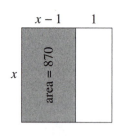

I took away my square-side from inside the area so that it was 14 30 [870]

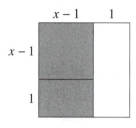

put down 1, the projection

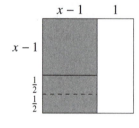

break off half of 1 [halve 1]

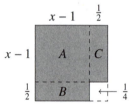

combine 0;30 and 0;30 [$\frac{1}{2} \times \frac{1}{2} = \frac{1}{4}$]

So, we see that the completed square has area

$$870 \quad + \quad \frac{1}{4} \quad = \quad 870\frac{1}{4}.$$

(shaded gnomon)   (added square)   (completed square)

Thus, one side of the completed square has length

$$\sqrt{870\frac{1}{4}} = 29\frac{1}{2}.$$

And, therefore, if we reconstruct the first rectangle

A
B
C

from the gnomon, we see that the length of one side of the original square (the vertical side of the rectangle) is

$$x = 29\frac{1}{2} + \frac{1}{2} = 30.$$

Put another way, in the previous examples, we were to find the length of the smaller square, A, so we subtracted from the length of the larger square. In this example, we know the side of the smaller square, and add to it to get the side of the larger square. □

Now You Try 20.5 Follow EXAMPLE 20.5 to solve the following. You may use Indo-Arabic numerals. Write your solutions in words as in the example.

1. (BM 13901 problem (xvi)) I took away [a third of the square-side] from inside the area so that (it was) 0;05. (Find the length of one side of the original square.)

2. (BM 13901 problem (iii)) I took away a third of the area. I added a third of the square-side to inside the area so that it was 0;20. (Find the length of one side of the original square. You will need to deal with the added given, namely, "I took away a third of the area.")

□

The difference between EXAMPLE 20.3 and EXAMPLE 20.5 is worth a little discussion before we move on.

The problem in EXAMPLE 20.3, "I summed the area and my square-side so that it was 0;45," can be described algebraically by the equation

$$x^2 + x = \frac{3}{4},$$

and the problem in EXAMPLE 20.5, "I took away my square-side from inside the area so that it was 14 30," can be described by the equation

$$x^2 - x = 870.$$

Today, we would view both equations to be special cases of the equation

$$x^2 + bx = c.$$

In both special cases, c is positive; however, because the Babylonians did not deal with negative numbers, the fact that $b = 1$ in EXAMPLE 20.3 and $b = -1$ in EXAMPLE 20.5 caused them to consider the two problems to be of different types (instead of their being special cases of the same type). This is why the two problems are not both solved with identical steps throughout.

With the advantage of symbolic algebra and the acceptance of negative numbers, we can solve the problems in both examples using exactly the same steps. Let us call this the "generalized Babylonian method of completing the square."[3]

Step	EXAMPLE 20.3	EXAMPLE 20.5
given total area (c)	$\frac{3}{4}$	870
put down b	1	-1
halve	$\frac{1}{2}$	$-\frac{1}{2}$
square	$\frac{1}{4}$	$\frac{1}{4}$
add given total area	1	$870\frac{1}{4}$
square root	1	$29\frac{1}{2}$

---

[3] This reference is for our benefit only. We know of no one else who refers to the "generalized Babylonian method of completing the square."

| subtract halved amount | $\frac{1}{2}$ | 30 |

Note that in the last step in EXAMPLE 20.5,

$$29\frac{1}{2} - \left(-\frac{1}{2}\right) = 30.$$

Solving the problems in both EXAMPLE 20.3 and EXAMPLE 20.5 using exactly the same steps was something that was not possible conceptually for the Babylonians—nor for some others, as we shall see.

Now You Try 20.6 Solve the following equations using the "generalized Babylonian method of completing the square."

1. $x^2 + 10x = 39$ 2. $x^2 - 6x = 16$

□

20.1.2   Arabic

Abu Ja'far Muḥammad ibn-Mūsā al-Khwārizmī[4] (ca. 780–850; FIGURE 20.1) is without question the best-known Arabic mathematician due to his two seminal works, *Kitāb al-jam'wal tafrīq bi ḥisāb al-Hind* (*Book on Addition and Subtraction after the Method of the Indians*)[5] and *Al-kitāb al-muḫtaṣar fī ḥisāb al-jabr wa'l muqābala* (*The Condensed Book on the Calculation of al-Jabr and al-Muqabala*).[6]

Figure 20.1: Al-Khwārizmī. (Source: Courtesy of Jeff Miller.)

In 830, the Abbasidian Caliph al-Ma'mūn (reigned 813–833) endowed the House of Wisdom [34] in Baghdad, which is often compared to the Museum of Alexandria. During al-Ma'mūn's reign, al-Khwārizmī became a member of the House of Wisdom, where he engaged in mathematics, astronomy, and geography. During that time, Babylonian, Greek, and Indian scientific and other texts were being translated into Arabic, and this influence on Arabic mathematics is apparent.

Al-Khwārizmī's *al-Hind*, written about 800, is noted for introducing the new Indian number system to the Arab world, which later entered Europe through Latin translations of the work (see SECTION 6.3). His

---

[4]Suzuki [112, p. 246] explains that " 'Abu Ja'far' means 'father of Ja'far'; 'Ibn Musa' is 'son of Musa' (the Arabic form of Moses); 'al Khwārizmī' is 'the man from Khwarizm' (a region near the Aral Sea, now in Uzbekistan and Kazakstan); thus, his name can be translated as 'Mohammed, father of Ja'far, son of Moses, the man from Khwarizm.' "

[5]The Arabic title is also translated variously as *Book of Addition and Subtraction According to the Hindu Calculation* [34, p.182] and *On the Indian Numbers* [112, p. 246], for example.

[6]The Arabic title is also translated variously as *Compendium on Calculation by Completion and Reduction* [10, p. 542] and *The Condensed Book of Completion and Restoration* [112, p. 247], for example.

al-jabr wa'l muqābala, written about 825, is noted for heavily influencing the continued development of the subject of algebra, in particular, quadratic equations, drawing from Babylonian, Greek, and Indian sources [34, p. 183]. In fact, it is from the title word *al-jabr* that the word "algebra" derives. Leonardo of Pisa, for example, was very much influenced by *al-jabr wa'l muqābala* and other Arabic mathematical works, such as those of Abū Kāmil and al-Karajī [21, 73], so much so that many of the problems and methods in his *Liber Abaci* were taken verbatim or nearly verbatim from them (although Leonardo did also introduce his own methods, as well). All of this paved the way for the solution of the general cubic equation later by Renaissance mathematicians, a high point in mathematics in that period.

Stedall [109, p. 35] tells us that "*al-jabr* is *completing* or *restoring*, used particularly in the sense of setting broken bones, while *al-muqābala* is *setting in opposition* or *balancing*. In al-Khwārizmī's text *Al-jabr wa'l muqābala*, *al-jabr* is used when a positive term is added in order to eliminate a negative quantity, while *al-muqābala* denotes the balancing of an equation by operating simultaneously on each side." In other words, in al-Khwārizmī's text, *al-jabr* refers to the transposition of terms that are subtracted in an equation, so that they become terms that are added, and *al-muqābala* refers to the cancellation of identical terms on both sides of an equation [73, p. 244]. Also, the words "algorism" and "algorithm" are derived from the Latin *algorismus*, which itself is a corruption of the name al-Khwārizmī from the titles of Latin translations of *Al-jabr wa'l muqābala*. The first translation into Latin was by John of Seville under the title *Liber Alghoarismi* (*Book of al-Khwārizmī*; ca. 1143). Another translation of al-Khwārizmī's algebra was by John of Sacrobosco under the title *Liber Algorismus* (*Book of al-Khwārizmī*; ca. 1240) [34, p. 184].

Arabic algebra at the time of al-Khwārizmī was rhetorical: no symbols were used, except perhaps for numerals. Today, a quadratic equation in one unknown x may be expressed generally as

$$ax^2 + bx + c = 0,$$

where a, b, and c are constants with $a \neq 0$ or, equivalently, as

$$x^2 + Bx + C = 0$$

after dividing through by a (which we may do because $a \neq 0$). This is, in fact, how al-Khwārizmī treats quadratic equations in which $a \neq 1$ [10, p. 543]:

> The solution is the same when two squares or three, or more or less are specified; you reduce them to one single square, and in the same proportion you reduce also the roots and simple numbers which are connected therewith.

We note that in general $B = \frac{b}{a}$ and $C = \frac{c}{a}$ may be *any* real numbers: positive, negative, or zero. However, like the Babylonians, Arabic mathematicians—despite having some of their mathematical roots reaching back to India—did not consider negative numbers or zero in their work. As a result, when it comes to quadratic equations, al-Khwārizmī considers five different cases in order to avoid negative numbers or zero as either coefficients or solutions [10, p. 543]:

> I observed that the numbers which are required in calculating by Completion and Reduction are of three kinds, namely, roots, squares, and simple numbers relative to neither root nor square.
>
> A root is any quantity which is to be multiplied by itself, consisting of units, or numbers ascending, or fractions descending. A square is the whole amount of the root multiplied by itself. A simple number is any number which may be pronounced without reference to a root or square.
>
> A number belonging to one of these three classes may be equal to a number of another class; you may say, for instance, "squares are equal to roots," or "squares are equal to numbers," or "roots are equal to numbers."[7]

[...]

---

[7]"Roots are equal to numbers" describes a linear equation, namely, $ax = b$.

I found that these three kinds, namely, roots, squares, and numbers, may be combined together, and thus three compound species arise; that is, "squares and roots equal to numbers"; "squares and numbers equal to roots"; "roots and numbers equal to squares."

These cases of "different" quadratic equations are listed in the following table. Note that, unlike the Babylonians, there is not attempt here to maintain dimensional homogeneity (see page 305).

al-Khwārizmī $(a > 0, b > 0, c > 0)$		Modern
Squares equal to roots	$ax^2 = bx$	$ax^2 - bx + 0 = 0$
Squares equal to numbers	$ax^2 = c$	$ax^2 + 0x - c = 0$
Squares and roots equal to numbers	$ax^2 + bx = c$	$ax^2 + bx - c = 0$
Squares and numbers equal to roots	$ax^2 + c = bx$	$ax^2 - bx + c = 0$
Roots and numbers equal to squares	$bx + c = ax^2$	$ax^2 - bx - c = 0$

Here is perhaps al-Khwārizmī's best-known example of solving quadratic equations. It is his first and simplest example.

Example 20.6 [10, p. 543]

Roots and Squares are equal to Numbers; for instance, "one square, and ten roots of the same, amount to thirty-nine dirhams"; that is to say, what must be the square which, when increased by ten of its own roots, amounts to thirty-nine? The solution is this: you halve the number of the roots, which in the present instance yields five. This you multiply by itself; the product is twenty-five. Add this to thirty-nine; the sum is sixty-four. Now take the root of this, which is eight, and subtract from it half the number of the roots, which is five; the remainder is three. This is the root of the square which you sought for; for the square itself is nine.

We note that a *dirham* is a "pure number," which distinguishes it from a square or a root. Thus, "one square" corresponds to $1x^2$, "ten roots" corresponds to $10x$, and "thirty-nine dirhams" corresponds to 39. Symbolically, then, the problem that al-Khwārizmī poses is this: Solve the equation

$$x^2 + 10x = 39.$$

However, not having symbolic algebra at his disposal, he solves the problem strictly using words.

Al-Khwārizmī's solution reminds us of the solution of the Babylonian problem BM 13901 (i) in EXAMPLE 20.3. We summarize the steps of al-Khwārizmī's solution.

Step	Result
given numbers	39
given roots	10
halve (the root)	5
square	25
add given numbers	64
square root	8
subtract halved amount	3

And al-Khwārizmī concludes that the solution is 3.

We check that 3 is, indeed, a solution of the quadratic equation $x^2 + 10x = 39$:

$$3^2 + 10(3) = 39. \quad \checkmark$$

But, just as the Babylonian scribe did not recognize a second, negative solution of the problem BM 13901 (i), al-Khwārizmī did not recognize a second, negative solution to this problem, namely, -13:

$$(-13)^2 + 10(-13) = 169 - 130 = 39. \quad \checkmark$$

The Babylonian scribe did not provide a justification for his solution, but as in the Greek tradition, al-Khwārizmī justified his solution using a geometrical argument. Now, since the steps of al-Khwārizmī's solution of this problem essentially are the same as the steps of the solution of BM 13901 (i), we may expect that al-Khwārizmī's geometrical justification would be similar to the one we proposed for BM 13901 (i) on page 308 and, indeed, one is; however, al-Khwārizmī also provides first another justification that uses a different geometrical figure ([34, pp. 202–203], [41, p. 424]), although it still clearly demonstrates the method of completing the square. We summarize the steps of al-Khwārizmī's first geometrical justification of the solution for this problem.

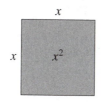

Begin with a square with area x^2.

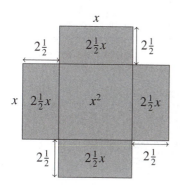

Add four rectangles, each with area two-and-a-half roots [$2\frac{1}{2}x$], so that the cross (shaded) has a total area of "one square, and ten roots of the same" [$x^2 + 10x$]. This is given to be equal to 39.

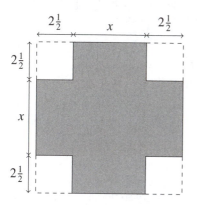

$$2\frac{1}{2} \times 2\frac{1}{2} = 6\frac{1}{4}$$

So, we see that the completed square has area

$$\underset{\text{(shaded cross)}}{39} + \underset{\text{(added squares)}}{(4 \times 6\frac{1}{4})} = \underset{\text{(completed square)}}{64}.$$

Thus, one side of the completed square has length

$$\sqrt{64} = 8.$$

And, therefore, we see that the length of one side of the original square is

$$x = 8 - (2 \times 2\tfrac{1}{2}) = 8 - 5 = 3.$$

Al-Khwārizmī then gives a second geometrical justification that is similar to the one we proposed for BM 13901 (i) on page 308.   □

Example 20.7 For quadratic equations in which the quadratic term is ax^2 with a different from one, al-Khwārizmī first divides the equation through by a. In this way, he avoids having to give a different solution for every value of a. This was an important step in abstraction and noteworthy in the development of algebra and mathematics in general. Al-Khwārizmī illustrates this with the following example [10, p. 544].

> For instance, "two squares and ten roots are equal to forty-eight dirhams"; that is to say, what must be the amount of two squares which, when summed up and added to ten times the root of one of them, make up a sum of forty-eight dirhams? You must at first reduce the two squares to one; and you know that one square of the two is the half of both. Then reduce everything mentioned in the statement to its half, and it will be the same as if the question had been, a square and five roots of the same are equal to twenty-four dirhams; or, what must be the amount of a square which, when added to five times its root, is equal to twenty-four dirhams?

He then proceeds to solve this reduced problem in the same way that he solves "one square, and ten roots of the same, amount to thirty-nine dirhams" (EXAMPLE 20.6).

> Now halve the number of the roots; the half is two and a half. Multiply that by itself; the product is six and a quarter. Add this to twenty-four; the sum is thirty dirhams and a quarter. Take the root of this; it is five and a half. Subtract from this the half of the number of the roots, that is two and a half; the remainder is three. This is the root of the square, and the square itself is nine.

Expressed symbolically, the original equation that al-Khwārizmī proposes to solve is

$$2x^2 + 10x = 48.$$

His first step is to divide through the equation by 2 to obtain the reduced equation

$$x^2 + 5x = 24.$$

He then solves the reduced equation, which is of the type "squares and roots equal to numbers," by completing the square and obtains the solution 3. We check that 3, indeed, is a solution of the reduced equation, and also that 3 is a solution of the original equation.

$$\text{reduced equation:} \quad 3^2 + 5(3) = 9 + 15 = 24 \qquad \checkmark$$
$$\text{original equation:} \quad 2(3)^2 + 10(3) = 18 + 30 = 48 \qquad \checkmark$$

However, as in the last example, al-Khwārizmī misses the second, negative solution, namely, -8:

$$2(-8)^2 + 10(-8) = 128 - 80 = 48. \qquad \checkmark$$

□

Now You Try 20.7 Solve the following quadratic equations. Write your solutions in words as al-Khwārizmī does. Each of these equations also has a second, negative solution; try to find the negative solution, as well.

1. One square, and twelve roots of the same, amount to forty-five.

2. Three squares and six roots are equal to three hundred sixty.

3. $x^2 + 3x = 18$ 4. $2x^2 + 10x = 48$ 5. $5x^2 + 4x = 1$

☐

We present another example of al-Khwārizmī's. This one is of the type "squares and numbers equal to roots."

Example 20.8 [10, p. 544]

Squares and Numbers are equal to Roots; for instance, "a square and twenty-one in numbers are equal to ten roots of the same square." That is to say, what must be the amount of a square, which, when twenty-one dirhams are added to it, becomes equal to the equivalent of ten roots of that square?

Al-Khwārizmī then proceeds to solve the equation. Note the difference in the steps from the last examples.

Solution: Halve the number of the roots; the half is five. Multiply this by itself; the product is twenty-five. Subtract from this the twenty-one which are connected with the square; the remainder is four. Extract its root; it is two. Subtract this from the half of the roots, which is five; the remainder is three. This is the root of the square which you required, and the square is nine. Or you may add the root to the half of the roots; the sum is seven; this is the root of the square which you sought for, and the square itself is forty-nine.

We summarize the steps of the solution. Compare the steps here with those in EXAMPLE 20.6.

Step	Operation	Result
given numbers		21
given roots		10
halve roots	$10 \div 2$	5
square	5^2	25
subtract given numbers	$25 - 21$	4
square root	$\sqrt{4}$	2
subtract from halved roots	$5 - 2$	3
add to halved roots	$5 + 2$	7

And al-Khwārizmī concludes that the solutions are 3 and 7.
Expressed in symbolically, the equation that al-Khwārizmī solves is

$$x^2 + 21 = 10x.$$

We check that 3 and 7 are, indeed, solutions of the equation.

$$3^2 + 21 = 30 \quad \text{and} \quad 10(3) = 30 \quad \checkmark$$
$$7^2 + 21 = 70 \quad \text{and} \quad 10(7) = 70 \quad \checkmark$$

Following his solution, al-Khwārizmī goes on to tell us that "squares and numbers equal to roots" is the only type that may admit two solutions:

When you meet with an instance which refers you to this case, try its solution by addition, and if that does not serve, then subtraction certainly will. For in this case both addition and subtraction may be employed, which will not answer in any other of the three cases in which the number of roots must be halved. And know, that, when in a question belonging to this case you have halved the number of roots connected with the square, then the instance is impossible; but if the product be equal to the dirhams by themselves, then the root of the square is equal to the half of the roots alone, without either addition or subtraction.

Finally, we find that al-Khwārizmī's geometrical justification for the case "squares and numbers equal roots" is not the same as that for "squares and roots equal numbers" in EXAMPLE 20.6. This reflects the fact that, geometrically, these are different types of quadratic equations, even though, algebraically, they are both special cases of one general quadratic equation. We show here only the accompanying figure [10, p. 545, figure 5.2], but leave out his explanation.

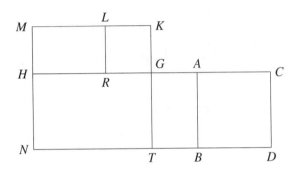

In the figure, x^2 is shown as the square with diagonal AD, so that CD has length x, and the rectangle with diagonal HB is taken to have area 21; thus, the rectangle with diagonal HD has area $x^2 + 21$. On the other hand, the side HC is taken to be 10, so that the rectangle with diagonal HD has area $10x$. Hence, the rectangle with diagonal HD represents "a square and twenty-one in numbers are equal to ten roots of the same square." Finally, both NT and MN are chosen to have length 5, so that the square with diagonal MT has area 25. From here, al-Khwārizmī demonstrates the validity of his method for solving the equation and arriving at the two solutions, namely, 3 and 7.   □

Now You Try 20.8 Solve the following quadratic equations. Write your solutions in words as in EXAMPLE 20.8.

1. A square and eighty in numbers are equal to eighteen roots.

2. One square and twelve in numbers are seven roots.

3. $5x^2 + 75 = 40x$ 4. $3x^2 + 1 = 4x$

                                                                                                          □

Remark 20.1 We point out three distinctions between the Babylonians and the Arabs at this point.

First, the Babylonians did not consider quadratic equations per se. They solved problems that are quadratic in nature, but they never wrote a quadratic or other equation. The Arabs, on the other hand, did write and solve quadratic equations, albeit not symbolically, for example, "one square, and ten roots of the same, amount to thirty-nine dirhams."

Second, the Babylonians did not combine numbers of different type or dimension. For example, consider the problem in BM 13901 (i) (EXAMPLE 20.3): "I summed the area and my square-side so that it was 0;45," in other words,

$$\text{area of a square + length of one side of the square} = \frac{45}{60} \text{ or } \frac{3}{4}.$$

or, if the length of one side of the square is x,

$$x^2 \text{ square units} + x \text{ units} = \frac{45}{60} \text{ or } \frac{3}{4} \text{ (no units).}$$

Thus, we see that not all the quantities in the problem are of the same type: area has square units, but length does not, for instance. So, the scribe's first instruction is to "put down 1, the projection," which converts the length into an area with the same numerical value,

$$x \text{ units} \cdot 1 \text{ unit} = x \text{ square units,}$$

so that the problem is recast as

$$x^2 \text{ square units} + x \text{ square units} = \frac{45}{60} \text{ or } \frac{3}{4} \text{ of a square unit.}$$

The Arabs, on the other hand, were comfortable combining numbers of different type. We see this in al-Khwārizmī's *al-jabr wa'l muqābala*: "I observed that the numbers which are required in calculating by Completion and Reduction are of three kinds, namely, roots, squares, and simple numbers relative to neither root or square....I found that these three kinds, namely, roots, squares, and numbers, may be combined together..." (see page 313).

Third, recall that a Babylonian scribe not only would not expect a negative number as a solution, but that he also would not consider quadratic problems that produce two positive solutions because of the misconception that what amounts to a single equation in one unknown cannot have more than one solution [73, p. 38]; in the latter case, a scribe would cast the problem in a way that would avoid this. Al-Khwārizmī, on the other hand, did treat single equations in one unknown that admit more than one positive solution. □

We present one last example of al-Khwārizmī's. In this example, he illustrates how to transform a quadratic equation that is not exactly one of his five types into one of those types (page 314).

Example 20.9 [2, pp. 41–42]

> If a person puts such a question to you as: "I have divided ten into two parts, and multiplying one of these by the other, the result was twenty-one;" then you know that one of the two parts is thing, and the other ten minus thing. Multiply, therefore, thing by ten minus thing; then you have ten things minus a square, which is equal to twenty-one.

Let us unravel this portion before we continue with the example.

Ten is divided into two parts: If x is one part ("one of the two parts is thing"), then the other part is $10 - x$ ("the other is ten minus thing") because the two parts have to add to 10, and $x + (10 - x) = 10$. Thus, multiplying x by $10 - x$, we obtain

$$x(10 - x) = 10x - x^2$$

("Multiply, therefore, thing by ten minus thing; then you have ten things minus a square"). Therefore, the algebraic equation is

$$10x - x^2 = 21$$

("then you have ten things minus a square, which is equal to twenty-one"). But $10x - x^2 = 21$ is not one of the five types. Hence, al-Khwārizmī continues by explaining how to transform this particular equation into one of the five types.

> Separate the square from the ten things, and add it to the twenty-one. Then you have ten things, which are equal to twenty-one dirhams and a square.

In other words, al-Khwārizmī tells us that

$$10x - x^2 = 21 \quad \text{is equivalent to} \quad \boxed{10x = 21 + x^2}$$

that *is* one of the five types, namely, "squares and numbers equal to roots." Al-Khwārizmī then solves the equation, $10x = 21 + x^2$, as before (see EXAMPLE 20.8), and concludes the example with the statement: "This is one of the problems which may be resolved by addition and subtraction."   □

Now You Try 20.9

1. Solve the following equations. Write your solutions in words as al-Khwārizmī does. Check your solutions.

   a) ([2, pp. 42–43]) "I have divided ten into two parts, and having multiplied each part by itself, I have subtracted the smaller from the greater, and the remainder was forty."

   b) ([2, pp. 43–44]) "I have divided ten into two parts, and having multiplied each part by itself, I have put them together, and have added to them the difference of the two parts previously to their multiplication, and the amount of all this is fifty-four."

                                                                                              □

20.1.3   Indian

Unlike Babylonian and Arabic mathematicians, for example, Chinese and Indian mathematicians were at ease with quantities that were negative or zero. For instance, the Chinese represented positive numbers on counting boards using red rods and negative numbers using black rods and easily manipulated them, and the Indians related to negative numbers as debts, and gave correct rules for operating with positive and negative numbers and zero. We take a look at two examples of quadratic equations from *Víja-gańita* (or *Bījagaṇita*) by the Indian mathematician Bhāskara II (1114–1185) in which negative quantities or zero appear. Before presenting the examples, however, we say something about the notation used by Indian mathematicians of that era.

To begin, Colebrooke [16, pp. x–xi] tells us:

> The Hindu algebraists use abbreviations and initials for symbols: they distinguish negative quantities by a dot, but have not any mark, besides the absence of the negative sign, to discriminate a positive quantity. No marks or symbols indicating operations of addition, or multiplication, &c. are employed by them: nor any announcing equality or relative magnitude (greater or less)....A fraction is indicated by placing the divisor under the dividend, but without a line of separation. The two sides of an equation are ordered in the same manner, one under the other....The symbols of unknown quantity are not confined to a single one: but extend to ever so great a variety of denominations: and the characters used are initial syllables of the names of colours, excepting the first, which is the initial of *yávat-távat*, as much as....Color therefore means unknown quantity, or the symbol of it: and the same *Sanscrĭt* word, *varńa*, also signifying a literal character, letters are accordingly employed likewise as symbols; either taken from the alphabet; or else initial syllables of words signifying the subjects of the problem....

Now, according to Cajori [31, p. 75], Brahmagupta (598–670) and later Indian mathematicians used the following abbreviations of the names of colors as variables:

- *ru* for *rupa*, the absolute number,

- *ya* for *yávat-távat*, the (first) unknown,

- *ca* for *calaca* (black), a second unknown,

- *ní* for *nílaca* (blue), a third unknown,

- *pí* for *pítaca* (yellow), a fourth unknown,

$$\vdots$$

- *c* for *caraní*, surd, or square root,

- *ya v* for x^2, the *v* being the contraction for *varga*, square number.

Thus, for example,[8]

$$ya\ v\ 1\ ya\ \dot{8}\ ru\ 16 \qquad \rightsquigarrow \qquad x^2 - 8x + 16,$$

where the dot above the 8 in "*ya* $\dot{8}$" indicates that the term is negative. In this way, Indian algebra was syncopated: symbols or abbreviations for variables and numerals were used, but otherwise operations and solutions were described using words or abbreviations. Finally, Bhāskara furnishes the following procedure for solving quadratic equations [16, pp. 207–210]:

> NEXT equation involving the square or other [power] of the unknown is propounded. [Its re-solution consists in] the elimination of the middle term, as teachers of the science dominate it. Here removal of one term, the middle one, in the square quantity, takes place: wherefore it is called elimination of the middle term. On this subject the following rule is delivered.
>
> 128–130. Rule: When a square and other [term] of the unknown is involved in the remainder; then, after multiplying both sides of the equation by an assumed quantity, something is to be added to them, so as the side may give a square-root. Let the root of the absolute number again be made equal to the root of the unknown: the value of the unknown is found from that equation. If the solution be not thus accomplished, in the case of cubes or biquadrates, this [value] must be elicited by the [calculator's] own ingenuity.
>
> [...]
>
> 131. ŚRÍD'HARA's rule on this point: Multiply both sides of the equation by a number equal to four times the [coefficient] of the square, and add to them a number equal to the square of the original [coefficient] of the unknown quantity. (Then extract the root.)

In other words, to solve the quadratic equation

$$ax^2 + bx = c,$$

multiply through the equation by $4a$ and add to both sides b^2:

$$4a^2x^2 + 4abx + b^2 = 4ac + b^2.$$

Observe that the expression on the left-hand side now factors as a perfect square:

$$(2ax + b)^2 = 4ac + b^2.$$

Therefore, we may take the square root on both sides to solve for the unknown, x.
We are now ready for the examples. Notice how Indian mathematics is written in prose.

Example 20.10 [16, pp. 211–212]

> 132. Example: The square-root of half the number of a swarm of bees is gone to a shrub of jasmin; and so are eight-ninths of the whole swarm: a female is buzzing to one remaining male, that is humming within a lotus, in which he is confined, having been allured to it by its fragrance at night. Say, lovely woman, the number of bees.

---

[8] We use the symbol $\rightsquigarrow$ to indicate a translation into modern notation.

Thus, the problem is to find the number of bees in a particular swarm given that this number is the sum of the square root of half the number of bees, eight-ninths the number of bees, and two bees ("a female is buzzing to one remaining male"). We may express the problem symbolically: If S denotes the number of bees in the swarm, then

$$S = \sqrt{\frac{1}{2}S} + \frac{8}{9}S + 2,$$

and now the problem is to solve the equation for S, the number of bees in the swarm. Bhāskara then proceeds to solve the problem in the following manner:

> Put the number of the swarm of bees *ya v* 2. The square-root of half this is *ya* 1. Eight-ninths of the whole swarm are *ya v* $\frac{16}{9}$. The sum of the square-root and fraction, added to the pair of bees specified, is equal to the amount of the swarm, namely *ya v* 2. Reducing the two sides of the equation to a common denomination, and dropping the denominator, the equation is $\frac{ya\ v\ 18\ ya\ 0\ ru\ 0}{ya\ v\ 16\ ya\ 9\ ru\ 18}$ and, subtraction being made, the two sides are $\frac{ya\ v\ 2\ ya\ 9\ ru\ 0}{ya\ v\ 0\ ya\ 0\ ru\ 18}$ Multiplying both these by eight, and adding the number eighty-one, and extracting both roots, the statement of them for an equation is $\frac{ya\ 4\ ru\ 9}{ya\ 0\ ru\ 15}$ Whence the value *yávat-távat* comes out 6. By substituting the square of this, the number of the swarm of bees is found 72.

Thus, we see that, because the problem calls for the square root of half the number of bees in the swarm, Bhāskara very cleverly lets $2x^2$ ("*ya v* 2") denote the number of bees in the swarm, for then the

$$\text{square root of half the number of bees} = \sqrt{\frac{1}{2}(2x^2)} = \sqrt{x^2} = x$$

(possibly positive or negative). This allows him to avoid having to deal with a square root explicitly in the remainder of his solution. So, with $2x^2$ denoting the number of bees in the swarm, Bhāskara obtains the equation

$$2x^2 = x + \frac{8}{9}(2x^2) + 2$$

or

$$2x^2 = x + \frac{16}{9}x^2 + 2.$$

At this point, Bhāskara seeks to derive an equation of the form $ax^2 + bx = c$ that he may solve using Śríd'hara's rule. Toward this end, his next step is to find a common denominator for the terms in the equation ("Reducing the two sides of the equation to a common denomination"). Note that it is implicit in Bhāskara's solution that he chooses to use 9 for the common denominator:

$$\frac{18x^2}{9} = \frac{9x + 16x^2 + 18}{9}.$$

Then Bhāskara clears denominators ("dropping the denominator") in the equation; explicitly, we would multiply both sides of the equation by 9. This yields, after rearranging the terms in the right-hand side,

$$18x^2 = 16x^2 + 9x + 18 \qquad \leftrightsquigarrow \qquad \frac{ya\ v\ 18\ ya\ 0\ ru\ 0}{ya\ v\ 16\ ya\ 9\ ru\ 18}.$$

Next, Bhāskara subtracts $16x^2$ and $9x$ from both sides of the equation ("subtraction being made") to obtain the equation

$$2x^2 - 9x = 18 \qquad \leftrightsquigarrow \qquad \frac{ya\ v\ 2\ ya\ \dot{9}\ ru\ 0}{ya\ v\ 0\ ya\ 0\ ru\ 18}.$$

Thus, Bhāskara derives an equation of the form $ax^2 + bx = c$ that he may solve using Śríd'hara's rule.

Continuing with his solution, Bhāskara follows Śríd'hara's rule (ŚR) and multiplies through the equation by 8 (ŚR: "a number equal to four times the [coefficient] of the square") and adds 81 to both sides of the equation (ŚR: "a number equal to the square of the original [coefficient] of the unknown quantity"). This gives

$$16x^2 - 72x + 81 = 225$$

or, after factoring the left-hand side,

$$(4x - 9)^2 = 225.$$

Having put the equation into this form, Bhāskara extracts the square root on both sides, obtaining the equation

$$4x - 9 = 15 \quad \rightsquigarrow \quad \begin{matrix} ya\ 4\ ru\ \dot{9} \\ ya\ 0\ ru\ 15 \end{matrix}.$$

He solves this linear equation to find that $x = 6$ ("the value *yávat-távat* comes out 6").

Finally, Bhāskara finds the number of bees, $2x^2$, in the swarm:

$$2(6^2) = 2 \cdot 36 = 72$$

("substituting the square of this, the number of the swarm of bees is found 72"). He concludes that there are 72 bees in the swarm.

We remark that Bhāskara finds that the solution of the equation $2x^2 - 9x = 18$ is 6, and, indeed,

$$2(6^2) - 9(6) = 72 - 54 = 18. \quad \checkmark$$

However, $-\frac{3}{2}$ is also a solution of the equation:

$$2\left(-\frac{3}{2}\right)^2 - 9\left(-\frac{3}{2}\right) = \frac{9}{2} + \frac{27}{2} = \frac{36}{2} = 18. \quad \checkmark$$

In fact, if we note that both $15^2 = 225$ and $(-15)^2 = 225$, then we see that extracting the square root on both sides of the equation

$$(4x - 9)^2 = 225$$

yields *two* linear equations, namely,

$$4x - 9 = 15 \quad \text{and} \quad 4x - 9 = -15$$

because both $15^2 = 225$ and $(-15)^2 = 225$: the former yields $x = 6$, and the latter $x = -\frac{3}{2}$. But, $-\frac{3}{2}$ would imply a fractional number of bees in the swarm:

$$2\left(-\frac{3}{2}\right)^2 = 2 \cdot \frac{9}{4} = \frac{9}{2}.$$

☐

Bhāskara's following additional instructions for solving quadratic equations apply to the next example [16, p. 208]:

> 130. If the root of the absolute side of the equation be less than the number, having the negative sign, comprised in the root of the side involving the unknown, then putting it negative or positive, a two-fold value is to be found of the unknown quantify: this [holds] in some cases.

Example 20.11 [16, pp. 216–217]

> 140. Example: The fifth part of the troop less three, squared, had gone to a cave; and one monkey was in sight, having climbed on a branch. Say how many they were?

Thus, the problem is to find the number of monkeys in a particular troop given that, if T denotes the number of monkeys in the troop,

$$T = \left(\frac{1}{5}T - 3\right)^2 + 1.$$

Bhāskara proceeds to solve the problem in this way:

> Here the troop is put *ya* 1. Its fifth part is *ya* $\frac{1}{5}$. Less three, it is *ya* $\frac{1}{5}$ *ru* $\overset{\cdot}{15}$. This squared is *ya* v $\frac{1}{25}$ *ya* $\overset{\cdot}{30}$ *ru* $\frac{225}{25}$. With the one seen ($\frac{25}{25}$), it is *ya* v $\frac{1}{25}$ *ya* $\overset{\cdot}{30}$ *ru* $\frac{250}{25}$. This is equal to the troop *ya* 1. Reducing these sides of equation to a common denomination, dropping the denominator, and making equal subtraction, the equation becomes $\begin{smallmatrix} ya\ v\ 1\ ya\ 55\ ru\ 0 \\ ya\ v\ 0\ ya\ 0\ ru\ 250 \end{smallmatrix}$ Multiplying by four, and adding a number equal to the square of fifty-five (3025), the roots extracted are $\begin{smallmatrix} ya\ 2\ ru\ 55 \\ ya\ 0\ ru\ 45 \end{smallmatrix}$ Here also a two-fold value is found as before, 50 and 5. But the second is in this case not to be taken: for it is incongruous. People do not approve of a negative absolute number.

Thus, we see that Bhāskara first lets x denote the number of monkeys in the troop ("the troop is put *ya* 1"). Then, "[t]he fifth part of the troop less three, squared" is

$$\left(\frac{1}{5}x - 3\right)^2 = \left(\frac{1}{5}x - \frac{15}{5}\right)^2 = \frac{1}{25}x^2 - \frac{30}{25}x + \frac{225}{25}$$

$$\rightsquigarrow \quad ya\ \frac{1}{5}\ ru\ \overset{\cdot}{15}\ \ \text{squared is}\quad ya\ v\ \frac{1}{25}\ ya\ \frac{30}{25}\ ru\ \frac{225}{25}.$$

Next, we add one or $\frac{25}{25}$ ("With the one seen ($\frac{25}{25}$)") and obtain

$$\frac{1}{25}x^2 - \frac{30}{25}x + \frac{250}{25} \quad\rightsquigarrow\quad ya\ v\ \frac{1}{25}\ ya\ \frac{30}{25}\ ru\ \frac{250}{25}.$$

This is equal to the number of monkeys in the troop, x ("This is equal to the troop *ya* 1"):

$$\frac{1}{25}x^2 - \frac{30}{25}x + \frac{250}{25} = x.$$

At this point, as in the last example, Bhāskara seeks to derive an equation of the form $ax^2 + bx = c$ that he may solve using Śríd'hara's rule (page 321). Toward this end, his next step is to find a common denominator for the terms in the equation ("Reducing these sides of equation to a common denomination"); he implicitly chooses to use 25:

$$\frac{x^2 - 30x + 250}{25} = \frac{25x}{25}.$$

He then clears denominators ("dropping the denominator"):

$$x^2 - 30x + 250 = 25x.$$

Now he subtracts 250 and 25x from both sides of the equation to obtain the equation ("making equal subtraction, the equation becomes")

$$x^2 - 55x = -250 \quad\rightsquigarrow\quad \begin{smallmatrix} ya\ v\ 1\ ya\ \overset{\cdot}{55}\ ru\ 0 \\ ya\ v\ 0\ ya\ 0\ ru\ 2\overset{\cdot}{50} \end{smallmatrix}.$$

Applying Śríd'hara's rule (ŚR), Bhāskara next multiplies through the equation by 4 (ŚR: "a number equal to four times the [coefficient] of the square") and adds the square of 55 or 3025 to both sides of the equation (ŚR: "a number equal to the square of the original [coefficient] of the unknown quantity"). This gives

$$4x^2 - 220x + 3025 = 2025$$

or, after factoring the left-hand side,

$$(2x - 55)^2 = 2025.$$

Now he extracts the square root on both sides ("the roots extracted are"):

$$2x - 55 = 45 \qquad \rightsquigarrow \qquad \begin{array}{l} ya\ 2\ ru\ \overset{.}{5}5 \\ ya\ 0\ ru\ 45 \end{array}.$$

However, because 45 on the right-hand side is less than 55 on the left-hand side, Bhāskara applies the rule (page 323),

> 130. If the root of the absolute side of the equation be less than the number, having the negative sign, comprised in the root of the side involving the unknown, then putting it negative or positive, a two-fold value is to be found of the unknown quantify...

that is to say, he recognizes that he now has to solve *two* linear equations, namely,

$$2x - 55 = 45 \quad \text{and} \quad 2x - 55 = -45,$$

since both $45^2 = 2025$ and $(-45)^2 = 2025$. He solves the two equations: "Here also a two-fold value is found as before, 50 and 5."

We check that 50 and 5 are, indeed, solutions of the equation $x^2 - 30x + 250 = 25x$:

$$50^2 - 30(50) + 250 = 1250 \quad \text{and} \quad 25(50) = 1250. \quad \checkmark$$
$$5^2 - 30(5) + 250 = 125 \quad \text{and} \quad 25(5) = 125. \quad \checkmark$$

However, the problem is not to solve a given equation, but to find the number of monkeys in a particular troop. Thus, it must be checked which of the solutions, 50 or 5, yields a reasonable result. On this point, Bhāskara tells us that 5 is *not* a solution of the problem: "the second is in this case not to be taken: for it is incongruous." The reason for this is the problem states that "[t]he fifth part of the troop less three ...had gone to a cave," and

$$\tfrac{1}{5}(5) - 3 = -4 :$$

a negative number of monkeys.

Therefore, the conclusion is that there are 50 monkeys in the troop. □

Now You Try 20.10 Complete the solution for each of the following problems. Write your solution in words following Bhāskara's examples; in particular, write any expressions or equations in the same way.

1. [16, p. 212]

> 133. Example: The son of PRĬT'HÁ, exasperated in combat, shot a quiver of arrows to slay CARŃA. With half his arrows he parried those of his antagonist; with four times the square-root of the quiver-full, he killed his horse; with six arrows he slew SALYA; with three he demolished the umbrella, standard and bow; and with one he cut off the head of the foe. How many were the arrows, which ARJUNA let fly?
>
> In this case put the number of the whole of the arrows *ya v* 1....

2. Solve the problem in EXAMPLE 20.11 after making the following change [16, footnote, p. 217]: " 'The fifth part of the troop taken from three' [instead of 'less three']."

3. [16, p. 215]

> 139. Example: The eighth part of a troop of monkeys, squared, was skipping in a grove and delighted with their sport. Twelve remaining were seen on the hill, amused with chattering to each other. How many were they in all?

In this case the troop of monkeys is put *ya* 1. The square of its eighth part, added to twelve, being equal to the whole troop, the two sides of the equation are....

□

Remark 20.2 One general trend in the evolution of problem solving throughout the history of mathematics, up until the seventeenth century in fact, was that problems were solved on a case-by-case basis, and very few general techniques were available for solving whole classes of problems. Take, for example, the Egyptian and Babylonian practice of presenting a method through the use of a specific example, a practice that is also seen in some Arabic works. Now, because a great deal of mathematics has its roots in practical problems, negative numbers were not used for the simple reason that no answer to a practical problem could be less that nothing; hence, if one of the numbers in an "equation" were what we would describe as negative, a new example with that number on the "opposite side" of the "equation" would have to be given.

The development of mathematics, like all sciences, is a slow process of continual revision. Over time, as practitioners began to try to streamline their work by the use of abbreviations, connections between types of problems were observed. This slowly led to grouping problems into cases, as is seen with the work of al-Khwārizmī, where only five methods instead of a multitude are needed. The Arabic scholars also took much of their knowledge from the Greeks. Thus the idea of geometrical proof and generalization became an important part of Arabic mathematics. They also borrowed some ideas from Indian sources, which allowed them to become gradually more comfortable with the concepts of zero and negative numbers.

Another reason for the difference in the Arabic and Indian methods is cultural. Ancient Indian society was much less tied to the here and now compared to the Mediterranean cultures. This may have allowed them to feel more comfortable with ideas that are not directly tied to the concrete, such as negative numbers, as seen in the work of Brahmagupta and Bhāskara.

All of these trends—the solution of practical problems using geometric methods, the gradual introduction of symbolism, and the generalization of method of solutions—were brought together by the Arabic scholars. Although they still wrote their mathematics in a rhetorical style, and then in a syncopated style, the mathematics they passed on to Western Europe set the stage for the gradual emergence of the completely symbolic methods that began to emerge in the Middle Ages and culminated in the Renaissance with the wholly symbolic mathematical notation we use today.   □

20.1.4 *The Quadratic Formula*

We now tie together the apparently different methods for solving quadratic equations that we have seen. We will look at the solutions of the problems in BM 13901 and al-Khwārizmī's *al-jabr wa'l muqābala* side by side because they are similar; then we will look at the solutions of the problems in Bhāskara's *Vījagaṅita*.

To be concrete, we first look at the problems in EXAMPLE 20.3 (page 305), in which we are essentially asked to solve the equation $x^2 + x = \frac{3}{4}$, EXAMPLE 20.6 (page 314), in which we are asked to solve the equation $x^2 + 10x = 39$, and EXAMPLE 20.8 (page 317), in which we are asked to solve $x^2 + 21 = 10x$.

EXAMPLE 20.3 (Babylonian)		
Step	$x^2 + x = \frac{3}{4}$	$x^2 + Bx = C$
given total area	$\frac{3}{4}$	C
put down 1	1	B
halve	$\frac{1}{2}$	$\frac{B}{2}$
square	$\frac{1}{4}$	$\frac{B^2}{4}$

add given total area	1	$\frac{B^2}{4} + C = \frac{B^2+4C}{4}$
square root	1	$\sqrt{\frac{B^2+4C}{4}} = \frac{\sqrt{B^2+4C}}{2}$
subtract halved amount	$\frac{1}{2}$	$\frac{\sqrt{B^2+4C}}{2} - \frac{B}{2}$

EXAMPLE 20.6 (Arabic)		
Step	$x^2 + 10x = 39$	$x^2 + Bx = C$
given numbers	39	C
given roots	10	B
halve	5	$\frac{B}{2}$
square	25	$\frac{B^2}{4}$
add given numbers	64	$\frac{B^2}{4} + C = \frac{B^2+4C}{4}$
square root	8	$\sqrt{\frac{B^2+4C}{4}} = \frac{\sqrt{B^2+4C}}{2}$
subtract halved amount	3	$\frac{\sqrt{B^2+4C}}{2} - \frac{B}{2}$

EXAMPLE 20.8 (Indian)		
Step	$x^2 + 21 = 10x$	$x^2 + C = Bx$
given numbers	21	C
given roots	10	B
halve	5	$\frac{B}{2}$
square	25	$\frac{B^2}{4}$
subtract given numbers	4	$\frac{B^2}{4} - C = \frac{B^2-4C}{4}$
square root	2	$\sqrt{\frac{B^2-4C}{4}} = \frac{\sqrt{B^2-4C}}{2}$
subtract from halved roots	3	$\frac{B}{2} - \frac{\sqrt{B^2-4C}}{2}$
add to halved roots	5	$\frac{B}{2} + \frac{\sqrt{B^2-4C}}{2}$

It is clear that the solutions of the equations in examples 20.3 and 20.6 are identical in form. This is not unexpected since the equations in both examples have the general form $x^2 + Bx = C$. Notice, however, that the solution of the equation in EXAMPLE 20.8, in which the equation has the similar general form $x^2 + C = Bx$, also has the same form. We summarize this observation.

Equation: $B > 0$ and $C > 0$			Solution
$x^2 + Bx = C$	or	$x^2 + Bx - C = 0$	$\frac{-B}{2} + \frac{\sqrt{B^2 + 4C}}{2}$
$x^2 + C = Bx$	or	$x^2 - Bx + C = 0$	$\frac{B}{2} - \frac{\sqrt{B^2 - 4C}}{2}$ $\frac{B}{2} + \frac{\sqrt{B^2 - 4C}}{2}$

Looking at the table, we see that the solutions for the first two examples do not subtract the square root. This is because that would yield a negative solution in that case and, if you recall, negative solutions were not accepted. Nevertheless, let us consider subtracting the square root, that is, let us consider

$$\frac{-B}{2} - \frac{\sqrt{B^2 + 4C}}{2}.$$

Then

- In EXAMPLE 20.3, in which $x^2 + x = \frac{3}{4}$, so that $B = 1$ and $C = \frac{3}{4}$, we find that

$$\frac{-B}{2} - \frac{\sqrt{B^2 + 4C}}{2} = \frac{-1}{2} - \frac{\sqrt{1^2 + 4(\frac{3}{4})}}{2} = \frac{-1}{2} - \frac{\sqrt{1+3}}{2} = \frac{-1}{2} - \frac{2}{2} = -\frac{3}{2},$$

and

$$\left(-\frac{3}{2}\right)^2 + \left(-\frac{3}{2}\right) = \frac{9}{4} - \frac{6}{4} = \frac{3}{4}. \quad \checkmark$$

- In EXAMPLE 20.6, in which $x^2 + 10x = 39$, so that $B = 10$ and $C = 39$, we find that

$$\frac{-B}{2} - \frac{\sqrt{B^2 + 4C}}{2} = \frac{-10}{2} - \frac{\sqrt{10^2 + 4(39)}}{2} = \frac{-10}{2} - \frac{\sqrt{256}}{2} = \frac{-10}{2} - \frac{13}{2} = -13,$$

and

$$(-13)^2 + 10(-13) = 169 - 130 = 39. \quad \checkmark$$

Thus, if we account for whether B and C are added or subtracted, we observe that the solutions of the equations in all three of the examples (20.3, 20.6, and 20.8) are, in fact, identical in form.

Since we have no difficulty with negative numbers or zero, ourselves, let us follow the steps in the solutions of the equations in these examples for the general quadratic equation,

$$ax^2 + bx + c = 0,$$

where a, b, and c are constants with $a \neq 0$ or, equivalently,

$$x^2 + Bx + C = 0$$

after dividing through the equation by a (which we may do because $a \neq 0$). We note that $B = \frac{b}{a}$ and $C = \frac{c}{a}$ may be *any* real numbers: positive, negative, or zero.

Before we continue, we pause to note that the equations in examples 20.3, 20.6, and 20.8 are, indeed, special cases of this general equation.

$x^2 + Bx + C = 0$		
Equation	B and C	
$x^2 + x = \frac{3}{4}$	$B = 1$	$C = -\frac{3}{4}$
$x^2 + 10x = 39$	$B = 10$	$C = -39$
$x^2 + 21 = 10x$	$B = -10$	$C = 21$

We now proceed to solve the equation $x^2 + Bx + C = 0$.

We follow the steps of the solution in EXAMPLE 20.8, in which we are asked to solve the equation $x^2 + 21 = 10x$, to solve the equation $x^2 + Bx + C = 0$ because the steps in that example lead to the two solutions: one that adds the square root and the other that subtracts the square root.

$x^2 + Bx + C = 0$	
Step	$x^2 + C = -Bx$
given numbers	C
given roots	$-B$
halve	$-\frac{B}{2}$
square	$\left(-\frac{B}{2}\right)^2 = \frac{B^2}{4}$
subtract given numbers	$\frac{B^2}{4} - C = \frac{B^2 - 4C}{4}$
square root	$\sqrt{\frac{B^2 - 4C}{4}} = \frac{\sqrt{B^2 - 4C}}{2}$
subtract from halved roots	$-\frac{B}{2} - \frac{\sqrt{B^2 - 4C}}{2}$
add to halved roots	$-\frac{B}{2} + \frac{\sqrt{B^2 - 4C}}{2}$

Hence, we see that the quadratic equation

$$x^2 + Bx + C = 0$$

admits two solutions, namely,

$$\frac{-B}{2} + \frac{\sqrt{B^2 - 4C}}{2} \quad \text{and} \quad \frac{-B}{2} - \frac{\sqrt{B^2 - 4C}}{2}.$$

Now, to find the solutions of the quadratic equation

$$ax^2 + bx + c = 0,$$

substitute $B = \frac{b}{a}$ and $C = \frac{c}{a}$ above. Doing so yields

$$\frac{-b}{2a} + \frac{\sqrt{b^2 - 4ac}}{2a} \quad \text{and} \quad \frac{-b}{2a} - \frac{\sqrt{b^2 - 4ac}}{2a}.$$

This is commonly called the *quadratic formula*.

Rule 20.1 (QUADRATIC FORMULA) Let a, b, and c be any real constants with $a \neq 0$. Then the quadratic equation

$$ax^2 + bx + c = 0$$

has solutions given by the formulas

$$\frac{-b}{2a} + \frac{\sqrt{b^2 - 4ac}}{2a} \quad \text{and} \quad \frac{-b}{2a} - \frac{\sqrt{b^2 - 4ac}}{2a}.$$

We may write the formulas compactly as

$$\frac{-b}{2a} \pm \frac{\sqrt{b^2 - 4ac}}{2a} \quad \text{or} \quad \frac{-b \pm \sqrt{b^2 - 4ac}}{2a}.$$

The quantity $b^2 - 4ac$ is called the *discriminant* of the quadratic equation. □

Think About It 20.1 How does the discriminant of a quadratic equation determine the number of real roots of the equation? □

Example 20.12 We solve the equations in examples 20.3, 20.6, and 20.8 using the quadratic formula. The key is to write the equations in the form $ax^2 + bx + c = 0$—that is, to set the equation equal to zero—before identifying the coefficients a, b, and c for the quadratic formula.

   You should compare the solutions found here to the solutions found in the respective examples.

1. EXAMPLE 20.3: $x^2 + x = \dfrac{3}{4}$

   First, rewrite the equation as $ax^2 + bx + c = 0$:

   $$x^2 + x - \frac{3}{4} = 0.$$

   Second, identify the coefficients a, b, and c:

   $$a = 1, \quad b = 1, \quad c = -\tfrac{3}{4}.$$

   Third, compute the discriminant, $b^2 - 4ac$:

   $$b^2 - 4ac = 1^2 - 4(1)\left(-\frac{3}{4}\right) = 1 + 3 = 4.$$

   Fourth, calculate the two solutions:

   $$\frac{-b}{2a} + \frac{\sqrt{b^2 - 4ac}}{2a} = \frac{-1}{2(1)} + \frac{\sqrt{4}}{2(1)} = \frac{-1}{2} + \frac{2}{2} = \frac{1}{2},$$

   $$\frac{-b}{2a} - \frac{\sqrt{b^2 - 4ac}}{2a} = \frac{-1}{2(1)} - \frac{\sqrt{4}}{2(1)} = \frac{-1}{2} - \frac{2}{2} = -\frac{3}{2}.$$

   Therefore, the solutions are $\frac{1}{2}$ and $-\frac{3}{2}$.

2. EXAMPLE 20.6: $x^2 + 10x = 39$

First, rewrite the equation as $ax^2 + bx + c = 0$:

$$x^2 + x - 39 = 0.$$

Second, identify the coefficients a, b, and c:

$$a = 1, \quad b = 10, \quad c = -39.$$

Third, compute the discriminant, $b^2 - 4ac$:

$$b^2 - 4ac = 10^2 - 4(1)(-39) = 100 + 156 = 256.$$

Fourth, calculate the two solutions:

$$\frac{-b}{2a} + \frac{\sqrt{b^2 - 4ac}}{2a} = \frac{-10}{2(1)} + \frac{\sqrt{256}}{2(1)} = \frac{-10}{2} + \frac{16}{2} = 3,$$

$$\frac{-b}{2a} - \frac{\sqrt{b^2 - 4ac}}{2a} = \frac{-10}{2(1)} - \frac{\sqrt{256}}{2(1)} = \frac{-10}{2} - \frac{16}{2} = -13.$$

Therefore, the solutions are 3 and -13.

3. EXAMPLE 20.8: $x^2 + 21 = 10x$

First, rewrite the equation as $ax^2 + bx + c = 0$:

$$x^2 - 10x + 21 = 0.$$

Second, identify the coefficients a, b, and c:

$$a = 1, \quad b = -10, \quad c = 21.$$

Third, compute the discriminant, $b^2 - 4ac$:

$$b^2 - 4ac = (-10)^2 - 4(1)(21) = 100 - 84 = 16.$$

Fourth, calculate the two solutions:

$$\frac{-b}{2a} + \frac{\sqrt{b^2 - 4ac}}{2a} = \frac{10}{2(1)} + \frac{\sqrt{16}}{2(1)} = \frac{10}{2} + \frac{4}{2} = 7,$$

$$\frac{-b}{2a} - \frac{\sqrt{b^2 - 4ac}}{2a} = \frac{10}{2(1)} - \frac{\sqrt{16}}{2(1)} = \frac{10}{2} - \frac{4}{2} = 3.$$

Therefore, the solutions are 7 and 3.

□

Now, what about the method in Bhāskara's *Víjagańita*? We apply his method (page 321) to the equation

$$ax^2 + bx + c = 0,$$

that is to say, more correctly, to the equation

$$ax^2 + bx = -c.$$

- "Multiply both sides of the equation by a number equal to four times the [coefficient] of the square":

$$4a(ax^2 + bx) = 4a(-c) \quad \text{or} \quad 4a^2x^2 + 4abx = -4ac.$$

- "and add to them a number equal to the square of the original [coefficient] of the unknown quantity":

$$(4a^2x^2 + 4abx) + b^2 = -4ac + b^2 \quad \text{or} \quad 4a^2x^2 + 4abx + b^2 = b^2 - 4ac.$$

- "Then extract the root."

First, observe that the left-hand side is a perfect square, to wit,

$$4a^2x^2 + 4abx + b^2 = (2ax + b)^2,$$

so that the equation

$$4a^2x^2 + 4abx + b^2 = b^2 - 4ac$$

may be rewritten as

$$(2ax + b)^2 = b^2 - 4ac.$$

Second, note that $(-r)^2 = r^2$, so that every positive real number R has two square roots: a positive square root and a negative square root, both with the same absolute value. (See CHAPTER 12 and SUBSECTION 20.2.2.) Hence, extracting the square root on both sides of the quadratic equation,

$$(2ax + b)^2 = b^2 - 4ac,$$

yields *two* linear equations, namely,

$$2ax + b = \sqrt{b^2 - 4ac} \quad \text{and} \quad -(2ax + b) = \sqrt{b^2 - 4ac},$$

or

$$2ax + b = \sqrt{b^2 - 4ac} \quad \text{and} \quad 2ax + b = -\sqrt{b^2 - 4ac}.$$

- Finally, solving the two linear equations for x, we find that the quadratic equation

$$ax^2 + bx + c = 0$$

has the two solutions

$$\frac{-b + \sqrt{b^2 - 4ac}}{2a} \quad \text{and} \quad \frac{-b - \sqrt{b^2 - 4ac}}{2a}.$$

In other words, we again have the quadratic formula (RULE 20.1).

Example 20.13 Solve the following equations using the quadratic formula. If the discriminant, $b^2 - 4ac$, is not a perfect square, leave the answer as a radical; if the discriminant is negative, then there is no real solution.

1. $x^2 + 3x = 18$

   First, rewrite the equation as
   $$x^2 + 3x - 18 = 0.$$

Second, identify the coefficients a, b, and c:

$$a = 1, \quad b = 3, \quad c = -18.$$

Third, compute the discriminant, $b^2 - 4ac$:

$$b^2 - 4ac = 3^2 - 4(1)(-18) = 9 + 72 = 81.$$

Fourth, calculate the two solutions:

$$\frac{-b + \sqrt{b^2 - 4ac}}{2a} = \frac{-3 + \sqrt{81}}{2(1)} = \frac{-3 + 9}{2} = \frac{6}{2} = 3,$$

$$\frac{-b - \sqrt{b^2 - 4ac}}{2a} = \frac{-3 - \sqrt{81}}{2(1)} = \frac{-3 - 9}{2} = \frac{-12}{2} = -6.$$

Therefore, the solutions are 3 and −6.

2. $4x^2 + 1 = 4x$

First, rewrite the equation as
$$4x^2 - 4x + 1 = 0.$$

Second, identify the coefficients a, b, and c:

$$a = 4, \quad b = -4, \quad c = 1.$$

Third, compute the discriminant, $b^2 - 4ac$:

$$b^2 - 4ac = (-4)^2 - 4(4)(1) = 16 - 16 = 0.$$

Fourth, calculate the two solutions:

$$\frac{-b + \sqrt{b^2 - 4ac}}{2a} = \frac{4 + \sqrt{0}}{2(4)} = \frac{4 + 0}{8} = \frac{1}{2},$$

$$\frac{-b - \sqrt{b^2 - 4ac}}{2a} = \frac{4 - \sqrt{0}}{2(4)} = \frac{4 - 0}{8} = \frac{1}{2}.$$

Therefore, there is only one solution, $\frac{1}{2}$. We say that $\frac{1}{2}$ is a *double root* of the equation.

3. $x^2 - 3x = 1$

First, rewrite the equation as
$$x^2 - 3x - 1 = 0.$$

Second, identify the coefficients a, b, and c:

$$a = 1, \quad b = -3, \quad c = -1.$$

Third, compute the discriminant, $b^2 - 4ac$:

$$b^2 - 4ac = (-3)^2 - 4(1)(-1) = 9 + 4 = 13.$$

Fourth, calculate the two solutions:

$$\frac{-b + \sqrt{b^2 - 4ac}}{2a} = \frac{3 + \sqrt{13}}{2(1)} = \frac{3 + \sqrt{13}}{2},$$

$$\frac{-b - \sqrt{b^2 - 4ac}}{2a} = \frac{3 - \sqrt{13}}{2(1)} = \frac{3 - \sqrt{13}}{2}.$$

Therefore, the solutions are $\frac{3+\sqrt{13}}{2}$ and $\frac{3-\sqrt{13}}{2}$.

4. $x^2 + 10 = 2x$

First, rewrite the equation as
$$x^2 - 2x + 10 = 0.$$

Second, identify the coefficients a, b, and c:
$$a = 1, \quad b = -2, \quad c = 10.$$

Third, compute the discriminant, $b^2 - 4ac$:
$$b^2 - 4ac = (-2)^2 - 4(1)(10) = 4 - 40 = -36.$$

The *discriminant is negative.*

Therefore, there is no real solution.

☐

Think About It 20.2 Why is there no real root of a quadratic equation when its discriminant is negative?
☐

Now You Try 20.11 Solve the following equations using the quadratic formula. If the discriminant is not a perfect square, leave the solution in radical form. This is an exact solution, as opposed to an approximate calculator answer.

1. $x^2 + 2x = 35$ 3. $2x^2 - 3x + 3 = 0$

2. $5x^2 + 75 = 40x$ 4. $x^2 + 25 = 10x$

5. $x^2 - x - 1 = 0$ (See NOW YOU TRY 19.11 on page 299.)

☐

We close our discussion of the method of completing the square and the quadratic formula for solving quadratic equations by mentioning the contribution of English mathematician Thomas Harriot (ca. 1560–1621). Whereas al-Khwārizmī's presentation of the method of completing the square was geometrical and Bhāskara's was rhetorical, Harriot presented the method of completing the square purely algebraically using only symbols. John Wallis (1616–1703) describes Harriot's "peculiar way of his own" thusly [109, pp. 115 and 242, footnote no. 87]:

To each part of his Quadratick Equation, $aa \pm 2ba = cc$; [Harriot] adds, the Square of half the Coefficient, bb, thereby making the Unknown part, a Compleat Square in Species equal to a Known Quantity.

$$aa \pm 2ba + bb = \pm cc + bb$$

And consequently, the Square Root of that, equal to the Square Root of this.

$$a \pm b = \sqrt{(\pm c + bb)}$$

which being known; the value of a is known also.

Moreover, Harriot recognized both negative real number solutions and complex number solutions without fanfare. (See SECTION 21.1 for an introduction to complex numbers.) Harriot is the first person known to have worked entirely symbolically. The shift from thinking geometrically to symbolically is one of the most notable features of seventeenth-century mathematics.

20.2 POLYNOMIAL EQUATIONS IN ONE VARIABLE

A *polynomial expression* of *degree n* in one variable, say, x, is an expression of the form

$$ax^n + bx^{n-1} + \cdots + cx^2 + dx + e,$$

where $a, b, \ldots, c, d$, and e are constants, $a \neq 0$, and n is a nonnegative integer. Here are a few examples.

Polynomial	Degree
$x^2 + 10x - 39$	2
$x^2 + 10x$	2
$x^2 - 39$	2
x^2	2
$10x - 39$	1
$10x$	1
-39	0
$x^3 - 5x + 6$	3
$2x^7 + x^6 - 3x^4 - 5x^3 + x^2 - 3x + 10$	7

It is common to order the terms of a polynomial so that the powers on the variable are in either descending or ascending order.

Now You Try 20.12

1. Identify the following expressions as either polynomials or not polynomials. Identify the degree of any polynomials.

   a) $-x^4 + 3x^3 + 2x^2 - 5x + 17$

   b) $3 + 4x - x^2$

   c) $x^3 + 2x^2 - \dfrac{5}{x} + 3$

   d) $5x - xy$

2. Write a polynomial expression of degree 35 and that has six terms.

☐

A *polynomial equation* of *degree n* is an equation of the form

$$ax^n + bx^{n-1} + \cdots + cx^2 + dx + e = 0,$$

where $a, b, \ldots, c, d$, and e are constants, $a \neq 0$, and n is a nonnegative integer. Here are a few examples.

Equation	Degree
$x^2 + 10x - 39 = 0$	2
$10x - 39 = 0$	1
$x^3 - 5x + 6 = 0$	3
$2x^7 + x^6 - 3x^4 - 5x^3 + x^2 - 3x + 10 = 0$	7

Remark 20.3 As an example, the equation

$$x^2 + 10x = 39$$

is also considered to be a polynomial equation of degree 2 because it is equivalent to the equation $x^2 + 10x - 39 = 0$. Likewise,

$$(2x - 1)(x^2 - 4) = 0$$

is considered to be a polynomial equation of degree 3 because it is equivalent to the equation $2x^3 - x^2 - 8x + 4 = 0$ after expanding.[9] Moreover, polynomial equations or expressions of degree 1 to 5 are referred to as:

> degree 1:   linear
> degree 2:   quadratic
> degree 3:   cubic
> degree 4:   quartic or biquadratic
> degree 5:   quintic

equations. We generally do not use special names for polynomial equations or expressions of higher degree.   □

Think About It 20.3 What does the prefix "quad" in the term *quadratic* represent? Give mathematical as well as nonmathematical examples of other words that use the prefix "quad." Given those examples, what degree would you expect a *quadratic* equation to be? Thinking as an ancient Greek, where mathematics has a geometric interpretation, what geometric shape would a "quadratic" be? What dimension does that shape have?   □

Now, solving a polynomial equation of degree n is intimately related to finding the nth root of a number. For example, solving an equation of degree 2 is related to finding the 2nd root or square root of a number (as we saw with the method of completing the square); solving an equation of degree 3 is related to finding the 3rd root or cube root of a number; solving an equation of degree 4 is related to finding the 4th root of a number; and so on.

We begin with a review of powers, then proceed to a general, although light, discussion of nth roots in SUBSECTION 20.2.2. After this, we turn to the practical problem of finding nth roots.

---

[9]We expand using the distributive law, $a(b + b) = ab + ac$:

$$(2x - 1)(x^2 - 4) = 2x(x^2 - 4) + (-1)(x^2 - 4) = 2x^3 - 8x - x^2 + 4 = 2x^3 - x^2 - 8x + 4.$$

20.2.1 Powers

Definition 20.1 Let n be a positive integer. Then the expression a^n is called a *power* and denotes the product of n factors of a, that is,

$$a^n = \underbrace{a \cdot a \cdot a \cdots a}_{n \text{ factors}}.$$

The expression a^n is read "the nth power of a" or "a raised to the nth power," or simply "a to the nth." The number a is called the *base* of the power and n is called the *exponent*. The operation indicated by a^n, namely, $a \cdot a \cdot a \cdots a$ (n factors), is called *exponentiation*. $\square$

Here are a few examples.

Expression	Read ...	Value
3^1	the 1st power of 3 or 3 to the first	3
3^2	the 2nd power of 3 or 3 to the second	$3 \cdot 3 = 9$
3^3	the 3rd power of 3 or 3 to the third	$3 \cdot 3 \cdot 3 = 27$
3^4	the 4th power of 3 or 3 to the fourth	$3 \cdot 3 \cdot 3 \cdot 3 = 81$
3^5	the 5th power of 3 or 3 to the fifth	$3 \cdot 3 \cdot 3 \cdot 3 \cdot 3 = 243$

Remark 20.4 The power a^2 is commonly read "a squared," and a^3 is commonly read "a cubed." $\square$

Think About It 20.4 Why would the second and third power be referred to as "square" and "cube," respectively? Why are there no corresponding names for higher powers? $\square$

We note that, for any positive integer n,

$$0^n = 0 \quad \text{and} \quad 1^n = 1.$$

Moreover, because the product of evenly many negative numbers is positive, and the product of oddly many negative numbers is negative, we also note that, for example,

$$(-2)^4 = (-2)(-2)(-2)(-2) = 16$$

and

$$(-2)^5 = (-2)(-2)(-2)(-2)(-2) = -32.$$

In general,

- If n is even, a^n is positive.

- If n is odd,

$$a^n \text{ is } \begin{cases} \text{positive} & \text{if } a \text{ is positive,} \\ \text{negative} & \text{if } a \text{ is negative.} \end{cases}$$

Because of this, the use of brackets is important in exponentiation whenever the base is negative and the exponent is even. We illustrate this using the base 6 and the exponent 4.

$$\begin{aligned}
\text{First:} \quad & 6^4 = 6 \cdot 6 \cdot 6 \cdot 6 = 1296 \\
\text{So:} \quad & -6^4 = -(6 \cdot 6 \cdot 6 \cdot 6) = -1296 \\
\text{But:} \quad & (-6)^4 = (-6)(-6)(-6)(-6) = 1296
\end{aligned}$$

At this point, we recall the rules for exponents.

Rule 20.2 (RULES FOR EXPONENTS) Let m and n be positive integers. Then,

	Rule	Example
(1)	$a^m a^n = a^{m+n}$	$a^2 \cdot a^4 = (a \cdot a)(a \cdot a \cdot a \cdot a) = a^6 = a^{2+4}$
(2)	$\dfrac{a^m}{a^n} = a^{m-n}$ if $m > n$	$\dfrac{a^5}{a^2} = \dfrac{\cancel{a} \cdot \cancel{a} \cdot a \cdot a \cdot a}{\cancel{a} \cdot \cancel{a}} = a^3 = a^{5-2}$
(3)	$(a^m)^n = a^{mn}$	$(a^2)^3 = (a \cdot a)(a \cdot a)(a \cdot a) = a^6 = a^{2 \cdot 3}$
(4)	$(ab)^n = a^n \cdot b^n$	$(a \cdot b)^2 = (a \cdot b)(a \cdot b) = (a \cdot a)(b \cdot b) = a^2 \cdot b^2$
(5)	$\left(\dfrac{a}{b}\right)^n = \dfrac{a^n}{b^n}$	$\left(\dfrac{a}{b}\right)^2 = \left(\dfrac{a}{b}\right)\left(\dfrac{a}{b}\right) = \dfrac{a \cdot a}{b \cdot b} = \dfrac{a^2}{b^2}$

☐

Now You Try 20.13 Simplify the following expressions using the rules for exponents.

1. $x^5 x^3$
2. $\dfrac{a^{11}}{a^5}$
3. $(x^5 y^3)^4$
4. $\dfrac{a^{11} b^6}{(a^5 b^3)^2}$

☐

Our notation for exponents is generally credited to René Descartes (1596–1650), an eminent French mathematician and philosopher who first published its use in his work *La géométrie* (*The Geometry*) [46] in 1637.[10] *La géométrie* is one of three essays that accompanies Descartes's magnum opus, *Discours de la méthode pour bien condiure sa raison et chercher la vérité dans les sciences* (*Discourse on the Method for Rightly Directing One's Reason and Searching for Truth in the Sciences*). About the same time,

---

[10] Across the English Channel, Thomas Harriot (ca. 1560–1621), in his *Treatise on equations* [110], in addition to using a variation of the symbol = for equals that Robert Recorde had introduced in *The Whetstone of Witte*, introduced the symbols < and > for inequality (less than and greater than, respectively), and the symbols ± and ∓ to handle several cases at once. But, according to Stedall [109, p. 90], "Harriot's most important innovation in notation was undoubtedly his use of ab to represent a multiplied by b, and consequently aa, aaa for what is now written a^2, a^3, etc." Stedall [109, p. 123] later goes on to say, "The only significant difference between modern notation and Harriot's is the use of superscripts for exponents. Terms such as $aaaabb$ seemed to beg for some kind of abbreviation, and Torporley in copying Harriot's manuscripts did indeed sometimes write a^{I}, a^{II}, a^{III}, a^{IV}, where Harriot had written a, aa, aaa, or $aaaa$."

Pierre de Fermat (1601–1665), another eminent French mathematician,[11] circulated his manuscript, *Ad locos planos et solidos isagoge* (*Introduction to Plane and Solid Loci*). With these works, both Descartes and Fermat (FIGURE 20.2) tied together in their own ways algebra and geometry to bring us the modern subject of *analytic geometry*, whereby geometric objects such as curves may be described and studied using equations, and vice versa. Analytic geometry turned out to be crucial for the later discovery of the calculus by Newton and Leibniz. However, because Descartes's work was published, albeit at first in French instead of the customary Latin, whereas Fermat's was only circulated as a manuscript, it was Descartes's work that firmly took hold [73]. (If Fermat's work had had more influence, perhaps we would today refer to the familiar graphing coordinate system as the "Fermatian coordinate system" instead of the "Cartesian coordinate system.")

(a) René Descartes.   (b) Pierre de Fermat.

Figure 20.2: René Descartes and Pierre de Fermat. (Source: (a) © Thinkstock. (b) Courtesy of Jeff Miller.)

Descartes used Indo-Arabic numerals for exponents written, as Cajori points out, in a position relative to the base as we do today for integer exponents greater than 2; for the second power, *aa* was often used instead of a^2 [31, pp. 205–208], presumably because not much space would be saved in writing or printing a^2 instead of *aa* and, at the time, it would have been easier to typeset *aa* than a^2. Before this, and even after, for it took some time for Descartes's notation to become widely accepted, a variety of notations for exponentiation were employed. Many people devised different ways of expressing powers, with some

---

[11]Fermat was a lawyer and government official in Toulouse, France. As Bell [9, chapter 4] puts it, Fermat was, "in the strictest sense of the word, so far as his science and mathematics were concerned, an amateur in the history of science, if not the very first." Nevertheless, Bell also asserts that Fermat was "the greatest mathematician of the seventeenth century" and was "*at least* Newton's equal *as a pure mathematician*." Among Fermat's mathematical accomplishments, we may count the creation of analytic geometry (independently of Descartes), his contribution to the evolution of the calculus, the creation of probability theory (together with Blaise Pascal (1623–1662)), and his investigations in number theory. Of the last, his most famous result is that which has come to be known as "Fermat's last theorem" (FLT). As Bell [9, p. 71] relates it:

> It was Fermat's custom in reading Bachet's *Diophantus* to record the results of his meditations in brief marginal notes in his copy. The margin was not suited for the writing out of proofs. Thus, in commenting on the eighth problem of the Second Book of Diophantus' Arithmetic, which asks for the solution in rational numbers (fractions or whole numbers) of the equation $x^2 + y^2 = a^2$, Fermat comments as follows:
>
> > "On the contrary, it is impossible to separate a cube into two cubes, a fourth power into two fourth powers, or, generally, any power above the second into two powers of the same degree: I have discovered a truly marvellous demonstration [of this general theorem] which this margin is too narrow to contain" (Fermat, *Oeuvres*, III, p. 241).

In other words, FLT states that no rational numbers x, y, and z exist that satisfy the equation $x^n + y^n = z^n$ for any integer $n > 2$. This was in 1637.

This tantalizing statement that Fermat had left in the margin enticed many mathematicians and amateurs to seek a proof of FLT. However, a proof of this simple-sounding statement eluded the greatest mathematicians until, in 1995, Andrew Wiles (b. 1953) finally published a several-hundred-page proof—358 years later!

inventing new symbols and others using words only. Cajori [31, pp. 335–360] provides an extensive discussion on powers, from which we present a few examples here.

The ancient Egyptians used a single symbol to denote the squaring of a number: ⌃, a pair of legs walking forward "signifying 'make in going,' that is, squaring the number" [31, p. 335]. This symbol used to denote the squaring of a number is found in problem 14 of the *Moscow Mathematical Papyrus* on finding the volume of a truncated pyramid. Interestingly, we also find in the *Rhind Mathematical Papyrus* the use of the symbol ⌃ for addition and the symbol ⌃ (a pair of legs walking away) for subtraction [31, p. 229].

Perhaps the first person to have used symbols to represent unknown quantities was the Greek mathematician Diophantus (ca. AD 250) in his work *Arithmetica* [31, pp. 71–74]. There we find that Diophantus uses the following symbols for an unknown quantity and powers of an unknown quantity.[12]

ς' or ς°'	undefined number (x)
Δ^Y	square (x^2)
K^Y	cube (x^3)
$\Delta^Y\Delta$	square-square (x^4)
ΔK^Y	square-cube (x^5)
$K^Y K$	cube-cube (x^6)

Note the additive principle in Diophantus's notation:

$$\Delta^Y\Delta = \text{square-square} = x^2 x^2 = x^{2+2} = x^4;$$
$$\Delta K^Y = \text{square-cube} = x^2 x^3 = x^{2+3} = x^5;$$
$$K^Y K = \text{cube-cube} = x^3 x^3 = x^{3+3} = x^6.$$

Later Greeks expressed numbers using the letters of the alphabet. TABLE 20.1 shows a few examples [31, p. 25]. To distinguish a numeral from a word, a bar may be written over the numeral. Thus, for instance, Cajori [31, p. 73] gives the example

$$K^Y \bar{a} \Delta^Y \overline{\iota\gamma} \varsigma \bar{\epsilon} \overset{\circ}{M} \bar{\beta},$$

where $\overset{\circ}{M}$ denotes units, as the Diophantine representation of the polynomial $x^3 + 13x^2 + 5x + 2$. Notice that the coefficient follows the unknown and that addition is understood. Diophantus indicated subtraction using the symbol ⋔ and collected all the terms to be subtracted at the end. Thus, for instance,

$$\Delta^Y\Delta \bar{\gamma} \Delta^Y \overline{\iota\gamma} \varsigma \bar{\epsilon} \pitchfork K^Y \bar{a} \overset{\circ}{M} \bar{\beta} \quad \text{is} \quad 3x^4 + 13x^2 + 5x - x^3 - 2.$$

Now You Try 20.14 Express the following polynomials using the Diophantine symbols.

1. $x^2 + 10x - 39$ 2. $x^3 - 5x + 6$ 3. $x^6 - 3x^4 - 5x^3 + x^2 - 3x + 10$

□

Indian mathematicians used a variety of symbols. Bhāskara II (1114–1185), in his work *Līlāvatī*, for example, alludes to raising numbers to powers greater than the third power using the words *varga* for the

---

[12]We introduce the symbols x, x^2, x^3, and so on, here and below for our benefit only. While Diophantus did use the symbols ς', Δ^Y, K^Y, and so on, in his syncopated algebra, he did not use letters in the way we do; and others, such as Indian and Arabic mathematicians, did not use any symbols at all in their rhetorical algebra.

α	β	γ	δ	ϵ	ι	κ	λ	μ	ν
1	2	3	4	5	10	20	30	40	50

ρ	σ	τ	υ	ϕ	$,\alpha$	$,\beta$	$,\gamma$	M	$\overset{\beta}{M}$
100	200	300	400	500	1000	2000	3000	10,000	20,000

Table 20.1: Greek alphabetic numerals.

square of a number (x^2) and *g'hana* for the cube of a number (x^3) in the following way (transliterated), as related by the sixteenth-century commentator Ganesa [31, p. 80]:

$$varga\text{-}varga = x^2 x^2 = x^4,$$
$$varga\text{-}g'hana \text{ or } g'hana\text{-}varga = x^2 x^3 \text{ or } x^3 x^2 = x^6,$$
$$g'hana\text{-}g'hana = x^3 x^3 = x^9.$$

Note the *multiplicative* principle in action so that, for instance, *varga-g'hana*—literally "$x^2 x^3$"—is $x^{2\cdot3} = x^6$, and not $x^{2+3} = x^5$ as we would have it. Of course, the multiplicative principle would never allow for one to raise a number to a prime power. One possible remedy would have been to have a different convention for naming numbers raised to prime powers; for example, Michael Psellus [31, p. 85], an eleventh-century Byzantine writer, in a letter[13] about Diophantus, relates how the Egyptians called x^6 the "cube-cube," but x^5 "the first undescribed" and x^7 "the second undescribed." (Note the Egyptian use of the additive property.) However, to raise a number to a prime power, Bhāskara instead dropped the multiplicative principle and turned to the additive principle by introducing the word *g'háta* ("product"): to raise a number to the fifth power, for example, the word *g'háta* is appended as follows:

$$varga\text{-}g'hana\text{-}g'háta = x^2 \cdot x^3 = x^{2+3} = x^5,$$

in which we see the additive principle now in action.

Arabic mathematicians generally followed the Indians in using the multiplicative principle to build up higher powers. One notable exception was the Arabic writer al-Karkhî of Baghdad (early eleventh century) [31, pp. 84–85]. In his work, the *Fakhrī*, al-Karkhî uses the word *mal* for the square of a number and *ka^{c}b* for the cube of a number. Then, following Diophantus in using the additive principle, al-Karkhî built up higher powers as follows:

$$m\bar{a}l\ m\bar{a}l = x^2 \cdot x^2 = x^{2+2} = x^4,$$
$$m\bar{a}l\ ka^c b = x^2 \cdot x^3 = x^{2+3} = x^5,$$
$$ka^c b\ ka^c b = x^3 \cdot x^3 = x^{3+3} = x^6,$$

and so on.

Leonardo of Pisa, in *Liber Abaci* [90], uses the word *radix* for x, *census* for x^2, and *cubus* for x^3, and goes on to form

$$census\ census \text{ for } x^4,$$
$$cubus\ cubus \text{ or } census\ census\ census \text{ for } x^6,$$
$$census\ census\ census\ census \text{ for } x^8,$$

reflecting the additive principle. Leonardo also refers to *radix census census* [31, pp. 89–91], presumably for x^5.

---

[13]Reproduced by Paul Tannery, *op. cit.*, Vol. II (1895), p. 37–42 .

Cajori states that Nicolas Chuquet (d. 1487), in his manuscript, *Triparty en la science des nombres* (1484), "elaborates the exponential notation to a completeness apparently never before dreamed of" and "was about one hundred and fifty years ahead of his time" [31, p. 100]. In *Triparty*, Chuquet brings us notation that comes close to what we use today, for example, writing "12^2." But, by "12^2." Chuquet does not mean 12×12; instead, he means $12x^2$. Indeed, we find that Chuquet writes, for example [31, p. 102],

$$.12^{\underline{o}} \text{ for } 12, \quad .12^1 \text{ for } 12x, \quad 12^2. \text{ for } 12x^2, \quad .12^3. \text{ for } 12x^3,$$

and so on. (Cf. Chuquet's use of the mark ○ to al-Khwārizmī's *dirham* to denote a "pure number"; page 314.) Of note is that Chuquet apparently reasoned that $x^0 = 1$.

As an example, using Chuquet's abbreviation of $\tilde{p}$ for *plus* (plus) and $\tilde{m}$ for *moins* (minus), he may have written

$$1^2. \text{ avec } \tilde{p}.10^1 \text{ monte tout } .39^{\underline{o}} \quad \text{for} \quad x^2 + 10x = 39.$$

Cajori [31, p. 100] posits that "had his work been printed at the time when it was written, it would, no doubt, have greatly accelerated the progress of algebra."

Now You Try 20.15 Use Chuquet's notation to express the following equations.

1. $x^2 - x = 1$ 2. $x^3 = 5x - 6$

□

In *Summa de arithmetica geometria proportioni et proportionalita* (1494), or the *Summa* for short, which was the common introduction to mathematics in Italy in the first half of the sixteenth century, Luca Pacioli (1445–1514) uses different notations for powers [31, pp. 106–110]. (A second edition of the *Summa* was published posthumously in 1523, making only changes in the spelling of some of the words [31]. See page 113.) In the *Summa*, we find the abbreviations *co.* (*cosa*) for x (*cosa* means "thing"), *ce.* (*censo*) for x^2, and *cu.* (*cubo*) for x^3. Then, to build up higher powers, Pacioli applies the multiplicative principle, for example, writing

$$\text{ce.ce. for } x^{2 \cdot 2} \text{ or } x^4,$$
$$\text{ce.cu. for } x^{2 \cdot 3} \text{ or } x^6,$$
$$\text{ce.ce.ce. for } x^{2 \cdot 2 \cdot 2} \text{ or } x^8.$$

(Compare Pacioli's convention with Leonardo of Pisa's.) For prime powers, Pacioli uses a naming scheme that is reminiscent of the Egyptians' (page 341):

$$p^{\underline{o}}r^{\underline{o}} \text{ (primo relato) for } x^5,$$
$$2^{\underline{o}}r^{\underline{o}} \text{ (secundo relato) for } x^7,$$
$$3^{\underline{o}}r^{\underline{o}} \text{ (terzo relato) for } x^{11}.$$

However, Pacioli also uses another notation for powers, namely, the symbol R that he also uses for roots. (In fact, Cajori tells us that Pacioli uses R five different ways!) For powers, Pacioli gives, for example,

$$R.p^{\underline{a}} \text{ (radix prima) for } x^0,$$
$$R.2^{\underline{a}} \text{ (radix secunda) for } x,$$
$$R.3^{\underline{a}} \text{ (radix terza) for } x^2.$$

Cajori [31, p. 109] gives this illustration:

$$R.5^{\underline{a}} \text{ via. } R.11^{\underline{a}} \text{ fa } R.15^{\underline{a}} \quad \text{for} \quad x^4 \times x^{10} = x^{14}.$$

Think About It 20.5 Explain Pacioli's naming system. □

Last, we spotlight Rafael Bombelli (1526–1572), a person to whom we shall return in CHAPTER 21, where we discuss cubic equations and complex numbers. Cajori [31, pp. 124–128] pronounces that an "important change in notation was made for the expression of powers which was new in Italian algebras" due to Bombelli. The notation Bombelli uses is similar to Chuquet's in that it expresses exponents without the base, which manages well as long as only one variable is expected. Using p. and m. for plus and minus, respectively, here is an example from page 251 of Bombelli's *L'algebra* (1572):

$$24.m.\overset{1}{20}.\text{ Equale à } 4.m.\overset{2}{16}.p.\overset{1}{16}. \quad \text{for} \quad 24 - 20x = 4x^2 - 16x + 16.$$

Many other conventions were used through the years, but we return now to our review of powers.

In 1867, Hermann Hankel (1839–1873) formulated his so-called *principle of permanence*, according to which "one tries to retain the validity of calculating rules, but to extend the concepts of the mathematical objects connected by them" [59, p. 50]. Thus, by Hankel's principle of permanence, we would want to allow in the rules for exponents (page 338) the rule

$$\frac{a^m}{a^n} = a^{m-n}$$

not only when $m > n$, but also when $m = n$ and $m < n$. Let us consider two concrete examples to see to what this may lead.

First, we take the case when $m = n$, say $m = n = 5$. Then,

$$\frac{a^5}{a^5} = a^{5-5} = a^0 \quad \text{and} \quad \frac{a^5}{a^5} = \frac{\cancel{a} \cdot \cancel{a} \cdot \cancel{a} \cdot \cancel{a} \cdot \cancel{a}}{\cancel{a} \cdot \cancel{a} \cdot \cancel{a} \cdot \cancel{a} \cdot \cancel{a}} = 1.$$

Second, we take the case when $m < n$, say $m = 2$ and $n = 5$. Then,

$$\frac{a^2}{a^5} = a^{2-5} = a^{-3} \quad \text{and} \quad \frac{a^2}{a^5} = \frac{\cancel{a} \cdot \cancel{a}}{\cancel{a} \cdot \cancel{a} \cdot a \cdot a \cdot a} = \frac{1}{a^3}.$$

Thus, we see that if we are to apply the principle of permanence, then we must allow for an exponent to be zero or negative in the following way.

Definition 20.2 Let n be an integer. If $a \neq 0$, then

$$a^0 = 1 \quad \text{and} \quad a^{-n} = \frac{1}{a^n}.$$

□

Here are a few examples.

Expression	Read ...	Value
3^0	the zeroth power of 3	1
-3^0	the opposite of the zeroth power of 3	-1
$(-3)^0$	the zeroth power of -3	1
3^{-1}	the negative-first power of 3	$\frac{1}{3}$
3^{-5}	the negative-fifth power of 3	$\frac{1}{3^5} = \frac{1}{243}$

Now RULE 20.2 (the rules for exponents) holds for all integer exponents, both positive, negative, and zero.

Think About It 20.6 What would 0^0 equal? Is there more than one logical answer?[14]   □

Example 20.14 Simplify the following expressions using the rules for exponents (page 338). Give your answers using positive exponents only.

1. $x^5 x^{-3}$

$$x^5 x^{-3} = x^{5+(-3)} = x^2$$

2. $\dfrac{a^{-11}}{a^{-5}}$

$$\frac{a^{-11}}{a^{-5}} = a^{-11-(-5)} = a^{-6} = \frac{1}{a^6}$$

3. $(x^{-5}y^3)^{-4}$

$$(x^{-5}y^3)^{-4} = (x^{-5})^{-4}(y^3)^{-4}$$
$$= x^{-5(-4)}y^{3(-4)}$$
$$= x^{20}y^{-12}$$
$$= \frac{x^{20}}{y^{12}}$$

4. $\left(\dfrac{a^{11}b^{-6}}{(a^{-5}b^3)^{-2}}\right)^{-3}$

To begin,

$$\left(\frac{a^{11}b^{-6}}{(a^{-5}b^{-3})^{-2}}\right)^{-3} = \frac{(a^{11}b^{-6})^{-3}}{((a^{-5}b^3)^{-2})^{-3}}.$$

Now, we simplify the numerator and the denominator separately.

First, the numerator:

$$(a^{11}b^{-6})^{-3} = (a^{11})^{-3}(b^{-6})^{-3} = a^{-33}b^{18}.$$

Second, the denominator:

$$((a^{-5}b^3)^{-2})^{-3} = (a^{-5}b^3)^6 = (a^{-5})^6(b^3)^6 = a^{-30}b^{18}.$$

Hence,

$$\left(\frac{a^{11}b^{-6}}{(a^{-5}b^{-3})^{-2}}\right)^{-3} = \frac{a^{-33}b^{18}}{a^{-30}b^{18}} = a^{-33-(-30)}b^{18-18}$$
$$= a^{-3}b^0$$
$$= \frac{b^0}{a^3}$$
$$= \frac{1}{a^3}.$$

□

---

[14]In order to avoid contradictions that may arise, mathematicians generally call 0^0 "indeterminate" or "undefined" instead of assigning 0^0 any *a priori* value.

Now You Try 20.16 Simplify the following expressions using the rules for exponents. Give your answers using positive exponents only.[15]

1. $x^{-5}x^3$

2. $\dfrac{a^{11}}{a^{-5}}$

3. $(x^{-5}y^3)^{-4}$

4. $\left(\dfrac{(a^{-11}b^6)^{-2}}{a^{-5}b^{-3}}\right)^3$

□

20.2.2 nth Roots

In mathematics, for a given operation, there may be a way to "undo" that operation that is called its inverse. Thus, the inverse of the operation addition is subtraction; the inverse of the operation multiplication is division; and the inverse of the operation exponentiation is the "extraction of roots."

Definition 20.3 Let n be a positive integer. If b is a nonnegative real number, then the expression $\sqrt[n]{b}$ denotes a nonnegative real number such that the nth power of $\sqrt[n]{b}$ is b, that is,

$$(\sqrt[n]{b})^n = b.$$

In other words,

$$\sqrt[n]{b} = a \quad \text{if} \quad a^n = b.$$

The expression $\sqrt[n]{b}$ is read "nth root of b." The number b is called the *radicand* or *argument* of the expression and n is called the *exponent* or *index* of the root. The operation indicated by $\sqrt[n]{b}$ is called *extraction of (nth) roots*. □

Defined in this way, extraction of roots and exponentiation are inverse operations if we restrict ourselves to positive numbers.

Remark 20.5 The expression $\sqrt{b}$ is understood to be $\sqrt[2]{b}$ with the index 2 suppressed. It is also common to write $\sqrt{}$, without the vinculum (horizontal bar), in place of $\sqrt{}$; the symbol is called a *surd* or *radical*. Last, a root is also called a radical. □

Here are a few examples.

Expression	Read	Value	Reason
$\sqrt{49}$	second root or square root of 49	7	$7^2 = 49$
$\sqrt[3]{343}$	third root or cube root of 343	7	$7^3 = 343$
$\sqrt[4]{81}$	fourth root of 81	3	$3^4 = 81$
$\sqrt[5]{243}$	fifth root of 243	3	$3^5 = 243$
$\sqrt[6]{64}$	sixth root of 64	2	$2^6 = 64$

We write

$$\sqrt{49} = 7, \quad \sqrt[3]{343} = 7,$$

$$\sqrt[4]{81} = 3, \quad \sqrt[5]{243} = 3, \quad \sqrt[6]{64} = 2.$$

(See CHAPTER 12.)

---

[15]There is often more than one way to start when simplifying these expressions; for example, we could simplify inside the grouping first instead of distributing the "outside" exponent first. As long as we follow the rules of algebra, we will get the same answer.

Now You Try 20.17 Simplify the following.

1. $\sqrt{400}$ 2. $\sqrt{40}$ 3. $\sqrt[3]{64}$

4. $\sqrt{\dfrac{1}{16}}$ 5. $\sqrt[3]{\dfrac{8}{27}}$ 6. $\sqrt[4]{\dfrac{81}{10\,000}}$

☐

We make the following observations.

1. Because a^n is positive if n is even, regardless of whether a is positive or negative, *every positive number has two real nth roots if n is even: a positive nth root and a negative nth root*, both with the same absolute value. For example, both 7 and -7 are square roots of 49 because both

$$7^2 = (7)(7) = 49 \quad \text{and} \quad (-7)^2 = (-7)(-7) = 49.$$

Likewise, both 3 and -3 are fourth roots of 81 because both

$$3^4 = (3)(3)(3)(3) = 81 \quad \text{and}$$
$$(-3)^4 = (-3)(-3)(-3)(-3) = 81.$$

And, also, both 2 and -2 are sixth roots of 64 because both

$$2^6 = 64 \quad \text{and} \quad (-2)^6 = 64.$$

But, in DEFINITION 20.3, $\sqrt[n]{b}$ is nonnegative. Therefore, we introduce the following notation.

> If a is negative and $a^n = b$ is positive, we write $a = -\sqrt[n]{b}$.

So, we would write, for example,

$$\sqrt{49} = 7, \qquad -\sqrt{49} = -7,$$
$$\sqrt[4]{81} = 3, \qquad -\sqrt[4]{81} = -3,$$
$$\sqrt[6]{64} = 2, \qquad -\sqrt[6]{64} = -2.$$

2. No negative number has a real nth root if n is even. For example, -4 does not have a real square root because, if b were a square root of -4, then $b^2 = -4$; however, b^2 cannot be negative. We say that an expression such as $\sqrt{-4}$ is *not a real number* or is *not real*. (See CHAPTER 21.)

3. For n odd,

$$a^n \text{ is } \begin{cases} \text{positive} & \text{if } a \text{ is positive,} \\ \text{negative} & \text{if } a \text{ is negative.} \end{cases}$$

For example, only 7 is a real cube root of 343, and only -7 is a real cube root of -343 because

$$7^3 = (7)(7)(7) = 343 \quad \text{and}$$
$$(-7)^3 = (-7)(-7)(-7) = -343.$$

Likewise, only 2 is a real fifth root of 32, and only -2 is a real fifth root of -32 because

$$2^5 = 32 \quad \text{and} \quad (-2)^5 = -32.$$

But, in DEFINITION 20.3, both b and $\sqrt[n]{b}$ are nonnegative, so strictly we would write, for example,

$$\sqrt[3]{343} = 7, \qquad -\sqrt[3]{-(-343)} = -7, \quad \text{and}$$
$$\sqrt[5]{32} = 2, \qquad -\sqrt[5]{-(-32)} = -2.$$

However, it is much more common to write for *odd* roots

$$\sqrt[3]{-343} = -7 \quad \text{instead of} \quad -\sqrt[3]{-(-343)} = -7,$$

and

$$\sqrt[5]{-32} = -2 \quad \text{instead of} \quad -\sqrt[5]{-(-32)} = -2,$$

even though this is not strictly correct by the definition.

Now You Try 20.18 Simplify the following.

1. $-\sqrt[5]{32}$ 2. $\sqrt[5]{-32}$ 3. $-\sqrt[5]{-32}$

4. $-\sqrt{400}$ 5. $\sqrt{-400}$ 6. $-\sqrt{-400}$

☐

Returning to our earlier examples, we see that by using the notion of inverses, we have:

$$(\sqrt{49})^2 = 49, \quad (\sqrt[3]{343})^3 = 343,$$
$$(\sqrt[4]{81})^4 = 81, \quad (\sqrt[5]{243})^5 = 243, \quad (\sqrt[6]{64})^6 = 64.$$

What may these suggest about Hankel's principle of permanence applied to the rules for exponents, in particular, to the rule

$$(a^m)^n = a^{mn} ?$$

Since

$$b^1 = b \quad \text{and} \quad mn = 1 \quad \text{if} \quad m = \tfrac{1}{n},$$

and both

$$(\sqrt[n]{b})^n = b \quad \text{and} \quad (b^{1/n})^n = b^{n/n} = b^1 = b,$$

to apply the principle of permanence, we must allow for an exponent to be a fraction in the following way.

Definition 20.4 Let m and n be integers, with n positive. Then, provided all the quantities are real,

$$b^{1/n} = \sqrt[n]{b}$$

and

$$b^{m/n} = (b^m)^{1/n} = \sqrt[n]{b^m}$$

or, equivalently,

$$b^{m/n} = (b^{1/n})^m = (\sqrt[n]{b})^m.$$

☐

Example 20.15

1. $\boxed{16^{1/2} = 4}$ because $16^{1/2} = \sqrt{16} = 4$ (for $4^2 = 16$).

2. $\boxed{125^{1/3} = 5}$ because $125^{1/3} = \sqrt[3]{125} = 5$ (for $5^3 = 125$).

3. $\boxed{(-32)^{1/5} = -2}$ because $(-32)^{1/5} = \sqrt[5]{-32} = -2$ (for $(-2)^5 = -32$).

4. $\boxed{64^{-1/2} = \frac{1}{8}}$ because $64^{-1/2} = \frac{1}{64^{1/2}} = \frac{1}{\sqrt{64}} = \frac{1}{8}$.

□

Now You Try 20.19 Simplify the following.

1. $169^{1/2}$

2. $81^{-1/4}$

3. $(-243)^{-1/5}$

□

Now that we have expressed roots as exponents, we have the following rules for extraction of roots that are analogous to the rules for exponents.

Rule 20.3 (Rules for extraction of roots) Let m and n be positive integers. Then, provided all the quantities are real,

	Rule	Analogous exponents rule
(1)	$\sqrt[m]{a}\sqrt[n]{a} = \sqrt[mn]{a^{m+n}}$	$a^m a^n = a^{m+n}$: $\quad a^{1/m}a^{1/n} = a^{(m+n)/(mn)}$
(2)	$\dfrac{\sqrt[m]{a}}{\sqrt[n]{a}} = \sqrt[mn]{a^{n-m}}$	$\dfrac{a^m}{a^n} = a^{m-n}$: $\quad \dfrac{a^{1/m}}{a^{1/n}} = a^{(n-m)/(mn)}$
(3)	$\dfrac{1}{\sqrt[n]{a}} = \sqrt[n]{\dfrac{1}{a}}$	$(a^m)^n = a^{mn}$: $\quad (a^{1/n})^{-1} = (a^{-1})^{1/n}$
(4)	$\sqrt[n]{ab} = \sqrt[n]{a} \cdot \sqrt[n]{b}$	$(ab)^n = a^n \cdot b^n$: $\quad (ab)^{1/n} = a^{1/n} \cdot b^{1/n}$
(5)	$\sqrt[n]{\dfrac{a}{b}} = \dfrac{\sqrt[n]{a}}{\sqrt[n]{b}}$	$\left(\dfrac{a}{b}\right)^n = \dfrac{a^n}{b^n}$: $\quad \left(\dfrac{a}{b}\right)^{1/n} = \dfrac{a^{1/n}}{b^{1/n}}$

□

Note that
$$\sqrt{a+b} \neq \sqrt{a} + \sqrt{b} \quad \text{and} \quad \sqrt{a-b} \neq \sqrt{a} - \sqrt{b}.$$
For example,
$$\sqrt{9+16} = \sqrt{25} = 5, \quad \text{but} \quad \sqrt{9} + \sqrt{16} = 3 + 4 = 7.$$

Remark 20.6 We are frequently cautioned to notice that "all quantities are real" because, otherwise, the rules for extraction of roots would imply, for example, that

$$\sqrt{16} = \sqrt{(-4)(-4)} = \underbrace{\sqrt{-4} \cdot \sqrt{-4}}_{\sqrt{-4} \text{ is not real}} = (\sqrt{-4})^2 = -4$$

when, in fact, $\sqrt{16} = 4$ (which is not negative). □

In summary, we have the following useful rule for simplifying roots.

Rule 20.4 Let n be a positive integer. Then, provided all quantities are real,

- If n is even,

$$\sqrt[n]{b^n} = |b| \quad \text{or} \quad (b^n)^{1/n} = |b|,$$

the absolute value of b.

- If n is odd,

$$\sqrt[n]{b^n} = b \quad \text{or} \quad (b^n)^{1/n} = b.$$

□

Example 20.16

$$\text{Even } n: \quad \sqrt[6]{(+3)^6} = \sqrt[6]{+729} = +3 = |+3|,$$

$$\sqrt[6]{(-3)^6} = \sqrt[6]{+729} = +3 = |-3|.$$

$$\text{Odd } n: \quad \sqrt[5]{(+3)^5} = \sqrt[5]{+243} = +3,$$

$$\sqrt[5]{(-3)^5} = \sqrt[5]{-243} = -3.$$

□

Now You Try 20.20 Simplify the following expressions using the rules for extraction of roots. Assume that all quantities are real.

1. $\sqrt[3]{x}\sqrt[3]{x}$ 2. $\dfrac{\sqrt[4]{a}}{\sqrt[5]{a}}$ 3. $\sqrt[4]{\sqrt[3]{x}\sqrt[5]{y}}$ 4. $\sqrt[6]{a^4}$

□

Example 20.17 Simplify the following expressions using the rules for exponents (page 338). Assume that all quantities are real. Give your answers using positive exponents only.

1. $x^{5/7} x^{-3/7}$

$$x^{5/7} x^{-3/7} = x^{5/7 + (-3/7)} = x^{2/7}$$

2. $\dfrac{a^{-11/2}}{a^{-5}}$

$$\frac{a^{-11/2}}{a^{-5}} = a^{-11/2 - (-5)} = a^{-1/2} = \frac{1}{a^{1/2}}$$

3. $(x^{-5/3}y^{3/2})^{-1/5}$

$$(x^{-5/3}y^{3/2})^{-1/5} = (x^{-5/3})^{-1/5}(y^{3/2})^{-1/5}$$

$$= x^{(-5/3)(-1/5)}y^{(3/2)(-1/5)}$$

$$= x^{1/3}y^{-3/10}$$

$$= \frac{x^{1/3}}{y^{3/10}}$$

4. $\left(\dfrac{a^{22}b^{-6}}{(a^{5/3}b^3)^{-2}}\right)^{-3/2}$

To begin,

$$\left(\frac{a^{22}b^{-6}}{(a^{5/3}b^3)^{-2}}\right)^{-3/2} = \frac{(a^{22}b^{-6})^{-3/2}}{((a^{5/3}b^3)^{-2})^{-3/2}}.$$

Now, we simplify the numerator and the denominator separately.

First, the numerator:

$$(a^{22}b^{-6})^{-3/2} = (a^{22})^{-3/2}(b^{-6})^{-3/2} = a^{-33}b^9.$$

Second, the denominator:

$$((a^{5/3}b^3)^{-2})^{-3/2} = (a^{5/3}b^3)^3 = (a^{5/3})^3(b^3)^3 = a^5b^9.$$

Hence,

$$\left(\frac{a^{22}b^{-6}}{(a^{5/3}b^3)^{-2}}\right)^{-3/2} = \frac{a^{-33}b^9}{a^5b^9} = a^{-33-5}b^{9-9}$$

$$= a^{-38}b^0$$

$$= \frac{b^0}{a^{38}}$$

$$= \frac{1}{a^{38}}.$$

□

Now You Try 20.21 Simplify the following expressions using the rules for exponents. Assume that all quantities are real. Give your answers using positive exponents only.

1. $x^{-5/4}x^{3/2}$ 2. $\dfrac{a^{11/3}}{a^{-5/3}}$ 3. $(x^{-5/7}y^{3/4})^{-4/5}$ 4. $\left(\dfrac{(a^{-1/2}b^6)^{-2/3}}{a^{-5/4}b^{-3/4}}\right)^3$

□

Finally, we may simplify square roots of numbers that are not perfect squares, cube roots of numbers that are not perfect cubes, and, in general, nth roots of numbers that are not nth powers in the sense demonstrated in the following example.

Example 20.18 Simplify the following radicals using the rules for extraction of roots. For each of the following, also use a calculator to approximate the root to four decimal places.

1. $\sqrt{18}$

We use the fact that $18 = 9 \cdot 2$ and that 9 is a perfect square: $9 = 3^2$:

$$\sqrt{18} = \sqrt{9 \cdot 2} = \sqrt{9} \cdot \sqrt{2} = 3\sqrt{2}.$$

To four decimal places, $\sqrt{18} \approx 4.2426$.

2. $\sqrt{75}$

In this case, $75 = 25 \cdot 3$ and 25 is a perfect square: $25 = 5^2$:

$$\sqrt{75} = \sqrt{25 \cdot 3} = \sqrt{25} \cdot \sqrt{3} = 5\sqrt{3}.$$

To four decimal places, $\sqrt{75} \approx 8.6603$.

3. $\sqrt[3]{16}$

This time we use the fact that $16 = 8 \cdot 2$ and that 8 is a perfect cube: $8 = 2^3$:

$$\sqrt[3]{16} = \sqrt[3]{8 \cdot 2} = \sqrt[3]{8} \cdot \sqrt[3]{2} = 2\sqrt[3]{2}.$$

To four decimal places, $\sqrt[3]{16} \approx 2.5198$.

4. $\sqrt[5]{320}$

For the fifth root, we seek a perfect fifth power: $320 = 32 \cdot 10$ and $32 = 2^5$. Thus,

$$\sqrt[5]{320} = \sqrt[5]{32 \cdot 10} = \sqrt[5]{32} \cdot \sqrt[5]{10} = 2\sqrt[5]{10}.$$

To four decimal places, $\sqrt[5]{320} \approx 3.1698$.

□

Now You Try 20.22 Simplify the following radicals using the rules for extraction of roots. For each of the following, also use a calculator to approximate the root to four decimal places.

1. $\sqrt{12}$ 　　　　　　2. $\sqrt{32}$ 　　　　　　3. $\sqrt[3]{32}$ 　　　　　　4. $\sqrt[4]{32}$

□

Now You Try 20.23 Solve the following equations using the quadratic formula (RULE 20.1 on page 330). If the discriminant is not a perfect square, leave the solution with a *simplified* radical (this would be an *exact solution*), and also use a calculator to approximate the solution to four decimal places.

1. $x^2 + 2x - 2 = 0$ 　　　　　　　　　　　　2. $2x^2 = \sqrt{10}x + 5$

□

The story of our use of the symbol $\sqrt[n]{\ }$ and $\sqrt{\ }$ for nth root is fascinating, but long, and so we mention only a few highlights that we glean from Cajori [31, pp. 360–379], who tells the greater story.

To begin, we find the use of the symbol $\ulcorner$ for square root in two Egyptian papyri that were found at Kahun. In India, Brahmagupta and others who followed him used the abbreviation c (*caraní*) for surd or square root:

$$c\ 18\ c\ 3 \quad \text{for} \quad \sqrt{18} + \sqrt{3}.$$

Cajori identifies four principal symbols that were used to denote roots in Europe since the twelfth century: R (*radix*), l (*latus*), $\sqrt{\ }$, and (most recently) fractional exponents. We discuss each of these symbols briefly.

According to Cajori [31, p. 361], when an Arabic copy of Euclid's *Elements* [53] was translated into Latin, the word *radix* was used for "square root," and this led to the use of the symbol R to denote "root"; however, in some other translations of Arabic into Latin, *radix* was used to denote an unknown, x, and so the symbol R was also used to denote x, albeit less commonly. One manner in which R was used to denote roots is found in Luca Pacioli's *Summa*. In fact, as you recall our mentioning, Pacioli uses R five different ways, one of which is to denote powers. Here are some of the many examples Cajori gives that illustrates the use of R in the *Summa* to denote roots.

$$R\ .200. \quad \text{for} \quad \sqrt{200},$$

$$R\ .cuba.\ de\ .64. \quad \text{for} \quad \sqrt[3]{64},$$

$$RR\ .120. \quad \text{for} \quad \sqrt[4]{120}.$$

And Pacioli writes, for example,

$$R\ .6.\tilde{m}.R.2. \quad \text{for} \quad \sqrt{6} - \sqrt{2}, \quad \text{and}$$

$$Rv.\ R.20\tfrac{1}{4}.\tilde{m}.\tfrac{1}{2}. \quad \text{for} \quad \sqrt{\sqrt{20\tfrac{1}{4}} - \tfrac{1}{2}},$$

where $Rv.$ denotes the square root of two or more terms.

The Latin *latus* means "side of a square" and was used to indicate "root" by the Roman Junius Nipsus of the second century A.D. [31, p. 364]. Of the several people Cajori mentions who used the symbol l for root is Peter Ramus (1515–1572).[16] As an example, Ramus writes

$$l\ 27\ ad\ l\ 12 \quad \text{for} \quad \sqrt{27} + \sqrt{12},$$

$$ll\ 32\ de\ ll\ 162 \quad \text{for} \quad \sqrt[4]{32} \text{ from } \sqrt[4]{162}, \quad \text{and}$$

$$lr.\ l112 - l76 \quad \text{for} \quad \sqrt{\sqrt{112} - \sqrt{76}},$$

where, Cajori states, "the r signifying here *residua*, or 'remainder,' and therefore $lr.$ signified the square root of the binomial difference." Cajori remarks that the use of the symbol l for calculating with roots never became popular.

Cajori [31, pp. 366–379] gives a detailed account of the symbol $\sqrt{\ }$, from which we mention here only a few points.

The symbol $\sqrt{\ }$ originated in Germany and is ultimately credited to Christoff Rudolff (ca. 1500–1545). We find the symbol in Rudolff's *Behend vnnd Hubsch Rechnung durch die kunstreichen regeln Algebre so gemeinlicklich die Coss genent werden* (Strassburg, 1525) [31, pp. 133–136], or *Coss* for short. Many, including the esteemed Leonhard Euler (1707–1783), have been of the opinion that $\sqrt{\ }$ derives from a deformed letter r for *radix*; in fact, Perez de Moya (1513–1597) of Spain uses the letter r for square root,

---

[16] *P. Rami Scholarvm mathematicarvm libri unus et triginti* (Basel, 1569), Lib. XXIV, p. 276, 277.

rr for fourth root (square root of the square root), and *rrr* for cube root. However, Cajori informs us that some German manuscript algebras, as well as the first printed algebras, suggest that the symbol $\sqrt{}$ derives from a dot. In support of this assertion, Cajori mentions that, in a volume of manuscripts housed in the Dresden Library, a manuscript written in Latin[17] specifies that one dot (.) be placed before the radicand to denote a square root, two dots (..) to denote a fourth root, three dots (...) to denote a cube root, and four dots (....) to denote a ninth root.[18] Unfortunately, no satisfactory explanation for this choice of notation has been discovered. Nevertheless, we may connect the dots to the codex marked *Codex Gotting. Philos. 30* that is kept at the University of Göttingen, which, according to Cajori, contains a letter written in Latin by Initius Algebras[19] probably before 1524. In there we find the symbol $\int$, a heavy point or dot with an upward stroke to the right, used for root:

$$\int_3 \quad \text{for square root,} \qquad \int c^e \quad \text{for cube root,}$$

and so on. Cajori gives the example

$$\int cs \overline{|8 + \int 22_3} \quad \text{for} \quad \sqrt{8 + \sqrt{22}},$$

where *cs* stands for *communis* and signals that we mean the root of the quantity marked by the gnomon.

So, did the symbol $\sqrt{}$ for root derive from a deformed letter *r* or from a dot? To help us form an opinion, Cajori [31, pp. 367–369] notes the following.

- Rudolff was familiar with the manuscript Vienna MS No. 5277, *Regule-Cose–uel Algobre–*, which states (translated from the Latin): "When $x^2 = \sqrt{x}$, erase the point before the *x* and multiply x^2 by itself, then things equal to each other are obtained." In other words, if we square both sides of the equation

$$x^2 = \sqrt{x}$$

we obtain the equation

$$x^4 = x,$$

in which we see that the square root has been "erased." Also, it is stated elsewhere in the manuscript: "by a point understand a root." Thus, Rudolff was aware of a precedent use of a point to mean the square root. (Interestingly, although the use of a point to mean the square root is described in the manuscript, Cajori reports that a point is never actually found used in this way in the manuscript.)

- In his work, the *Coss*, Rudolff refers to the use of a *Punkt* (point) with regard to a root, but he uses, as Cajori describes it, "a mark with a very short heavy downward stroke (almost a point), followed by a straight line or stroke, slanting upward...."

- Johannes Widmann, in his work, *Behende vnnd hübsche Rechnũg auff allen Kauffmanschafften*, writes R and *ra* for root even though he is familiar of the use of a point for root from the Dresden Library manuscript.

---

[17]M. Cantor, *Vorles. über Geschichte der Mathematik*, Vol. II (2. Aufl., 1900), p. 241.

[18]If you do not see the logic in the progression of the number of dots, then you stand in the company of Cajori, who opines [31, p. 366]:

Evidently this notation is not a happy choice. If one dot meant square root and two dots meant square root of square root (i.e., $\sqrt{\sqrt{}}$), then three dots should mean square root of square root of square root, or eighth root.

[19]*Initius Algebras: Algebrae Arabis Arithmetici viri clarrisimi Liber ad Ylem geometram magistrum suum.* Cajori [31, p. 367] states that this was published by M. Curtze in *Abhandlungen zur Geschichte der mahematischen Wissenschaften*, Heft XIII (1902), pp. 435–611, and that notations are explained in the introduction, pp. 443–448. Curtze's publication in German is available online at <http://books.google.com/books?id=o4NsAAAAMAAJ>. See also the review by David Eugene Smith that appears in the *Bulletin of the American Mathematical Society*, Volume 9, Number 7 (1903), <http://www.ams.org/bull/1903-09-07/>.

Where do these observations leave us? To tie the ends together, Cajori quotes a remark of H. Wieleitner:[20] "The dot appears at times in manuscripts as an abbreviation for the syllable *ra*." So, it seems that
the dot may have stood for "*ra*," which had been used for root, and that may have led to Rudolff's deformed
letter *r* ($\sqrt{}$) for root. However, Wieleitner continues in his remark: "Whether the dot used in the Dresden
manuscript represents this normal abbreviation for *radix* does not appear to have been specially examined."
Alas, then, the question of the origin of the symbol $\sqrt{}$ is not settled.

Nevertheless, the use of Rudolff's basic symbol $\sqrt{}$ for root spread. The convention for marking the
index of the root, however, remained in flux. Our modern convention of writing the index in the opening
of the symbol $\sqrt{}$ was suggested by Albert Girard (1595–1632) around 1629—his writing $\sqrt[3]{}$, for instance,
although he also writes $\sqrt{}\sqrt{}$ for $\sqrt[4]{}$—but it was not until the eighteenth century that this convention gradually
became widely accepted. In the intervening time, some of the different conventions used include

- Simon Stevin in 1585:

$$\sqrt{} \quad \text{for square root,} \quad \sqrt{}\sqrt{} \quad \text{for fourth root,} \quad \sqrt{}\sqrt{}\sqrt{} \quad \text{for eighth root,}$$
$$\sqrt{}③ \quad \text{for cube root,} \quad \mathcal{W}③ \quad \text{for ninth root.}$$

- Christophorus Dibuadius in 1605:

$$\sqrt{},\ \sqrt{}Q,\ \sqrt{}_3 \quad \text{for square root,}$$
$$\sqrt{}C,\ \sqrt{}c,\ \sqrt{}c^e \quad \text{for cube root,}$$
$$\sqrt{}\sqrt{},\ \sqrt{}QQ,\ \sqrt{}_{33} \quad \text{for fourth root.}$$

- William Oughtred in 1631:

$$\sqrt{}qu \quad \text{for square root,} \quad \sqrt{}qq \quad \text{for fourth root,} \quad \sqrt{}qe \quad \text{for fifth root,}$$
$$\sqrt{}c \quad \text{for cube root,} \quad \sqrt{}cc \quad \text{for sixth root,} \quad \sqrt{}ccc \quad \text{for ninth root,}$$
$$\sqrt{}cccc,\ \sqrt{}[12],\ \sqrt{}\boxed{12} \quad \text{for twelfth root.}$$

- John Wallis in 1655:

$$\sqrt{},\ \sqrt{}^2 \quad \text{for square root,} \quad \sqrt{}^3 \quad \text{for cube root,} \quad \sqrt{}^4 \quad \text{for fourth root.}$$

Last, Descartes promoted the use of a vinculum (horizontal bar) for aggregation or grouping, for
example,
$$\sqrt{aa + bb},$$
while Leibniz promoted the use of brackets because of easier typesetting, for example,
$$\sqrt{}(aa + bb).$$

Both of these conventions are commonly used today, although, as usual, a variety of conventions was used
before we settled on these two.

---

[20]H. Wieleitner, *Die Sieben Rechnungsarten* (Leipzig-Berlin, 1912), p. 49.

The use of fractions to denote roots was first suggested by Nicole Oresme (1323–1349), a bishop in Normandy, in his manuscript *Algorismus proportionum* (*Algorithm of Ratios*) [31, pp. 91–93], although the notation he uses does not resemble our exponential notation. Cajori gives the example

$$\boxed{\begin{array}{cc} 1 & . & p \\ 2 & . & 2 \end{array}} \quad \text{for } 2^{\frac{1}{2}},$$

and that he reads "*medietas [proportionis] duplae*." To make sense of Oresme's notation, we first need to understand how he expresses ratios. Oresme writes, for example,

$$\boxed{\begin{array}{c} 1 \\ 3 \end{array}} \quad \text{for } \tfrac{1}{3} \text{ and } \quad \boxed{\begin{array}{ccc} p & . & 1 \\ 1 & . & 3 \end{array}} \quad \text{for } 1\tfrac{1}{3}.$$

So, in his notation for $2^{\frac{1}{2}}$, $\frac{p}{2}$ represents the base 2, and the "exponent" $\frac{1}{2}$ is written to its left. To be sure, Cajori provides another example:

$$\boxed{\begin{array}{ccccc} 1 & . & p & . & 1 \\ 4 & . & 2 & . & 2 \end{array}}$$

for $(2\frac{1}{2})^{\frac{1}{4}}$, and that he reads "*quarta pars [proportionis] duplae sesquialterae*." But, as we learn from Cajori:

> Oresme expresses in words, "…. proponatur proportio, que sit due tertia quadruple; et quia duo est numerator, ipsa erit vna tertia quadruple duplicate, seu sedecuple," i.e., $4^{\frac{2}{3}} = (4^2)^{\frac{1}{3}} = 16^{\frac{1}{3}}$. Oresme writes also: "Sequitur quod .*a*. moueatur velocius .*b*. in proportione, que est medietas proportionis .50. ad .49.," which means, "the velocity of a : velocity of $b = \sqrt{50} : \sqrt{49}$," the word *medietas* meaning "square root."

Thus, we see that Oresme does not use his "fractional exponent" notation for roots in his computations.

On page 6 in his *Œuvres* (*Arithmetic*), Simon Stevin instructs [31, p. 158]: "$\frac{3}{2}$ en un circle seroit le charactere de racine quarrée de ③, par ce que telle $\frac{3}{2}$ en circle multipliée en soy donne produict ③ et ainsi des autres," in other words, $\frac{3}{2}$ written in a circle is to stand for $x^{3/2}$.

As a last example, Cajori cites the following from page 332 of John Wallis's *Treatise of algebra* [31, p. 217]:

$$\overline{c^5 + c^4x - x^5}\Big|^{\frac{1}{5}} \quad \text{for} \quad \sqrt[5]{c^5 + c^4x - x^5}.$$

To bring this to a close, recall that it was Descartes who brought us our modern exponential notation: a noteworthy achievement. On the contrary, Cajori laments Descartes's contribution toward notation for roots [31, p. 375]:

> Great as were Descartes' services toward perfecting algebraic notation, he missed a splendid opportunity of rendering a still greater service. Before him Oresme and Stevin had advanced the concept of fractional as well as integral exponents. If Descartes, instead of extending the application of the radical sign $\sqrt{}$ by adding to it the vinculum, had discarded the radical sign altogether and had introduced the notation for fractional as well as integral exponents, then it is conceivable that the further use of radical signs would have been discouraged and checked; it is conceivable that the unnecessary duplication in notation, as illustrated by $b^{\frac{3}{4}}$ and $\sqrt[4]{b^3}$, would have been avoided; it is conceivable that generations upon generations of pupils would have been saved the necessity of mastering the operations with two difficult notations when one alone (the exponential) would have answered all purposes. But Descartes missed this opportunity, as did later also I. Newton who introduced the notation of the fractional exponent, yet retained and used radicals.

20.3   CONTINUED FRACTIONS

As the name implies, *continued fractions* are fractions that have some sort of ongoing or continued structure: a continued fraction is an expression that has a fraction inside a fraction inside a fraction, and so on, so to speak. For example, the sequence of numbers $\phi_0, \phi_1, \phi_2, \phi_3, \ldots$, given by

$$\phi_0 = 1,$$

$$\phi_1 = 1 + \frac{1}{1},$$

$$\phi_2 = 1 + \cfrac{1}{1 + \cfrac{1}{1}},$$

$$\phi_3 = 1 + \cfrac{1}{1 + \cfrac{1}{1 + \cfrac{1}{1}}},$$

$$\phi_4 = 1 + \cfrac{1}{1 + \cfrac{1}{1 + \cfrac{1}{1 + \cfrac{1}{1}}}},$$

and so on, is a sequence of continued fractions.

Definition 20.5 An expression of the form

$$a_1 + \cfrac{b_1}{a_2 + \cfrac{b_2}{a_3 + \cdots}},$$

is called a *continued fraction*. If all the *b* equal 1, then the expression

$$a_1 + \cfrac{1}{a_2 + \cfrac{1}{a_3 + \cdots}},$$

is called a *simple continued fraction*. A continued fraction may be "finite" (terminates) or "infinite" (does not terminate).   □

Continued fractions (or their equivalent before our modern style of writing fractions evolved) have been found in writings of the Greeks and Indians. They became more prevalent in Western Europe in the late medieval and Renaissance periods, finding full fruition with the work of Leonhard Euler (1707–1783) in the eighteenth century. We learn from Olds [91] that the modern theory of continued fractions can be traced back to Rafael Bombelli (1526–1572). Olds attributes the earliest important step in the conception of continued fractions to Euclid (ca. 323–285 BC), for he notes that Euclid's algorithm in essence converts a common fraction into a continued fraction. Prior to this, Olds says that "the idea of a continued fraction are somewhat confused, for many ancient arithmetical results are suggestive of these fractions, but there was no systematic development of the subject."

The first systematic work done on continued fractions is that of John Wallis (1616–1703). For example, he expressed

$$\frac{4}{\pi} = \frac{3 \times 3 \times 5 \times 5 \times 7 \times 7 \times 9 \times \cdots}{2 \times 4 \times 4 \times 6 \times 6 \times 8 \times 9 \times \cdots}$$

that Lord Brouncker (1620–1684), the first president of the Royal Society, reworked into a true continued fraction

$$\frac{4}{\pi} = 1 + \cfrac{1^2}{2 + \cfrac{3^2}{2 + \cfrac{5^2}{2 + \cfrac{7^2}{2 + \cdots}}}};$$

From this beginning, great mathematicians such as Euler, Lambert (1728–1777), Lagrange (1736–1823), and many others developed the theory as we know it today. In particular, Euler's great memoir, *De Fractionibus Continuis* (1737), laid the foundation for the modern theory.

Continued fractions play an important role in present-day mathematics. They constitute an important tool for new discoveries in the theory of numbers and in the field of Diophantine approximations. There is the important generalization of continued fractions called the analytic theory of continued fractions, an extensive area for present and future research. In the computer field, continued fractions are used to give approximations to various complicated functions to give rapid numerical results valuable to scientists and to those working in applied mathematical fields.

In this section, we consider only simple continued fractions. Returning to our introductory sequence, simplifying each continued fraction, we see that

$$\phi_0 = 1 = \frac{1}{1},$$

$$\phi_1 = 1 + \frac{1}{\phi_0} = \frac{2}{1},$$

$$\phi_2 = 1 + \frac{1}{\phi_1} = 1 + \frac{1}{2} = \frac{3}{2},$$

$$\phi_3 = 1 + \frac{1}{\phi_2} = 1 + \frac{2}{3} = \frac{5}{3},$$

$$\phi_4 = 1 + \frac{1}{\phi_3} = 1 + \frac{3}{5} = \frac{8}{5},$$

and so on.

Now You Try 20.24 Write the continued fractions for the next numbers $\phi_5, \phi_6, \ldots, \phi_{10}$ of the sequence given above, and simplify each expression. □

Let us consider the decimal fraction representations of the numbers $\phi_0, \phi_1, \ldots, \phi_{10}$:

$$\phi_0 = 1, \qquad \phi_1 = 2, \qquad \phi_2 = \frac{3}{2} = 1.5,$$

$$\phi_3 = \frac{5}{3} = 1.\overline{6}, \qquad \phi_4 = \frac{8}{5} = 1.6, \qquad \phi_5 = \frac{13}{8} = 1.625,$$

$$\phi_6 = \frac{21}{13} = 1.\overline{615\,384},$$

$$\phi_7 = \frac{34}{21} = 1.\overline{619\,0476},$$

$$\phi_8 = \frac{55}{34} = 1.6\overline{17\,647\,058\,823\,529\,41},$$

$$\phi_9 = \frac{89}{55} = 1.6\overline{18},$$

$$\phi_{10} = \frac{144}{89} = 1.617\,977\,528\,0\overline{89}.$$

We plot these values below.

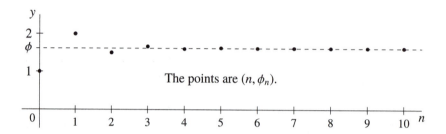

The points are (n, ϕ_n).

Observe that the values $\phi_0, \phi_1, \ldots, \phi_{10}$ appear to be settling down upon some value ϕ. What is this value?

To begin, notice that the numbers $\phi_0, \phi_1, \phi_2, \ldots$ become more like the infinite continued fraction

$$\phi = 1 + \cfrac{1}{1 + \cfrac{1}{1 + \cfrac{1}{1 + \cfrac{1}{1 + \cfrac{1}{1 + \cdots}}}}}$$

as we progress through the sequence. Next observe that ϕ is "self-similar" in the sense that the expression in the dashed box shown below is exactly ϕ itself:

$$\phi = 1 + \cfrac{1}{1 + \cfrac{1}{1 + \cfrac{1}{1 + \cfrac{1}{1 + \cfrac{1}{1 + \cdots}}}}}$$

Thus, we see that

$$\phi = 1 + \frac{1}{\phi}.$$

To solve this equation for ϕ, we clear the denominators by multiplying by ϕ to obtain the equation

$$\phi^2 = \phi + 1,$$

then apply the quadratic formula to obtain the solutions

$$\phi_+ = \frac{1 + \sqrt{5}}{2} \quad \text{and} \quad \phi_- = \frac{1 - \sqrt{5}}{2}.$$

Hence, because $\phi > 0$, we conclude that

$$\phi = \frac{1 + \sqrt{5}}{2},$$

that is,

$$1 + \cfrac{1}{1 + \cfrac{1}{1 + \cfrac{1}{1 + \cfrac{1}{1 + \cfrac{1}{1 + \cdots}}}}}$$

$$= \frac{1 + \sqrt{5}}{2} = 1.618\,033\,988\,749\ldots.$$

Think About It 20.7 Where have we seen the number $\frac{1+\sqrt{5}}{2}$ before? □

20.3.1 Finite Simple Continued Fractions

We shall return to the sequence of numbers $\phi_0, \phi_1, \phi_2, \ldots$, as well as the "limit" ϕ of the sequence, at the end of this section to explore some unexpected and interesting mathematical connections. For now, we turn our attention to *finite* simple continued fractions. We begin with an example.

Example 20.19 The continued fraction

$$2 + \cfrac{1}{1 + \cfrac{1}{3 + \cfrac{1}{3}}}$$

$$= 2 + \cfrac{1}{1 + \cfrac{1}{\frac{10}{3}}}$$

$$= 2 + \cfrac{1}{1 + \cfrac{3}{10}}$$

$$= 2 + \cfrac{1}{\frac{13}{10}}$$

$$= 2 + \frac{10}{13}$$

$$= 2\tfrac{10}{13} \quad \text{or} \quad \frac{36}{13}.$$

To express the common fraction $\frac{36}{13}$ as a continued fraction, we would reverse the steps. □

Example 20.20 We express $\frac{9}{24}$ as a continued fraction. Note that in each step, the remaining fraction is less than 1.

$$\frac{9}{24} = \frac{1}{\frac{24}{9}} = \frac{1}{2 + \frac{6}{9}} = \frac{1}{2 + \frac{2}{3}} = \frac{1}{2 + \frac{1}{\frac{3}{2}}} = \frac{1}{2 + \cfrac{1}{1 + \cfrac{1}{2}}}$$

Therefore,

$$\frac{9}{24} = \frac{1}{2 + \cfrac{1}{1 + \cfrac{1}{2}}}.$$

Now You Try 20.25 Express the following as common fractions or mixed numerals.

1. $2 + \cfrac{1}{7 + \cfrac{1}{1 + \cfrac{1}{8}}}$

2. $\cfrac{1}{3 + \cfrac{1}{1 + \cfrac{1}{4 + \cfrac{1}{2}}}}$

Now You Try 20.26 Express the following as finite simple continued fractions.

1. $\dfrac{22}{7}$

2. $1.414 = 1\frac{414}{1000}$

Think About It 20.8 How is the process of expressing a common fraction as a finite simple continued fraction related to Euclid's algorithm?

20.3.2   Infinite Simple Continued Fractions

We restrict our discussion to only those infinite simple continued fractions that are *periodic*, that is, have a repeating pattern: either *purely periodic* or *periodic after some point*. For example, the continued fractions

$$1 + \cfrac{1}{1 + \cfrac{1}{1 + \cfrac{1}{1 + \cfrac{1}{1 + \cfrac{1}{1 + \cdots}}}}} \qquad \text{and} \qquad 1 + \cfrac{1}{2 + \cfrac{1}{1 + \cfrac{1}{1 + \cfrac{1}{2 + \cfrac{1}{1 + \cdots}}}}}$$

are purely periodic, and the continued fractions

$$1 + \cfrac{1}{2 + \cfrac{1}{2 + \cfrac{1}{2 + \cfrac{1}{2 + \cfrac{1}{2 + \cdots}}}}} \qquad \text{and} \qquad 2 + \cfrac{1}{7 + \cfrac{1}{1 + \cfrac{1}{8 + \cfrac{1}{1 + \cfrac{1}{8 + \cdots}}}}}$$

are periodic after some point.

We have already seen that

$$\phi = 1 + \cfrac{1}{1 + \cfrac{1}{1 + \cfrac{1}{1 + \cfrac{1}{1 + \cfrac{1}{1 + \cdots}}}}}$$

$$= \frac{1 + \sqrt{5}}{2}$$

by solving the equation

$$\phi = 1 + \frac{1}{\phi}$$

or, after clearing denominators, the equation

$$\phi^2 = \phi + 1.$$

Let us try a similar approach to determine the value, if it exists, of the continued fraction

$$1 + \cfrac{1}{2 + \cfrac{1}{2 + \cfrac{1}{2 + \cfrac{1}{2 + \cfrac{1}{2 + \cdots}}}}}.$$

Example 20.21 Let

$$\alpha = 1 + \cfrac{1}{2 + \cfrac{1}{2 + \cfrac{1}{2 + \cfrac{1}{2 + \cdots}}}} = 1 + \cfrac{1}{1 + 1 + \cfrac{1}{2 + \cfrac{1}{2 + \cfrac{1}{2 + \cdots}}}}$$

$$= 1 + \frac{1}{1 + \alpha}.$$

Clearing denominators in the equation

$$\alpha = 1 + \frac{1}{1 + \alpha},$$

we obtain

$$\alpha + \alpha^2 = 1 + \alpha + 1.$$

After simplifying and using the quadratic formula, we find that the solutions of this equation are

$$\alpha_+ = \sqrt{2} \quad \text{and} \quad \alpha_- = -\sqrt{2}.$$

Hence, because $\alpha > 0$, we conclude that

$$\alpha = \sqrt{2},$$

that is,

$$1 + \cfrac{1}{2 + \cfrac{1}{2 + \cfrac{1}{2 + \cfrac{1}{2 + \cfrac{1}{2 + \cdots}}}}} = \sqrt{2}.$$

□

Example 20.22 Express $\sqrt{5}$ as a simple continued fraction.

To begin, we note that $\sqrt{5}$ is between 2 and 3 (because $2^2 = 4$ and $3^2 = 9$), so that $\sqrt{5} = 2 +$ something small. Let

$$\sqrt{5} = 2 + \frac{1}{x},$$

where $\frac{1}{x}$ is the "something small." Then,

$$\sqrt{5} - 2 = \frac{1}{x}, \quad \text{so that} \quad x = \frac{1}{\sqrt{5} - 2} = -\frac{1}{2 - \sqrt{5}}.$$

We "rationalize the denominator" to find that

$$x = -\frac{1}{2 - \sqrt{5}} = -\frac{1}{2 - \sqrt{5}} \cdot \frac{2 + \sqrt{5}}{2 + \sqrt{5}} = -\frac{2 + \sqrt{5}}{4 - 5} = 2 + \sqrt{5},$$

rationalize the denominator

so that

$$\sqrt{5} = 2 + \frac{1}{x} = 2 + \frac{1}{2 + \sqrt{5}}.$$

Thus, by repeated substitution of $\sqrt{5}$, we find that

$$\sqrt{5} = 2 + \cfrac{1}{2 + \sqrt{5}}$$

$$= 2 + \cfrac{1}{2 + \left(2 + \cfrac{1}{2 + \sqrt{5}}\right)}$$

$$= 2 + \cfrac{1}{2 + 2 + \cfrac{1}{2 + \left(2 + \cfrac{1}{2 + \sqrt{5}}\right)}}$$

$$= 2 + \cfrac{1}{2 + 2 + \cfrac{1}{2 + 2 + \cfrac{1}{2 + \left(2 + \cfrac{1}{2 + \cdots}\right)}}}$$

$$= 2 + \cfrac{1}{4 + \cfrac{1}{4 + \cfrac{1}{4 + \cdots}}}.$$

To check, we determine the value of the continued fraction. Toward this end, let

$$\alpha = 2 + \cfrac{1}{4 + \cfrac{1}{4 + \cfrac{1}{4 + \cdots}}} = 2 + \cfrac{1}{2 + 2 + \cfrac{1}{4 + \cfrac{1}{4 + \cdots}}}$$

$$= 2 + \cfrac{1}{2 + \alpha}.$$

Clearing denominators in the equation

$$\alpha = 2 + \cfrac{1}{2 + \alpha},$$

we obtain

$$2\alpha + \alpha^2 = 4 + 2\alpha + 1.$$

After rearranging the equation and using the quadratic formula, we find that the solutions of this equation are

$$\alpha_+ = \sqrt{5} \quad \text{and} \quad \alpha_- = -\sqrt{5}.$$

Hence, because $\alpha > 0$, we conclude that

$$\alpha = \sqrt{5},$$

that is,

$$2 + \cfrac{1}{4 + \cfrac{1}{4 + \cfrac{1}{4 + \cdots}}}$$

$$= \sqrt{5}.$$

□

An important step in expressing $\sqrt{5}$ as a simple continued fraction was to "rationalize the denominator" in the expression

$$\frac{1}{2 - \sqrt{5}}.$$

We did this by using the fact that

$$(a - b)(a + b) = a^2 - b^2,$$

the difference of squares factoring, so that

$$(2 - \sqrt{5})(2 + \sqrt{5}) = 2^2 - (\sqrt{5})^2 = 4 - 5 = -1,$$

which is no longer irrational (it has been "rationalized"). Thus,

$$-\frac{1}{2 - \sqrt{5}} \cdot \frac{2 + \sqrt{5}}{2 + \sqrt{5}} = -\frac{2 + \sqrt{5}}{4 - 5} = -\frac{2 + \sqrt{5}}{-1} = 2 + \sqrt{5}$$

As another example, we rationalize the denominator in the expression

$$\frac{1}{3 + \sqrt{15}}.$$

Again, using the fact that $(a - b)(a + b) = a^2 - b^2$, we want to multiply the irrational denominator $3 + \sqrt{15}$ by $3 - \sqrt{15}$, for

$$(3 + \sqrt{15})(3 - \sqrt{15}) = 3^2 - (\sqrt{15})^2 = 9 - 15 = -6,$$

which is no longer irrational. Thus,

$$\frac{1}{3 + \sqrt{15}} = \frac{1}{3 + \sqrt{15}} \cdot \frac{3 - \sqrt{15}}{3 - \sqrt{15}} = \frac{3 - \sqrt{15}}{9 - 15} = -\frac{3 - \sqrt{15}}{6}.$$

rationalize the denominator

Definition 20.6 Let p, s, and q be real numbers with $q \geq 0$. The *conjugate* of $p + s\sqrt{q}$ is the number $p - s\sqrt{q}$. We say that $p + s\sqrt{q}$ and $p - s\sqrt{q}$ are a *conjugate pair*.   □

Thus, the expressions $3 + \sqrt{15}$ and $3 - \sqrt{15}$ are a conjugate pair: $3 - \sqrt{15}$ is the conjugate of $3 + \sqrt{15}$, and also $3 + \sqrt{15}$ is the conjugate of $3 - \sqrt{15}$.

Now You Try 20.27 Rationalize the denominators in the following.

1. $\dfrac{1}{3 - \sqrt{10}}$

2. $\dfrac{1}{2 + \sqrt{6}}$

3. $\dfrac{1}{-1 - \sqrt{5}}$

□

Example 20.23 We return to the expression for ϕ. Express $\frac{1+\sqrt{5}}{2}$ as a simple continued fraction.

To begin, we note that $\sqrt{5}$ is between 2 and 3 (because $2^2 = 4$ and $3^2 = 9$), so that

$$\frac{1 + 2}{2} < \frac{1 + \sqrt{5}}{2} < \frac{1 + 3}{2} \quad \text{or} \quad \tfrac{3}{2} < \frac{1 + \sqrt{5}}{2} < 2.$$

Consequently, $\frac{1+\sqrt{5}}{2} = 1 + \text{something small}$. Let

$$\frac{1 + \sqrt{5}}{2} = 1 + \frac{1}{x},$$

where $\frac{1}{x}$ is the "something small." Then,

$$\frac{1 + \sqrt{5}}{2} - 1 = \frac{1}{x} \quad \text{or} \quad \frac{-1 + \sqrt{5}}{2} = \frac{1}{x}, \quad \text{so that} \quad x = \frac{2}{-1 + \sqrt{5}}.$$

We rationalize the denominator using the conjugate of $-1 + \sqrt{5}$ to find that

$$x = \underbrace{\frac{2}{-1 + \sqrt{5}} = \frac{2}{-1 + \sqrt{5}} \cdot \frac{-1 - \sqrt{5}}{-1 - \sqrt{5}}}_{\text{rationalize the denominator}} = \frac{2(-1 - \sqrt{5})}{1 - 5} = \frac{-1 - \sqrt{5}}{-2} \quad \text{or} \quad \frac{1 + \sqrt{5}}{2},$$

so that

$$\frac{1 + \sqrt{5}}{2} = 1 + \frac{1}{x} = 1 + \frac{1}{\frac{1+\sqrt{5}}{2}}.$$

By repeated substitution of $\frac{1+\sqrt{5}}{2}$, we find that

$$\frac{1 + \sqrt{5}}{2} = 1 + \frac{1}{\frac{1+\sqrt{5}}{2}}$$

$$= 1 + \frac{1}{1 + \frac{1}{\frac{1+\sqrt{5}}{2}}}$$

$$= 1 + \frac{1}{1 + \frac{1}{1 + \frac{1}{\frac{1+\sqrt{5}}{2}}}}$$

$$= 1 + \frac{1}{1 + \frac{1}{1 + \frac{1}{1 + \frac{1}{1 + \cdots}}}}.$$

$\square$

Which is exactly as we expected.

Now You Try 20.28 Express the following as simple continued fractions. Check by determining the values of the continued fractions.

1. $\sqrt{10}$ 2. $\sqrt{6}$

$\square$

We now look at continued fractions that are periodic after some point.

Example 20.24 To find the value of the continued fraction

$$2 + \cfrac{1}{7 + \cfrac{1}{1 + \cfrac{1}{8 + \cfrac{1}{1 + \cfrac{1}{8 + \cdots}}}}},$$

if it exists, let

$$\alpha = 2 + \cfrac{1}{7 + \cfrac{1}{\beta}},$$

where

$$\beta = 1 + \cfrac{1}{8 + \cfrac{1}{1 + \cfrac{1}{8 + \cdots}}}$$

$$= 1 + \cfrac{1}{8 + \cfrac{1}{\beta}}.$$

Clearing denominators in the equation

$$\beta = 1 + \cfrac{1}{8 + \cfrac{1}{\beta}},$$

we obtain

$$8\beta^2 + \beta = 8\beta + 1 + \beta \quad \text{or} \quad 8\beta^2 - 8\beta - 1 = 0.$$

(Check it.) Using the quadratic formula, we find that the solutions are

$$\beta_+ = \frac{8 + \sqrt{96}}{16} \quad \text{and} \quad \beta_- = \frac{8 - \sqrt{96}}{16}.$$

Because $\beta > 0$, we conclude that

$$\beta = \frac{8 + \sqrt{96}}{16} = \frac{2 + \sqrt{6}}{4}.$$

Thus,

$$\alpha = 2 + \cfrac{1}{7 + \cfrac{1}{\beta}} = 2 + \cfrac{1}{7 + \cfrac{1}{\dfrac{2 + \sqrt{6}}{4}}} = 2 + \cfrac{1}{7 + \dfrac{4}{2 + \sqrt{6}}}$$

$$= 2 + \cfrac{1}{\dfrac{18 + 7\sqrt{6}}{2 + \sqrt{6}}}$$

$$= 2 + \frac{2 + \sqrt{6}}{18 + 7\sqrt{6}}$$

$$= \frac{38 + 15\sqrt{6}}{18 + 7\sqrt{6}}.$$

(Check it.) Finally, we rationalize the denominator using the conjugate of $18 + 7\sqrt{6}$ to find that

$$\alpha = \frac{38 + 15\sqrt{6}}{18 + 7\sqrt{6}} = \frac{38 + 15\sqrt{6}}{18 + 7\sqrt{6}} \cdot \frac{18 - 7\sqrt{6}}{18 - 7\sqrt{6}}$$

rationalize the denominator

$$= \frac{684 + 270\sqrt{6} - 266\sqrt{6} - 105(6)}{324 - 49(6)}$$

$$= \frac{54 + 4\sqrt{6}}{30}$$

$$= \frac{27 + 2\sqrt{6}}{15},$$

that is,

$$2 + \cfrac{1}{7 + \cfrac{1}{1 + \cfrac{1}{8 + \cfrac{1}{1 + \cfrac{1}{8 + \cdots}}}}}$$

$$= \frac{27 + 2\sqrt{6}}{15}.$$

□

Now You Try 20.29 Find the values of the following continued fractions if they exist.

1. $1 + \cfrac{1}{2 + \cfrac{1}{1 + \cfrac{1}{1 + \cfrac{1}{2 + \cfrac{1}{1 + \cdots}}}}}$

2. $1 + \cfrac{1}{1 + \cfrac{1}{2 + \cfrac{1}{1 + \cfrac{1}{2 + \cfrac{1}{1 + \cdots}}}}}$

□

Think About It 20.9 In the examples above, we said to "find the value if it exists." When or why would the value not exist?

□

We end this subsection with a subtlety about "infinite sums" for which we have so far turned a blind eye, namely, that an "infinite sum" may or may not represent a unique number.[21] For example, consider the *alternating* "infinite sum"

$$1 - 1 + 1 - 1 + 1 - 1 + 1 - 1 + \cdots.$$

Now notice that

$$(1 - 1) + (1 - 1) + (1 - 1) + (1 - 1) + \cdots = 0,$$

but that

$$1 + (-1 + 1) + (-1 + 1) + (-1 + 1) + \cdots = 1.$$

Thus, the alternating "infinite sum" does not represent a unique number.

As another example, consider the following two "infinite sums" of decreasing terms:

$$1 + \frac{1}{2} + \frac{1}{3} + \frac{1}{4} + \frac{1}{5} + \cdots \quad \text{and} \quad 1 + \frac{1}{2} + \frac{1}{4} + \frac{1}{8} + \frac{1}{16} + \cdots.$$

We examine the first "infinite sum"

$$1 + \frac{1}{2} + \frac{1}{3} + \frac{1}{4} + \frac{1}{5} + \cdots,$$

which is called the *harmonic series*. Observe that

$$\frac{1}{3} > \frac{1}{4}, \quad \text{so that} \quad \frac{1}{3} + \frac{1}{4} > \frac{1}{4} + \frac{1}{4} = \frac{1}{2},$$

$$\frac{1}{5} > \frac{1}{6} > \frac{1}{7} > \frac{1}{8}, \quad \text{so that} \quad \frac{1}{5} + \frac{1}{6} + \frac{1}{7} + \frac{1}{8} > 4 \cdot \frac{1}{8} = \frac{1}{2},$$

$$\frac{1}{9} > \frac{1}{10} > \frac{1}{11} > \frac{1}{12} > \frac{1}{13} > \frac{1}{14} > \frac{1}{15} > \frac{1}{16},$$

$$\text{so that} \quad \frac{1}{9} + \frac{1}{10} + \frac{1}{11} + \frac{1}{12} + \frac{1}{13} + \frac{1}{14} + \frac{1}{15} + \frac{1}{16} > 8 \cdot \frac{1}{16} = \frac{1}{2},$$

and so on. Thus, the harmonic series

$$1 + \frac{1}{2} + \underbrace{\frac{1}{3} + \frac{1}{4}}_{> \frac{1}{4} + \frac{1}{4} = \frac{1}{2}} + \underbrace{\frac{1}{5} + \frac{1}{6} + \frac{1}{7} + \frac{1}{8}}_{> \frac{1}{8} + \frac{1}{8} + \frac{1}{8} + \frac{1}{8} = \frac{1}{2}} + \cdots > 1 + \frac{1}{2} + \frac{1}{2} + \frac{1}{2} + \cdots$$

becomes increasingly greater and greater, so that the harmonic series does not represent a number at all.

Next, we examine the second "infinite sum,"

$$1 + \frac{1}{2} + \frac{1}{4} + \frac{1}{8} + \frac{1}{16} + \cdots,$$

that is an example of a *geometric series*.[22] We claim that this geometric series represents a unique number; moreover, we claim that it represents the number 2. To see this, consider an interval of length 2 marked as shown below.

---

[21] In mathematical parlance, we say that an "infinite sum" (which we have not defined precisely) may or may not *converge*.

[22] In general, a geometric series is an "infinite sum" of the form $a + ar + ar^2 + ar^3 + ar^4 + \cdots$. Here, $a = 1$ and $r = \frac{1}{2}$:

$$1 + \frac{1}{2} + \left(\frac{1}{2}\right)^2 + \left(\frac{1}{2}\right)^3 + \left(\frac{1}{2}\right)^4 + \cdots = 1 + \frac{1}{2} + \frac{1}{4} + \frac{1}{8} + \frac{1}{16} + \cdots.$$

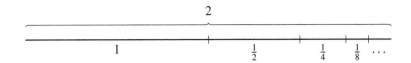

Thus, we see that the geometric series

$$1 + \frac{1}{2} + \frac{1}{4} + \frac{1}{8} + \frac{1}{16} + \cdots$$

represents the number 2. The above visualization is not a proof that the sum is 2. Providing a rigorous proof would take us too far afield.

Think About It 20.10 At each step in the diagram above, how much of the line is left compared to the step before? At any step, will the amount covered go past 2? Where have we seen this idea before? □

Therefore, to be careful, we should always wonder about whether an "infinite sum" represents a unique number.

20.3.3 The Number ϕ

Recall that the number

$$\phi = \frac{1 + \sqrt{5}}{2}$$

is the "limit" of the sequence of numbers $\phi_0, \phi_1, \phi_2, \phi_3, \ldots$, given by the finite simple continued fractions

$$\phi_0 = 1,$$

$$\phi_1 = 1 + \frac{1}{1},$$

$$\phi_2 = 1 + \cfrac{1}{1 + \cfrac{1}{1}},$$

$$\phi_3 = 1 + \cfrac{1}{1 + \cfrac{1}{1 + \cfrac{1}{1}}},$$

$$\phi_4 = 1 + \cfrac{1}{1 + \cfrac{1}{1 + \cfrac{1}{1 + \cfrac{1}{1}}}},$$

and so on. That is, if we plot the values

$$\phi_0 = 1 = \frac{1}{1},$$

$$\phi_1 = 1 + \frac{1}{\phi_0} = \frac{2}{1},$$

$$\phi_2 = 1 + \frac{1}{\phi_1} = 1 + \frac{1}{2} = \frac{3}{2},$$

$$\phi_3 = 1 + \frac{1}{\phi_2} = 1 + \frac{2}{3} = \frac{5}{3},$$

$$\phi_4 = 1 + \frac{1}{\phi_3} = 1 + \frac{3}{5} = \frac{8}{5},$$

and so on, they approach the value ϕ. (See page 358.) Now, notice that for each term ϕ_i in the sequence, the numerator of ϕ_i is the sum of the numerator and the denominator of the preceding term ϕ_{i-1}, and the denominator of ϕ_i is the numerator of ϕ_{i-1}. For example, the numerator of ϕ_1 is the sum of the numerator and the denominator of ϕ_0.

| $\phi_0 = \dfrac{1}{1}$ | $\phi_1 = \dfrac{1+1}{1} = \dfrac{2}{1}$ | $\phi_2 = \dfrac{2+1}{2} = \dfrac{3}{2}$ | $\phi_3 = \dfrac{3+2}{3} = \dfrac{5}{3}$ | $\phi_4 = \dfrac{5+3}{5} = \dfrac{8}{5}$ |

Now You Try 20.30 Continuing in the same way, write the next terms of the sequence $\phi_5, \phi_6, \ldots, \phi_{10}$ as common fractions. ☐

Thus, if we let $f_{-1} = 1$, let f_0 equal the numerator of ϕ_0, let f_1 equal the numerator of ϕ_1, let f_2 equal the numerator of ϕ_2, and so on, we obtain the following sequence of numbers.

f_{-1}	f_0	f_1	f_2	f_3	f_4	f_5	f_6	f_7	f_8	f_9	f_{10}	$\cdots$
1	1	2	3	5	8	13	21	34	55	89	144	$\cdots$

Think About It 20.11 What pattern do you notice in the sequence of numbers? ☐

The sequence of numbers follows the *recursive formula*

$$f_{-1} = 1, \quad f_0 = 1, \quad \text{and}$$
$$f_i = f_{i-2} + f_{i-1} \quad \text{for } i = 1, 2, 3, \ldots.$$

For example, beginning with

$$f_{-1} = 1, \quad f_0 = 1,$$

we find that

$$f_1 = f_{-1} + f_0 = 1 + 1 = 2,$$
$$f_2 = f_0 + f_1 = 1 + 2 = 3,$$
$$f_3 = f_1 + f_2 = 2 + 3 = 5,$$
$$f_4 = f_2 + f_3 = 3 + 5 = 8,$$

and so on. This sequence of numbers,

$$1, \ 1, \ 2, \ 3, \ 5, \ 8, \ 13, \ 21, \ 34, \ 55, \ 89, \ 144, \ldots,$$

is known as the *Fibonacci sequence* because it appears in the 1202 book *Liber Abaci* [90] by Leonardo of Pisa, popularly known as Fibonacci. Leonardo developed the sequence that is named after him by thinking about how a population of rabbits would increase over time. This is his famous "rabbit problem" (see pages 107 and 508). It is important to note that Leonardo developed the sequence, but did not look at the ratios of its terms. The connection between the Fibonacci sequence and ϕ was found later. Thus, we have presented it here in the reverse order from its historical development. And because of this, we have numbered our terms in the sequence beginning with f_{-1}. When you encounter Fibonacci's sequence in other contexts, the standard enumeration begins with f_0. The Fibonacci sequence not only shows up in the most unexpected places in nature—in the branching of trees, the spirals of a pine cone, and the alignment of the seeds in the head of a sunflower—it also leads the so-called golden ratio or golden section (see NOW YOU TRY 19.11 on page 299).

Consider the sequence of rectangles shown in FIGURE 20.3 that are obtained by placing together squares the length of whose sides are the terms of the Fibonacci sequence. Let us agree to call each of these rectangles a *Fibonacci rectangle*.

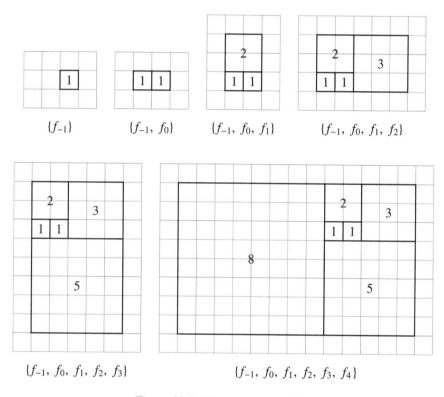

Figure 20.3: Fibonacci rectangles.

Now You Try 20.31 Draw the Fibonacci rectangles corresponding to $\{f_{-1}, f_0, f_1, f_2, f_3, f_4, f_5\}$ and $\{f_{-1}, f_0, f_1, f_2, f_3, f_4, f_5, f_6\}$ following the pattern shown above. □

Now consider the sequence of rational numbers in which each term is the ratio of the longer side to the shorter side of the corresponding rectangle in the sequence of Fibonacci rectangles, and observe that the sequence of ratios is

$$\{f_{-1}\}: \qquad \frac{1}{1} = \phi_0,$$

$$\{f_{-1}, f_0\}: \qquad \frac{2}{1} = \phi_1,$$

$$\{f_{-1}, f_0, f_1\}: \qquad \frac{3}{2} = \phi_2,$$

$$\{f_{-1},\ f_0,\ f_1,\ f_2\}: \qquad \frac{5}{3} = \phi_3,$$

$$\{f_{-1},\ f_0,\ f_1,\ f_2,\ f_3\}: \qquad \frac{8}{5} = \phi_4,$$

$$\{f_{-1},\ f_0,\ f_1,\ f_2,\ f_3,\ f_4\}: \qquad \frac{13}{8} = \phi_5,$$

$$\{f_{-1},\ f_0,\ f_1,\ f_2,\ f_3,\ f_4,\ f_5\}: \qquad \frac{21}{13} = \phi_6,$$

$$\{f_{-1},\ f_0,\ f_1,\ f_2,\ f_3,\ f_4,\ f_5,\ f_6\}: \qquad \frac{34}{21} = \phi_7,$$

and so on. We see that the "limit" of the ratio of the longer side to the shorter side of the sequence of Fibonacci rectangles is the number

$$\phi = \frac{1 + \sqrt{5}}{2}$$

(see page 358), the so-called *golden ratio* or *golden section*.

> A rectangle for which the ratio of the longer side to the shorter side is the golden ratio is called a *golden rectangle*.

Thus, we see that the "limit" of the sequence of Fibonacci rectangles is a golden rectangle.

Now consider a rectangle with base b and height h such that, if a square is removed from the rectangle, the ratio of the longer side to the shorter side of the remaining rectangle equals the ratio of the longer side to the shorter side of the original rectangle. This means that

$$\frac{h}{b - h} = \frac{b}{h}.$$

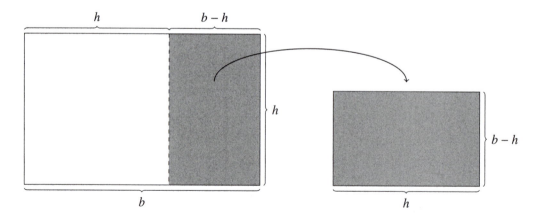

Cross multiplying leads to a quadratic equation in b, namely,

$$h^2 = b(b - h) \quad \text{or} \quad b^2 - hb - h^2 = 0.$$

The positive solution of this equation, which can be found using the quadratic formula (page 330), is

$$b = h\left(\frac{1 + \sqrt{5}}{2}\right).$$

(Check it.) Hence, the ratio

$$\frac{b}{h} = \frac{h\left(\frac{1+\sqrt{5}}{2}\right)}{h} = \frac{1+\sqrt{5}}{2} = \phi,$$

the golden ratio.

> A rectangle is a golden rectangle if, when a square is removed from the rectangle, the ratio of the longer side to the shorter side of the remaining rectangle equals the ratio of the longer side to the shorter side of the original rectangle.

Finally, when a square is removed from a golden rectangle, the remaining rectangle is another golden rectangle. And if a square is removed from this smaller golden rectangle, the remaining rectangle is yet another golden rectangle, and so on. Consequently, a golden rectangle contains infinitely many ever smaller golden rectangles. Moreover, these nested golden rectangles contain a spiral called the *golden spiral*; see FIGURE 20.4.

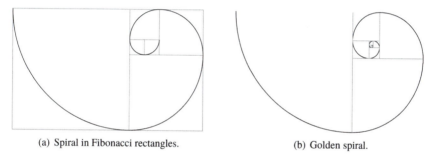

(a) Spiral in Fibonacci rectangles. (b) Golden spiral.

Figure 20.4: A spiral drawn inside Fibonacci rectangles and a golden spiral drawn inside golden rectangles.

Another geometric figure in which the golden ratio appears is the pentagram or the regular pentagon with all its diameters. We see in FIGURE 20.5 that the pentagram contains in its center another pentagram, which in turn contains in its center another pentagram, and so on. (This is akin to the fact that a golden rectangle contains a smaller golden rectangle—after we remove a square—which in turn contains another even smaller golden rectangle, and so on.) Now, because $\triangle ACD$ is similar to $\triangle DEC'$, which in turn is similar to $\triangle AD'C'$, we see that $AD : DC'$ as $DC' : AC'$. If we let $AD = x$ and $DC' = y$, it follows that

$$\frac{x}{y} = \frac{y}{x-y} \quad \text{or} \quad x^2 - yx - y^2 = 0$$

after cross multiplying. As we noted earlier, the positive solution of this quadratic equation, which can be found using the quadratic formula, is

$$x = y\left(\frac{1+\sqrt{5}}{2}\right),$$

so that the ratio

$$AD : DC' = \frac{y\left(\frac{1+\sqrt{5}}{2}\right)}{y} = \frac{1+\sqrt{5}}{2} = \phi,$$

the golden ratio. Note that $\triangle ACD$ is an isosceles triangle with base angle 72°. For this reason, an isosceles triangle with base angle 72° is called a *golden triangle*.

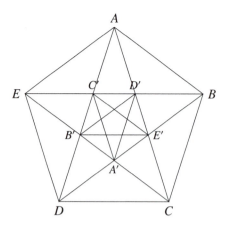

Figure 20.5: Nested pentagrams.

Think About It 20.12 How can you form a golden rectangle from a pentagram or a golden triangle? □

Recall that the pentagram was a sacred symbol of the Pythagoreans, and we now have seen some reasons why it would be considered mystical. As we remarked earlier, some scholars believe that because the pentagram contains the golden ratio, the number $\sqrt{5}$ was the first incommensurable (or irrational number) to be discovered, while other scholars maintain that $\sqrt{2}$ was the first; for example, see Von Fritz [123] for the former and Heath [68, pp. 206–207] for the latter.

We have only scratched the surface of elementary number theory as seen by the ancients. The larger field of number theory, both historically and currently, is a vast and beautiful area of mathematics. It has appeal both from the simplicity and creativity of the mathematics alone, which has inspired generations of mathematicians, and the ever-unfolding uses of the results applied to everything from computing to biology.

Chapter 21

Cubic Equations and Complex Numbers

Luca Pacioli declared at the end of his *Summa de arithmetica geometria proportioni et proportionalita* (1494) that solving the cubic equation would be as impossible as solving the ancient problem of squaring a circle[1] [21, p. 319]. This is not to say that no one had solved any cubic equations before then. Indeed, people had solved cubic equations, but the equations were specific ones, and each required its own method of solution.[2] What had not yet been found was a *general method* of solution using radicals for any cubic equation,

$$x^3 + bx^2 + cx + d = 0,$$

where b, c, and d are any real numbers, like completing the square for solving any quadratic equation. It was such a general method that Pacioli declared would be impossible. What Pacioli apparently could not imagine was that a general method for solving not only any cubic equation, but also any biquadratic equation, would not only be found, but that finding it would be just around the corner.

Within twenty years of the *Summa*, the first crack was made by fellow Italian Scipione del Ferro (1465–1526) of the University of Bologna, who found a method for solving the "cube and first power equal to the number,"

$$x^3 + px = q,$$

where p and q are any positive real numbers. This was, indeed, a monumental step because, as Girolamo Cardano (or Cardan; 1501–1576) would show some years later, using the change of variable $x = y - \frac{b}{3}$ in the general cubic equation $x^3 + bx^2 + cx + d = 0$ leads to the reduced cubic equation $y^3 + py + q = 0$. (We leave it to you as an exercise to show this.) Then, once having solved the reduced equation for y, we may recover the solution of the general equation using $x = y - \frac{b}{3}$.[3]

The solutions of the cubic equation and of the quartic equation were milestones in the history of algebra, and were certainly pinnacle achievements in mathematics during the Renaissance. The principal players were the Italian mathematicians del Ferro, Niccolò Tartaglia [4] (1500–1557), Cardano, Ludovico Ferrari (1522–1565), and Rafael Bombelli (1526–1572) [21, p. 317]. In steps, del Ferro, Tartaglia, and Cardano found a method for solving the cubic, and Ferrari, a pupil of Cardano, found a method for solving the quartic (or biquadratic); the solutions appear in Cardano's seminal work, *Artis magnae, sive de regulis*

---

[1]The three famous construction problems of Greek antiquity are, by the use of a straightedge and compasses only, the squaring (or quadrature) of a circle, the duplication of a cube, and the trisection of an angle [21, pp. 122–128].

[2]For example, Diophantus (ca. AD 250) proposes in problem 17 of Book VI of the *Arithmetica* to "Find a right triangle such that the area added to the hypotenuse gives a square, while the perimeter gives a cube." This led him to the cubic equation $x^3 + x = 4x^2 + 4$, for which he states that 4 is a solution without indicating how he found it [21, pp. 317–318]. For another example, cubic equations appear in Chinese mathematics as long ago as the seventh century, and Jia Xian (eleventh century) developed a numerical method for solving cubics [41, p. 414].

[3]Cardano neither wrote one general cubic equation, $x^3 + bx^2 + cx + d = 0$, nor used explicitly the change of variable $x = y - \frac{b}{3}$, because the coefficients in his equations were never negative. Thus, Cardano had to express a different rule for each case.

[4]Born Niccolò Fontana, he was better known during his lifetime as Tartaglia, which means "stammerer," a moniker he received because of the speech impediment that was wrought by the sword of a French soldier to his jaw as a child during the sacking of his hometown of Brescia in 1512.

algebraicis (*The Great Art, or the Rules of Algebra*; 1545), commonly referred to as the *Ars magna* [36].[5] Cardano provides the following account in the opening chapter of the *Ars magna* [36, pp. 7–9]:

CHAPTER I

On Double Solutions in Certain Types of Cases

1. This art originated with Mahomet the son of Moses the Arab [that is, al-Khwārizmī]. Leonardo of Pisa is a trustworthy source for this statement....

In our own days Scipione del Ferro of Bologna has solved the case of the cube and first power equal to a constant [$x^3 + px = q$], a very elegant and admirable accomplishment. Since this art surpasses all human subtlety and the perspicuity of mortal talent and is a truly celestial gift and a very clear test of the capacity of men's minds, whoever applies himself to it will believe that there is nothing that he cannot understand. In emulation of him, my friend Niccolò Tartaglia of Brescia, wanting not to be outdone, solved the same case when he got into a contest with his [Scipione's] pupil, Antonio Maria Fior, and, moved by my entreaties, gave it to me. For I had been deceived by the words of Luca Pacioli, who denied that any more general rule could be discovered than his own. Notwithstanding the many things which I had already discovered, as is well known, I had despaired and had not attempted to look any further. Then, however, having received Tartaglia's solution and seeking for the proof of it, I came to understand that there were a great many other things that could also be had. Pursuing this thought and with increased confidence, I discovered these others, partly by myself and partly through Lodovico Ferrari, formerly my pupil. Hereinafter those things which have been discovered by others have their names attached to them; those to which no name is attached are mine. The demonstrations, except for the three by Mahomet and the two by Lodovico, are all mine.

It turns out that "imaginary numbers" (square roots of negative numbers; see SECTION 21.1) sometimes appear in the solution of the cubic when one uses what is now often called "Cardan's cubic formula," even when the solutions were known to be real. This was something that Cardano was unable to resolve [21, p. 326]. It would be left to Bombelli, who accepted the existence of imaginary numbers and was proficient at operating with them, to resolve this apparent impasse [21, p. 327].

Although Cardano was very good about acknowledging in the *Ars magna* those before him who contributed toward the solution of the cubic, his account gives no hint of the bitter feud between him and Tartaglia that was to ensue. Here is the story that unfolded.

Publishing one's new results quickly is a way of life for many mathematicians today, but that was not so during the time that the solution of the cubic equation was unfolding in Italy. In those days in Europe, mathematicians built their reputations by challenging each other to public contests. In a contest between two mathematicians, each would propose to the other a set of problems of an agreed-upon number to be solved within a selected amount of time, perhaps some number of weeks or months, with the person who solves the greater number of problems correctly being the winner; the reward might be a large prize, or a university appointment, or the patronage of a wealthy nobleman. Clearly, it was to a contestant's advantage to propose problems that his opponent could not solve. To ensure this, a mathematician typically guarded any new discoveries very closely, even to his deathbed. And so it was with del Ferro, who did not disclose his secret for solving the "cube and first power equal to the number," $x^3 + px = q$, until he transmitted it to his pupils Annibale della Nave (1500–1558) and Antonio Maria Fiore (first half of the sixteenth century) at his death.

About a decade after del Ferro's death, in 1535, Tartaglia burst onto the cubic scene with the announcement that he had discovered the solution to the "cube and square equal to the number," that is, to equations of the form $x^3 + px^2 = q$. This prompted Fiore, who believed Tartaglia to have been bluffing and who felt secure in his own knowledge of the solution of the "cube and first power equal to the number," to challenge Tartaglia to a public problem-solving contest. Each contestant proposed 30 problems to the other and had 50 days to solve as many of the problems as he could. All of the 30 problems that Fiore had proposed to

---

[5]Because the unknown in equations was called *res* in Latin or *cosa* in Italian or *Coss* in German (literally translated as the "thing"), algebra was also referred to as the "cossic art." On the other hand, the "lesser art" referred to arithmetic.

Tartaglia were of the form $x^3 + px = q$, which, after having worked doggedly, Tartaglia finally discovered how to solve just before the end of the 50 days. Consequently, as Burton [21, p. 321] tells us, "Within two hours, Tartaglia had reduced all 30 problems posed to him to particular cases of the equation $x^3 + px = q$, for which he knew the answer. Of the problems he himself put to Fiore, the latter failed to master a single one (most of which led to equations of the form $x^3 + px^2 = q$)." Fiore suffered a humiliating defeat.

Before long, Cardano received news in Milan of Tartaglia's feat and immediately set out to obtain the secret of the cubic solution from Tartaglia. For his part, Cardano offered to publish Tartaglia's solution in a work, giving full credit to Tartaglia. Tartaglia, however, refused to divulge his secret, saying that he would publish it in his own work at the right time. Up to this point in the story, Oystein Ore [36, p. x] asserts, "the facts seem clear, but for the subsequent events we are dependent almost exclusively upon Tartaglia's printed accounts, which by no stretch of the imagination can be regarded as objective." It seems that Tartaglia soon had a change of heart, perhaps in the hope of securing a favor from Cardano, and provided Cardano with his solution of the cubic; but this was only after Cardano "swore a most solemn oath, by the Sacred Gospels and his word as a gentleman, never to publish the method, and he pledged by his Christian faith to put it down in cipher, so that it would be unintelligible to anyone after his death" [36, p. x]. Yet, as we know, Cardano did publish the solution of the cubic in his *Ars magna*.

It seems that, after pledging to Tartaglia that he would never divulge the secret of Tartaglia's method, Cardano and his young pupil Ferrari, who had been a servant in Cardano's household, discovered that del Ferro had originally solved the "cube and first power equal to the number." After having confirmed this by inspecting the papers of the late del Ferro, Cardano apparently no longer felt bound to his oath of secrecy to Tartaglia because it was del Ferro, and not Tartaglia, who was the first discoverer. Nevertheless, Cardano gives credit to Tartaglia in the *Ars magna*; and, nevertheless, Tartaglia was furious that Cardano had broken his oath. This was followed by bitter exchanges between Tartaglia and Cardano, with Ferrari being Cardano's champion. Ore relates what ensues [36, pp. x–xii]:

> After the publication of the *Ars Magna*, Tartaglia's rage knew no bounds. Already the following year he published another book *Quesiti et Inventioni Diverse* in which he exposed Cardano's perfidy, reproducing the correspondence between them, and giving a word-for-word account of their conversations. On February 10, 1574, Ferrari responded with a printed *cartello*, a challenge to Tartaglia to meet him in a dispute on almost any scientific topic, for a prize of up to 200 scudi. "This I have proposed to make known, that you have written things which false and unworthily slander the above-mentioned Signor Gerolamo (Cardano), compared to whom you are hardly worth mentioning."
>
> [...]
>
> ...The meeting took place in a church in Milan before a great audience on August 10, 1548. We have no account of the proceedings, except for a few statements from Tartaglia to the effect that the meeting broke up when the supper hour drew near....
>
> From various inferences there does not seem to be much doubt that Ferrari was declared the winner. He received numerous distinguished offers afterwards, among them a request to serve as the tutor for the emperor's son.

In the end, it appears that Tartaglia's fears about not receiving credit for the solution of the cubic have come to pass, for, although Cardano clearly gives Tartaglia credit in the *Ars magna*, the cubic formula is today commonly attributed to Cardano without any mention of Tartaglia (nor of del Ferro nor Ferrari).

21.1 COMPLEX NUMBERS

Even as late as the seventeenth century, prominent mathematicians like René Descartes referred to positive roots of an equation as "true" roots and negative roots as "false" roots; Descartes called roots that contained a square root of a negative number "imaginary" [46, pp. 159, 175]. By this time, however, negative numbers were beginning to gain full acceptance in Europe, even though they had long ago been accepted in India and China, but it would be somewhat longer (not until the nineteenth century) before imaginary or complex numbers would enjoy an equivalent standing.

Even though square roots of negative numbers had made an appearance by the sixteenth century—Cardano mentions them in the *Ars magna*, for example—they were treated only as a convenient contrivance and not really as numbers; mathematicians often referred to them as "sophistic," "impossible," "imaginary," and "useless" [11, p. 143]. When confronted with an equation like $x^2 + 1 = 0$ or $x^2 + 2x + 2 = 0$, one could simply say that the equation has no solution. What forced the matter was the discovery of the cubic formula (RULE 21.6 on page 405), for it turns out that the formula expresses some real-number solutions of cubic equations using square roots of negative numbers: a real puzzler! (See page 407 for an example.) Thus, these "sophistic," "impossible," "imaginary," and "useless" objects no longer could be ignored.

We introduce complex numbers in this section.

Definition 21.1 The *imaginary unit*, denoted i, is the number such that its square equals -1:

$$i^2 = -1.$$

$\square$

Definition 21.2 If b is a real number and i is the imaginary unit, then

$$b\text{i} \quad \text{or} \quad i b$$

is called a *pure imaginary number*. If a and b are real numbers and i is the imaginary unit, then

$$a + b\text{i} \quad \text{or} \quad a + i b$$

is called a *complex number*.

In the set of complex numbers, addition and multiplication are defined as follows: If $a + b\text{i}$ and $c + d\text{i}$ are complex numbers, then

addition: $(a + b\text{i}) + (c + d\text{i}) = (a + c) + (b + d)\text{i}$;

multiplication: $(a + b\text{i})(c + d\text{i}) = (ac - bd) + (ad + bc)\text{i}$.

If $z = a + b\text{i}$ is a complex number, then a is called the *real part* of z and b is called the *imaginary part* of z. These are denoted $\mathfrak{R}(z) = a$ and $\mathfrak{I}(z) = b$. Two complex numbers are equal if their real and imaginary parts are equal. $\square$

Remark 21.1

- The set of (pure) real numbers may be considered a subset of the set of complex numbers because, for example, the real number a corresponds to the complex number $a + 0\text{i}$. Similarly, the set of pure imaginary numbers may be considered a subset of the set of complex numbers because, for example, the pure imaginary number $b\text{i}$ corresponds to the complex number $0 + b\text{i}$.

- The addition of two complex numbers is consistent with *combining like terms*. Moreover, because addition of real numbers is commutative, we see that

$$(c + d\text{i}) + (a + b\text{i}) = (c + a) + (d + b)\text{i}$$
$$= (a + c) + (b + d)\text{i} = (a + b\text{i}) + (c + d\text{i}),$$

that is to say, addition of complex numbers is also commutative.

- The multiplication of two complex numbers is consistent with *distributing*:

$$(a + b\text{i})(c + d\text{i}) = ac + ad\text{i} + bc\text{i} + bd\text{i}^2 \qquad (\text{use } i^2 = -1)$$
$$= ac + (ad + bc)\text{i} - bd$$
$$= (ac - bd) + (ad + bc)\text{i}.$$

In particular,
$$a(c + di) = ac + adi.$$

Moreover, because addition and multiplication of real numbers are commutative, we see that

$$(c + di)(a + bi) = (ca - db) + (cb + da)i$$
$$= (ac - bd) + (ad + bc)i = (a + bi)(c + di),$$

that is to say, multiplication of complex numbers is also commutative.

• As with commutativity, it follows from the associativity of addition and multiplication of real numbers that the addition and multiplication of complex numbers are also associative: If u, v, and w are complex numbers, then

$$(u + v) + w = u + (v + w) \quad \text{and} \quad (uv)w = u(vw).$$

• The set of complex numbers also obeys the distributive law: If u, v, and w are complex numbers, then

$$u(v + w) = uv + uw \quad \text{and} \quad (v + w)u = vu + wu.$$

□

Example 21.1 Let $z = 2 + 5i$ and $w = 3 - 2i$. Find $3z$, $z + w$, $z - 2w$, and zw.

• $3z$
$$3z = 3(2 + 5i) = 6 + 15i$$

• $z + w$
$$z + w = (2 + 5i) + (3 - 2i) = (2 + 3) + (5 - 2)i = 5 + 3i$$

• $z - 2w$

$$z - 2w = (2 + 5i) - 2(3 - 2i)$$
$$= (2 + 5i) + (-6 + 4i)$$
$$= (2 - 6) + (5 + 4)i$$
$$= -4 + 9i$$

• zw

$$zw = (2 + 5i)(3 - 2i)$$
$$= (2)(3) + (2)(-2i) + (5i)(3) + (5i)(-2i)$$
$$= 6 - 4i + 15i - 10i^2 \qquad\qquad (\text{use } i^2 = -1)$$
$$= 6 + (-4 + 15)i + 10$$
$$= 16 + 11i$$

□

Now You Try 21.1 Let $u = -3 + 2i$, $v = 4 - i$, and $w = 3 + 5i$. Find $-5u$, $u + v$, $2u - 4v$, and uv. Show that $u(v + w) = uv + uw$. □

Figure 21.1: Leonhard Euler. (Source: Courtesy of Jeff Miller.)

The convention of using i for $\sqrt{-1}$ is due to Leonhard Euler (1707–1783), a notation that he adopted late in his life [15, p. 442].[6] (See FIGURE 21.1.) Euler was a remarkable and an extremely influential mathematician who contributed significantly to many areas of mathematics, from analysis to graph theory to topology [3]. He was also remarkably prolific, writing an average of 800 pages a year, often with his children playing around him. In fact, Bell [9, p. 139] tells us that "Euler wrote his great memoirs as easily as a fluent writer composes a letter to an intimate friend. Even total blindness during the last seventeen years of his life did not retard his unparalleled productivity; indeed, if anything, the loss of his eyesight sharpened Euler's perceptions in the inner world of his imagination." Euler's works,[7] which number 866, have been estimated to be able to fill about seventy-five large volumes [15, p. 440].

To get a sense of how mathematicians thought of imaginary numbers in the eighteenth century, we turn to Euler's *Elements of Algebra*, in which he writes [54, pp. 40–41]:

> 143. And, since all numbers which it is possible to conceive, are either greater or less than 0, or are 0 itself, it is evident that we cannot rank the square root of a negative number amongst possible numbers, and we must therefore say that is an impossible quantity. In this manner we are led to the idea of numbers, which from their nature are impossible; and therefore they are usually called *imaginary quantities*, because they exist merely in the imagination.

> 144. All such expressions, as $\sqrt{-1}$, $\sqrt{-2}$, $\sqrt{-3}$, $\sqrt{-4}$, &c. are consequently impossible, or imaginary numbers, since they represent roots of negative quantities; and of such numbers we may truly assert that they are neither nothing, nor greater than nothing, nor less than nothing; which necessarily constitutes them imaginary, or impossible.

> 145. But notwithstanding this, these numbers present themselves to the mind; they exist in our imagination, and we still have a sufficient idea of them; since we know that by $\sqrt{-4}$ is meant a number which, multiplied by itself, produces −4; for this reason also, nothing prevents us from making use of these imaginary numbers and employing them in calculation.

However, the following paragraphs in the *Elements of Algebra* show that Euler is not quite consistent with how he treats these "impossible" numbers:

---

[6]Euler introduced or popularized many of the notations we use in mathematics today. For example [15, p. 442],

- e: For "that number whose hyperbolic logarithm = 1"; perhaps because e is the first letter of "exponential." We shall encounter e in CHAPTER 24.

- π: For the ratio of the circumference of a circle to its diameter; Euler did not introduce π, but he popularized it.

- $\sum$: For summation.

- $f(x)$: For a function of x.

[7]See "The Euler Archive: The works of Leonhard Euler online" <http://eulerarchive.maa.org>.

146. The first idea that occurs on the present subject is, that the square of $\sqrt{-3}$, for example, or the product of $\sqrt{-3}$ by $\sqrt{-3}$, must be -3; that the product of $\sqrt{-1}$ by $\sqrt{-1}$, is -1; and, in general, that by multiplying $\sqrt{-a}$ by $\sqrt{-a}$, or by taking the square of $\sqrt{-a}$ we obtain $-a$.

147. Now, as $-a$ is equal to $+a$ multiplied by -1, and as the square root of a product is found by multiplying together the roots of its factors, it follows that the root of a times -1, or $\sqrt{-a}$, is equal to $\sqrt{a}$ times $\sqrt{-1}$; but $\sqrt{a}$ is a possible or real number, consequently the whole impossibility of an imaginary quantity may be always reduced to $\sqrt{-1}$; for this reason $\sqrt{-4}$ is equal to $\sqrt{4}$ multiplied by $\sqrt{-1}$, or equal to $2\sqrt{-1}$, because $\sqrt{4}$ is equal to 2; likewise -9 is reduced to $\sqrt{9} \times \sqrt{-1}$, or $3\sqrt{-1}$; and $\sqrt{-16}$ is equal to $4\sqrt{-1}$.

148. Moreover, as $\sqrt{a}$ multiplied by $\sqrt{b}$ makes $\sqrt{ab}$ we shall have $\sqrt{6}$ for the value of $\sqrt{-2}$ multiplied by $\sqrt{-3}$; and $\sqrt{4}$ or 2, for the value of the product of $\sqrt{-1}$ and $\sqrt{-4}$. Thus we see that two imaginary numbers, multiplied together, produce a real, or possible one.

But, on the contrary, a possible number, multiplied by an impossible number, gives always an imaginary product: thus $\sqrt{-3}$ by $\sqrt{+5}$, gives $\sqrt{-15}$.

Thus, we see that, on the one hand, Euler states in ¶146 that

$$\sqrt{-3} \cdot \sqrt{-3} = -3,$$

but, on the other hand, he states in ¶148 that

$$\sqrt{-2} \cdot \sqrt{-3} = \sqrt{(-2)(-3)} = \sqrt{+6};$$

however, if we apply the reasoning Euler uses in the latter to the former, we would find that

$$\sqrt{-3} \cdot \sqrt{-3} = \sqrt{(-3)(-3)} = \sqrt{+9} = 3, \text{ not } -3.$$

Nevertheless, Euler acknowledges the usefulness of imaginary numbers and declares that "the calculation of imaginary quantities is of the greatest importance":

151. It remains for us to remove any doubt, which may be entertained concerning the utility for the numbers of which we have been speaking; for those numbers being impossible, it would not be surprising if they were thought entirely useless, and the object only of an unfounded speculation. This, however, would be a mistake; for the calculation of imaginary quantities is of the greatest importance, as questions frequently arise, of which we cannot immediately say whether they include any things real and possible, or not; but when the solution of such a question leads to imaginary numbers, we are certain that what is required is impossible.

Remark 21.2 Assuming that a and b are positive, and using the associative and commutative laws, what Euler should have noted is that

$$\sqrt{-a} \cdot \sqrt{-b} = (\sqrt{a} \cdot \sqrt{-1})(\sqrt{b} \cdot \sqrt{-1}) = (\sqrt{a} \cdot \sqrt{b})(\sqrt{-1} \cdot \sqrt{-1})$$

$$= (\sqrt{a} \cdot \sqrt{b}) \cdot (-1) = \sqrt{ab} \cdot (-1) = -\sqrt{ab}.$$

□

We now move to division. In order to divide by a complex number, we must know what is its reciprocal.

Definition 21.3 A *reciprocal* or *multiplicative inverse* of a complex number z is a complex number w such that

$$zw = 1 \quad \text{and} \quad wz = 1.$$

□

Thus, for example, a reciprocal of i is $-i$ because[8]

$$i(-i) = -i^2 = -(-1) = 1 \quad \text{and, likewise,} \quad -i \cdot i = 1.$$

In general, a reciprocal of $z = a + bi$ is a complex number $w = x + yi$ such that

$$zw = 1 \text{ or } 1 + 0i \quad \text{and} \quad wz = 1 \text{ or } 1 + 0i.$$

Now, since $wz = zw$ (multiplication of complex numbers is commutative) and

$$zw = (a + bi)(x + yi) = (ax - by) + (ay + bx)i,$$

$zw = 1$ and $wz = 1$ mean that

$$(ax - by) + (ay + bx)i = 1 + 0i.$$

Hence, for $w = x + yi$ to be a reciprocal of $z = a + bi$, we see that $\mathcal{R}(w) = x$ and $\mathfrak{I}(w) = y$ must satisfy the system of equations

$$ax - by = 1,$$
$$bx + ay = 0.$$

Example 21.2 Find a reciprocal of z.

1. $z = 2 + 5i$

   To find a reciprocal $w = x + yi$, solve the system

   $$2x - 5y = 1,$$
   $$5x + 2y = 0.$$

   a) Using the method of elimination to eliminate the y terms here, multiply the first equation by 2 and the second equation by 5:

   $$4x - 10y = 2,$$
   $$25x + 10y = 0.$$

   b) Add the equations to eliminate the y terms:

   $$29x = 2, \quad \text{so} \quad x = \frac{2}{29}.$$

   c) Substitute $x = \frac{2}{29}$ into the equation $5x + 2y = 0$ and solve for y:

   $$5\left(\frac{2}{29}\right) + 2y = 0$$

   $$\frac{10}{29} + 2y = 0$$

---

[8] We note that Euler incorrectly states that the reciprocal of i is itself, i [54, p. 41]:

149. It is the same with regard to division; for $\sqrt{a}$ divided by $\sqrt{b}$ making $\sqrt{\frac{a}{b}}$, it is evident that $\sqrt{-4}$ divided by $\sqrt{-1}$ will make $\sqrt{+4}$ or 2; that $\sqrt{+3}$ divided by $\sqrt{-3}$ will give $\sqrt{-1}$; and that 1 divided by $\sqrt{-1}$ gives $\sqrt{\frac{+1}{-1}}$, or $\sqrt{-1}$; because 1 is equal to $\sqrt{+1}$.

$$2y = -\frac{10}{29}$$

$$y = -\frac{5}{29}.$$

Thus, $w = \frac{2}{29} - \frac{5}{29}i$. We check that, in fact, $zw = 1$:

$$zw = (2 + 5i)\left(\frac{2}{29} - \frac{5}{29}i\right)$$

$$= (2)\left(\frac{2}{29}\right) + (2)\left(-\frac{5}{29}i\right) + (5i)\left(\frac{2}{29}\right) + (5i)\left(-\frac{5}{29}i\right)$$

$$= \frac{4}{29} - \frac{10}{29}i + \frac{10}{29}i - \frac{25}{29}i^2 \qquad \text{(use } i^2 = -1\text{)}$$

$$= \frac{4}{29} + \frac{25}{29}$$

$$= \frac{29}{29} = 1.$$

Therefore, a reciprocal of $z = 2 + 5i$ is $\boxed{w = \frac{2}{29} - \frac{5}{29}i.}$

2. $z = \dfrac{2}{5}i$

We note that $\frac{2}{5}i = 0 + \frac{2}{5}i$. Thus, to find a reciprocal $w = x + yi$, we solve the system

$$0x - \frac{2}{5}y = 1,$$

$$\frac{2}{5}x + 0y = 0.$$

From the first equation, we see that $y = -\frac{5}{2}$; and from the second equation, we see that $x = 0$. Thus, $w = 0 - \frac{5}{2}i$ or $-\frac{5}{2}i$. We check that, in fact, $zw = 1$:

$$zw = \left(\frac{2}{5}i\right)\left(-\frac{5}{2}i\right) = -i^2 = -(-1) = 1.$$

Therefore, a reciprocal of $z = \frac{2}{5}i$ is $\boxed{w = -\frac{5}{2}i.}$

3. $z = \dfrac{2}{5}$

Since $z = \frac{2}{5}$ is a real number, we expect $w = \frac{5}{2}$ to be its reciprocal. Let us see if the system of equations bears this out. We note that $\frac{2}{5} = \frac{2}{5} + 0i$. Thus, to find a reciprocal $w = x + yi$ of the complex number $z = \frac{2}{5} + 0i$, we solve the system

$$\frac{2}{5}x - 0y = 1,$$

$$0x + \frac{2}{5}y = 0.$$

From the first equation, we see that $x = \frac{5}{2}$; and from the second equation, we see that $y = 0$. Thus, in fact, $w = \frac{5}{2} + 0i$ or $\frac{5}{2}$, so that a reciprocal of $z = \frac{2}{5}$ is $\boxed{w = \frac{5}{2}}$ just as we had expected.

4. $z = 0$

Since $z = 0$ is a real number, we expect that its reciprocal is undefined. Let us see if the system of equations bears this out. We note that $0 = 0 + 0i$. Thus, to find a reciprocal $w = x + yi$ of the complex number $z = 0 + 0i$, we solve the system

$$0x - 0y = 1,$$
$$0x + 0y = 0.$$

We see from the first equation that there is no solution of the system. Therefore, we conclude that the complex number $z = 0 + 0i$ does not have a reciprocal, just as we had expected.

$\square$

Instead of solving a system of equations each time we seek a reciprocal of a complex number, let us solve the system once for the general complex number $z = a + bi$.

To find a reciprocal $w = x + yi$ of the nonzero complex number $z = a + bi$, where not both $a = 0$ and $b = 0$, we solve the system

$$ax - by = 1,$$
$$bx + ay = 0.$$

1. Multiply through the first equation by a and the second equation by b so that we may eliminate the y terms here:

$$a^2x - aby = a,$$
$$b^2x + aby = 0.$$

2. Add the equations to eliminate the y terms:

$$(a^2 + b^2)x = a, \quad \text{so} \quad x = \frac{a}{a^2 + b^2}.$$

3. Substitute $x = \dfrac{a}{a^2 + b^2}$ into the equation $bx + ay = 0$ and solve for y:

$$b\left(\frac{a}{a^2 + b^2}\right) + ay = 0$$

$$\frac{ab}{a^2 + b^2} + ay = 0$$

$$ay = -\frac{ab}{a^2 + b^2}$$

$$y = -\frac{b}{a^2 + b^2}.$$

Thus,

$$w = \frac{a}{a^2 + b^2} - \frac{b}{a^2 + b^2}i.$$

We check that, in fact, $zw = 1$:

$$zw = (a + bi)\left(\frac{a}{a^2 + b^2} - \frac{b}{a^2 + b^2}i\right)$$

$$= \frac{a^2}{a^2 + b^2} - \frac{ab}{a^2 + b^2}i + \frac{ab}{a^2 + b^2}i - \frac{b^2}{a^2 + b^2}i^2 \qquad \text{(use } i^2 = -1)$$

$$= \frac{a^2}{a^2 + b^2} + \frac{b^2}{a^2 + b^2}$$

$$= \frac{a^2 + b^2}{a^2 + b^2} = 1.$$

Therefore, a reciprocal of $z = a + bi$ is

$$w = \frac{a}{a^2 + b^2} - \frac{b}{a^2 + b^2}i.$$

In fact, this is the only reciprocal of z, so we may say rightly that it is *the* reciprocal of z. For, if w_1 and w_2 were reciprocals of z, we would find that

$$w_1 = w_1 \cdot 1 = w_1(zw_2) = (w_1z)w_2 = 1 \cdot w_2 = w_2,$$

that is, in fact, $w_1 = w_2$ are one and the same.

Rule 21.1 Let $z = a + bi$ be a nonzero complex number, that is, not both $a = 0$ and $b = 0$. The reciprocal of z is

$$z^{-1} = \frac{1}{z} = \frac{a}{a^2 + b^2} - \frac{b}{a^2 + b^2}i.$$

☐

Now You Try 21.2 Use RULE 21.1 to find the reciprocal of $z = a + bi$.

1. $z = 2 + 5i$ 2. $z = -3 + 2i$ 3. $z = \dfrac{2}{5}i$ 4. $z = \dfrac{2}{5}$

☐

Instead of memorizing RULE 21.1, there is another way to find the reciprocal of a complex number—one that extends nicely to division by a complex number—namely, by using its complex conjugate.

Definition 21.4 Let $z = a + bi$. The *complex conjugate* of z is the number $\bar{z} = a - bi$. ☐

(Recall DEFINITION 20.6 in which we defined $p - s\sqrt{q}$ to be the conjugate of $p + s\sqrt{q}$. Hence, it is consistent to define $a - bi = a - b\sqrt{-1}$ to be the complex conjugate of $a + bi = a + b\sqrt{-1}$.)

Here are a few examples.

- $\overline{4 + i} = 4 - i$
- $\overline{3i} = -3i$
- $\overline{10 - 5i} = 10 + 5i$
- $\overline{-2} = -2$

Notice that $\overline{-2} = -2$. This is because $\overline{-2} = \overline{-2 + 0i} = -2 - 0i = -2$. In fact, a number $z = a + bi$ is a real number if and only if $\bar{z} = z$.

Think About It 21.1 By simply changing the sign on the imaginary term, what does this accomplish in the multiplication? Pick a complex number and multiply by its complex conjugate to see what happens. □

We now make the observation that, if $z = a + bi$, then

$$\frac{1}{z} = \frac{1}{z} \cdot \frac{\bar{z}}{\bar{z}}$$ (Why?)

$$= \frac{1}{a + bi} \cdot \frac{a - bi}{a - bi}$$

$$= \frac{a - bi}{a^2 - abi + abi - b^2 i^2}$$

$$= \frac{a - bi}{a^2 + b^2}$$

$$= \frac{a}{a^2 + b^2} - \frac{b}{a^2 + b^2} i,$$

which is precisely RULE 21.1 for the reciprocal of a complex number. Thus, we have the following rule for dividing by a complex number.

Rule 21.2 Let z and w be complex numbers. To divide z by w, compute

$$\frac{z}{w} = \frac{z}{w} \cdot \frac{\bar{w}}{\bar{w}}.$$

□

Example 21.3 Divide 3 by $2 - i\sqrt{2}$.

$$\frac{3}{2 - i\sqrt{2}} = \frac{3}{2 - i\sqrt{2}} \cdot \frac{2 + i\sqrt{2}}{2 + i\sqrt{2}}$$

$$= \frac{3(2 + i\sqrt{2})}{4 + i2\sqrt{2} - i2\sqrt{2} - 2i^2}$$

$$= \frac{3(2 + i\sqrt{2})}{4 + 2}$$

$$= \frac{3(2 + i\sqrt{2})}{6}$$

$$= \frac{2 + i\sqrt{2}}{2}$$

$$= 1 + i\frac{\sqrt{2}}{2}$$

Therefore,

$$3 \div (2 - i\sqrt{2}) = \frac{3}{2 - i\sqrt{2}} = 1 + i\frac{\sqrt{2}}{2}.$$

□

Now You Try 21.3 Use RULE 21.2 to divide z by w.

1. $z = 1$, $w = 2 + 5i$

2. $z = 4 + i$, $w = -3 + 2i$

3. $z = -10 + 5i$, $w = 2i$

4. $z = -1 + 4i$, $w = 2$

☐

We now return our attention briefly to nth roots of real numbers that we discussed in SUBSECTION 20.2.2. To begin, recall that every positive real number has two square roots, namely, a positive square root and a negative square root (see CHAPTER 12). If we allow ourselves to roam into the land of complex numbers, however, then it turns out that every *negative* nonzero real number also has two square roots. In fact, if we allow ourselves complex numbers, then every nonzero real number not only has two square roots, but also three cube roots, four fourth roots, five fifth roots, and, in general, n nth roots!

Example 21.4 The number 1 has two square roots, namely, 1 and -1, for both

$$1^2 = 1 \quad \text{and} \quad (-1)^2 = 1.$$

The number -1 also has two square roots, namely, i and $-i$, for both

$$i^2 = -1 \quad \text{and} \quad (-i)^2 = i^2 = -1.$$

Similarly, -25 has two square roots, namely, 5i and $-5i$, for both

$$(5i)^2 = 25i^2 = -25 \quad \text{and} \quad (-5i)^2 = 25i^2 = -25.$$

☐

At this point, we recall DEFINITION 20.3 and notation 1, in which we declared that

- If a is a real number that is not negative and $a^2 = b$, then $a = \sqrt{b}$.
- If a is a real number that is negative and $a^2 = b$, then $a = -\sqrt{b}$.

In the same way, we introduce the following notation.

If b is a real number that is not negative, then we write

$$\sqrt{-b^2} = bi,$$

called the *principal square root*. If b is a real number that is negative, then we write

$$bi = -\sqrt{-b^2}.$$

Thus,

$$\sqrt{1} = 1, \quad \sqrt{-1} = i, \quad -\sqrt{1} = -1, \quad -\sqrt{-1} = -i,$$
$$\sqrt{25} = 5, \quad \sqrt{-25} = 5i, \quad -\sqrt{25} = -5, \quad -\sqrt{-25} = -5i.$$

Now You Try 21.4

1. Find both square roots of the following.

a) 49   b) −36   c) −30   d) −64

2. Show that the two square roots of the imaginary unit i are

$$\frac{\sqrt{2}}{2} + i\frac{\sqrt{2}}{2} \quad \text{and} \quad -\frac{\sqrt{2}}{2} - i\frac{\sqrt{2}}{2}.$$

Hint: Find the squares of the numbers.

3. Solve the following equations using the quadratic formula (RULE 20.1 on page 330). If the discriminant is not a perfect square, express the radical as simply as possible (page 350); if the discriminant is negative, express the solution as a complex number.

a) $x^2 + 2x + 2 = 0$ b) $2x^2 = \sqrt{10}x - 5$

□

Example 21.5 A cube root of 1 is a number b such that $b^3 = 1$. What are the cube roots of 1?

We claim that 1 has *three* cube roots altogether, namely,

$$\omega_0 = 1, \quad \omega_1 = -\frac{1}{2} + i\frac{\sqrt{3}}{2}, \quad \text{and} \quad \omega_2 = -\frac{1}{2} - i\frac{\sqrt{3}}{2}.$$

(The symbol ω is the lowercase Greek letter *omega*.) To verify this, we must check that $\omega_0^3 = 1$, $\omega_1^3 = 1$, and $\omega_2^3 = 1$.

- $\omega_0 = 1$

$$\omega_0^3 = 1^3 = 1$$

- $\omega_1 = -\frac{1}{2} + i\frac{\sqrt{3}}{2}$

$$\omega_1^3 = \left(-\frac{1}{2} + i\frac{\sqrt{3}}{2}\right)^3$$

$$= \left(-\frac{1}{2} + i\frac{\sqrt{3}}{2}\right)^2\left(-\frac{1}{2} + i\frac{\sqrt{3}}{2}\right)$$

$$= \left(\frac{1}{4} - i\frac{\sqrt{3}}{4} - i\frac{\sqrt{3}}{4} + \frac{3}{4}i^2\right)\left(-\frac{1}{2} + i\frac{\sqrt{3}}{2}\right)$$

$$= \left(\frac{1}{4} - i\frac{2\sqrt{3}}{4} - \frac{3}{4}\right)\left(-\frac{1}{2} + i\frac{\sqrt{3}}{2}\right)$$

$$= \left(-\frac{2}{4} - i\frac{2\sqrt{3}}{4}\right)\left(-\frac{1}{2} + i\frac{\sqrt{3}}{2}\right)$$

$$= \left(-\frac{1}{2} - i\frac{\sqrt{3}}{2}\right)\left(-\frac{1}{2} + i\frac{\sqrt{3}}{2}\right)$$

$$= \frac{1}{4} - i\frac{\sqrt{3}}{4} + i\frac{\sqrt{3}}{4} - \frac{3}{4}i^2$$

$$= \frac{1}{4} + \frac{3}{4}$$

$$= 1$$

- $\omega_2 = -\dfrac{1}{2} - i\dfrac{\sqrt{3}}{2}$

We leave it to you as an exercise to show that, indeed,

$$\omega_2^3 = \left(-\frac{1}{2} - i\frac{\sqrt{3}}{2}\right)^3 = 1.$$

Thus, we see that 1 has *three* cube roots altogether, one real and two complex, namely,

$$\omega_0 = 1, \quad \omega_1 = -\frac{1}{2} + i\frac{\sqrt{3}}{2}, \quad \text{and} \quad \omega_2 = -\frac{1}{2} - i\frac{\sqrt{3}}{2}.$$

Observe that $\omega_2 = \overline{\omega_1}$: $-\frac{1}{2} - i\frac{\sqrt{3}}{2}$ is the complex conjugate of $-\frac{1}{2} + i\frac{\sqrt{3}}{2}$. □

Think About It 21.2 Do you think that a real number could have *exactly one* nth root that is complex, and with all its other *nth* roots real? □

Remark 21.3

1. The nth roots of 1 are also called the *nth roots of unity* ("unity" being 1).

2. For any real number, nth roots come in "complex conjugate pairs." What this means is that if $a + bi$ is an nth root of a real number x, then $\overline{a + bi} = a - bi$ is also an nth root of x.

3. We have not said how to find nth roots that are complex numbers, which is not too hard with a little trigonometry, for example, but that is beyond our scope here. □

Now You Try 21.5

1. Verify that $-\frac{1}{2} - i\frac{\sqrt{3}}{2}$ is a cube root of unity.

2. Verify that the three cube roots of 27 are

$$3\omega_0, \quad 3\omega_1, \quad \text{and} \quad 3\omega_2,$$

where ω_0, ω_1, and ω_2 are the cube roots of unity given in EXAMPLE 21.5. Identify the complex conjugate pairs.

3. Verify that

$$\omega_0 = 1, \quad \omega_1 = i, \quad \omega_2 = -1, \quad \text{and} \quad \omega_3 = -i$$

are the four fourth roots of unity. Identify the complex conjugate pairs.

4. Find the four fourth roots of 16. Identify the complex conjugate pairs.

□

The absolute value of a real number tells us the "size" of the number. This is often described as "the distance from the number to zero on the number line." We would like to define the absolute value of a complex number similarly to tell us the "size" of a complex number.

Definition 21.5 Let $z = a + bi$ be a complex number. The *absolute value* of z is

$$|z| = \sqrt{a^2 + b^2}.$$

□

Think About It 21.3 How would you graph complex numbers? Of what theorem does $|z| = \sqrt{a^2 + b^2}$ remind you? Geometrically, what does $\sqrt{a^2 + b^2}$ represent?   □

Remark 21.4

- The absolute value of z is also called the *magnitude* of z or the *modulus* of z or the *norm* of z.

- If $z = a + bi$, then

$$z\bar{z} = (a + bi)(a - bi) = a^2 + b^2,$$

  so we could have defined the absolute value of z by

$$|z| = \sqrt{z\bar{z}}.$$

- For a complex number z, just as with real numbers, $|z| \geq 0$, with $|z| = 0$ if and only if $z = 0$.

                                                                                                        □

Example 21.6 Let $u = 1$, $v = -1$, $w = i$, and $z = -\frac{1}{2} + i\frac{\sqrt{3}}{2}$. Find $|u|$, $|v|$, $|w|$, and $|z|$.

- $u = 1$

  We note that $u = 1 + 0i$, so

$$|u| = |1 + 0i| = \sqrt{1^2 + 0^2} = \sqrt{1} = 1.$$

- $v = -1$

  We note that $v = -1 + 0i$, so

$$|v| = |-1 + 0i| = \sqrt{(-1)^2 + 0^2} = \sqrt{1} = 1.$$

- $w = i$

  We note that $w = 0 + 1i$, so

$$|w| = |0 + 1i| = \sqrt{0^2 + 1^2} = \sqrt{1} = 1.$$

- $z = -\dfrac{1}{2} + i\dfrac{\sqrt{3}}{2}$

$$|z| = \left| -\frac{1}{2} + i\frac{\sqrt{3}}{2} \right| = \sqrt{\left(-\frac{1}{2}\right)^2 + \left(\frac{\sqrt{3}}{2}\right)^2} = \sqrt{\frac{1}{4} + \frac{3}{4}} = \sqrt{1} = 1$$

                                                                                                        □

Now You Try 21.6 Find the absolute value of the following complex numbers.

1. 5   2. 5i   3. $-1 + 3i$ 4. $\dfrac{\sqrt{2}}{2} - i\dfrac{\sqrt{2}}{2}$

                                                                                                        □

Figure 21.2: Carl Friedrich Gauss. (Source: Courtesy of Jeff Miller.)

Observe that there are only two real numbers x such that $|x| = 1$, namely, $x = 1$ and $x = -1$. However, EXAMPLE 21.6 shows four complex numbers z such that $|z| = 1$. In fact, there is an infinitude of complex numbers z such that $|z| = 1$. More generally, there are only two real numbers x such that $|x| = r$ for any positive real number r, namely, $x = r$ and $x = -r$; however, there is an infinitude of complex numbers z such that $|z| = r$. How can this be?

Think About It 21.4 What is the geometric interpretation of the absolute value for real numbers? How does that imply that $|a| = |-a|$? How would you interpret the absolute value of a complex number so that there is an infinitude of complex numbers that all have the same absolute value? □

This leads us to our last topic on complex numbers: their geometric representation.

Because a complex number has a real part and an imaginary part, we cannot plot them on a number line. But we may treat a complex number as an ordered pair that can be plotted in the "complex plane," a plane that has a pure real axis (the x axis) and a pure imaginary axis (the iy axis). The idea to represent complex numbers in a plane occurred to three persons from three different parts of Europe all about the turn of the nineteenth century: Caspar Wessel (1745–1818), a Norwegian surveyor and cartographer; Jean Robert Argand (1768–1822), a French-Swiss bookkeeper; and Carl Friedrich Gauss (1777–1855), arguably the greatest of all German mathematicians [21, pp. 628–631]. Both Wessel and Argand proposed using line segments or "vectors" to represent complex numbers. Wessel presented his idea first to the Royal Academy of Sciences of Denmark in 1797, and then in a paper published in the *Philosophical Transactions* of the Academy in 1799. Argand published his idea in a booklet titled *Essai sur une manière de représenter les quantités imaginaires dans les constructions géométriques* that was privately printed in 1806. However, perhaps because neither Wessel nor Argand was well known in mathematics, their ideas went largely overlooked. It was not until Gauss (FIGURE 21.2) proposed representing a complex number as a point in the plane that the plane geometric interpretation took hold. Apparently, Gauss had used his idea implicitly in his 1799 doctoral dissertation on the fundamental theorem of algebra; it was not until 1831 before Gauss's idea was described publicly in a commentary on his paper *Theoria Residuorum Biquadraticorum* [21, p. 631]. Today, just as the rectangular coordinate plane (the xy plane, where x and y are real numbers) is often called the *Cartesian plane* (after René Descartes), the complex plane is often called the *Gaussian plane*.

Example 21.7 The following illustrates how we may plot complex numbers $x + iy$ in the complex plane. Note that the horizontal axis is the real axis (x axis), and the vertical axis is the imaginary axis (iy axis).

$$5 \leftrightarrow (5, 0) \qquad -2i \leftrightarrow (0, -2)$$
$$3 + 4i \leftrightarrow (3, 4) \qquad -5 + 4i \leftrightarrow (-5, 4)$$
$$-3 - 5i \leftrightarrow (-3, -5) \qquad 4 - 3i \leftrightarrow (4, -3)$$

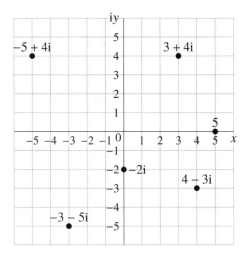

Remark 21.5 Note that the imaginary axis in the complex plane should be labeled with pure imaginary numbers, i, 2i, −i, −2i, and so on. Throughout this book, however, in order to keep the graphs from being too cluttered, we label the imaginary axis with numbers 1, 2, −1, −2, and so on, instead. Nevertheless, keep in mind that the imaginary axis is a pure imaginary number line.

Now You Try 21.7 Plot the following complex numbers as points in the complex plane.

$$-3, \quad 4, \quad 3 - i, \quad -4 + 5i, \quad 2 + 2i, \quad -4 - 3i$$

Now, recall that the absolute value of a real number x is the distance between x and 0 (the origin) on the real number line. For example, |3| is the distance between 3 and 0 on the number line, and |−5| is the distance between −5 and 0 on the number line.

In the same way, the absolute value of a complex number z is the distance between z and 0 or $0 + 0i$ (the origin) in the complex plane.

Think About It 21.5 How many points are the same distance from a given fixed point in the plane? What is the locus of points (that is, what curve is traced by the points) that are the same distance from a given fixed point in the plane?

Example 21.8 Consider the complex numbers $u = 5$, $v = -5$, $w = -5i$, and $z = 3 + 4i$. We plot numbers in the complex plane and find their absolute values.

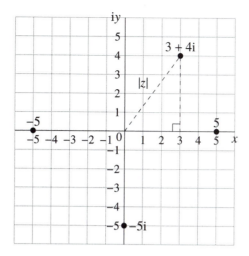

We see that the distance from $u = 5$ (or $5 + 0i$) to 0 (or $0 + 0i$) in the complex plane is 5 (along the real or x axis), and that a direct computation yields

$$|u| = |5 + 0i| = \sqrt{5^2 + 0^2} = \sqrt{25} = 5.$$

Likewise, we see that the distance from $v = -5$ to 0 in the complex plane is 5, and that a direct computation yields

$$|v| = |-5 + 0i| = \sqrt{(-5)^2 + 0^2} = \sqrt{25} = 5.$$

We also see that the distance from $w = -5i$ to 0 is 5 (along the imaginary or iy axis), and that a direct computation yields

$$|w| = |0 + 5i| = \sqrt{0^2 + 5^2} = \sqrt{25} = 5.$$

To find the distance from $z = 3 + 4i$ to 0, we draw a right triangle as shown above. Then, $|z| = |3 + 4i|$ (the distance from $z = 3 + 4i$ to 0) is the length of the hypotenuse of the right triangle with legs 3 and 4. Thus, we apply Pythagoras's theorem (page 216):

$$|z|^2 = 3^2 + 4^2 = 25,$$

so that

$$|z| = \sqrt{25} = 5.$$

On the other hand, a direct computation yields

$$|z| = |3 + 4i| = \sqrt{3^2 + 4^2} = \sqrt{25} = 5,$$

which we see is precisely the application of Pythagoras's theorem, above.

In summary, we find that

$$|5| = 5, \quad |-5| = 5, \quad |5i| = 5, \quad |3 + 4i| = 5.$$

In fact, any complex number that lies on the circle of radius 5 centered at the origin in the complex plane has absolute value 5. See FIGURE 21.3.

$\square$

Now You Try 21.8

1. Find four more complex numbers with absolute value 5. Hint: Use symmetry.

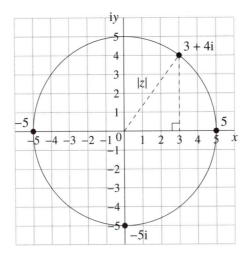

Figure 21.3: Any complex number that has absolute value 5 lies on the circle of radius 5 centered at the origin in the complex plane.

2. Let $z = -2 + iy$. Find y so that $|z| = 5$. Plot z in the complex plane.

3. True or false: If $z = x + iy$ and $|z| = 5$, then $|x| \le 5$ and $|y| \le 5$. Explain why.

☐

Example 21.9 Consider the complex numbers $u = 4$, $v = 4i$, and $w = -2\sqrt{2} - i2\sqrt{2}$. We find their absolute values.

$$|u| = |4 + 0i| = \sqrt{4^2 + 0^2} = \sqrt{16} = 4,$$

$$|v| = |0 + 4i| = \sqrt{0^2 + 4^2} = \sqrt{16} = 4,$$

$$|w| = \left|-2\sqrt{2} - i2\sqrt{2}\right| = \sqrt{(-2\sqrt{2})^2 + (-2\sqrt{2})^2} = \sqrt{8 + 8} = \sqrt{16} = 4.$$

We plot the numbers in the complex plane. Notice that the numbers lie on the circle of radius 4 centered at the origin. In fact, every complex number that has absolute value 4 lies on the circle of radius 4 centered at the origin in the complex plane.

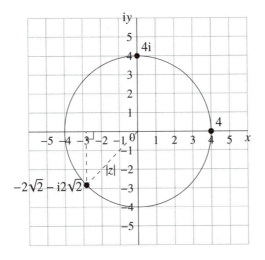

☐

Now You Try 21.9

1. Let $z = 5i$, $u = -4 + 3i$, $v = 4 - 3i$, and $w = 3 + 4i$.

   a) Plot z, u, v, and w in the complex plane and show that

   $$|z| = |u| = |v| = |w|.$$

   b) Find two more complex numbers with the same absolute value, and show that all of these complex numbers lie on the circle of radius $|z|$ centered at the origin in the complex plane.

2. Let $z = 5 + 12i$.

   a) Plot z in the complex plane and find $|z|$.

   b) Find seven other complex numbers that have the same absolute value and show that they all lie on the circle of radius $|z|$ centered at the origin in the complex plane.

3. Let $z = x + 24i$. Find x so that $|z| = 25$.

$\square$

21.2 SOLVING CUBIC EQUATIONS AND THE CUBIC FORMULA

To avoid having coefficients that are either zero or negative numbers, in chapters XI to XXIII of the *Ars magna* [36, pp. 3–4] Cardano lists thirteen cases of the *general cubic equation*,

$$x^3 + bx^2 + cx + d = 0,$$

where b, c, and d are real numbers.

Cardano $(p, q, a, b, c > 0)$	
Cube and first power equal to the number	$x^3 + px = q$
Cube equal to the first power and number	$x^3 = px + q$
Cube and number equal to the first power	$x^3 + q = px$
Cube equal to the square and number	$x^3 = bx^2 + d$
Cube and square equal to the number	$x^3 + bx^2 = d$
Cube and number equal to the square	$x^3 + d = bx^2$
Cube, square, and first power equal to the number	$x^3 + bx^2 + cx = d$
Cube and first power equal to the square and number	$x^3 + cx = bx^2 + d$
Cube and square equal to the first power and number	$x^3 + bx^2 = cx + d$
Cube equal to the square, first power, and number	$x^3 = bx^2 + cx + d$
Cube and number equal to the square and first power	$x^3 + d = bx^2 + cx$
Cube, first power, and number equal to the square	$x^3 + cx + d = bx^2$
Cube, square, and number equal to the first power	$x^2 + bx^2 + d = cx$

The first three cases, "cube and first power equal to the number," "cube equal to the first power and number," and "cube and number equal to the first power," are special cases of what is called the *reduced cubic equation* or the *depressed cubic equation,*

$$x^3 + px + q = 0,$$

where p and q are real numbers. The formula that is commonly referred to as "Cardan's cubic formula" is a formula for solving the reduced cubic equation (one in which the squared term is missing). To use this formula to solve a general cubic equation, one first makes the change of variable $x = y - \frac{b}{3}$:

$$x^3 + bx^2 + cx + d = 0 \quad \longrightarrow \quad y^3 + py + q = 0,$$

where

$$p = c - \frac{b^2}{3} \quad \text{and} \quad q = \frac{2b^3}{27} - \frac{bc}{3} + d.$$

The formula is then used to solve the reduced equation for y. Finally, the solutions of the general equation are given again by the change of variable $x = y - \frac{b}{3}$. This is, indeed, how Cardano solves all the cases of the cubic equation. We leave it to you to verify that the above substitutions work.

In treating the different cases of the cubic in chapters XI to XXIII of the *Ars magna*, with the exception of chapter XVI, Cardano first provides a proof of the method ("demonstratio"), then he states the rule ("regula") for solving the particular case, and follows the rule with one or more examples ("exemplum"); in chapter XVI, Cardano states the rule before providing a proof of the method. His proofs are in the Greek geometrical tradition. To ease the reading in the following examples, we write, for instance, $x^3 + 6x = 20$ where Cardano writes "a cube and 6 things equal 20" ("cubus & 6 positiones, æquantur 20"), and

$$\sqrt[3]{\sqrt{108} + 10} - \sqrt[3]{\sqrt{108} - 10}$$

instead of Cardano's

"$Rv : cub : R\,108\,p : 10\,m : Rv : cubica\,R\,108\,m : 10$".

Rule 21.3 *Ars magna* chapter XI, "On the Cube and First Power Equal to the Number" [36, pp. 96–101],[9] that is, $x^3 + px = q$:

RULE

Cube one-third the coefficient of x; add to it the square of one-half the constant of the equation; and take the square root of the whole. You will duplicate this, and to one of the two you add one-half the number you have already squared and from the other you subtract one-half the same. You will then have a *binomium* and its *apotome*. Then, subtracting the cube root of the *apotome* from the cube root of the *binomium* the remainder [or] that which is left is the value of x.

Cardano illustrates the preceding rule by using it to solve the equation $x^3 + 6x = 20$.

Example 21.10 (*Ars magna* chapter XI [36, p. 99])

---

[9] According to Witmer [36, p. xxii], while he has attempted to preserve the rhetorical form of Cardano's *Ars magna*, he also has not hesitated to use modern terminology and modern symbolism to ease the reading of the translation. For example, Witmer uses "first power" where Cardano writes *res* (Latin), that is translated literally as "thing," to refer to the unknown (x).

For example,

$$x^3 + 6x = 20.$$

Cube 2, one-third of 6, making 8; square 10, one-half the constant; 100 results. Add 100 and 8, making 108, the square root of which is $\sqrt{108}$. This you will duplicate: to one add 10, one-half the constant, and from the other subtract the same. Thus you will obtain the *binomium* $\sqrt{108} + 10$ and its *apotome* $\sqrt{108} - 10$. Take the cube roots of these. Subtract [the cube root of the] *apotome* from that of the *binomium* and you will have the value of x:

$$\sqrt[3]{\sqrt{108} + 10} - \sqrt[3]{\sqrt{108} - 10}$$

□

Note that 2 is a solution of the equation: $2^3 + 6 \cdot 2 = 20$. We leave it for you as an exercise to show that, in fact, $\sqrt[3]{\sqrt{108} + 10} - \sqrt[3]{\sqrt{108} - 10} = 2$.

Example 21.11 We use Cardano's rule to solve the equation

$$x^3 + 12x = 10.$$

1. "Cube one-third the coefficient of x":

    The coefficient of x is 12; one-third of 12 is 4; the cube of 4 is 64.

2. "add to it the square of one-half the constant of the equation":

    The constant is 10; one-half of 10 is 5; the square of 5 is 25; the sum of 64 (from the previous step) and 25 is 89.

3. "and take the square root of the whole":

    The square root of the sum is $\sqrt{89}$.

4. "You will duplicate this, and to one of the two you add one-half the number you have already squared and from the other you subtract one-half of the same. You will then have a *binomium* and its *apotome*":

    Take two $\sqrt{89}$'s, that is, take one $\sqrt{89}$ and another $\sqrt{89}$; to one add 5 (one-half of the constant) and to the other subtract 5 to obtain

    $$\sqrt{89} + 5 \quad (binomium) \text{ and } \quad \sqrt{89} - 5 \quad (apotome).$$

5. "subtracting the cube root of the *apotome* from the cube root of the *binomium* the remainder [or] that which is left is the value of x":

    The cube root of the *binomium* is $\sqrt[3]{\sqrt{89} + 5}$ and the cube root of its *apotome* is $\sqrt[3]{\sqrt{89} - 5}$; the difference between the cube roots is the solution x, that is,

    $$x = \sqrt[3]{\sqrt{89} + 5} - \sqrt[3]{\sqrt{89} - 5}.$$

We check that $\sqrt[3]{\sqrt{89} + 5} - \sqrt[3]{\sqrt{89} - 5}$ is, indeed, a solution of the equation $x^3 + 12x = 10$. Toward this end, we use the identity

$$(a - b)^3 = a^3 - 3a^2b + 3ab^2 - b^3,$$

with $a = \sqrt[3]{\sqrt{89} + 5}$ and $b = \sqrt[3]{\sqrt{89} - 5}$.

To begin, we compute

$$a^3 = \left(\sqrt[3]{\sqrt{89} + 5}\right)^3 = \sqrt{89} + 5,$$

$$a^2 = \left(\sqrt[3]{\sqrt{89} + 5}\right)^2 = \sqrt[3]{(\sqrt{89} + 5)^2} = \sqrt[3]{114 + 10\sqrt{89}},$$

$$a^2b = \sqrt[3]{114 + 10\sqrt{89}} \cdot \sqrt[3]{\sqrt{89} - 5}$$

$$= \sqrt[3]{(114 + 10\sqrt{89})(\sqrt{89} - 5)}$$

$$= \sqrt[3]{64\sqrt{89} + 320}$$

$$= \sqrt[3]{64(\sqrt{89} + 5)}$$

$$= \sqrt[3]{64} \cdot \sqrt[3]{\sqrt{89} + 5} = 4\sqrt[3]{\sqrt{89} + 5},$$

$$b^3 = \left(\sqrt[3]{\sqrt{89} - 5}\right)^3 = \sqrt{89} - 5,$$

$$b^2 = \left(\sqrt[3]{\sqrt{89} - 5}\right)^2 = \sqrt[3]{(\sqrt{89} - 5)^2} = \sqrt[3]{114 - 10\sqrt{89}},$$

$$ab^2 = \sqrt[3]{\sqrt{89} + 5} \cdot \sqrt[3]{114 - 10\sqrt{89}}$$

$$= \sqrt[3]{(\sqrt{89} + 5)(114 - 10\sqrt{89})}$$

$$= \sqrt[3]{64\sqrt{89} - 320}$$

$$= \sqrt[3]{64(\sqrt{89} - 5)}$$

$$= \sqrt[3]{64} \cdot \sqrt[3]{\sqrt{89} - 5} = 4\sqrt[3]{\sqrt{89} - 5}.$$

Thus, we find that

$$(a - b)^3 = a^3 - 3a^2b + 3ab^2 - b^3$$

$$= (\sqrt{89} + 5) - 3 \cdot 4\sqrt[3]{\sqrt{89} + 5} + 3 \cdot 4\sqrt[3]{\sqrt{89} - 5} - (\sqrt{89} - 5)$$

$$= 10 - 12\sqrt[3]{\sqrt{89} + 5} + 12\sqrt[3]{\sqrt{89} - 5},$$

and, therefore,

$$\left(\sqrt[3]{\sqrt{89}+5}-\sqrt[3]{\sqrt{89}-5}\right)^3+12\left(\sqrt[3]{\sqrt{89}+5}-\sqrt[3]{\sqrt{89}-5}\right)$$

$$=\left(10-12\sqrt[3]{\sqrt{89}+5}+12\sqrt[3]{\sqrt{89}-5}\right)+12\left(\sqrt[3]{\sqrt{89}+5}-\sqrt[3]{\sqrt{89}-5}\right)$$

$$= 10. \quad \checkmark$$

☐

Example 21.12 Use Cardano's rule to solve the equation $2x^3 + 12x = 40$.

We first put the equation in the form $x^3 + px = q$ by dividing through the equation by 2:

$$2x^3 + 12x = 40 \quad\Longrightarrow\quad x^3 + 6x = 20.$$

We may now apply Cardano's rule. This equation was solved in EXAMPLE 21.10. ☐

Now You Try 21.10

1. Verify that $\sqrt[3]{\sqrt{108}+10}-\sqrt[3]{\sqrt{108}-10}$ is a solution of $x^3 + 6x = 20$.

2. Solve the following equations using Cardano's rule.

   a) (*Ars magna* chapter XI [36, pp. 99–100]) $x^3 + 3x = 10$
   b) (*Ars magna* chapter XI [36, p. 100]) $x^3 + 6x = 2$
   c) $3x^3 + 18x = 30$

☐

We now move on to Cardano's second case.

Rule 21.4 *Ars magna* chapter XII, "On the Cube Equal to the First Power and Number" [36, pp. 102–103], that is, $x^3 = px + q$:

RULE

The rule, therefore, is: When the cube of one-third the coefficient of x is not greater than the square of one-half the constant of the equation, subtract the former from the latter and add the square root of the remainder to one-half the constant of the equation and, again, subtract it from the same half, and you will have, as was said, a *binomium* and its *apotome*, the sum of the cube roots of which constitutes the value of x.

☐

Example 21.13 (*Ars magna* chapter XII [36, p. 103])

For example,

$$x^3 = 6x + 40.$$

Raise 2, one-third the coefficient of x, to the cube, which makes 8; subtract this from 400, the square of 20, one-half the constant, making 392; the square root of this added to 20 makes $20 + \sqrt{392}$, and subtracted from 20 makes $20 - \sqrt{392}$; the sum of the cube roots of these, $\sqrt[3]{20 + \sqrt{392}} + \sqrt[3]{20 - \sqrt{392}}$, is the value of x.

☐

Note that 4 is a solution of the equation: $4^3 = 6 \cdot 4 + 40$. We leave it for you as an exercise to show that, in fact, $\sqrt[3]{20 + \sqrt{392}} + \sqrt[3]{20 - \sqrt{392}} = 4$.

Example 21.14 We use Cardano's rule to solve the equation

$$x^3 = 3x + 6.$$

1. "the cube of one-third the coefficient of x ... the square of one-half the constant of the equation":

   The coefficient of x is 3; one-third of 3 is 1; the cube of 1 is 1. The constant is 6; one-half of 6 is 3; the square of 3 is 9.

2. "subtract the former from the latter":

   The difference between 9 and 1 is 8.

3. "and add the square root of the remainder to one-half the constant of the equation and, again, subtract it from the same half, and you will have, as was said, a *binomium* and its *apotome*":

   The square root of 8 is $\sqrt{8}$. Add $\sqrt{8}$ to 3 (one-half the constant) and subtract $\sqrt{8}$ from 3 to obtain

   $$3 + \sqrt{8} \quad (binomium) \text{ and } \quad 3 - \sqrt{8} \quad (apotome).$$

4. "the sum of the cube roots of which constitutes the value of x":

   The cube root of the *binomium* is $\sqrt[3]{3 + \sqrt{8}}$ and the cube root of its *apotome* is $\sqrt[3]{3 - \sqrt{8}}$; the sum of the cube roots is the solution x, that is,

   $$x = \sqrt[3]{3 + \sqrt{8}} + \sqrt[3]{3 - \sqrt{8}}.$$

□

Now You Try 21.11

1. Verify that $\sqrt[3]{20 + \sqrt{392}} + \sqrt[3]{20 - \sqrt{392}}$ is a solution of $x^3 = 6x + 40$.

2. Verify that $\sqrt[3]{3 + \sqrt{8}} + \sqrt[3]{3 - \sqrt{8}}$ is a solution of $x^3 = 3x + 6$.

3. Solve the following equations using Cardano's rule.

   a) (*Ars magna* chapter XII [36, p. 103]) $x^3 = 6x + 6$

   b) $x^3 = 3x + 4$ c) $x^3 = 3x + 2$ d) $x^3 = 18x + 35$

□

Think About It 21.6 In the rule to solve $x^3 = px + q$, Cardano states: "When the cube of one-half the coefficient of x is not greater than the square of one-half the constant of the equation...." What would happen otherwise? □

Now we look at Cardano's third case.

Rule 21.5 *Ars magna* chapter XIII, "On the Cube and Number Equal to the First Power" [36, pp. 104–109], that is, $x^3 + q = px$:

RULE

The rule, therefore, is: When the cube and the constant are equal to the first power, find the solution for the cube equal to the same number of y's and the same constant; take three times the square of one-half of this and subtract it from the coefficient of the first power; and the square root of the remainder added to or subtracted from one-half the solution for the cube equal to y plus the constant gives the solution for the cube and constant equal to x.

☐

Example 21.15 (*Ars magna* chapter XIII [36, p. 103])

Example:
$$x^3 + 3 = 8x.$$

Solving
$$y^3 = 8y + 3$$

according to the preceding rule, I obtain 3. The square of one-half of this is $2\frac{1}{4}$, which multiplied by 3 is $6\frac{3}{4}$. Subtracting this from 8, the coefficient of x, leaves $1\frac{1}{4}$, the square root of which added to or subtracted from $1\frac{1}{2}$, which is one-half the solution for the cube equal to the first power and constant, gives both solutions which were being sought. One is $1\frac{1}{2} + \sqrt{1\frac{1}{4}}$, the other $1\frac{1}{2} - \sqrt{1\frac{1}{4}}$.

☐

Remark 21.6 In the preceding example, even though 3 is a solution of $y^3 = 8y + 3$, Cardano could not have used RULE 21.4 to solve the equation as he claims because it is not true that "the cube of one-third the coefficient of x [here y] is not greater than the square of one-half the constant of the equation":

$$\left(\frac{8}{3}\right)^3 = \frac{512}{27}, \quad \left(\frac{3}{2}\right)^2 = \frac{9}{4}, \quad \text{and} \quad \frac{512}{27} \not< \frac{9}{4}.$$

How, then, did Cardano solve the equation? The answer may be found later in the same chapter, where Cardano writes [36, p. 103]:

> When the cube of one-third the coefficient of x is greater than the square of one-half the constant of the equation, which happens whenever the constant is less than three-fourths of this cube or when two-thirds of the coefficient of x multiplied by the square root of one-third the same number is greater than the constant of the equation, then the solution of this can be found by the aliza problem which is discussed in the book of geometrical problems. But if you wish to avoid such a difficulty you may, for the most part, be satisfied by chapter XXV of this work.

Turning to chapter XXV of the *Ars magna*, "On Imperfect and Particular Rules," we find [36, p. 160]:

> 2. When the cube is equal to the first power and constant, find two numbers whose product is the constant and one of which is the square root of the sum of the other and the coefficient of x. The square root is the value.

We apply this rule to the equation $y^3 = 8y + 3$:

1. "find two numbers whose product is the constant":

   The constant is 3, and
   $$3 = 1 \cdot 3, \quad \text{so } \underline{1} \text{ and } \underline{\underline{3}}.$$

2. "and one of which is the square root of the sum of the other and the coefficient of x":

   The coefficient of x is 8, and

   $$\underline{\underline{3}} = \sqrt{1 + 8}.$$

3. "The square root is the value":

   $$y = \underline{\underline{3}}.$$

We point out that we are not sure that this is how Cardano, in fact, proceeded.   □

   In the following example, the solutions of the equation are complex numbers, $a + bi$, where i is the imaginary unit ($i^2 = -1$).

Example 21.16 We use Cardano's rule to solve the equation

$$x^3 + 35 = 18x.$$

1. "find the solution for the cube equal to the same number of y's and the same constant":

   $$y^3 = 18y + 35.$$

2. Solve the new equation in y following RULE 21.4.

   a) "the cube of one-third the coefficient of x [here y] ... the square of one-half the constant of the equation":

      The coefficient of y is 18; one-third of 18 is 6; the cube of 6 is 216. The constant is 35; one-half of 35 is $\frac{35}{2}$; the square of $\frac{35}{2}$ is $\frac{1225}{4}$.

   b) "subtract the former from the latter":

      The difference between $\frac{1225}{4}$ and 216 is $\frac{361}{4}$.

   c) "and add the square root of the remainder to one-half the constant of the equation and, again, subtract it from the same half, and you will have, as was said, a *binomium* and its *apotome*":

      The square root of $\frac{361}{4}$ is $\frac{19}{2}$. Add $\frac{19}{2}$ to $\frac{35}{2}$ (one-half the constant) and subtract $\frac{19}{2}$ from $\frac{35}{2}$ to obtain

      $$\frac{35}{2} + \frac{19}{2} = 27 \quad (binomium) \text{ and } \quad \frac{35}{2} - \frac{19}{2} = 8 \quad (apotome).$$

   d) "the sum of the cube roots of which constitutes the value of x [here y]":

      The cube root of the *binomium* is $\sqrt[3]{27} = 3$ and the cube root of its *apotome* is $\sqrt[3]{8} = 2$; the sum of the cube roots is the solution y, that is,

      $$y = 3 + 2 = 5.$$

3. "take three times the square of one-half of this [that is, y]":

   $$3\left(\frac{5}{2}\right)^2 = \frac{75}{4}.$$

4. "and subtract it from the coefficient of the first power":

   $$18 - \frac{75}{4} = -\frac{3}{4}.$$

5. "and the square root of the remainder added to or subtracted from one-half the solution for the cube equal to y plus the constant gives the solution for the cube and constant equal to x":

$$\sqrt{-\frac{3}{4}} = \frac{\sqrt{3}}{2}i,$$

$$x = \frac{5}{2} + \frac{\sqrt{3}}{2}i \quad \text{and} \quad x = \frac{5}{2} - \frac{\sqrt{3}}{2}i.$$

☐

Now You Try 21.12

1. Verify that $1\frac{1}{2} + \sqrt{1\frac{1}{4}}$ and $1\frac{1}{2} - \sqrt{1\frac{1}{4}}$ are solutions of the equation $x^3 + 3 = 8x$.

2. Verify that $\frac{5}{2} \pm \frac{\sqrt{3}}{2}i$ are solutions of the equation $x^3 + 35 = 18x$.

3. Solve the following equations using Cardano's rule.

   a) $x^3 + 91 = 36x$

   b) $x^3 + 72 = 12\sqrt[3]{4}x.$

☐

Remark 21.7 Cardano would not have expressed the solutions of the equation $x^3 + 35 = 18x$ in the preceding example using the imaginary unit, i; instead, he would have written simply

$$\frac{5}{2} + \sqrt{-\frac{3}{4}} \quad \text{and} \quad \frac{5}{2} - \sqrt{-\frac{3}{4}}.$$

Imaginary numbers were not understood during his day beyond their being a convenient contrivance. Nevertheless, Cardano did address the appearance of the square root of negative numbers in chapter XXXVII of the *Ars magna*. In "Rule II" of that chapter, Cardano considers a problem that is equivalent to solving the equation $x(10 - x) = 30$ or $x(10 - x) = 40$ [36, p. 219]:

> The second species of negative assumption involves the square root of a negative. I will give an example: If it should be said, Divide 10 into two parts the product of which is 30 or 40, it is clear that this case is impossible. Nevertheless, we will work thus: We divide 10 into two equal parts, making each 5. These we square, making 25. Subtract 40, if you will, from the 25 thus produced ... leaving a remainder of -15, the square root of which added to or subtracted from 5 gives parts the product of which is 40. These will be $5 + \sqrt{-15}$ and $5 - \sqrt{-15}$.

Cardano then goes on to give a geometrical demonstration of the solution, in which he says:

> ... you will have to imagine $\sqrt{-15}$ ... and you will have that which you seek, namely $5 + \sqrt{25 - 40}$ and $5 - \sqrt{25 - 40}$, or $5 + \sqrt{-15}$ and $5 - \sqrt{-15}$. Putting aside the mental tortures involved, multiply $5 + \sqrt{-15}$ by $5 - \sqrt{-15}$, making $25 - (-15)$ which is $+15$. Hence this product is $40....$ So progresses arithmetic subtlety the end of which, as is said, is as refined as it is useless.

☐

Think About It 21.7 Attempt to solve the equations $x(10 - x) = 30$ and $x(10 - x) = 40$ by graphing. How do the graphs show that there are no real solutions? Now use the quadratic formula to solve the two equations. Why are there no real solutions?

☐

Let us see if Cardano's rules for solving $x^3 + px = q$, $x^3 = px + q$, and $x^3 + q = px$, where p and q are positive, are related in any way. We apply the three rules to the respective equations with arbitrary positive p and q.

Cube and First Power Equal to the Number	
Step	$x^3 + px = q$
cube one-third the coefficient of x	$(\frac{p}{3})^3$
add it to the square of one-half the constant	$(\frac{q}{2})^2 + (\frac{p}{3})^3$
take the square root of the whole	$\sqrt{(\frac{q}{2})^2 + (\frac{p}{3})^3}$
add one-half the number already squared (*binomium*)	$\sqrt{(\frac{q}{2})^2 + (\frac{p}{3})^3} + \frac{q}{2}$
subtract one-half the same (*apotome*)	$\sqrt{(\frac{q}{2})^2 + (\frac{p}{3})^3} - \frac{q}{2}$

subtracting the cube root of the *apotome* from the cube root of the *binomium* ...is the value of x:

$$x = \sqrt[3]{\sqrt{\left(\frac{q}{2}\right)^2 + \left(\frac{p}{3}\right)^3} + \frac{q}{2}} - \sqrt[3]{\sqrt{\left(\frac{q}{2}\right)^2 + \left(\frac{p}{3}\right)^3} - \frac{q}{2}}.$$

Cube Equal to the First Power and Number	
Step	$x^3 = px + q$
cube of one-third the coefficient of x	$(\frac{p}{3})^3$
square of one-half the constant	$(\frac{q}{2})^2$
subtract the former from the latter	$(\frac{q}{2})^2 - (\frac{p}{3})^3$
add the square root of the remainder to one-half the constant	$\frac{q}{2} + \sqrt{(\frac{q}{2})^2 - (\frac{p}{3})^3}$
and, again, subtract it from the same half	$\frac{q}{2} - \sqrt{(\frac{q}{2})^2 - (\frac{p}{3})^3}$

as was said, a *binomium* and its *apotome*, the sum of the cube roots of which constitutes the value of x:

$$x = \sqrt[3]{\frac{q}{2} + \sqrt{\left(\frac{q}{2}\right)^2 + \left(\frac{p}{3}\right)^3}} + \sqrt[3]{\frac{q}{2} - \sqrt{\left(\frac{q}{2}\right)^2 + \left(\frac{p}{3}\right)^3}}.$$

Cube and Number Equal to the First Power	
Step	$x^3 + q = px$
find the solution for the cube equal to the same number of y's and the same constant, namely, $y^3 = py + q$: $$y = \sqrt[3]{\frac{q}{2} + \sqrt{\left(\frac{q}{2}\right)^2 + \left(\frac{p}{3}\right)^3}} - \sqrt[3]{\frac{q}{2} - \sqrt{\left(\frac{q}{2}\right)^2 + \left(\frac{p}{3}\right)^3}}.$$	
take three times the square of one-half of this and subtract it from the coefficient of the first power	$p - 3(\frac{y}{2})^2$
the square root of the remainder added to or subtracted from one-half the solution [for $y^3 + py = q$] gives the solution [for $x^3 + q = px$]: $$x = \frac{y}{2} \pm \sqrt{p - 3\left(\frac{y}{2}\right)^2}.$$	

Hence, we see that, with p and q positive, the solutions of the "cube and first power equal to the number" and the "cube equal to the first power and number" appear to be very similar, but the solution of the "cube and number equal to the first power" appears to be quite different from the first two. It may surprise you that it can be shown that all three solutions can be obtained from a single formula if we allow p and q to be any real number. This single formula is referred to as the cubic formula or Cardano's (or Cardan's) cubic formula.

Rule 21.6 (CUBIC FORMULA) Let p and q be any real numbers. Then the reduced or depressed cubic equation

$$x^3 + px + q = 0$$

has solutions x_1, x_2, and x_3 given by

$$x_1 = r + s, \quad x_2 = r\omega_1 + s\omega_2, \quad x_3 = r\omega_2 + s\omega_1,$$

where

$$r = \sqrt[3]{-\frac{q}{2} + \sqrt{\frac{q^2}{4} + \frac{p^3}{27}}}, \quad s = \sqrt[3]{-\frac{q}{2} - \sqrt{\frac{q^2}{4} + \frac{p^3}{27}}},$$

$$\omega_1 = -\frac{1}{2} + \frac{\sqrt{3}}{2}i, \quad \omega_2 = -\frac{1}{2} - \frac{\sqrt{3}}{2}i.$$

The quantity $\frac{q^2}{4} + \frac{p^3}{27}$ is called the *discriminant* of the cubic equation. □

Remark 21.8

1. In the cubic formula, ω_1 and ω_2 are the two complex cube roots of unity (see page 388).

2. The discriminant $b^2 - 4ac$ of the quadratic equation $ax^2 + bx + c = 0$ determines the number of real roots of the equation: there are two distinct real roots if $b^2 - 4ac > 0$; one (double) real root if $b^2 - 4ac = 0$; and no real root if $b^2 - 4ac < 0$. In the same way, the discriminant $\frac{q^2}{4} + \frac{p^3}{27}$ of the

cubic equation $x^3 + px + q = 0$ determines the number of real roots of the equation: if p and q are not both zero, then there is one real root if $\frac{q^2}{4} + \frac{p^3}{27} > 0$; two distinct real roots (one is a double root) if $\frac{q^2}{4} + \frac{p^3}{27} = 0$; and three distinct real roots if $\frac{q^2}{4} + \frac{p^3}{27} < 0$. The last case in which a cubic equation has three distinct real roots is referred to as the "irreducible" case.

3. The cubic formula is commonly attributed to Cardano and referred to as "Cardan's cubic formula," even though del Ferro, Tartaglia, Cardano, and Ferrari all had a hand in the solution of the cubic.

□

Example 21.17 We solve the equations in examples 21.10, 21.13, and 21.15 using the cubic formula. The key is to write the equations in the form $x^3 + px + q = 0$—that is, to set the equation equal to zero—before identifying the coefficients p and q for the cubic formula.

1. EXAMPLE 21.10: $x^3 + 6x = 20$.

   First, we write the equation in the form $x^3 + px + q = 0$ and identify p and q:

   $$x^3 + 6x - 20 = 0, \quad \text{so} \quad p = 6 \text{ and } q = -20.$$

   Second, we compute $\frac{q}{2}$, $\frac{q^2}{4}$, and $\frac{p^3}{27}$:

   $$\frac{q}{2} = \frac{-20}{2} = -10, \quad \frac{q^2}{4} = \frac{400}{4} = 100, \quad \frac{p^3}{27} = \frac{216}{27} = 8.$$

   Third, we compute r and s:

   $$r = \sqrt[3]{-\frac{q}{2} + \sqrt{\frac{q^2}{4} + \frac{p^3}{27}}} = \sqrt[3]{10 + \sqrt{108}},$$

   $$s = \sqrt[3]{-\frac{q}{2} - \sqrt{\frac{q^2}{4} + \frac{p^3}{27}}} = \sqrt[3]{10 - \sqrt{108}}.$$

   Fourth, we compute $x_1 = r + s$, $x_2 = r\omega_1 + s\omega_2$, and $x_3 = r\omega_2 + s\omega_1$, where $\omega_1 = -\frac{1}{2} + \frac{\sqrt{3}}{2}i$ and $\omega_2 = -\frac{1}{2} - \frac{\sqrt{3}}{2}i$:

   $$x_1 = r + s = \sqrt[3]{10 + \sqrt{108}} + \sqrt[3]{10 - \sqrt{108}},$$

   $$x_2 = r\omega_1 + s\omega_2$$
   $$= -\frac{1}{2}(r + s) + i\frac{\sqrt{3}}{2}(r - s)$$
   $$= -\frac{1}{2}\left(\sqrt[3]{10 + \sqrt{108}} + \sqrt[3]{10 - \sqrt{108}}\right)$$
   $$+ i\frac{\sqrt{3}}{2}\left(\sqrt[3]{10 + \sqrt{108}} - \sqrt[3]{10 - \sqrt{108}}\right),$$

   $$x_3 = r\omega_2 + s\omega_1$$
   $$= -\frac{1}{2}(r + s) - i\frac{\sqrt{3}}{2}(r - s)$$

$$= -\frac{1}{2}\left(\sqrt[3]{10 + \sqrt{108}} + \sqrt[3]{10 - \sqrt{108}}\right)$$

$$- i\frac{\sqrt{3}}{2}\left(\sqrt[3]{10 + \sqrt{108}}\right) - \left(\sqrt[3]{10 - \sqrt{108}}\right).$$

Therefore, we see that the only real root is x_1, and this is, in fact, the root found using Cardano's rule in EXAMPLE 21.10:

$$x_1 = \sqrt[3]{10 + \sqrt{108}} + \sqrt[3]{10 - \sqrt{108}}$$

$$= \sqrt[3]{\sqrt{108} + 10} - \sqrt[3]{\sqrt{108} - 10};$$

as we remarked earlier, this is 2 in disguise, a fact that we leave to you to show as an exercise. The two other roots, x_2 and x_3, are complex and simplify to $-1 + 3i$ and $-1 - 3i$, respectively. Note that the discriminant $\frac{q^2}{4} + \frac{p^3}{27} = 108 > 0$.

2. EXAMPLE 21.13: $x^3 = 6x + 40$.

First, we write the equation in the form $x^3 + px + q = 0$ and identify p and q:

$$x^3 - 6x - 40 = 0, \quad \text{so} \quad p = -6 \text{ and } q = -40.$$

Second, we compute $\frac{q}{2}$, $\frac{q^2}{4}$, and $\frac{p^3}{27}$:

$$\frac{q}{2} = \frac{-40}{2} = -20, \quad \frac{q^2}{4} = \frac{1600}{4} = 400, \quad \frac{p^3}{27} = \frac{-216}{27} = -8.$$

Third, we compute r and s:

$$r = \sqrt[3]{-\frac{q}{2} + \sqrt{\frac{q^2}{4} + \frac{p^3}{27}}} = \sqrt[3]{20 + \sqrt{392}},$$

$$s = \sqrt[3]{-\frac{q}{2} - \sqrt{\frac{q^2}{4} + \frac{p^3}{27}}} = \sqrt[3]{20 - \sqrt{392}}.$$

Fourth, we compute $x_1 = r + s$, $x_2 = r\omega_1 + s\omega_2$, and $x_3 = r\omega_2 + s\omega_1$, where $\omega_1 = -\frac{1}{2} + \frac{\sqrt{3}}{2}i$ and $\omega_2 = -\frac{1}{2} - \frac{\sqrt{3}}{2}i$:

$$x_1 = r + s = \sqrt[3]{20 + \sqrt{392}} + \sqrt[3]{20 - \sqrt{392}},$$

$$x_2 = r\omega_1 + s\omega_2$$

$$= -\frac{1}{2}(r + s) + i\frac{\sqrt{3}}{2}(r - s)$$

$$= -\frac{1}{2}\left(\sqrt[3]{20 + \sqrt{392}} + \sqrt[3]{20 - \sqrt{392}}\right)$$

$$+ i\frac{\sqrt{3}}{2}\left(\sqrt[3]{20 + \sqrt{392}} - \sqrt[3]{20 - \sqrt{392}}\right),$$

$$x_3 = r\omega_2 + s\omega_1$$

$$= -\frac{1}{2}(r+s) - i\frac{\sqrt{3}}{2}(r-s)$$

$$= -\frac{1}{2}\left(\sqrt[3]{20+\sqrt{392}} + \sqrt[3]{20-\sqrt{392}}\right)$$

$$\quad - i\frac{\sqrt{3}}{2}\left(\sqrt[3]{20+\sqrt{392}} - \sqrt[3]{20-\sqrt{392}}\right).$$

Therefore, we see that the only real root is x_1, and this is, in fact, the root found using Cardano's rule in EXAMPLE 21.13; as we remarked earlier, this is 4 in disguise, a fact that we leave to you to show as an exercise. The other two roots, x_2 and x_3, are complex and simplify to $-2 + i\sqrt{6}$ and $-2 - i\sqrt{6}$, respectively. Note that the discriminant $\frac{q^2}{4} + \frac{p^3}{27} = 392 > 0$.

3. EXAMPLE 21.15: $x^3 + 3 = 8x$.

First, we write the equation in the form $x^3 + px + q = 0$ and identify p and q:

$$x^3 - 8x + 3 = 0, \quad \text{so} \quad p = -8 \text{ and } q = 3.$$

Second, we compute $\frac{q}{2}$, $\frac{q^2}{4}$, and $\frac{p^3}{27}$:

$$\frac{q}{2} = \frac{3}{2}, \quad \frac{q^2}{4} = \frac{9}{4}, \quad \frac{p^3}{27} = \frac{-512}{27}.$$

Third, we compute r and s:

$$r = \sqrt[3]{-\frac{q}{2} + \sqrt{\frac{q^2}{4} + \frac{p^3}{27}}} = \sqrt[3]{-\frac{3}{2} + i\sqrt{\frac{1805}{108}}},$$

$$s = \sqrt[3]{-\frac{q}{2} - \sqrt{\frac{q^2}{4} + \frac{p^3}{27}}} = \sqrt[3]{-\frac{3}{2} - i\sqrt{\frac{1805}{108}}}.$$

Fourth, we compute $x_1 = r + s$, $x_2 = r\omega_1 + s\omega_2$, and $x_3 = r\omega_2 + s\omega_1$, where $\omega_1 = -\frac{1}{2} + \frac{\sqrt{3}}{2}i$ and $\omega_2 = -\frac{1}{2} - \frac{\sqrt{3}}{2}i$:

$$x_1 = r + s = \sqrt[3]{-\frac{3}{2} + i\sqrt{\frac{1805}{108}}} + \sqrt[3]{-\frac{3}{2} - i\sqrt{\frac{1805}{108}}},$$

$$x_2 = r\omega_1 + s\omega_2$$

$$= -\frac{1}{2}(r+s) + i\frac{\sqrt{3}}{2}(r-s)$$

$$= -\frac{1}{2}\left(\sqrt[3]{-\frac{3}{2} + i\sqrt{\frac{1805}{108}}} + \sqrt[3]{-\frac{3}{2} - i\sqrt{\frac{1805}{108}}}\right)$$

$$\quad + i\frac{\sqrt{3}}{2}\left(\sqrt[3]{-\frac{3}{2} + i\sqrt{\frac{1805}{108}}} - \sqrt[3]{-\frac{3}{2} - i\sqrt{\frac{1805}{108}}}\right),$$

$$x_3 = r\omega_2 + s\omega_1$$

$$= -\frac{1}{2}(r + s) - i\frac{\sqrt{3}}{2}(r - s)$$

$$= -\frac{1}{2}\left(\sqrt[3]{-\frac{3}{2} + i\sqrt{\frac{1805}{108}}} + \sqrt[3]{-\frac{3}{2} - i\sqrt{\frac{1805}{108}}}\right)$$

$$- i\frac{\sqrt{3}}{2}\left(\sqrt[3]{-\frac{3}{2} + i\sqrt{\frac{1805}{108}}} - \sqrt[3]{-\frac{3}{2} - i\sqrt{\frac{1805}{108}}}\right).$$

Now, in EXAMPLE 21.15, Cardano finds the roots of the equation $x^3 + 3 = 8x$ to be $1\frac{1}{2} \pm \sqrt{1\frac{1}{4}}$, but none of x_1, x_2, or x_3 that we found using the cubic formula appears to be either of the roots that Cardano finds. Indeed, we see that x_1, x_2, and x_3 all contain the imaginary number $\sqrt{-\frac{1805}{108}} = i\sqrt{\frac{1805}{108}}$. It turns out that using some very clever algebra to simplify x_1, x_2, and x_3 reveals that, in fact,

$$x_1 = 1\frac{1}{2} + \sqrt{1\frac{1}{4}}, \quad x_2 = -3, \quad \text{and} \quad x_3 = 1\frac{1}{2} - \sqrt{1\frac{1}{4}}.$$

Therefore, the equation has three real roots. Note that the discriminant $\frac{q^2}{4} + \frac{p^3}{27} = -\frac{1805}{108} < 0$.

□

Now You Try 21.13

1. Let

$$x_2 = -\frac{1}{2}\left(\sqrt[3]{\left\{-\frac{3}{2} + i\sqrt{\frac{1805}{108}}\right\}} + \sqrt[3]{\left\{-\frac{3}{2} - i\sqrt{\frac{1805}{108}}\right\}}\right)$$

$$+ i\frac{\sqrt{3}}{2}\left(\sqrt[3]{\left\{-\frac{3}{2} + i\sqrt{\frac{1805}{108}}\right\}} - \sqrt[3]{\left\{-\frac{3}{2} - i\sqrt{\frac{1805}{108}}\right\}}\right)$$

and

$$x_3 = -\frac{1}{2}\left(\sqrt[3]{\left\{-\frac{3}{2} + i\sqrt{\frac{1805}{108}}\right\}} + \sqrt[3]{\left\{-\frac{3}{2} - i\sqrt{\frac{1805}{108}}\right\}}\right)$$

$$- i\frac{\sqrt{3}}{2}\left(\sqrt[3]{\left\{-\frac{3}{2} + i\sqrt{\frac{1805}{108}}\right\}} - \sqrt[3]{\left\{-\frac{3}{2} - i\sqrt{\frac{1805}{108}}\right\}}\right).$$

Given that

$$\sqrt[3]{-\frac{3}{2} + i\sqrt{\frac{1805}{108}}} + \sqrt[3]{-\frac{3}{2} - i\sqrt{\frac{1805}{108}}} = 1\frac{1}{2} + \sqrt{1\frac{1}{4}}$$

and

$$\sqrt[3]{-\frac{3}{2} + i\sqrt{\frac{1805}{108}}} - i\sqrt[3]{\frac{3}{2} - \sqrt{\frac{1805}{108}}} = i\left(\sqrt{6\frac{3}{4}} - \sqrt{\frac{5}{12}}\right),$$

show that

a) $x_2 = -3$.

b) $x_3 = 1\frac{1}{2} - \sqrt{1\frac{1}{4}}$.

2. Show that $2, -1 + 3i$ and $-1 - 3i$ are roots of the equation $x^3 + 6x = 20$ in EXAMPLE 21.10.

3. Show that $4, -2 + i\sqrt{6}$ and $-2 - i\sqrt{6}$ are roots of the equation $x^3 = 6x + 40$ in EXAMPLE 21.13.

4. Use the discriminant to determine the number of distinct real roots of each equation. Then use the cubic formula (RULE 21.6) to solve the following equations.

a) $x^3 + 12x = 10$.

b) $x^3 = 3x + 6$.

c) $x^3 + 35 = 18x$.

□

When the discriminant of the cubic is negative, the cubic formula produces cube roots of complex numbers. Curiously, it is precisely in this case that we have three distinct real roots of the cubic equation. (Otherwise, we have only one or two distinct real roots, depending on whether the discriminant is positive or zero, respectively.) The case when the discriminant of the cubic is negative has come to be known as the *irreducible case*. Cardano did not pursue the irreducible case of the cubic. It would be left to Rafael Bombelli of Bologna to pick up the torch on this matter. In his seminal work *L'Algebra* [14], first printed in 1569, Bombelli writes [76, pp. 14–15]:[10]

> ...I have found another kind of cubic root of a polynomial which is very different from the others. This [cubic root] arises in the chapter dealing with the equation of the kind $x^3 = px + q$, when $p^3/27 > q^2/4$, as we will show in that chapter. This kind of square root has in its calculation [algorismo] different operations than the others and has a different name. Since when $p^3/27 > q^2/4$, the square root of their difference can be neither positive nor negative, therefore I will call it "more than minus" when it should be added and "less than minus" when it should be subtracted. This operation is extremely necessary, more than the other cubic roots of polynomia, which come up when we treat equations of the kind $x^4 + ax^3 + b$ or $x^4 + ax + b$ or $x^4 + ax^3 + ax + b$. Because in solving these equations, the cases in which we obtain this [new] kind of root are many more than the cases in which we obtain the other kind. [This new kind of root] will seem to most people more sophistic than real. This was the opinion I held, too, until I found its geometrical proof (as it will be shown in the proof given in the above mentioned chapter on the plane). I will first treat multiplication, giving the law of plus and minus:[11]

$$(+)(+i) = +i$$
$$(-)(+i) = -i$$
$$(+)(-i) = -i$$
$$(-)(-i) = +i$$
$$(+i)(+i) = -$$
$$(+i)(-i) = +$$
$$(-i)(+i) = +$$
$$(-i)(-i) = -$$

---

[10]La Nave and Mazur have shortened the translation by using modern algebraic notation.

[11]In a footnote, La Nave and Mazur provide this more literal translation of Bombelli's text:

Plus times more than minus makes more than minus.

Minus times more than minus makes less than minus.

Plus times less than minus makes less than minus.

Minus times less than minus makes more than minus.

More than minus times more than minus makes minus.

More than minus times less than minus makes plus.

Less than minus times more than minus makes plus.

Less than minus times less than minus makes minus.

Notice that this kind of root of polynomials cannot be obtained if not together with its conjugate. For instance, the conjugate of $\sqrt[3]{2 + i\sqrt{2}}$ will be $\sqrt[3]{2 - i\sqrt{2}}$. It has never happened to me to find one of these kinds of cubic root without its conjugate. It can also happen that the second quantity [inside the cubic root] is a number and not a root (as we will see in solving equations). Yet, [even if the second quantity is a number], an expression like $\sqrt[3]{2} + 2i$ cannot be reduced to only one monomial, despite the fact that both 2 and 2i are numbers.

The famous example of Bombelli's is this [21, pp. 326–328]:

$$x^3 = 15x + 4.$$

A direct substitution shows that 4 is a solution of the equation, yet the cubic formula produces

$$x = \sqrt[3]{2 + 11i} + \sqrt[3]{2 - 11i}.$$

Bombelli's genius was to presume that the two cube roots are, themselves, complex conjugates, that is to say,

$$\sqrt[3]{2 + 11i} = a + bi \quad \text{and} \quad \sqrt[3]{2 - 11i} = a - bi.$$

(There was no reason at that time to think that the cube root of a complex number may be a complex number.) He then proceeded to show that, in fact,

$$\sqrt[3]{2 + 11i} = 2 + i \quad \text{and} \quad \sqrt[3]{2 - 11i} = 2 - i,$$

so that

$$\sqrt[3]{2 + 11i} + \sqrt[3]{2 - 11i} = (2 + i) + (2 - i) = 4.$$

About all this, Bombelli says interestingly [21, p. 328]:

> It was a wild thought in the judgement of many; and I too for a long time was of the same opinion. The whole matter seemed to rest on sophistry rather than on truth. Yet I sought so long, until I actually proved this to be case.

To show that $\sqrt[3]{2 + 11i} = 2 + i$, Bombelli first cubes both sides of the equation

$$\sqrt[3]{2 + 11i} = a + bi.$$

This yields

$$2 + 11i = (a + bi)^3$$
$$= a^3 + 3a^2 \cdot (bi) + 3a \cdot (bi)^2 + (bi)^3$$
$$= a^3 + 3a^2bi - 3ab^2 - b^3i$$
$$= (a^3 - 3ab^2) + i(3a^2b - b^3).$$

Thus, by equating real and imaginary parts, we have the following system of equations for a and b:

$$a^3 - 3ab^2 = 2,$$
$$3a^2b - b^3 = 11.$$

Now, if we guess that $a = 2$, substitution into the system yields $b = 1$. Therefore, we conclude that

$$\sqrt[3]{2 + 11i} = 2 + i.$$

Similarly, by cubing both sides of the equation

$$\sqrt[3]{2 - 11i} = a - bi,$$

we are led to the conclusion that

$$\sqrt[3]{2 - 11i} = 2 - i.$$

Example 21.18 Had we been stubborn and used RULE 21.4 to solve the equation $y^3 = 8y + 3$ in EXAMPLE 21.15 (see REMARK 21.6 on page 401), we would have obtained the solution

$$y = \sqrt[3]{\frac{3}{2} + \sqrt{-\frac{1805}{108}}} + \sqrt[3]{\frac{3}{2} - \sqrt{-\frac{1805}{108}}} = \sqrt[3]{\frac{3}{2} + i\sqrt{\frac{1805}{108}}} + \sqrt[3]{\frac{3}{2} - i\sqrt{\frac{1805}{108}}}.$$

We show that, in fact, $y = 3$.

We use Bombelli's trick and let

$$\sqrt[3]{\frac{3}{2} + i\sqrt{\frac{1805}{108}}} = a + ib \quad \text{and} \quad \sqrt[3]{\frac{3}{2} - i\sqrt{\frac{1805}{108}}} = a - ib.$$

Cubing both sides of the equation $\sqrt[3]{\frac{3}{2} + i\sqrt{\frac{1805}{108}}} = a + ib$, we find that

$$\frac{3}{2} + i\sqrt{\frac{1805}{108}} = (a + ib)^3$$
$$= a^3 + i3a^2b - 3ab^2 - ib^3$$
$$= (a^3 - 3ab^2) + i(3a^2b - b^3).$$

Thus,

$$a^3 - 3ab^2 = \frac{3}{2},$$

$$3a^2b - b^3 = \sqrt{\frac{1805}{108}}.$$

To solve this system, we suppose that $a = \frac{3}{2}$. Then,

$$a^3 - 3ab^2 = \left(\frac{3}{2}\right)^3 - 3\left(\frac{3}{2}\right)b^2,$$

so that $a^3 - 3ab^2 = \frac{3}{2}$ is

$$\left(\frac{3}{2}\right)^3 - 3\left(\frac{3}{2}\right)b^2 = \frac{3}{2}$$
$$\frac{27}{8} - \frac{9}{2}b^2 = \frac{3}{2}$$
$$b^2 = \frac{5}{12}$$
$$b = \pm\sqrt{\frac{5}{12}}.$$

We choose $b = \sqrt{\frac{5}{12}}$ and use this, together with $a = \frac{3}{2}$, in the second equation, namely, $3a^2b - b^3 = \sqrt{\frac{1805}{108}}$; we find that

$$3a^2b - b^3 = 3 \cdot \frac{9}{4} \cdot \sqrt{\frac{5}{12}} - \left(\sqrt{\frac{5}{12}}\right)^3 = \frac{76}{12}\sqrt{\frac{5}{12}} = \sqrt{\frac{76^2}{12^2} \cdot \frac{5}{12}} = \sqrt{\frac{1805}{108}}. \quad \checkmark$$

Hence, we see that

$$\sqrt[3]{\frac{3}{2} + i\sqrt{\frac{1805}{108}}} = a + ib = \frac{3}{2} + i\sqrt{\frac{5}{12}};$$

similarly,

$$\sqrt[3]{\frac{3}{2} - i\sqrt{\frac{1805}{108}}} = a - ib = \frac{3}{2} - i\sqrt{\frac{5}{12}}.$$

Therefore,

$$y = \sqrt[3]{\frac{3}{2} + i\sqrt{\frac{1805}{108}}} + \sqrt[3]{\frac{3}{2} - i\sqrt{\frac{1805}{108}}} = \left(\frac{3}{2} + i\sqrt{\frac{5}{12}}\right) + \left(\frac{3}{2} - i\sqrt{\frac{5}{12}}\right) = 3.$$

□

Example 21.19 Find the square roots of i.

To begin, we use Bombelli's trick and assume that the square roots of i are $a + bi$, for example, $\sqrt{i} = a + bi$. Then, squaring both sides of the equation, we find that

$$\begin{aligned} i &= (a + bi)^2 \\ &= a^2 + abi + abi + b^2i^2 \\ &= (a^2 - b^2) + 2abi. \end{aligned}$$

Thus, because $i = 0 + 1i$, it follows that

$$a^2 - b^2 = 0,$$
$$2ab = 1.$$

To solve this system, we note that the first equation implies that

$$a^2 = b^2 \quad \Longrightarrow \quad a = \pm b.$$

Now, the second equation implies that

$$ab = \frac{1}{2},$$

which is positive, so that a and b are either both positive or both negative. Hence, it cannot be that $a = -b$, and it must be that $a = b$. Substituting $a = b$ into the equation $ab = \frac{1}{2}$, we find that

$$b^2 = \frac{1}{2} \quad \Longrightarrow \quad b = \pm\frac{1}{\sqrt{2}}.$$

Finally, since $a = b$, we conclude that the square roots of i are

$$\frac{1}{\sqrt{2}} + i\frac{1}{\sqrt{2}} \quad \text{and} \quad -\frac{1}{\sqrt{2}} - i\frac{1}{\sqrt{2}}.$$

□

Now You Try 21.14

1. We found that the square roots of i are $\frac{1}{\sqrt{2}} + i\frac{1}{\sqrt{2}}$ and $-\frac{1}{\sqrt{2}} - i\frac{1}{\sqrt{2}}$. Rationalize the denominators.

2. Use Bombelli's trick to find the square roots of $-4i$.

□

Chapter 22

Polynomial Equations

Al-Khwārizmī, as we saw in EXAMPLE 20.8 on page 317, solves the equation $x^2 + 21 = 10x$ ("a square and twenty-one in numbers are equal to ten roots") by completing the square to find that the roots or solutions are 3 and 7.

Observe that

$$x = 3 \quad \text{implies} \quad x - 3 = 0, \quad \text{so that} \quad (x-3)b = 0b = 0$$

for any number b. Similarly,

$$x = 7 \quad \text{implies} \quad x - 7 = 0, \quad \text{so that} \quad a(x-7) = a0 = 0$$

for any number a. Thus, if we assume that either $x = 3$ or $x = 7$, then

$$(x-3)(x-7) = 0.$$

Now, expanding the left-hand side of this equation yields the equation

$$x^2 - 10x + 21 = 0 \quad \text{or} \quad x^2 + 21 = 10x.$$

Hence, we see that we have come full circle: By completing the square, we find that the roots of the equation $x^2 + 21 = 10x$ are 3 and 7; and 3 and 7 lead to the factors $x - 3$ and $x - 7$ that generate the equation $x^2 + 21 = 10x$.

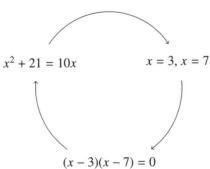

$$x^2 + 21 = 10x \qquad\qquad x = 3, x = 7$$

$$(x-3)(x-7) = 0$$

Remark 22.1 Thomas Harriot (ca. 1560–1621) and René Descartes (1596–1650) were the first known to have systematically written equations as "something $= 0$" [11, p. 108], and in their theory of equations they both set the factored form of polynomials equal to zero. In this way, they would already know the roots of the equation so that, upon expanding the expression, they would better be able to study the relation between the roots of the equation and the coefficients of the polynomial expression. Factoring was not a method for solving equations. This is contrary to how we generally teach the solving of polynomial equations today, particularly quadratic equations, that is, by factoring the polynomial expression after having set it equal to zero. □

Example 22.1 To find a polynomial equation that has roots 4 and 11, we note that

$$x = 4 \quad \text{implies} \quad x - 4 = 0, \quad \text{so that} \quad (x - 4)b = 0b = 0$$

for any number b, and that

$$x = 11 \quad \text{implies} \quad x - 11 = 0, \quad \text{so that} \quad a(x - 11) = a0 = 0$$

for any number a. Thus, if we assume that either $x = 4$ or $x = 11$, then

$$(x - 4)(x - 11) = 0.$$

Hence, expanding the left-hand side of this equation, we obtain the quadratic equation

$$x^2 - 15x + 44 = 0.$$

We check that 4 and 11 are roots of the quadratic equation.

$$x = 4: \quad 4^2 - 15 \cdot 4 + 44 = 16 - 60 + 44 = 0. \qquad \checkmark$$
$$x = 11: \quad 11^2 - 15 \cdot 11 + 44 = 121 - 165 + 44 = 0. \qquad \checkmark$$

□

Example 22.2 To find a polynomial equation that has roots 0, -2 and 5, we note that if we assume that $x = 0$, $x = -2$, or $x = 5$, then

$$x(x + 2)(x - 5) = 0$$

is an equation whose roots are 0, -2, and 5. Hence, expanding the left-hand side of this equation, we obtain the cubic equation

$$x^3 - 3x^2 - 10x = 0.$$

We check that 0, -2, and 5 are roots of the cubic equation.

$$x = 0: \quad 0^3 - 3 \cdot 0^2 - 10 \cdot 0 = 0 - 0 - 0 = 0. \qquad \checkmark$$
$$x = -2: \quad (-2)^3 - 3(-2)^2 - 10(-2) = -8 - 12 + 20 = 0. \qquad \checkmark$$
$$x = 5: \quad 5^3 - 3 \cdot 5^2 - 10 \cdot 5 = 125 - 75 - 50 = 0. \qquad \checkmark$$

□

22.1   RELATION BETWEEN ROOTS AND COEFFICIENTS

That we may generate a polynomial equation with known roots, instead of finding the roots of a given equation, is a marvelous observation, for it allows us to study the relations between the roots of a polynomial equation and its coefficients. For example, suppose that a polynomial equation has roots a and b. Then,

$$(x - a)(x - b) = 0 \quad \text{leads to} \quad x^2 - (a + b)x + ab = 0,$$

so we see that the equation we obtained with the two roots, a and b, has degree 2; moreover, the coefficient of the linear term (the x term) is the opposite of the sum of the roots, and the constant term is the product of the roots.

Example 22.3 The roots of $x^2 - 15x + 44 = 0$ are 4 and 11, and

$$x^2 - 15x + 44 = x^2 - (4 + 11)x + (4)(11).$$

As a second example, the roots of $x^2 + 10x - 39 = 0$ are 3 and -13, for

$$3^2 - 10 \cdot 3 - 39 = 0 \quad \text{and} \quad (-13)^2 - 10(-13) - 39 = 0,$$

and

$$x^2 + 10x - 39 = x^2 - (3 + (-13))x + (3)(-13).$$

□

Example 22.4 Let $a = \sqrt{2}$ and $b = \sqrt{8}$. Then, because $\sqrt{8} = 2\sqrt{2}$,

$$a + b = \sqrt{2} + \sqrt{8} = \sqrt{2} + 2\sqrt{2} = 3\sqrt{2},$$
$$ab = \sqrt{2} \cdot \sqrt{8} = \sqrt{16} = 4.$$

Thus, $\sqrt{2}$ and $\sqrt{8}$ are roots of the equation

$$x^2 - \underbrace{(\sqrt{2} + \sqrt{8})}_{a+b}x + \underbrace{\sqrt{2} \cdot \sqrt{8}}_{ab} = 0 \quad \text{or} \quad x^2 - 3\sqrt{2}\,x + 4 = 0.$$

We check:

$$x = \sqrt{2}: \quad (\sqrt{2})^2 - 3\sqrt{2} \cdot \sqrt{2} + 4 = 2 - 6 + 4 = 0. \qquad \checkmark$$
$$x = \sqrt{8}: \quad (\sqrt{8})^2 - 3\sqrt{2} \cdot \sqrt{8} + 4 = 8 - 12 + 4 = 0. \qquad \checkmark$$

□

Example 22.5 Let $a = -3 + 2i$ and $b = -3 - 2i$, where i is the imaginary unit ($i^2 = -1$). Then,

$$a + b = (-3 + 2i) + (-3 - 2i) = -6,$$
$$ab = (-3 + 2i)(-3 - 2i) = 9 - 4i^2 = 13.$$

Thus, $-3 + 2i$ and $-3 - 2i$ are roots of the equation

$$x^2 - \underbrace{((-3 + 2i) + (-3 - 2i))}_{a+b}x + \underbrace{(-3 + 2i)(-3 - 2i)}_{ab} = 0$$

or

$$x^2 + 6x + 13 = 0.$$

□

Now You Try 22.1

1. Verify that $-3 + 2i$ and $-3 - 2i$ are roots of the equation $x^2 + 6x + 13 = 0$.

2. Show that -7 and -3 are roots of the equation $x^2 + 10x + 21 = 0$. Express the coefficients of the linear term and the constant term in the equation in terms of its roots.

□

Think About It 22.1 Having made the observations above, what two questions would you ask yourself if you were given a quadratic equation such as $x^2 + 7x + 10 = 0$ to solve? □

Now suppose that a polynomial equation has roots a, b, and c. Then,

$$(x - a)(x - b)(x - c) = 0$$

leads to

$$x^3 - (a + b + c)x^2 + (ab + ac + bc)x - abc = 0,$$

so we see that the equation we obtained that has the three roots, a, b, and c, has degree 3; moreover, the coefficient of the quadratic term (the x^2 term) is the opposite of the sum of the roots, the coefficient of the linear term is the sum of the product of the roots taken two at a time, and the constant term is the opposite of the product of all the roots.

Think About It 22.2 Do you see a pattern emerging? What might the polynomial equation with four roots, say a, b, c, and d, look like? Verify your conjecture. □

Now You Try 22.2

1. Verify that

$$(x - a)(x - b)(x - c) = x^3 - (a + b + c)x^2 + (ab + ac + bc)x - abc.$$

2. The roots of the equation $x^3 - 3x^2 - 10x = 0$ are 0, -2, and 5. Express the coefficients of the quadratic and linear terms and the constant term in the equation in terms of its roots.

3. The roots of $x^3 + 4x^2 + x - 6 = 0$ are 1, -2, and -3. Express the coefficients of the quadratic and linear terms and the constant term in the equation in terms of its roots.

□

Example 22.6 We have observed that if a and b are roots of the equation $x^2 + Bx + C = 0$, then

$$B = -(a + b) \quad \text{and} \quad C = ab.$$

Thus, if $a = -7$ and $b = 5$, for instance, then

$$B = -(-7 + 5) = 2 \quad \text{and} \quad C = (-7)(5) = -35,$$

so that -7 and 5 are roots of the equation $x^2 + 2x - 35 = 0$. We check:

$$x = -7 : \quad (-7)^2 + 2(-7) - 35 = 49 - 14 - 35 = 0. \quad \checkmark$$
$$x = 5 : \quad \quad 5^2 + 2 \cdot 5 - 35 = 25 + 10 - 35 = 0. \quad \checkmark$$

□

Example 22.7 If a and b are roots of the equation $x^2 + Bx + C = 0$, then

$$a + b = -B \quad \text{and} \quad ab = C.$$

Suppose that a and b are solutions of the equation

$$x^2 - x - 6 = 0.$$

Since $B = -1$ and $C = -6$, we find that

$$a + b = -B = 1 \quad \text{and} \quad ab = C = -6.$$

So we need two numbers whose sum is 1 and whose product is -6. We start by considering the integer factors of -6, namely, ± 1, ± 2, ± 3, and ± 6.

a	b	$a + b$	ab
$+1$	-6	-5	-6
-1	$+6$	$+5$	-6
$+2$	-3	-1	-6
-2	$+3$	$+1$	-6

Therefore, we choose $a = -2$ and $b = 3$, so that $a + b = 1$, $ab = -6$, and

$$(x - (-2))(x - 3) = 0 \quad \text{or} \quad (x + 2)(x - 3) = 0.$$

Moreover,

$$(x + 2)(x - 3) = x^2 - x - 6.$$

We check that -2 and 3 are roots of the equation $x^2 - x - 6 = 0$:

$$x = -2 : \quad (-2)^2 - (-2) - 6 = 4 + 2 - 6 = 0. \qquad \checkmark$$
$$x = 3 : \quad \ \ 3^2 - 3 - 6 = 9 - 3 - 6 = 0. \qquad \checkmark$$

□

Now You Try 22.3 Suppose that a and b are roots of the equation $x^2 - 8x + 12 = 0$, so that

$$ab = 12 \quad \text{and} \quad a + b = -(-8).$$

Find a and b. □

22.2   VIÈTE AND HARRIOT

Thomas Harriot (ca. 1560–1621), in his *Treatise on equations*,[1] was one of the first mathematicians to study the relation between the roots of polynomial equations and the coefficients of their terms. (See FIGURE 22.1.) But this story more properly begins with a Frenchman named François Viète.

Figure 22.1: Thomas Harriot. (Source: Courtesy of Jeff Miller.)

---

[1] *Treatise on equations* is the title given by Jacqueline Stedall [110, p. 6] to Harriot's undated material on the structure of polynomial equations. Stedall's edition is the first complete version of Harriot's work on equations.

François Viète (1540–1603),[2] who often published his mathematics in Latin under the name Franciscus Vieta or Francisci Vietæ, was a lawyer by trade. He was born in Fontenay-le-Comte and, at a young age, studied with the Franciscan monks at their cloister in Fontenay. At the age of 18, Viète left for the University of Poitiers where he received his bachelor's degree in law, and subsequently began working as a lawyer.

Throughout his professional career in law, Viète had several prominent clients. Indeed, Viète's experience led to his appointment as *maître des réquètes* (master of requests) at the court and a member of the privy council of Henri III in 1580 until sometime around the end of 1584 or the beginning of 1585. One (mathematically) interesting episode in Viète's service to Henri III was his working as a cryptanalyst (code breaker), deciphering coded messages between the king's enemies. In fact, Viète became so good at it that some "denounced him by saying that the decipherment could only have been the product of sorcery and necromancy" [122, p. 3].

Henri III was killed in 1589 and was succeeded by Henri IV. By this time, Viète had become one of the most influential persons at the court, and in 1594 he was summoned to be a privy councilor to Henri IV. Not long after, Viète was apparently overtaken by ill health. He was then sent on a mission to Poitou that allowed him to live in his hometown of Fontenay until his retirement in 1602. He died the next year.

Witmer tells us that there were two distinct periods to Viète's mathematical life.[3] The first began around the time Viète began his full-time employment as a lawyer for the Soubise family in 1564 and ended in 1571 when the royal printer, Jean Mettayer, made his press available for the publication of Viète's *Canon Mathematicus* and *Universalium Inspectionum Liber Singularis* (both published in 1579). These two works consisted mainly of trigonometric tables on an ambitious scale. The second period began around 1584 and it was during this period that Viète produced his most influential work.

Viète is perhaps best remembered for his contribution to algebraic notation, providing in his seminal work *In artem analyticem isagoge* (*Introduction to the Analytic Art*; 1591) [120], or the *Isagoge* for short, "for the first time in algebra a clear-cut distinction between the important concept of a parameter and the idea of an unknown quantity" [15, p. 304], and for his systematic study of equations. As Boyer [15, p. 301] tells us:

> Letters had indeed been used to represent magnitudes known or unknown, since the days of Euclid, and Jordanus had done this freely; but there had been no way of distinguishing magnitudes assumed to be known from those unknown quantities that are to be found. [For example, one could not write a general quadratic equation because one could not tell in the equation $x^2 + xy = z$ (using modern notation), say, which letters represent quantities assumed to be known, and which represent quantities assumed to be unknown.] Here Viète introduced a convention as simple as it was fruitful. He used a vowel [A, E, I, O, U, or Y] to represent the quantity in algebra that was assumed to be unknown or undetermined and a consonant [B, G, D,...] to represent a magnitude or number assumed to be known or given.

So, using Viète's convention, it would be clear in the equation $A^2 + BA = G$—more likely to have been expressed verbally as "*A squared* plus *B* times *A* is equal to *G plane*" with the homogeneity of the terms[4]

---

[2] We have drawn sparsely from Witmer's introduction to his translation of Viète's *The Analytic Art* [122, pp. 1–10]. In the footnote on page 1, Witmer states:

> There is no full-scale biography of Viète of which I am aware. This and what follows are pieced together largely from the sketches by Frederic Ritter, *François Viète, Inventeur de l'Algebre Moderne* (Paris, 1895) and Joseph E. Hofmann's introduction to the facsimile reprint of Viète's *Opera Mathematica* (Frans van Schooten, originally Leyden, 1646; reprint Georg Olms Verlag, Hldesheim, 1970). A note appended to the Ritter piece says that he had prepared a complete biography, intended to accompany a translation into French of Viète's complete works, which would run to 350 pages. The whereabouts of this manuscript is unknown.

[3] Viète spent his nonprofessional life dedicated to mathematics. Chronologically, Viète (1540–1603) places after the notable Italian Girolamo Cardano (1501–1575) and before his also notable fellow countrymen René Descartes (1596–1650) and Pierre de Fermat (1601–1665).

[4] By "homogeneity of the terms" is meant that all the terms should have the same geometric dimension. Thus, because both

maintained—that the unknown quantity is A, while B and G are assumed to be known.[5] As Witmer puts it, "Viète's new way of doing it had the great advantage of making more visible the operations which went into building up or solving a complex series of terms. This is due more to his substitution of letters for the givens—a substitution which, as far as is known, was Viète's own contribution—than to the use of letters to represent the variables ... [for] numerical coefficients tend, first, to obscure the generality of what is being proposed and, second, to merge with each other as letters do not when a given expression is subjected to processing" [122, p. 5].

Yet, despite Viète's revolutionary advance in symbolic algebraic notation, he was hampered by the fact that he continued to use essentially the syncopated algebra of his predecessors. We see this in the following example [109, p. 8] translated from Viète's *Isagoge*:

If to $\dfrac{A\ plane}{B}$ there should be added $\dfrac{Z\ squared}{G}$, the sum will be

$$\dfrac{G \text{ times } A\ plane + B \text{ times } Z\ squared}{B \text{ times } G}.$$

Across the English Channel, Thomas Harriot (ca. 1560–1621), who had studied Viète, was to change the face of algebra. In his own *Treatise on equations*, Harriot, in addition to using a variation of the symbol = for equals that Robert Recorde had introduced in *The Whetstone of Witte*, introduced the symbols < and > for inequality, and the symbols ± and ∓ to handle several cases at once. But, according to Stedall [109, p. 90], "Harriot's most important innovation in notation was undoubtedly his use of ab to represent a multiplied by b, and consequently aa, aaa for what is now written a^2, a^3, etc." (Harriot followed Viète and used vowels for unknowns, but changed to using lowercase letters instead of capitals.) And with this, Harriot spawned purely symbolic algebra. So, keeping with using vowels for the unknowns (lowercase instead, but still maintaining homogeneity of the terms[6]), Harriot would write the equation from Viète's *Isagoge* above thus [109, p. 11]:

$$\frac{ac}{b} + \frac{zz}{g} = \frac{acg + bzz}{bg}.$$

As another example [109, p. 11]:

Viète: If B times G should be divided by $\dfrac{A\ plane}{D}$, both magnitudes having been multiplied by D, the

result will be $\dfrac{B \text{ times } G \text{ times } D}{A\ plane}.$

Harriot: $\dfrac{bg}{\frac{ac}{d}} = \dfrac{bgd}{ac}$

"A squared" and "B times A" are 2-dimensional (both are areas), "A squared plus B times A" is set equal to "G plane," which is 2-dimensional, and not set equal to "G," which is 1-dimensional. In this way, every term in the equation "A squared plus B times A is equal to G plane" has dimension 2 (every term is an area).

[5]Our convention of using the latter letters of the alphabet to represent unknown quantities and earlier letters to represent parameters—writing, for example, $x^2 + bx = c$, with x assumed to be unknown and b and c assumed to be known—was introduced by Descartes without comment in his 1637 work *La géométrie* [46]. It is also in this work that Descartes introduced our notation for exponents. See page 339.

[6]Note that Harriot writes "ac" (fully symbolic), for example, where Viète writes "A plane" (syncopated).

And here is one more example to illustrate the advantage of Harriot's purely symbolic algebra. Here is shown the rule for moving terms from one side of an equation to the other, which Viète called *antithesis*, terminology that Harriot kept [109, p. 11]:

Viète: *A squared* minus *D plane* is supposed equal to *G squared* minus *B* times *A*. I say that *A squared* plus
   B times *A* is equal to *G squared* plus *D plane* and that by this transposition and under opposite signs of
   conjunction the equation is not changed.

Harriot:

$$\text{Suppose} \quad aa - dc = gg - ba$$
$$\text{I say that} \quad aa + ba = gg + dc \quad \text{by } antithesis.$$

As Stedall trumpets, "the lucidity and economy of Harriot's notation is obvious: Easy to read and easy to use, it reveals algebraic structure and acts as an aid to thinking in a way that Viète's verbal descriptions can not" [110, p. 11]. With this innovative notation, Harriot was one of the first mathematicians to study the relation between the roots of polynomial equations and the coefficients of their terms.

Nothing is known of Harriot's early background.[7] What we know is that Harriot's name appears in the Oxford University Register in 1577, so that he was likely born around 1560, and that he graduated in 1580. We also know that by the time of his death in 1621, Harriot had written over four thousand manuscript sheets in the areas in which he worked: navigation, astronomy, optics, geometry, and algebra; however, he had never published any of his work. The only book that Harriot published in his lifetime was *A briefe and true report of the new found land of Virginia* (1588) following his employment by Walter Raleigh from 1585 to 1586 as a navigator and scientist on a voyage to North America. Of the thousands of manuscript sheets that Harriot left upon his death, about 140 sheets constitute his *Treatise on equations* [110, p. 3].

Around the early 1590s, Sir Henry Percy, the ninth earl of Northumberland, became a lifelong patron and benefactor of Harriot. Even during the time that Percy was imprisoned in the Tower of London (1605–1621), Harriot continued to be maintained at the Earl's London home, Syon House at Isleworth in Middlesex. Indeed, most of Harriot's scientific and mathematical friends were in some way connected to the Earl. One who was particularly close to Harriot was Nathaniel Torporley (1564–1632) and, according to Stedall [110, pp.4–5], "it was almost certainly through Torporley that Harriot acquired his detailed knowledge of Viète's mathematics: there is, for example, among Harriot's papers a sheet headed: 'A proposition of Vietas delivered by Mr. Thorperly'."

Stedall dates Harriot's *Treatise on equations* indirectly, saying: "much of it arises so directly from *De potestatum resolutione* that it was probably written shortly after that book appeared in 1600" [110, p. 7]. *De numerosa potestatum ad exegesin resolutione*[8] [121] was Viète's treatment on numerical methods developed by twelfth-century Arabic mathematicians for solving equations, which was the first such exposition in Europe [110, p. 6]. Some of Harriot's material on equations was edited and published after his death by Walter Warner in 1631 under the title *Artis analyticae praxis* (*The practice of the analytic art*) [106], that has come to be referred to commonly as the *Praxis* for short. Harriot had appointed Torporley in his will "to be Overseer of my Mathematical Writings to be received of my Executors to peruse and order and to separate the chief of them from my waste papers, to the end that after he doth understand them he may make use in penning such doctrine that belongs unto them for public uses as it shall be thought Convenient by my Executors and him selfe" [110, p. 18]. For uncertain reasons, however, Torporley was relieved of this duty by Harriot's Executors. Following Torporley's relief, what emerged was the *Praxis*, "but the editors selected and reordered Harriot's work in such a way that the *Praxis* often bears little resemblance to the manuscripts, and fails to do full justice to the quality and originality of

---

[7]We have drawn from the Stedall's introduction to Harriot's *Treatise on equations* [110]. See also the Seltman and Goulding's introduction to their translation of the *Artis Analyticae Praxis* [106].

[8]First published in 1600 under this title, this work was retitled *De numerosa potestatum purarum, atque adfectarum ad exegesin resolutione tractus* when it was placed in Schooten's 1646 collected works of Viète.

Harriot's insights" [110, p. 3]. Yet it was the *Praxis* by which Harriot's esteem as a mathematician has been judged by and large for many years.

This is not to say, however, that the *Praxis*, regarded on its own, is not an important work. Whereas Harriot had not published his *Treatise on equations*, the *Praxis*, from its first printing, was readily available (at least in Britain) [105, p. 73] and, thus, disseminated widely Harriot's algebraic notation that presented true symbolic algebra for the first time. Further, despite some shortcomings of the *Praxis* compared with *Treatise on equations*, Seltman [105, p.76] posits: "The mathematics of the [original] papers [of Harriot's] may be more sophisticated than that of the *Praxis* but the latter is more accessible. We have the wonderful calculative logic of Harriot's purely symbolic notation combined with a clear and straightforward presentation. In one way the *Praxis* has the edge on the papers insofar as the mode of thinking implied by the manipulations is closer to modern mathematical thinking than that of Harriot himself. Think of the facility in algebraic manipulation that a large number of mathematicians would acquire through this—and there were many copies available."

We illustrated earlier how Harriot's symbolic algebra makes it easier to see the structure of an equation than does Viète's syncopated algebra. In addition, whereas Viète did not consider negative solutions [122, p. 8], Harriot considered not only negative solutions, but imaginary ones, as well [109, p. 118]. And whereas Viète proceeded with his theory of equations in terms of proportions, Harriot proceeded in terms of setting factors equal to zero. For example, we find in Sheet $d.1$ of Harriot's *Treatise on equations* (Add MS 6783, f. 183) [110, pp. 15, 126] that Harriot concludes from

$$(a - b)(a - c) = aa - ba - ca + bc = \text{oo},$$

where a is the unknown and zero is written as oo to maintain homogeneity, that "$bc = ba + ca - aa$ and we will have $a = b$ and $a = c$ if b and c are unequal, but nothing other than b and c."

On this last point, we note that Descartes also proceeded with his theory of equations in *La géométrie* (1637) in a way similar to Harriot's. For example, we find in the Third Book of *La géométrie* [46, p. 159] that Descartes concludes from

$$(x - 2)(x - 3) = x^2 - 5x + 6 = 0$$

that "[$x^2 - 5x + 6 = 0$, or $x^2 = 5x - 6$] is an equation in which x has the value 2 and at the same time x has the value 3." As another example, we find also in the Third Book of *La géométrie* [46, p. 167] the equation

$$
\begin{array}{r}
z^4 - 16z^3 + 96zz - 256z + 256 \\
+16z^3 - 192zz + 768z - 1024 \\
+ 71zz - 568z + 1136 \\
- 4z + 16 \\
- 420 \\
\hline
z^4 \qquad - 25zz - 60z - 36 = 0
\end{array}
$$

that is very similar in procedure and layout to this example in Harriot's manuscript (Add MS 6783 f. 130) [110, p. 258]:

$$eeee - 96ee + 512e = +72$$
$$+ 49ee - 392e \quad + 768$$
$$+ 6e \quad - 784$$
$$+ 24$$
$$eeee - 47ee + 126e = +80$$

All this led, Stedall [109, pp. 111–112], [110, pp. 27–29] tells us, to a little controversy over priority, with Harriot's English supporters suspecting that Descartes must have seen, or at least known of, Harriot's manuscript before writing *La géométrie*. The matter has not yet been settled.

Nevertheless, all this not withstanding, Descartes did make at least two contributions to algebra in *La géométrie*, namely, the introduction of the modern notation for exponents and the move away from the requirement of maintaining homogeneity in equations.

22.3   ZEROS OF A POLYNOMIAL

As we have seen, the roots or solutions of a polynomial equation are inextricably related to the factors of the polynomial, as well as to it coefficients. The roots of a polynomial equation that is set equal to zero are called the zeros of the polynomial expression.

Definition 22.1 A number is called a *zero of a polynomial* (or, more generally, a zero of a function) if the evaluation of the polynomial at that number yields zero. ☐

For example, 3 is a zero of the polynomial $x^2 - x - 6$, but 5 is not:

$$x = 3 : \qquad 3^2 - 3 - 6 = 0,$$

but

$$x = 5 : \qquad 5^2 - 5 - 6 = 14 \quad (\neq 0).$$

In fact, from EXAMPLE 22.7, we know that the roots of the equation $x^2 - x - 6 = 0$ are 3 and −2, so that the zeros of the polynomial $x^2 - x - 6$ are 3 and −2.[9]

22.3.1   *Factoring*

Theorem 22.1 (FACTOR THEOREM) *If α is a zero of the polynomial $p(x)$, then the polynomial factors as*

$$p(x) = (x - \alpha)q(x),$$

where $q(x)$ is a polynomial of one degree lower than $p(x)$, and conversely. ☐

Example 22.8

1. We have already seen that $\alpha = 3$ is a zero of the polynomial $x^2 - x - 6$, and that

$$\underbrace{x^2 - x - 6}_{p(x)} = (\underbrace{x - 3}_{x-\alpha})(\underbrace{x + 2}_{q(x)}).$$

   Note that $q(x)$ is one degree lower than $p(x)$: The degree of $q(x) = x + 2$ is 1, and the degree of $p(x) = x^2 - x - 6$ is 2.

---

[9]Equations have roots and functions have zeros.

Likewise, $\alpha = -2$ is a zero of $x^2 - x - 6$ and

$$\underbrace{x^2 - x - 6}_{p(x)} = (\underbrace{x - (-2)}_{x-\alpha})(\underbrace{x - 3}_{q(x)}) \quad \text{or} \quad (x + 2)(x - 3).$$

Note that $q(x)$ is one lower than $p(x)$: The degree of $q(x) = x - 3$ is 1, and the degree of $p(x) = x^2 - x - 6$ is 2.

2. The number $\alpha = 2$ is a zero of $x^3 - 8$, for

$$2^3 - 8 = 8 - 8 = 0;$$

hence, $x^3 - 8 = (x - 2)q(x)$. To find $q(x)$, we divide $x^3 - 8$ by $x - 2$. Here is one method of doing polynomial long division.[10]

- Draw boxes with the divisor written in decreasing order on the left side.

- The goal is to fill in the boxes with terms that combine to produce the dividend, $x^3 - 8$. Write the highest-order term of the dividend, x^3, in the top box of the first column (top left box).

- Ask, "What times x is x^3?" and write the quotient, x^2, above the first column of boxes.

- Multiply x^2 by the divisor to fill in the boxes in the first column.

---

[10]There are different ways to layout polynomial long division. You may use a different layout. One common method looks almost identical to long division of numbers.

	x^2			
x	x^3			
-2	$-2x^2$			

- Now the terms inside the boxes combine to $x^3 - 2x^2$, but the dividend is $x^3 - 8$. To cancel the $-2x^2$ inside the boxes, because $-2x^2$ is not a term in the dividend, we must add $2x^2$; we write this in the top box of the second column.

	x^2			
x	x^3	$2x^2$		
-2	$-2x^2$			

- Ask, "What times x is $2x^2$?" and write the quotient, $2x$, above the second column of boxes.

	x^2	$2x$		
x	x^3	$2x^2$		
-2	$-2x^2$			

- Multiply $2x$ by the divisor to fill in the boxes in the second column.

	x^2	$2x$		
x	x^3	$2x^2$		
-2	$-2x^2$	$-4x$		

- Now the terms inside the boxes combine to $x^3 - 4x$, but the dividend is $x^3 - 8$. To cancel the $-4x$ inside the boxes, because $-4x$ is not a term in the dividend, we must add $4x$; we write this in the top box of the third column.

	x^2	$2x$		
x	x^3	$2x^2$	$4x$	
-2	$-2x^2$	$-4x$		

- Ask, "What times x is $4x$?" and write the quotient, 4, above the third column of boxes.

	x^2	$2x$	4	
x	x^3	$2x^2$	$4x$	
-2	$-2x^2$	$-4x$		

- Multiply 4 by the divisor to fill in the boxes in the third column.

	x^2	$2x$	4	
x	x^3	$2x^2$	$4x$	
-2	$-2x^2$	$-4x$	-8	

- Now the terms inside the boxes combine to $x^3 - 8$, which is exactly the dividend; hence, the remainder is zero, which we may write in the top box of the fourth column if we choose.

	x^2	$2x$	4	
x	x^3	$2x^2$	$4x$	0
-2	$-2x^2$	$-4x$	-8	

Thus, we see that

$$\underbrace{x^3 - 8}_{p(x)} = (\underbrace{x - 2}_{x-\alpha})(\underbrace{x^2 + 2x + 4}_{q(x)}).$$

Note that $q(x)$ is one degree lower than $p(x)$: The degree of $q(x) = x^2 + 2x + 4$ is 2, and the degree of $p(x) = x^3 - 8$ is 3.

3. The number $\alpha = -1$ is a zero of $2x^2 + 5x + 3$, for

$$2(-1)^2 + 5(-1) + 3 = 2 - 5 + 3 = 0;$$

hence, because $x - (-1) = x + 1$, the polynomial $2x^2 + 5x + 3 = (x + 1)q(x)$. To find $q(x)$, we divide $2x^2 + 5x + 3$ by $x + 1$.

- Draw boxes with the divisor written in decreasing order on the left side.

x				
1				

- The goal is to fill in the boxes with terms that combine to the dividend, $2x^2 + 5x + 3$. Write the highest-order term of the dividend, $2x^2$, in the top box of the first column (top left box).

x	$2x^2$			
1				

- Ask, "What times x is $2x^2$?" and write the quotient, $2x$, above the first column of boxes.

	$2x$			
x	$2x^2$			
1				

- Multiply $2x$ by the divisor to fill in the boxes in the first column.

	$2x$			
x	$2x^2$			
1	$2x$			

- Now the terms inside the boxes combine to $2x^2 + 2x$, but the dividend is $2x^2 + 5x + 3$. To obtain $5x$ inside the boxes, because $5x$ is a term in the dividend, we must add $3x$; we write this in the top box of the second column.

	$2x$			
x	$2x^2$	$3x$		
1	$2x$			

- Ask, "What times x is $3x$?" and write the quotient, 3, above the second column of boxes.

	$2x$	3		
x	$2x^2$	$3x$		
1	$2x$			

- Multiply $3x$ by the divisor to fill in the boxes in the second column.

	$2x$	3		
x	$2x^2$	$3x$		
1	$2x$	3		

- Now the terms inside the boxes combine to $2x^2 + 5x + 3$, which is exactly the dividend; hence, the remainder is zero, which we may write in the top box of the third column if we choose.

	$2x$	3		
x	$2x^2$	$3x$	0	
1	$2x$	3		

Thus, we see that

$$\underbrace{2x^2 + 5x + 3}_{p(x)} = (\underbrace{x + 1}_{x-\alpha})(\underbrace{2x + 3}_{q(x)}).$$

Note that $q(x)$ is one lower degree than $p(x)$: The degree of $q(x) = 2x + 3$ is 1, and the degree of $p(x) = 2x^2 + 5x + 3$ is 2.

$\square$

Now You Try 22.4 Perform the following polynomial divisions.

1. $(x^2 + 9x + 20) \div (x + 5)$

2. $(x^2 - 7x - 30) \div (x + 3)$

3. $\dfrac{3x^3 - 14x^2 - 3x + 44}{x - 4}$

4. $\dfrac{x^2 - 11x + 28}{x - 2}$ (There is a nonzero remainder.)

$\square$

Example 22.9 Factor the polynomial $x^2 - 2$.

Since $(\sqrt{2})^2 = 2$, we see that $\sqrt{2}$ is a zero of the polynomial:

$$(\sqrt{2})^2 - 2 = 2 - 2 = 0.$$

Hence, the polynomial $x^2 - 2$ factors as

$$x^2 - 2 = (x - \sqrt{2})q(x).$$

To find $q(x)$, we divide $x^2 - 2$ by $x - \sqrt{2}$.

	x	$\sqrt{2}$		
x	x^2	$\sqrt{2}x$	0	
$-\sqrt{2}$	$-\sqrt{2}x$	-2		

Thus,

$$q(x) = x + \sqrt{2},$$

and the given quadratic polynomial factors as

$$x^2 - 2 = (x - \sqrt{2})(x + \sqrt{2}).$$

We observe that this is just the difference of squares factoring, to wit,

$$x^2 - y^2 = (x - y)(x + y),$$

with $y = \sqrt{2}$.   □

Example 22.10 Factor the polynomial $x^2 + 2x - 1$.

Recall that if $x^2 + Bx + C = (x - a)(x - b)$, then $a + b = -B$ and $ab = C$. Since there are not two integers a and b such that $a + b = -2$ and $ab = -1$, we are unable to factor the $x^2 + 2x - 1$ in the usual way; however, if we are able to find a zero of $x^2 + 2x - 1$, say α, then, in fact, the polynomial would factor as

$$x^2 + 2x - 1 = (x - \alpha)q(x).$$

Let us try to find a zero of the polynomial by solving the equation

$$x^2 + 2x - 1 = 0.$$

Since the equation is in the form $ax^2 + bx + c = 0$, we apply the quadratic formula to solve it. Toward this end, we identify a, b, and c:

$$a = 1, \quad b = 2, \quad c = -1.$$

Next, we compute the discriminant:

$$b^2 - 4ac = 2^2 - 4(1)(-1) = 4 + 4 = 8.$$

Thus, we find that the solutions of the equation are

$$x = \frac{-b}{2a} \pm \frac{\sqrt{b^2 - 4ac}}{2a} = \frac{-2}{2} \pm \frac{\sqrt{8}}{2}$$

$$= \frac{-2}{2} \pm \frac{2\sqrt{2}}{2}$$

$$= -1 \pm \sqrt{2}.$$

Hence, the zeros of $x^2 + 2x - 1$ are

$$\alpha_1 = -1 + \sqrt{2} \quad \text{and} \quad \alpha_2 = -1 - \sqrt{2},$$

so that the polynomial factors as either

$$x^2 + 2x - 1 = (x - \alpha_1)q_1(x) = (x + 1 - \sqrt{2})q_1(x)$$

or

$$x^2 + 2x - 1 = (x - \alpha_2)q_2(x) = (x + 1 + \sqrt{2})q_2(x).$$

Consider the factorization

$$x^2 + 2x - 1 = (x + 1 - \sqrt{2})q_1(x).$$

To find $q_1(x)$, we divide $x^2 + 2x - 1$ by $x + 1 - \sqrt{2}$.

	x	$1 + \sqrt{2}$	
x	x^2	$(1 + \sqrt{2})x$	0
$1 - \sqrt{2}$	$(1 - \sqrt{2})x$	-1	

Note that the boxes give us

$$x^2 + (1 - \sqrt{2})x + (1 + \sqrt{2})x - 1 = x^2 + 2x - 1.$$

Thus,

$$q_1(x) = x + 1 + \sqrt{2},$$

and the given quadratic polynomial factors as

$$x^2 + 2x - 1 = (x + 1 - \sqrt{2})(x + 1 + \sqrt{2}).$$

No wonder we were unable to factor $x^2 + 2x - 1$ in the usual way!

We leave it to you as an exercise to check that the factorization is correct by expanding the right-hand side. Also, note that we could just as well have found $q_2(x)$, instead of $q_1(x)$, to factor the given quadratic polynomial. □

Think About It 22.3 In the last example, the factors match the roots found by using the quadratic formula, so the long division was not necessary. Is this always the case? □

Example 22.11 Factor the polynomial $2x^2 + 2x - 1$.

Again, we are unable to factor the polynomial in the usual way; however, if we are able to find a zero of $2x^2 + 2x - 1$, say α, then, in fact, the polynomial would factor as

$$2x^2 + 2x - 1 = (x - \alpha)q(x).$$

Let us try to find a zero of the polynomial by solving the equation

$$2x^2 + 2x - 1 = 0.$$

Since the equation is in the form $ax^2 + bx + c = 0$, we apply the quadratic formula to solve it.[11] Toward this end, we identify a, b, and c:

$$a = 2, \quad b = 2, \quad c = -1.$$

Next, we compute the discriminant:

$$b^2 - 4ac = 2^2 - 4(2)(-1) = 4 + 8 = 12.$$

---

[11] A zero can sometimes be found by inspection, particularly if it is a small integer.

Thus, we find that the solutions of the equation are

$$x = \frac{-b}{2a} \pm \frac{\sqrt{b^2 - 4ac}}{2a} = \frac{-2}{4} \pm \frac{\sqrt{12}}{4}$$

$$= \frac{-2}{4} \pm \frac{2\sqrt{3}}{4}$$

$$= \frac{-1}{2} \pm \frac{1}{2}\sqrt{3}.$$

Hence, the zeros of $2x^2 + 2x - 1$ are

$$\alpha_1 = \frac{-1}{2} + \frac{1}{2}\sqrt{3} = \frac{-1 + \sqrt{3}}{2} \quad \text{and} \quad \alpha_2 = \frac{-1}{2} - \frac{1}{2}\sqrt{3} = \frac{-1 - \sqrt{3}}{2},$$

so that the polynomial factors as either

$$2x^2 + 2x - 1 = (x - \alpha_1)q_1(x) = \left(x - \frac{-1 + \sqrt{3}}{2}\right)q_1(x)$$

or

$$2x^2 + 2x - 1 = (x - \alpha_2)q_2(x) = \left(x - \frac{-1 - \sqrt{3}}{2}\right)q_2(x).$$

Consider the factorization

$$2x^2 + 2x - 1 = \left(x - \frac{-1 + \sqrt{3}}{2}\right)q_1(x) = \left(x + \frac{1 - \sqrt{3}}{2}\right)q_1(x).$$

To find $q_1(x)$, we divide $2x^2 + 2x - 1$ by $x + \frac{1-\sqrt{3}}{2}$.

	$2x$	$1 + \sqrt{3}$		
x	$2x^2$	$(1 + \sqrt{3})x$	0	
$\frac{1-\sqrt{3}}{2}$	$(1 - \sqrt{3})x$	-1		

Note that the boxes give us

$$2x^2 + (1 - \sqrt{3})x + (1 + \sqrt{3})x - 1 = 2x^2 + 2x - 1.$$

Thus,

$$q_1(x) = 2x + 1 + \sqrt{3},$$

and the given quadratic polynomial factors as

$$2x^2 + 2x - 1 = \left(x + \frac{1 - \sqrt{3}}{2}\right)(2x + 1 + \sqrt{3}).$$

We leave it to you to check that the factorization is correct by expanding the right-hand side. Also, as we noted in the last example, we could just as well have found $q_2(x)$, instead of $q_1(x)$, to factor the given quadratic polynomial.

Lastly, observe in this example that, even though the zeros of $2x^2 + 2x - 1$ found by the quadratic formula are $\alpha_1 = \frac{-1+\sqrt{3}}{2}$ and $\alpha_2 = \frac{-1-\sqrt{3}}{2}$,

$$(x - \alpha_1)(x - \alpha_2) = \left(x - \frac{-1 + \sqrt{3}}{2}\right)\left(x - \frac{-1 - \sqrt{3}}{2}\right) \neq 2x^2 + 2x - 1.$$

Thus, we really do have to divide to find the factor $q(x)$.   □

Think About It 22.4 In the last example, why did the factors not match the roots that were found using the quadratic formula? How is the last example different from the one before it? □

Example 22.12 Factor the polynomial $x^3 + 4x^2 + x - 6$ given that 1 is a zero of the polynomial.

Because 1 is a zero of $x^3 + 4x^2 + x - 6$,

$$1^3 + 4 \cdot 1^2 + 1 - 6 = 1 + 4 + 1 - 6 = 0,$$

the polynomial factors as

$$x^3 + 4x^2 + x - 6 = (x - 1)q(x).$$

To find $q(x)$, we divide $x^3 + 4x^2 + x - 6$ by $x - 1$.

	x^2	$5x$	6	
x	x^3	$5x^2$	$6x$	0
-1	$-x^2$	$-5x$	-6	

Thus,

$$q(x) = x^2 + 5x + 6,$$

and a factoring of the given cubic polynomial is

$$x^3 + 4x^2 + x - 6 = (x - 1)(x^2 + 5x + 6).$$

In this case, we find that we may factor the quadratic factor, $x^2 + 5x + 6$, in the usual way:

$$x^2 + 5x + 6 = (x + 2)(x + 3).$$

Therefore, a complete factoring of the given cubic polynomial is

$$x^3 + 4x^2 + x - 6 = (x - 1)(x + 2)(x + 3).$$

□

Now You Try 22.5

1. Verify that $x^2 + 2x - 1 = (x + 1 - \sqrt{2})(x + 1 + \sqrt{2})$.

2. Show by expanding the right-hand side that, even though $\frac{-1+\sqrt{3}}{2}$ and $\frac{-1-\sqrt{3}}{2}$ are zeros of $2x^2 + 2x - 1$,

$$2x^2 + 2x - 1 \neq \left(x - \frac{-1 + \sqrt{3}}{2}\right)\left(x - \frac{-1 - \sqrt{3}}{2}\right);$$

however, the quadratic polynomial does factor as

$$2x^2 + 2x - 1 = 2\left(x - \frac{-1 + \sqrt{3}}{2}\right)\left(x - \frac{-1 - \sqrt{3}}{2}\right).$$

3. Verify that $\alpha = \frac{-1-\sqrt{3}}{2}$ is a zero of $2x^2 + 2x - 1$, then divide $2x^2 + 2x - 1$ by $x - \alpha$ to factor the polynomial.

4. Verify that -4 is a zero of the polynomial $x^3 - 3x^2 - 22x + 40$, then factor the polynomial completely.

5. A polynomial $ax^n + bx^{n-1} + \cdots + c$ is called a *monic polynomial* if $a = 1$. For example, the polynomial $x^3 - 3x^2 - 22x + 40$ is a monic polynomial, but $2x^3 - 6x^2 - 44x + 80$ is not.

    a) Find a monic polynomial of degree 3 that has zeros 7, 3, and -5.

    b) Find a monic polynomial of degree 2 that has zeros 7i and -7i.

    c) Find a monic polynomial of least degree that has zeros $2 + 3$i and 5 that has real coefficients.

□

Remark 22.2 The examples illustrate that $x^2 + bx + c = (x - \alpha_1)(x - \alpha_2)$ if α_1 and α_2 are roots of the equation $x^2 + bx + c = 0$, but that $ax^2 + bx + c \neq (x - \alpha_1)(x - \alpha_2)$ even if α_1 and α_2 are roots of the equation $ax^2 + bx + c = 0$; instead, $ax^2 + bx + c = a(x - \alpha_1)(x - \alpha_2)$ if α_1 and α_2 are roots of the equation $ax^2 + bx + c = 0$.

□

22.3.2 Descartes's Rule of Signs

The Third Book of Descartes's *La géométrie* contains this curious statement about the number of real roots of a polynomial equation [46, p. 160]:[12]

> An equation can have as many true roots as it contains changes of sign, from $+$ to $-$ or from $-$ to $+$; and as many false [negative] roots as the number of times two $+$ signs or two $-$ signs are found in succession.

Descartes goes on to give this example to illustrate the rule:

> Thus, in the last equation [$x^4 - 4x^3 - 19x^2 + 106x - 120 = 0$], since $+x^4$ is followed by $-4x^3$, giving a change of sign from $+$ to $-$, and $-19x^2$ is followed by $+106x$ and $+106x$ by -120, giving two more changes, we know there are three true [positive] roots; and since $-4x^3$ is followed by $-19x^2$ there is one false [negative] root.

Note that Descartes's rule of signs tells us only the *possible number* of positive and negative real roots of a polynomial equation. So, in his example above, there may be *as many as* three positive real roots (0, 1, 2, or 3 roots) and, in fact, there are 3; and there may be *as many as* one negative real root (0 or 1 root) and, in fact, there is one. As another example, consider the equation

$$x^2 + 4x + 4 = 0.$$

Descartes's rule of signs tells us that there may be as many as two negative real roots (0, 1, or 2 roots) because $+x^2$ is followed by $+4x$ and $+4x$ is followed by $+4$, so that two $+$ signs are found in succession twice; and there is no positive real root because there is no change of sign, from $+$ to $-$ or from $-$ to $+$. In fact, solving the equation shows that it has only one negative root (a double root), namely, -2:

$$0 = x^2 + 4x + 4 = (x + 2)(x + 2).$$

Now You Try 22.6

1. Descartes claims that the equation $x^4 - 4x^3 - 19x^2 + 106x - 120 = 0$ has three positive ("true") roots and one negative ("false") root.

    a) Verify that 2, 3, 4 are positive roots of the equation, and that -5 is a negative root.

    b) Factor the expression $x^4 - 4x^3 - 19x^2 + 106x - 120$ on the left-hand side of the equation into a product of linear factors. What does this say about the total number of roots of the equation?

2. Use Descartes's rule of signs to determine the possible number of positive and negative real roots of the following equations. You do not have to solve the equations.

---

[12] A footnote on page 160 informs us: "This is the well known 'Descartes's Rule of Signs.' It was known however, before his time, for Harriot had given it in his *Artis analyticae praxis*, London, 1631. Cantor says Descartes may have learned it from Cardan's writings, but was the first to state it as a general rule."

a) $x^2 - 2x - 2 = 0$

c) $x^3 - 3x^2 - 18x + 40 = 0$

b) $x^4 + 7x^3 + 5x^2 - 31x - 30 = 0$

d) $x^3 - 8 = 0$

□

Think About It 22.5 Here is a modern presentation of Descartes's rule of signs [59, p. 122]:

[T]he number of positive zeros of a polynomial is equal to the number of its sign changes, or differs from it by a positive even number. The number of negative zeros is obtained similarly from the number of sign changes in the sequence of coefficients of the polynomial $f(-x)$.

For example, let
$$f(x) = x^3 - 7x + 6$$
Then we see that f has two sign changes, so that f has 2 positive zeros or none (because the number of possible zeros differs by a positive even integer). Also,
$$f(-x) = -x^3 + 7x + 6$$
has one sign change, so that $f(x) = 0$ has 1 negative zero.

Apply Descartes's rule of signs that is given in *La géométrie* to the equation $f(x) = 0$, that is, to
$$x^3 - 7x + 6 = 0.$$

Do you arrive at the same conclusion? Reconcile any differences. (You may verify that $\alpha = 1$ is a zero of f, then factor f to determine all of the zeros.)   □

We conclude this subsection with the following interesting result that is also found in the Third Book of *La géométrie* [46, pp. 160, 163]:

It is also easy to transform an equation so that all the roots that were false shall become true roots, and all those that were true shall become false. This is done by changing the signs of the second, fourth, sixth, and all even terms, leaving unchanged the signs of the first, third, fifth, and other odd terms. Thus, if instead of
$$+x^4 - 4x^3 - 19x^2 + 106x - 120 = 0$$
[that we recall has three true roots, namely, 2, 3, and 4, and one false root, namely, −5] we write
$$+x^4 + 4x^3 - 19x^2 - 106x - 120 = 0$$
we get an equation having one true root, 5, and three false roots, 2, 3, and 4 [in absolute value].

Example 22.13 Beginning with
$$(x - 1)(x + 2)(x + 3) = 0$$
and expanding the left-hand side, we see that the roots of the equation
$$x^3 + 4x^2 + x - 6 = 0$$
are 1, −2, and −3. Now we change the signs of the second and fourth terms, and leave the signs of the others unchanged, to obtain the equation
$$x^3 - 4x^2 + x + 6 = 0.$$

We check that now the roots are −1, 2, and 3:

$$x = -1: \quad (-1)^3 - 4(-1)^2 + (-1) + 6 = -1 - 4 - 1 + 6 = 0. \quad \checkmark$$
$$x = 2: \quad 2^3 - 4 \cdot 2^2 + 2 + 6 = 8 - 16 + 2 + 6 = 0. \quad \checkmark$$
$$x = 3: \quad 3^3 - 4 \cdot 3^2 + 3 + 6 = 27 - 36 + 3 + 6 = 0. \quad \checkmark$$

□

Now You Try 22.7

1. Verify that the roots of the equation

$$x^5 - 7x^4 + 3x^3 + 43x^2 - 28x - 60 = 0$$

   are -2, -1, 2, 3, and 5. Now change the signs of the terms in the equation appropriately so that the roots of the new equation are 2, 1, -2, -3, and -5. Verify this.

2. Verify that the roots of the equation
$$x^3 - 7x + 6 = 0$$

   are -3, 1, and 2. Now change the signs of the terms in the equation appropriately so that the roots of the new equation are 3, -1, and -2. Verify this.

$\square$

22.4   THE FUNDAMENTAL THEOREM OF ALGEBRA

The fundamental theorem of algebra speaks to the number of solutions of a polynomial equation. The result is something that you probably have noticed when solving equations and, perhaps, even hypothesized.

The first explicit statement of the theorem, without a proof, was by Albert Girard (1595–1632) in his work *Invention nouvelle en l'algèbre* (*A New Discovery in Algebra*, 1629) [73, p. 446]:

> THEOREM. Every algebraic equation ... admits of as many solutions as the denomination of the highest quantity indicates....

By the "denomination of the highest quantity" is meant the degree of the equation, so Girard asserts that a polynomial of degree n has n solutions. Katz [73] tells us that Girard realized that some solutions may be repeated (multiplicity greater than one) and some may be imaginary ("impossible"). According to Katz [73, p. 447], "In answer to the anticipated question of the value of these impossible solutions, Girard answered that 'they are good for three things: for the certainty of the general rule, for being sure that there are no other solutions, and for its utility.' " Katz indicates that Girard provides the example $x^4 + 3 = 4x$ that has solutions 1, 1, $-1 + i\sqrt{2}$, and $-1 - i\sqrt{2}$.

Some eight years later, Descartes, in his Third Book of *La géométrie*, would enunciate essentially Girard's theorem, writing [46, p. 159]:

> Every equation can have as many distinct roots (values of the unknown quantity) as the number of dimensions of the unknown quantity in the equation.

Katz [73, p. 447] remarks that Descartes uses the phrase "can have" instead of Girards "admits of" regarding the number of roots or solutions because, unlike Girard, Descartes considers only distinct roots and, at least earlier on, does not consider imaginary roots.

Today, the fundamental theorem of algebra may be stated as follows:

Theorem 22.2 (FUNDAMENTAL THEOREM OF ALGEBRA) *Let $a_0, a_1, \ldots, a_{n-1}$ be complex numbers. If $n \geq 1$, then the equation*
$$x^n + a_{n-1}x^{n-1} + \cdots + a_2x^2 + a_1x + a_0 = 0$$

has at least one complex number solution.

$\square$

The fundamental theorem of algebra together with the factor theorem (page 424) imply that every polynomial equation of degree n has exactly n complex number solutions. Note that the coefficients and solutions may be real numbers (because $a + 0i = a$), as well, and that the solutions may be repeated.

Gauss was the first to prove the fundamental theorem of algebra; he did so in his doctoral thesis of 1799. Katz [73, p. 738] informs us that "Gauss was so intrigued with the fundamental theorem—that every polynomial $p(x)$ with real coefficients has a real or complex root—that he published four different proofs of it, in 1799, 1815, 1816, and 1848. Each proof used in some form or other the geometric interpretation of complex numbers, although in the first three proofs Gauss hid this notion by considering the real and imaginary parts of the numbers separately.... It was only in his final proof in 1848 that Gauss believed mathematicians would be comfortable enough with the geometric interpretation of complex numbers for him to use it explicitly. In fact, in that proof, similar to his first one, he even permitted the coefficients of the polynomial to be complex." As Crowe [42, p. 9] remarks, however, "Ironically Gauss himself did not accept the geometrical representation of complex numbers as a sufficient justification for them."

Chapter 23

Rule of Three

The *rule of three* is a method for solving proportion problems such as: *If, since your last fill up, it took 3 gallons of gasoline to fill up your car after having driven 80 miles, how many miles would you be able to drive on 12 gallons?* To answer the question, we may set up a proportion equation, such as

$$\frac{x \text{ miles}}{12 \text{ gallons}} = \frac{80 \text{ miles}}{3 \text{ gallons}}.$$

Solving for x (by cross multiplying), we find that $x = 320$. Thus, we conclude that we would be able to drive 320 miles on 12 gallons of gasoline. We use a proportion (or ratios) to solve this type of problem because the assumption is being made that one quantity increases or decreases proportionally as the other quantity increases or decreases. However, before we came to write equations, the answer likely would have been gotten using the "rule of three." (There were also employed through the ages rules of five, seven, and so on, for different proportion problems.)

23.1 CHINA

Our first examples are from Chinese mathematics and taken from the *Nine Chapters*. Examples of the *rule of three* are found in chapter 2 of the *Nine Chapters*, "*Su mi*" ("Millet and Rice"). The chapter begins with the following table and a statement of the rule of three; bold text is commentary added to the *Nine Chapters* [44, p. 241].

Liu: For barter trade and exchange of goods.

Millet and Rice Exchange Rule			
millet rate	50	cooked imperial millet	42
hulled millet	30	soya beans	45
milled millet	27	small beans	45
highly milled millet	24	sesame seed	45
imperial millet	21	wheat	45
fine crushed wheat	13 $\frac{1}{2}$	paddy	60
coarse crushed wheat	54	fermented beans	63
cooked hulled millet	75	porridge	90
cooked milled millet	54	cooked soya beans	103 $\frac{1}{2}$
cooked highly milled millet	48	malt	175

> **Liu: The various exchange rates in the table are in proportion. They can be mutually converted by selecting the corresponding rates. Reduce whenever reducible. The [Exchange] Rule holds in other cases.**

Rule of Three: Take the given number to multiply the sought rate. [The product] is the dividend. The given rate is the divisor. Divide.

> **Liu: This is a general rule. [The concept of] rate can be applied in various problems in the *Nine Chapters*. As the saying goes: "Knowing the past one can predict the future. Shown one corner [of a square], one can infer the other three." This Rule can resolve complicated and tricky problems and overcome the barriers between different quantities [of different items]. Take the rates from the given situation, clearly distinguishing their positions in the array, [then] homogenize and uniformize[1] [the columns]. All of these ultimately depend on this Rule.**

To fix ideas, we look at two examples in the *Nine Chapters* that use the rule of three.

Example 23.1 *Nine Chapters* chapter 2: "Millet and Rice" [44, p. 242]

1. Now millet, 1 *dou*, is required as hulled millet. Tell: how much is obtained? Answer: As hulled millet: 6 *sheng*. Method: Taking millet is required as hulled millet, multiply by 3, divide by 5.

First, we note that *dou* and *sheng* are units of capacity (volume) [108, pp. 9–10], and that there are 10 *sheng* in 1 *dou*.

Second, referring to the "Exchange Rule" table (page 440), we find that the exchange rates are

$$\text{given rate}: \quad \text{millet } 50,$$
$$\text{sought rate}: \quad \text{hulled millet } 30.$$

We see that both rates have a common factor of 10, so, following Liu's advice to "Reduce whenever reducible," we reduce the rates:

$$\text{given rate}: \quad \text{millet } 5,$$
$$\text{sought rate}: \quad \text{hulled millet } 3.$$

Last, the given number is 1 *dou* or 10 *sheng*. Thus, applying the rule of three, we find that the

$$[\text{sought result}] = \text{given number} \times \text{sought rate} \div \text{given rate}$$
$$= 10 \times 3 \div 5 = 6.$$

Therefore, we conclude that the sought result is 6 *sheng*.   □

Example 23.2 *Nine Chapters* chapter 2: "Millet and Rice" [108, p. 144]

[Problem 2]

Now 2 *dou* 1 *sheng* millet is required as milled millet. Tell: how much is obtained.

Answer: As milled millet 1 *dou* $1\frac{17}{50}$ *sheng*.

Method: Taking millet is required as milled millet, multiply by 27, divide by 50.

---

[1]By "uniformize" is meant to find a common denominator for a collection of fractions, and by "homogenize" is meant to build those fractions to have that common denominator [108, pp. 50, 72–73]. Thus, for example, the fractions $\frac{1}{2}$, $\frac{2}{3}$, and $\frac{3}{4}$ can be uniformized using the common denominator 12, in which case the fractions are homogenized by the numerators 6, 8, and 9, giving the equivalent fractions $\frac{6}{12}$, $\frac{8}{12}$, and $\frac{9}{12}$. Or we could uniformize the denominators to 24, in which case the numerators homogenize to 12, 16, and 18, giving the equivalent fractions $\frac{12}{24}$, $\frac{16}{24}$, and $\frac{18}{24}$, instead. Shen et al. [108, p. 50] tell us that "The notion of greatest common divisor and least common multiple, although used in China, did not play as large a rôle as they do in the West."

Referring to the "Exchange Rule" table, we find that the exchange rates are

$$\text{given rate}: \quad \text{millet } 50,$$
$$\text{sought rate}: \quad \text{milled millet } 27.$$

We see that both rates do not have a common factor, so we do not reduce the rates.

Now, the given number is 2 *dou* 1 *sheng* or $(2 + \frac{1}{10})$ *dou* or $\frac{21}{10}$ *dou*. Thus, applying the rule of three, we find that the

$$[\text{sought result}] = \text{given number} \times \text{sought rate} \div \text{given rate}$$

$$= \frac{21}{10} \times 27 \div 50 = \frac{567}{500}.$$

So, we conclude that the sought result is $\frac{567}{500}$ *dou* or $1\frac{67}{500}$ *dou*. Finally, using that there are 10 *sheng* in 1 *dou*, we convert $\frac{67}{500}$ *dou* to *sheng*:

$$\frac{67}{500} \times 10 = \frac{670}{500} = 1\frac{17}{50}.$$

Therefore, in *dou* and *sheng*, the sought result is 1 *dou* $1\frac{17}{50}$ *sheng*. □

Now You Try 23.1 Solve the following problems using the rule of three. Identify the given number, sought rate, and the given rate. Give your answer in *sheng* or in *dou* and *sheng* as appropriate.

1. (*Nine Chapters* chapter 2 problem 3 [108, p. 144]) Now 4 *dou* 5 *sheng* millet is required as highly milled millet. Tell: how much is obtained?

2. (*Nine Chapters* chapter 2 problem 9 [44, p. 242]) Now 8 *dou* 6 *sheng* of millet is required as cooked highly milled millet. Tell: How much is obtained?

□

Remark 23.1 First, Dauben [44] remarks that the commodity rice is conspicuously missing from the "Exchange Rule" table even though the title of the chapter includes rice. He explains that this is because, although *mi* is today commonly translated as rice, the word *mi* had a broader meaning in ancient China. Indeed, Dauben suggests that a better translation of the chapter's title may be "grains." For a discussion, see [108, pp. 142–143]. Second, Dauben points out that, although this chapter in the *Nine Chapters* begins with a table of grain equivalencies, the chapter is not only about that; the chapter includes other commodities such as bricks, bamboo, silk, and so on. Moreover, not all of the problems in this chapter are solved using the rule of three; in fact, the last nine problems of the chapter (problems 38–46) do not involve proportions. Last, Shen et al. [108, p. 144] compare these "exchange rule" problems in the *Nine Chapters* to the bread-and-beer problems in the Rhind Mathematical Papyrus (see page 442) and make the comment that "[i]t is interesting that even 4000 years ago the ancient Egyptians had corresponding solutions to solve the same daily life problem. However, they had not formulated a rule (Rule of Three). Otherwise, they would have got the answer more conveniently." □

The rule of three was not used only by early Chinese mathematicians. Indeed, the rule of three (and the inverse) appears to have been a standard mathematical technique that was employed by many different peoples in many ages. For example, Cooke [41, pp. 134–135] tells us that the ancient Egyptians used the rule of three to solve some proportion problems and provides an example from the Ahmose Papyrus (Rhind Mathematical Papyrus): "The Rule of Three procedure is invoked in problem 73, which asks how many loaves of 15-*pesu* bread are required to provide the same amount of grain as 100 loaves of 10-*pesu* bread. The answer is found by dividing 100 by 10, then multiplying by 15, which is precisely the Rule of Three."[2] Shen et al. [108, pp. 135–139] give a good summary of the use of the rule of three by different peoples.

---

[2]A *pesu* is a rate that is [41] "defined as the number of loaves of bread or jugs of beer obtained from one *hekat* of grain. A hekat was slightly larger than a gallon, 4.8 liters to be precise." According to Imhausen [70, p. 38], *Rhind* problem 73 is one of the group of

23.2   INDIA

Our next examples of the rule of three are taken from Indian mathematics. Bhāskara I composed his *Bhāṣya* (big commentary) on the work *Āryabhatīya* of Āryabhata in AD 629, and it is here that we find one of the most flowery descriptions of the rule of three.

Rule 23.1 (RULE OF THREE) *Āryabhatīya* of Āryabhata with commentary of Bhāskara I (bold text is Bhāskara's commentary) [95, p. 414]

> **In order to explain the Rule of Three, he says an āryā and a half:**
>
> 2.26. Now, when one has multiplied that fruit quantity of the Rule of Three by the desire quantity, what has been obtained from that divided by the measure should be this fruit of the desire.

□

Remark 23.2 The "desire quantity" is not the quantity sought. The quantity sought is the "fruit of the desire." So, we see that the rule of three stated here is

$$\text{fruit of the desire} = \text{fruit} \times \text{desire} \div \text{measure}.$$

Comparing this to the rule of three in the *Nine Chapters* (page 440),

$$[\text{sought result}] = \text{given number} \times \text{sought rate} \div \text{given rate}.$$

we see that

Āryabhatīya	$\leftrightarrow$	*Nine Chapters*
fruit of the desire	$\leftrightarrow$	[sought result]
fruit quantity	$\leftrightarrow$	given number
desire quantity	$\leftrightarrow$	sought rate
measure	$\leftrightarrow$	given rate

□

To try to understand Āryabhata's prescription for the rule of three, we continue with Bhāskara's commentary [95]:

> **An example:**
>
> **1. I have bought five *palas* of sandalwood for nine *rūpakas*. How much sandalwood, then, should be obtained for one *rūpaka*? [...]**
>
> **Setting down: 9   5   1**
>
> **Procedure: Since five *palas* of sandalwood [have been obtained] with nine *rūpakas*, nine is the measure quantity, five is the fruit quantity. Since "how much [has been obtained] with one *rūpaka*?" [is the question], one is the desire quantity. The fruit quantity multiplied by that desire quantity which is one, 5, should be divided by the measure quantity which is nine, $\frac{5}{9}$.**

---

so-called bread-and-beer problems (nine in the *Rhind* and eleven in the *Moscow Mathematical Papyrus*). Apparently, the production of bread and beer was a significant part of ancient Egyptian civilization. Evidence of this can be found in both archaeological artifacts and in scenes of daily life that are depicted on tomb walls [70]. We note, however, that not all of the bread-and-beer problems were solved using the rule of three.

Note that the "fruit quantity" is the *same type* of quantity as the "fruit of the desire," and the "measure" is the *same type* of quantity as the "desire quantity." Indeed, the twelfth-century Indian mathematician Bhāskara II says so plainly in his work *Līlāvatī* [95, p. 456]:

Now, the rule of operation for the [rule] of three quantities, in one verse:

73. The [given] amount and the desired [amount], [being of] the same type, are [written] in the first and last [positions, respectively]. The result of that [given amount], [being of] the same type, is [put] in the middle. That, multiplied by the desired [amount] and divided by [the given amount in] the first [position], is the result of the desired [amount]. In the inverse [rule of three quantities], the procedure is reversed.

Observe that Bhāskara II also instructs us to list the quantities in the order "measure-fruit-desire" just as Āryabhata does.

Example 23.3 If 100 widgets cost $36, how much would 30 widgets cost?

Since "how much would 30 widgets cost?" is the question, "how much [how many dollars]" is the "fruit of the desire" and 30 (widgets) is the "desire quantity." Thus, 100 (widgets; same *type* as the "desire quantity") is the "measure" and 36 (dollars; same *type* as the "fruit of the desire") is the "fruit quantity."

$$\text{measure-fruit-desire:} \quad 100 \quad 36 \quad 30$$

Thus, the
$$\text{fruit of the desire} = \text{fruit} \times \text{desire} \div \text{measure} = 36 \times 30 \div 100 = 10\tfrac{4}{5}.$$

Therefore, we conclude that it would cost $10\tfrac{4}{5}$ or $10.80. ☐

Example 23.4 According to your grandmother, the classic grape-milk drink is made with $1\tfrac{1}{2}$ fluid ounces of grape juice and $6\tfrac{1}{2}$ fluid ounces of whole milk. How many fluid ounces of grape juice must you mix with 8 fluid ounces (one cup) of whole milk to make the classic grape-milk drink?

Since "how many fluid ounces of grape juice must you mix with 8 fluid ounces (one cup) of whole milk ... ?" is the question, "how many fluid ounces of grape juice" is the "fruit of the desire" and 8 (fluid ounces of milk) is the "desire quantity." Thus, $1\tfrac{1}{2}$ (fluid ounces of grape juice) is the "fruit quantity" (because it is the same *type*, grape juice, as the "fruit of the desire") and $6\tfrac{1}{2}$ (fluid ounces of milk) is the "measure" (because it is the same *type*, milk, as the "desire quantity").

$$\text{measure-fruit-desire:} \quad 6\tfrac{1}{2} \quad 1\tfrac{1}{2} \quad 8$$

Thus, the
$$\text{fruit of the desire} = \text{fruit} \times \text{desire} \div \text{measure} = 1\tfrac{1}{2} \times 8 \div 6\tfrac{1}{2} = 1\tfrac{11}{13}.$$

Therefore, we conclude that we would need $1\tfrac{11}{13}$ fluid ounces of grape juice. ☐

Think About It 23.1 How would you use algebra to solve the problems in the last two examples? ☐

Now You Try 23.2 Solve the following problems using the rule of three. Identify the "fruit of the desire," the "fruit quantity," the "desire quantity," and the "measure ."

1. If, since your last fill up, it took 3 gallons of gasoline to fill up your car after having driven 80 miles, how many miles would you be able to drive on 12 gallons? Compare the solution using the rule of three here to the solution using a proportion equation that was presented on page 439.

2. According to your grandmother, an 8-fluid ounce cup of the classic grape-milk drink is made with $1\tfrac{1}{2}$ fluid ounces of grape juice and $6\tfrac{1}{2}$ fluid ounces of whole milk. How many fluid ounces of grape juice and whole milk must you mix to make 12 fluid ounces of the classic grape-milk drink?

3. (*Līlāvatī* v. 74 [95]) If two and a half *palas* of saffron are obtained for three-sevenths of a *niṣka*, tell me at once, best of merchants, how much of that [can be bought] with nine *niṣkas*.

$\square$

The rule of three, namely,

$$\text{fruit of the desire} = \text{fruit} \times \text{desire} \div \text{measure},$$

is used when one expects the "measure" and the "fruit" to be in the same ratio as the "desire" and the "fruit of the desire." We illustrate this in the following table. Here, we let

- M denote "measure,"
- F denote "fruit,"

- x denote "desire,"
- y denote "fruit of the desire."

Then, the rule of three is used when one expects

$$M : F \quad \text{as} \quad x : y.$$

	M	F	$M : F$	x	y	$x : y$
EXAMPLE 23.3	100	36	$100 : 36 = \mathbf{25 : 9}$	30	$10\frac{4}{5}$	$30 : 10\frac{4}{5} = \mathbf{25 : 9}$
EXAMPLE 23.4	$6\frac{1}{2}$	$1\frac{1}{2}$	$6\frac{1}{2} : 1\frac{1}{2} = \mathbf{13 : 3}$	8	$1\frac{11}{13}$	$8 : 1\frac{11}{13} = \mathbf{13 : 3}$

However, in certain problems, we may expect the "desire" and the "fruit" to be in the same ratio as the "measure" and the "fruit of the desire":

$$x : F \quad \text{as} \quad M : y.$$

In this case, Bhāskara advises that one uses the *inverse rule of three* [95, p. 456]:

> Now the rule of operation in the inverse [rule of] three quantities:

> 77. [Sometimes] decrease in the result occurs when there is increase in the desired [quantity], or increase when decrease. So the inverse [rule of] three quantities should be known by those who understand calculation.

> When there is decrease in the result, when there is increase in the desired [quantity], or increase in the result in the case of decrease [in the desired quantity], then the inverse [rule of] three quantities [is used].

He goes on to illustrate when the inverse rule of three is to be used, and follows this with an example that illustrates the rule [95]:[3]

> That is as follows:

> 78. In the case of the cost of living beings [according to their] age, and the weight and alloy of gold, and the subdivision of amounts, the [rule of] three quantities should be inverted.

> An example involving the price with respect to the age of a living being:

> 79. If a woman [slave] sixteen years old is bought for [a price of] 32, what is [the price of one] twenty years old? [If] an ox after two years of labor is bought for four *niṣkas*, then what is [the price of one] after six years of labor?

> Statement: 16   32   20. The result is $25\frac{3}{5}$ *niṣkas*.

> Statement of the second [problem]: 2   4   6. The result is $1\frac{1}{3}$ *niṣkas*. [...]

---

[3]When reading historical material, we may come across practices that today we would find to be unacceptable. In these instances, we should be mindful of the time and place in which they occurred.

Before we try to understand how to apply the inverse rule of three, let us make the following observations.

- In Bhāskara's examples, the value (of a slave or draft animal) decreases as the age (of the slave or draft animal) increases.

- The quantities are still listed in the order "measure-fruit-desire."

- The "desire" and the "fruit" are in the same ratio as the "measure" and the "fruit of the desire," that is, if we again let M denote the "measure," F denote the "fruit," x denote the "desire," and y denote the "fruit of the desire," we see that

$$x : F \quad \text{as} \quad M : y.$$

	x	F	$x : F$	M	y	$M : y$
Slave example	20	32	$20 : 32 = 5 : 8$	16	$25\frac{3}{5}$	$16 : 25\frac{3}{5} = 5 : 8$
Ox example	6	4	$6 : 4 = 3 : 2$	2	$1\frac{1}{3}$	$2 : 1\frac{1}{3} = 3 : 2$

Bhāskara's examples give us the general inverse rule of three.

Rule 23.2 (INVERSE RULE OF THREE) Let the "measure," "desire quantity," and "fruit quantity" be given. Then, the

$$\text{fruit of the desire} = \text{fruit} \times \text{measure} \div \text{desire}.$$

□

Example 23.5 In the year 2005 you paid \$10,500 for a BMW R1150RS model motorcycle that originally sold new in 2002. For how much could you have expected to sell the motorcycle in 2009?

Since "for how much could you have expected to sell the motorcycle in 2009?" is the question, "for how much [how many dollars]" is the "fruit of the desire" and 7 (years, the age of the motorcycle) is the "desire quantity." Thus, 10,500 (dollars) is the "fruit quantity" and 3 (years) is the "measure."

measure-fruit-desire: 3   10,500   7

Now, because we expect the price of the motorcycle to *decrease* with age, we apply the *inverse* rule of three:

$$\text{fruit of the desire} = \text{fruit} \times \text{measure} \div \text{desire} = 10,500 \times 3 \div 7 = 4500.$$

Therefore, you could have expected to sell the motorcycle for \$4,500.

□

Think About It 23.2 Since the price decreases as the age increases, this is an inverse proportion. How would you use algebra to solve the problem in the last example?

□

Now You Try 23.3 Solve the following problems using the inverse rule of three. Identify the "fruit of the desire," the "fruit quantity," the "desire quantity," and the "measure."

1. (Francès Pellos, a fifteenth-century nobleman from Nice [39, p. 106])[4]   Similarly, if three and a half are worth 6 and a half, how much is 4 and a third worth?

2. (*Līlāvatī* v. 77 [34, p. 222])   If a gadyánaca of gold of the touch of ten may be had for one *nishca* [of silver], what weight of gold of fifteen touch may be bought for the same price?

□

---

[4]From *Compendion de l'abaco* (1492), after the edition by R. Laffont, Montepellier: Editions de la Revue des Langues Romanes, 1967, pp. 104, 181. From the French translation of the Occitan by M. Guillemot.

23.3   MEDIEVAL EUROPE

As a last example of the rule of three, we turn to the twelfth-century Italian mathematician Leonardo of Pisa, who presents the rule of three in his 1202 work, *Liber Abaci* (see page 107). Goetzmann [61, p. 130] provides an example of the rule of three from *Liber Abaci* chapter 8, "On Finding the Value of Merchandise by the Principal Method":

> The first problem Fibonacci addresses is how to determine an unknown price from a given quantity of merchandise when the price per unit is known: suppose 100 rolls cost 40 lira, how much would five rolls cost? He offers the solution through a diagram $\left[\begin{smallmatrix} 40 & 100 \\ ? & 5 \end{smallmatrix}\right]$ and the solution as $(40 * 5)/100$.
>
> This simple solution is generally called the "Rule of Three" and is one of the oldest algebraic tools in mathematics.

Goetzmann goes on to say that "Leonardo uses the Rule of Three with increasingly complex quantities and currencies, applying it to examples drawn from trade around the Mediterranean," and offers the following example from *Liber Abaci*:

> Or for traders of raw cotton here is a useful problem:
>
> One has near Sicily a certain ship laden with 11 hundredweights and 47 rolls of cotton, and one wishes to convert them to packs; because $^1/_3$ 1 hundredweights of cotton ... is one pack, then four hundredweight of cotton are 3 packs and four rolls of cotton are 3 rolls of a pack; you write down in the problem the 11 hundredweights and 47 rolls, that is 1147 rolls below the 4 rolls of cotton and you will multiply the 1147 by 3 and you divide by the 4; the quotient will be $^1/_4$ 860 rolls of a pack.

Following Leonardo's instructions, one would construct the diagram $\left[\begin{smallmatrix} 3 & 4 \\ ? & 1147 \end{smallmatrix}\right]$, from which it would follow that the solution is $1147 \times 3 \div 4 = 860\frac{1}{4}$ packs.

Now You Try 23.4 Use Leonardo's method to solve the following problems. (See NOW YOU TRY 23.2.)

1. If, since your last fill up, it took 3 gallons of gasoline to fill up your car after having driven 80 miles, how many miles would you be able to drive on 12 gallons?

2. According to your grandmother, an 8-fluid ounce cup of the classic grape-milk drink is made with $1\frac{1}{2}$ fluid ounces of grape juice and $6\frac{1}{2}$ fluid ounces of whole milk. How many fluid ounces of grape juice and whole milk must you mix to make 12 fluid ounces of the classic grape-milk drink?

$\Box$

Let us now turn to the question, Why do the rule of three and the inverse rule of three work? To fix ideas, we answer the question using the language of Āryabhata: "fruit quantity," "desire quantity," "measure," and "fruit of the desire."

To begin, the rule of three or the inverse rule of three may be applied to a problem only if

1. there are three related given quantities and we are to find a fourth related unknown quantity, and

2. two of the given quantities are of the same type, and the third given quantity and the unknown quantity are of another same type.

Now, when we expect the "measure" and the "fruit" to be in the same ratio as the "desire quantity" and the "fruit of the desire," then the quantities in the problem satisfy the proportion

measure : fruit   as   desire : fruit of the desire

or

$$\frac{\text{measure}}{\text{fruit}} = \frac{\text{desire}}{\text{fruit of the desire}}.$$

Solving the proportion equation for "fruit of the desire" (by cross multiplying), we find that

$$\text{fruit of the desire} = \frac{\text{fruit} \times \text{desire}}{\text{measure}},$$

which is precisely the rule of three (RULE 23.1 on page 442).

On the other hand, when we expect the "desire" and the "fruit" to be in the same ratio as the "measure" and the "fruit of the desire," then the quantities in the problem satisfy the (inverse) proportion

$$\text{desire : fruit} \quad \text{as} \quad \text{measure : fruit of the desire}$$

or

$$\frac{\text{desire}}{\text{fruit}} = \frac{\text{measure}}{\text{fruit of the desire}}.$$

Solving the proportion equation for "fruit of the desire," we find that

$$\text{fruit of the desire} = \frac{\text{fruit} \times \text{measure}}{\text{desire}},$$

which is precisely the inverse rule of three (RULE 23.2 on page 445).

We note that with the rule of three or the inverse rule of three there is no need to write a proportion equation; all that is needed is to multiply and divide appropriately to obtain the answer.

23.4   THE RULE OF THREE IN FALSE POSITION

We look at the use of the rule of three in solving problems by the method of false position.

Plofker [95, p. 454] provides the following example from the work *Līlāvatī* of Bhāskara II. Bhāskara calls the method *ista-karma*, which means "operating with a trial number" [39].

Example 23.6 *Līlāvatī* of Bhāskara II

Before the example, Bhāskara gives the method:

Now, the rule of operation for methods of assumption [of some arbitrary quantity], with reduction of given [quantities] and remainders and simplification of fractional differences, in one verse:

51. As in the statement of the example, any desired number is multiplied, divided, or decreased and increased by fractions. The given quantity, multiplied by the desired [number] and divided by the [result], is the [required] quantity; [that] is called the operation with an assumed [quantity].

52. What quantity, multiplied by five, diminished by its own one-third, divided by ten, increased by one-third, one-half, and one-fourth of the [orginal] quantity, is seventy minus two?

Statement: The multiplier is 5; its own part is negative, $\frac{1}{3}$; [the proportion] subtracted, $\frac{1}{3}$; the divisor, 10; parts of the quantity are added, $\frac{1}{3}, \frac{1}{2}, \frac{1}{4}$; the given [answer] is 68.

The quantity assumed here is 3. [It is] multiplied by five (15), decreased by its own third part (10), [and] divided by ten (1). [It is] added to the third, half, and quarter ($\frac{3}{3}, \frac{3}{2}, \frac{3}{4}$) of the quantity assumed here (3); the result is $\frac{17}{4}$. The given [answer], 68, multiplied by the assumed [number], is divided by that. The resulting quantity is 48.

He then recapitulates the method:

So however the [unknown] quantity in an example [when] multiplied or divided by something, or decreased or increased by a part of the quantity, [becomes] the given [answer], that is the way that [some] imagined given quantity [is transformed] via the operation in the [above-]mentioned explanation. Then divide the given [answer], multiplied by the assumed quantity, by whatever [result] is obtained. The result [of that] is the [desired] quantity. [...]

To solve the problem, Bhāskara proceeds as follows:

Guess:	3
Work:	five times the quantity: $5 \times 3 = \underline{15}$ a third of this is removed: $\underline{15} - \frac{1}{3} \times \underline{15} = \underline{\underline{10}}$ of this one tenth: $\frac{1}{10} \times \underline{\underline{10}} = \underset{\smile}{1}$ to this its one-third, one-half, and one-fourth (of the quantity): $\quad \underset{\smile}{1} + \frac{1}{3} \times 3 + \frac{1}{2} \times 3 + \frac{1}{4} \times 3 = \frac{17}{4}$ or $4\frac{1}{4}$, but the desired result is $70 - 2$ or 68
Adjust:	rule of three: $68 \times 3 \div \frac{17}{4} = 48$
Answer:	48

It is interesting to compare Bhāskara's method to that of the ancient Egyptians. We restate the problem in EXAMPLE 23.6 along the lines of the ancient Egyptian texts:

A quantity, five times the quantity, a third of this is removed, and of this one-tenth, and to this (of the original) one-third, one-half, and one-fourth (is added) so that 70 less 2 results.

The Egyptian scribe A'hmosè may then have solved this problem using the method of false position as follows (see EXAMPLE 19.5):

Guess:	12
Work:	five times the quantity: $5 \times 12 = \underline{60}$ a third of this is removed: $\underline{60} - \frac{1}{3} \times \underline{60} = \underline{\underline{40}}$ of this one tenth: $\frac{1}{10} \times \underline{\underline{40}} = \underset{\smile}{4}$ to this its one-third, one-half, and one-fourth (of the quantity): $\quad \underset{\smile}{4} + \frac{1}{3} \times 12 + \frac{1}{2} \times 12 + \frac{1}{4} \times 12 = 17$, but the desired result is $70 - 2$ or 68
Divide:	$68 \div \underline{17} = \underset{\smile}{4}$
Multiply:	$\underset{\smile}{4} \times 12 = 48$
Answer:	48

Although the method is the same, notice the difference between the Indian mathematician Bhāskara II and the Egyptian scribe A'hmosè, for example, in their respective facilities for using numbers. Regarding this, Chabert et al. [39, p. 98] point out that "In contrast to the Ancient Egyptians, for example, who almost invariably used unit fractions, Bhāskara uses general fractions. Nor does he try as much as possible, to operate with integers. Here [EXAMPLE 23.6], he takes 3 as the supposed value and obtains the result 17/4...."

Observe further that, although essentially the same, the last steps that Bhāskara takes in finding the solution in EXAMPLE 23.6 differ from what A'hmosè may have taken: Bhāskara uses the rule of three (one step), whereas A'hmosè may have done this (two steps):

Divide:	$68 \div \frac{17}{4} = 16$
Multiply:	$16 \times 3 = 48$

(Of course, A'hmosè would never have written the non-unit fraction $\frac{17}{4}$.)

As another example, Francès Pellos, a fifteenth-century nobleman from Nice, also uses the rule of three in conjunction with false position. This example is from his *le Compendion de l'abaco* (*Compendium of the abacus*; 1492), the first published Occitan arithmetic [39]. (Occitan is in what is now the southwestern region of France.)

Example 23.7 Pellos (in [39, p. 107])

Similarly, find a number such that if it is divided by 7 it comes to 3 and a half. Do thus: choose as you wish 14, which you divide by 7, it comes to 2. Then, say this, if 2 comes from 14, what does 3 and a half come from? Use, as you know, the rule of three and you will find 24 and a half. And this is the required number.

In summary, the problem is to find a number such that one-seventh the number results in $3\frac{1}{2}$. Pellos proceeds as follows:

Guess:	14
Divide:	$14 \div 7 = \underline{2}$, but the desired result is $3\frac{1}{2}$
Rule of three:	$3\frac{1}{2} \times \mathbf{14} \div \underline{2} = 24\frac{1}{2}$
Answer:	$24\frac{1}{2}$

□

Now You Try 23.5

1. Verify the arithmetic in examples 23.6 and 23.7.

2. Solve the problem in EXAMPLE 23.6 using the method of false position, but, unlike Bhāskara, try as much as possible to operate with integers. (Hint: Guess a number that is a common multiple of the denominators.)

□

Finally, we return to the following problem from *Liber Abaci* (see page 289):

Indeed the value of one hundredweight, namely 100 rolls, is 13 pounds, and it is sought how much 1 roll is worth.

We already showed how Leonardo solved this problem using his second method of elchataym. We now show how Leonardo solved this problem using his first method of elchataym in which he used the rule of three.

The way Leonardo solves this problem using his first method of elchataym is to make two guesses for the value of one roll, namely, 1 soldo and 2 soldi. He finds that 1 soldo for one roll leads to 100 soldi or 5 pounds for 100 rolls, which is 8 pounds short of the given 13 pounds for 100 rolls; he also finds that 2 soldi for one roll leads to 200 soldi or 10 pounds for 100 rolls, which is 3 pounds short of the given 13 pounds for 100 rolls. He then notes that by increasing his guess by 1 pound or 12 denari (the difference between his two guesses of 2 soldi and 1 soldo), he has decreased the shortage from the given 13 pounds for 100 rolls by 5 pounds (the difference between the two shortages of 8 pounds and 3 pounds). Thus, Leonardo poses the question, "what shall I add to the price of the same roll [2 soldi for one roll] in order to decrease the difference of 3 pounds that resulted from the second position to the true price of the same hundredweight?" To answer this question, he applies the rule of three:

> You therefore multiply the extreme numbers, and divide by the middle one ... namely the 12 by the 3, and you divide by the 5 ... the quotient is $\frac{1}{5}$ 7 denari which is added to the 2 soldi....

Leonardo, therefore, concludes that if the value of 100 rolls is 13 pounds, then one roll is worth 2 soldi and $\frac{1}{5}$ 7 denari.

The following figure illustrates Leonardo's first solution. Here, the horizontal axis represents the price for one roll in denari, and the vertical axis represents the price for 100 rolls in soldi.

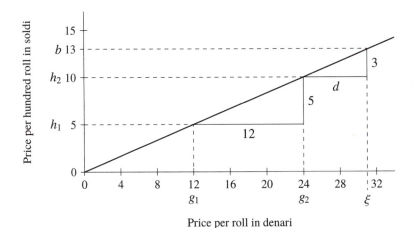

Price per roll in denari

We see that Leonardo has set up two similar triangles, for which he is solving for the difference d between the second guess of 2 soldi or 24 denari and the sought value of ξ denari for one roll. The two similar triangles lead to the proportion equation

$$\frac{d}{3} = \frac{12}{5},$$

so that

$$d = \frac{12 \times 3}{5} = 7\frac{1}{5}.$$

Therefore, $\xi = 24 + 7\frac{1}{5}$ or $31\frac{1}{5}$ denari, or 2 soldi and $7\frac{1}{5}$ denari.

23.5   DIRECT VARIATION, INVERSE VARIATION, AND MODELING

The rule of three is related to the modern notions of *direct variation* or *direct proportionality*, and the inverse rule of three is related to *inverse variation* or *inverse proportionality*. Let us begin by defining these notions, and then see how they are related to the rule of three and the inverse rule of three.

Definition 23.1 A number y *varies (directly)* as a number x, or y is *(directly) proportional* to x, if

$$y = kx$$

for some constant k that is called the *constant of (direct) proportionality*. □

Example 23.8 Suppose that y varies (directly) as x and $y = 9$ when $x = 6$. Find the constant of proportionality and write the variation equation.

The number y varies as x, so

$$y = kx.$$

Thus, using the given values of y and x, we have

$$9 = k(6).$$

Solving for k, we find that the constant of proportionality is

$$k = \frac{3}{2}$$

and, therefore, the variation equation is

$$y = \frac{3}{2}x.$$

□

Example 23.9 When traveling at a constant speed (rate), the distance traveled is (directly) proportional to the amount of time traveled.

Suppose that, on one trip, you drove 300 miles in 6 hours at a constant speed. Traveling at that same constant speed, how long would it take you to drive 240 miles?

Let d denote the number of miles and t denote the number of hours driven. Then, d is proportional to t, so

$$d = kt,$$

Thus, using the given values of d and t, $d = 300$ when $t = 6$, we have

$$300 = k(6).$$

Solving for k, we find that the constant of proportionality is

$$k = 50$$

and, hence, the variation equation is

$$d = 50t.$$

Finally, to find how long it would take to drive 240 miles at the same constant speed, we substitute $d = 240$ into the variation equation and solve for t:

$$240 = 50t, \quad \text{so that} \quad t = \frac{240}{50} = 4.8.$$

Therefore, it would take 4.8 hours or 4 hours and 48 minutes.[5] □

---

[5]If we include the units of miles for distance and hours for time, we see that

$$d \text{ miles} = 50 \text{ units} \cdot t \text{ hours}.$$

Now, dividing both sides by t hours, we see that the constant of proportionality $k = 50$ has units of miles per hour:

$$\frac{d \text{ miles}}{t \text{ hours}} = 50 \text{ units} \quad \text{or} \quad \frac{4 \text{ miles}}{t \text{ hours}} = 50 \text{ miles per hour.}$$

Thus, the constant of proportionality here is a speed or rate. In fact, you may have recognized this problem as a "distance $=$ rate$\times$time" problem, for which we would usually write $d = rt$ instead of $d = kt$.

Now You Try 23.6

1. Write an equation that relates x and y if y is proportional to x and $y = 18$ when $x = 10$.

2. Driving at a constant speed, the distance it takes to bring a vehicle to a stop at a particular breaking effort varies as the square of the speed.

   Suppose that you find it takes B feet to bring your car to a stop from a constant speed of 20 miles per hour. How many feet would it take to bring your car to a stop in the same conditions from a constant speed of 40 miles per hour?

   □

Definition 23.2 A number y *varies inversely* as a number x, or y is *inversely proportional* to x, if

$$y = \frac{k}{x}$$

for some constant k that is called the *constant of inverse proportionality*.   □

Example 23.10 Suppose that y is inversely proportional to x and $y = 9$ when $x = 6$. Find the constant of inverse proportionality and write the inverse variation equation.

The number y is inversely proportional to x, so

$$y = \frac{k}{x}.$$

Thus, using the given values of y and x, we have

$$9 = \frac{k}{6}.$$

Solving for k, we find that the constant of inverse proportionality is

$$k = 54$$

and, therefore, the inverse variation equation is

$$y = \frac{54}{x}.$$

□

Example 23.11 The time it takes to travel a fixed distance varies inversely as the average speed (constant rate) of travel. In other words, as your speed increases, your travel time decreases.

Suppose that, on one occasion, you drove from home to work in 20 minutes at an average speed of 30 miles per hour. How long would it take you to drive home from work if your average speed would be 20 miles per hour?

Let t denote the number of hours driven and r denote the average speed (constant rate) in miles per hour driven. Note that 20 minutes equal $\frac{1}{3}$ hour.[6] Then, t varies inversely as r, so

$$t = \frac{k}{r}.$$

---

[6]It is important that the units match. Here, the time should be in hours if the average speed is in miles per *hour*.

Thus, using the given values of t and r, $t = \frac{1}{3}$ when $r = 30$, we have

$$\frac{1}{3} = \frac{k}{30}.$$

Solving for k, we find that the constant of inverse proportionality is

$$k = 10$$

and, hence, the inverse variation equation is

$$t = \frac{10}{r}.$$

Finally, to find the amount of time it would take you to drive home at an average speed of 20 miles per hour, we substitute $r = 20$ into the inverse variation equation and solve for r:

$$t = \frac{10}{20}, \quad \text{so that} \quad t = \frac{1}{2}.$$

Therefore, it would take you $\frac{1}{2}$ hour or 30 minutes. □

Now You Try 23.7

1. Write an equation that relates x and y if y varies inversely as x and $y = 18$ when $x = 10$.

2. The intensity of light is inversely proportional to the square of the distance from the light source.

   Suppose that you notice that the intensity of light from your desk lamp is I lumens when the light bulb is two feet away from you. Find the intensity of light from the desk lamp if it is moved so that the light bulb is now four feet away from you.

   □

Notice that the equation $y = kx$ is a linear equation, so that its graph is a straight line,[7] whereas the equation $y = \frac{k}{x}$ is a nonlinear equation, so that its graph is not a straight line. FIGURE 23.1 shows two typical graphs with $k = 2$. We remark that, in the case of direct variation,

- if k is positive, y increases at a constant rate as x increases,

- if k is negative, y decreases at a constant rate as x increases,

in fact, at the rate k that is the slope of the line. However, in the case of inverse variation,

- if k is positive, y decreases at a varying rate as x increases,

- if k is negative, y increases at a varying rate as x increases,

in fact, at the rate $-\frac{k}{x^2}$, so that y decreases or increases more slowly as x increases.

Now let us see how the rule of three and the inverse rule of three are related to direct proportionality and inverse proportionality. Toward this end, let us define the following symbols: Let

- M denote "measure,"

- F denote "fruit,"

- x denote "desire,"

- y denote "fruit of the desire."

---

[7]In ancient times, "line" could refer to a straight line or a curve.

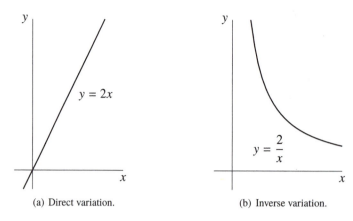

(a) Direct variation.   (b) Inverse variation.

Figure 23.1: Direction variation and inverse variation.

Then the rule of three may be expressed as

$$y = \frac{Fx}{M}.$$

Now, rearrange the factors F and M,

$$y = \frac{F}{M}x,$$

and note that $\frac{F}{M}$ is a constant. So, if we let $k = \frac{F}{M}$, then the rule of three transforms into the variation equation

$$y = kx,$$

where $k = \frac{F}{M}$ is the constant of proportionality.

Similarly, the inverse rule of three may be expressed as

$$y = \frac{FM}{x}.$$

So, if this time we let $k = FM$, then the inverse rule of three transforms into the inverse variation equation

$$y = \frac{k}{x},$$

where $k = FM$ is the constant of inverse proportionality.

To firm up ideas, let us take another look at the problem in EXAMPLE 23.5, but now from the point of view of variation:

> In the year 2005 you paid $10,500 for a BMW R1150RS model motorcycle that originally sold new in 2002. For how much could you have expected to sell the motorcycle in 2009?

Let y denote the value in dollars of the motorcycle when it is x years old. Then the problem may be summarized in the following table.

Year	x (years old)	y (dollars)
2005	3	10,500
2009	7	?

Now, let us assume that y is inversely proportional to x, so that

$$y = \frac{k}{x}$$

for some constant k. We use the given information, $(x, y) = (3, 10{,}500)$, to find k, the constant of inverse proportionality:

$$10{,}500 = \frac{k}{3} \quad \text{implies that} \quad k = 10{,}500 \times 3 = 31{,}500.$$

Therefore, we find that the value in dollars, y, of the motorcycle is related to its age in years, x, by the equation

$$y = \frac{31{,}500}{x}.$$

We use this equation to complete the table and, hence, answer the question in the problem.

To find for how much you could have expected to sell the motorcycle in the year 2009, let $x = 7$ in the equation. This gives

$$y = \frac{31{,}500}{7} = 4500.$$

So, you could have expected to sell the motorcycle for $4,500.

Year	x (years old)	y (dollars)
2005	3	10,500
2009	7	4500

Note that the answer here agrees with the answer we obtained before using the inverse rule of three, but now we have an equation, namely,

$$y = \frac{31{,}500}{x},$$

that we may use to analyze our solution. To begin, we graph the equation (by finding several solutions, plotting the solutions as points, and connecting the points with a smooth curve).

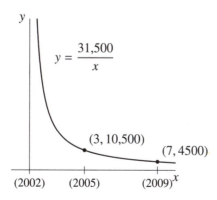

Observe that the graph cries out one glaring shortcoming with the inverse variation model for the value of the motorcycle: In the year 2002, when $x = 0$, the value in dollars, y, of the motorcycle was infinite!

Let us try to construct a different model for the value of the motorcycle. For this, we need a little more given information. Let us suppose that you found that the 2008 "blue book" value of the motorcycle was $4,200. Now we have the following given information.

Year	x (years old)	y (dollars)
2005	3	10,500
2008	6	4200
2009	7	?

The simplest model that we can construct with this (new) given information is a *linear* model, that is, an equation whose graph is a line. To begin, we compute the slope of the line using the points $(3, 10{,}500)$ and $(6, 4200)$:

$$\frac{\Delta y}{\Delta x} = \frac{4200 - 10{,}500}{6 - 3} = -2100.$$

Now, let (x, y) be *any other point* on the line.[8] Then, computing the slope of the line using the point $(3, 10{,}500)$ and (x, y), we find that

$$\frac{\Delta y}{\Delta x} = \frac{y - 10{,}500}{x - 3}.$$

Since the line has only one slope, this implies that

$$\frac{y - 10{,}500}{x - 3} = -2100$$

or (by cross multiplying)

$$y - 10{,}500 = -2100(x - 3) \qquad \text{(expand the right-hand side)}$$
$$y - 10{,}500 = -2100x + 6300 \qquad \text{(add 10,500 to both sides)}$$
$$y = -2100x + 16{,}800.$$

Thus, a linear model for the value in dollars, y, of the motorcycle when it is x years old is

$$y = -2100x + 16{,}800.$$

First, let us check that the model produces the given information.

$$x = 3: \quad y = -2100(3) + 16{,}800 = 10{,}500 \qquad \checkmark$$
$$x = 6: \quad y = -2100(6) + 16{,}800 = 4200 \qquad \checkmark$$

Now, using this model, we find that, when $x = 7$ (in the year 2009),

$$y = -2100(7) + 16{,}800 = 2100.$$

So, using the linear model based on the (new) given information, you could have expected to sell the motorcycle for \$2,100 (considerably less than \$4,500, the value that was produced by the inverse variation model).

Now You Try 23.8 Construct a linear model for the preceding problem, but use the points $(6, 4200)$ and (x, y) instead to compute the slope of the line the second time. All of the other steps would be the same. Do you get the same equation as when we used the points $(3, 10{,}500)$ and (x, y)? □

---

[8] Alternatively, we may use the slope-intercept form of an equation for a line directly, namely, $y = ax + b$, where a is the slope, (x, y) is a point on the line, and $(0, b)$ is the y intercept. In this instance, we substitute $a = -2100$ and $(x, y) = (3, 10{,}500)$ into $y = ax + b$ and solve for b:

$$10{,}500 = -2100(3) + b \qquad \text{(multiply: } -2100(3) = -6300)$$
$$10{,}500 = -6300 + b \qquad \text{(add 6300 to both sides)}$$
$$16{,}800 = b.$$

Thus, we obtain the equation $y = -2100x + 16{,}800$.

But is the linear model better than the inverse variation model? To help ourselves decide, we graph both equations and compare the graphs in FIGURE 23.2. Observe that the linear model is better than the inverse variation model in one way: In the year 2002, when $x = 0$, the value in dollars, y, of the motorcycle is not infinite. In fact, we can compute the value of the motorcycle when it was new:

$$x = 0: \quad y = -2100(0) + 16{,}800 = 16{,}800.$$

So, according to the linear model, the motorcycle cost $16,800 when it was new. However, observe that the linear model is worse than the inverse variation model in one way: The value of the motorcycle becomes negative in the future in the linear model, but it is always positive (and decreasing) in the inverse variation model.

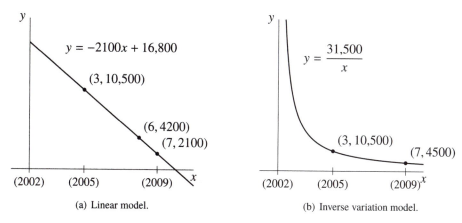

(a) Linear model.   (b) Inverse variation model.

Figure 23.2: Linear model versus inverse variation model of the same problem.

Therefore, we see that each model, the inverse variation model and the linear model, has aspects that we may like and aspects that we may not like. Such is mathematical modeling: It is the art of finding a happy compromise.

We finish this discussion by constructing one more model for the value of the motorcycle. In this model, we seek to keep the decreasing characteristic of the inverse variation model, but also seek to set the value of the motorcycle when it was new to be $16,800.

Year	x (years old)	y (dollars)
2002	0	16,800
2005	3	10,500
2009	7	?

To keep the decreasing characteristic of the inverse variation model, we begin by supposing that

$$y = \frac{k}{x}$$

for some constant k. However, we see immediately that the right-hand side is undefined when $x = 0$. Two ways to overcome this are

1. Let x begin with 1 instead of zero, that is, let $x = 1$ in the year 2002, $x = 4$ in the year 2005, $x = 8$ in the year 2009, and so on.

2. Shift $x \mapsto x - a$ for some number a, that is, suppose that

$$y = \frac{k}{x-a}.$$

If we choose the latter, shift $x \mapsto x - a$, then it remains for us to find the constants k and a.
   Using the given information, we find that

$$(x, y) = (0, 16{,}800) \quad \Longrightarrow \quad 16{,}800 = \frac{k}{-a},$$

$$(x, y) = (3, 10{,}500) \quad \Longrightarrow \quad 10{,}500 = \frac{k}{3-a}.$$

Thus, to find the constants k and a, we must solve the system of equations

$$16{,}800 = -\frac{k}{a}, \quad 10{,}500 = \frac{k}{3-a}.$$

We leave the solving of this system to you as an exercise. You should find that the solution is $k = 84{,}000$
and $a = -5$, so that the model for the value in dollars, y, of the motorcycle when it is x years old is

$$y = \frac{84{,}000}{x+5}.$$

First, let us check that the model produces the given information.

$$x = 0: \quad y = \frac{84{,}000}{0+5} = 16{,}800 \quad \checkmark$$

$$x = 3: \quad y = \frac{84{,}000}{3+5} = 10{,}500 \quad \checkmark$$

Now, using this model, we find that, when $x = 7$ (in the year 2009),

$$y = \frac{84{,}000}{7+5} = 7000.$$

So, using the "shifted inverse variation" model, you could have expected to sell the motorcycle for $7,000.
Here is the graph of the model.

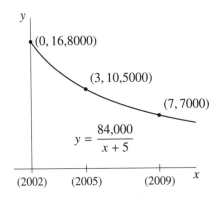

Now You Try 23.9 In the year 2005 you paid $10,500 for a BMW R1150RS model motorcycle.

1. Let y denote the value in dollars of the motorcycle and let x denote a number of years with $x = 1$ in 2002, $x = 4$ in 2005, $x = 8$ in 2009, and so on. Suppose that the value of the motorcycle when it was new was \$16,800.

   a) Construct an inverse variation model,

   $$y = \frac{k}{x},$$

   using the given information and graph the model.

Year	x (years old)	y (dollars)
2002	1	16,800
2005	4	10,500

   b) Using the model, find the value of the motorcycle in 2008. In 2009. In 2012.

   c) How do these values compare to the values obtained using the "shifted inverse variation" model

   $$y = \frac{84,000}{x + 5},$$

   where here $x = 0$ in 2002, $x = 3$ in 2005, $x = 7$ in 2009, and so on.

2. Let y denote the value in dollars of the motorcycle when it is x years old. Suppose that you found that the 2008 "blue book" value of the motorcycle was \$4,200.

   a) Construct a "shifted inverse variation" model,

   $$y = \frac{k}{x - a},$$

   using the given information and graph the model.

Year	x (years old)	y (dollars)
2005	3	10,500
2008	6	4200

   b) Using this model, how much did the motorcycle cost when it was new?

   c) Using this model, for how much could you have expected to sell the motorcycle in 2009? In 2012?

   □

Chapter 24

Logarithms

It is often necessary to multiply two large numbers such as 4,030,185 by 5,518,112. Prior to the mid-twentieth century, this would be done by hand. Try it for yourself. Multiply 4,030,185 by 5,518,112 by hand and note: How much time did it take? How many steps did it take? How sure of the answer are you?

Before the advent of modern computing devices, "simple" calculations that involved multiplication, division, or the extraction of roots of large numbers would be very time-consuming processes with significant potential for errors, especially when they had to be completed quickly. For example, in the field of sea navigation during the mid- to latter centuries of the second millennium, calculations with large numbers were required to find one's location at sea; however, if it would take the better part of an hour to complete the calculations, particularly multiplication, one's ship would have moved quite a distance from its location by the time that location had been determined. Thus, it was essential that navigators could complete such calculations both quickly and accurately. A better method had to be found. Enter John Napier and Henry Briggs in the late sixteenth century.

We begin by getting a feel for what Napier and Briggs developed.

Let b be a positive real number, and let a, L_1, L_2, L_3, and L_4 be real numbers. Now, if we suppose that

$$\frac{ab^{L_1}}{ab^{L_2}} = \frac{ab^{L_3}}{ab^{L_4}},$$

then

$$b^{L_1 - L_2} = b^{L_3 - L_4},$$

so that

$$L_1 - L_2 = L_3 - L_4.$$

To say this another way, we introduce the notation[1]

$$\boxed{\mathcal{L}(x) = y \quad \text{if and only if} \quad x = ab^y}$$

and let

$$X = ab^{L_1}, \quad Y = ab^{L_2}, \quad U = ab^{L_3}, \quad V = ab^{L_4}.$$

Then

$$\frac{X}{Y} = \frac{U}{V} \quad \Longrightarrow \quad \mathcal{L}(X) - \mathcal{L}(Y) = \mathcal{L}(U) - \mathcal{L}(V).$$

---

[1]The notation $\mathcal{L}(x)$ is function notation (read "$\mathcal{L}$ of x"), and is not the product of $\mathcal{L}$ and x. Recall that a function is a set of ordered pairs (x, y) for which every value x is paired with one and only one value y. Here, the ordered pairs $(x, y) = (x, \mathcal{L}(x))$.

In particular, we see that

$$\frac{XY}{Y} = \frac{X}{1} \quad \Longrightarrow \quad \mathcal{L}(XY) - \mathcal{L}(Y) = \mathcal{L}(X) - \mathcal{L}(1)$$

or, equivalently,

$$\mathcal{L}(XY) = \mathcal{L}(X) + \mathcal{L}(Y) - \mathcal{L}(1).$$

Let us see how we can use this relation. The first examples use small numbers, for which the method is overkill, but numbers of six or more digits are encountered routinely in astronomy and in navigation, for example. It is in those applications that the method earns its keep.

Example 24.1 We are interested in values $N = ab^L$, so that $\mathcal{L}(N) = L$. In this example, we take $a = 10^2$ and $b = 0.9$, so that $N = 10^2(0.9)^L$, and compute $N = 10^2(0.9)^L$ for different positive integers L. For example,[2]

$$L = 0 \quad \Longrightarrow \quad N = 10^2(0.9)^0 = 100(1) = 100,$$
$$L = 1 \quad \Longrightarrow \quad N = 10^2(0.9)^1 = 100(0.9) = 99,$$
$$L = 2 \quad \Longrightarrow \quad N = 10^2(0.9)^2 = 100(0.81) = 81,$$
$$L = 3 \quad \Longrightarrow \quad N = 10^2(0.9)^3 = 100(0.729) = 72.9,$$

and so on. TABLE 24.1 lists the values of N for $L = 0, 1, 2, \ldots, 45$.[3]

Now, consider the product of 12 and 5. We know that $12 \cdot 5 = 60$, but let us try to find the product using the values in the table. For this, recall that $\mathcal{L}(N) = L$ and

$$\mathcal{L}(XY) = \mathcal{L}(X) + \mathcal{L}(Y) - \mathcal{L}(1).$$

Here, $X = 12$ and $Y = 5$, so that

$$\mathcal{L}(12 \cdot 5) = \mathcal{L}(12) + \mathcal{L}(5) - \mathcal{L}(1).$$

When we look in the table, however, we do not find $N = 12$ nor $N = 5$ nor $N = 1$; the closest we find are[4]

$$N = 12.157\,665\,46, \quad N = 5.233\,476\,33, \quad \text{and} \quad N = 1.077\,526\,366.$$

Hence, we may only *approximate* the product using the table:

$$\begin{aligned}\mathcal{L}(12 \cdot 5) &= \mathcal{L}(12) + \mathcal{L}(5) - \mathcal{L}(1) \\ &\approx \mathcal{L}(12.157\,665\,46) + \mathcal{L}(5.233\,476\,33) - \mathcal{L}(1.077\,526\,366) \\ &= 20 + 28 - 43 \\ &= 5.\end{aligned}$$

Looking at the table again, we see that

$$5 = \mathcal{L}(59.049).$$

Therefore, we conclude that

$$12 \cdot 5 \approx 59.049 \quad \text{or} \quad 59$$

when rounded to the nearest one.   □

---

[2]Computed using the 10-digit-display Texas Instruments TI-30X II S calculator.

[3]A sequence of numbers $a, a + d, a + 2d, a + 3d, \ldots$ is called an *arithmetic progression* with (constant) difference d. A sequence of numbers $a, ar, ar^2, ar^3, \ldots$ is called a *geometric progression* with (constant) ratio r. Thus, in the table, the sequence of numbers L is an arithmetic progression with $a = 0$ and difference $d = 1$, and the sequence of numbers N is a geometric progression with $a = 10^2$ and ratio $r = 0.9$. Moreover, the table defines a one-to-one correspondence between the terms of the arithmetic progression and the terms of the geometric progression.

[4]One may also apply a method called *linear interpolation* to "read between the lines" for those numbers N that are not listed in the table, but we shall not do so here.

L	N	L	N	L	N
0	100.	16	18.53020189	31	3.815204245
1	90.	17	16.6771817	32	3.43368382
2	81.	18	15.00946353	33	3.090315438
3	72.9	19	13.50851718	34	2.781283894
4	65.61	20	12.15766546	35	2.503155505
5	59.049	21	10.94189891	36	2.252839954
6	53.1441	22	9.847709022	37	2.027555959
7	47.82969	23	8.86293812	38	1.82400363
8	43.046721	24	7.976644308	39	1.642320327
9	38.7420489	25	7.178979877	40	1.478088294
10	34.86784401	26	6.461081889	41	1.330279465
11	31.38105961	27	5.8149737	42	1.197251518
12	28.24295365	28	5.23347633	43	1.077526366
13	25.41865828	29	4.710128697	44	0.96977373
14	22.87679245	30	4.239115828	45	0.872796357
15	20.58911321		.		.

Table 24.1: $N = 10^2(0.9)^L$

Now You Try 24.1 Let $\mathcal{L}(x) = y$ if and only if $x = ab^y$, and recall that

$$\mathcal{L}(XY) = \mathcal{L}(X) + \mathcal{L}(Y) - \mathcal{L}(1).$$

1. Use the table for $N = 10^2(0.9)^L$ to find or approximate the following products.

   a) $15 \cdot 4$ b) $20 \cdot 3$ c) $2 \cdot 3 \cdot 10$ d) 3^4

2. Show that

$$\mathcal{L}\left(\frac{X}{Y}\right) = \mathcal{L}(X) - \mathcal{L}(Y) + \mathcal{L}(1).$$

   (Hint: Begin with $\frac{X/Y}{1} = \frac{X}{Y}$.) Now use this relation and the table for $N = 10^2(0.9)^L$ to find or approximate $63 \div 7$.

3. Find a relation for $\mathcal{L}(X^2)$, $\mathcal{L}(X^3)$, and in general $\mathcal{L}(X^k)$ for any positive integer k.

□

Example 24.2 In this example, we consider the problem of finding the *mean proportional* between (or the *middle proportional* to)[5] two positive numbers a and b, that is, to find a number x such that

$$a : x = x : b \quad \text{or} \quad \frac{a}{x} = \frac{x}{b}.$$

---

[5]The term "mean proportional" was used widely until around the 1920s, when the term "geometric mean" surpassed it in usage; the term "middle proportional" apparently never enjoyed much use, although it can be found in the literature. A comparison of their usage can be found using Google books "Ngram Viewer" <http://books.google.com/ngrams>.

The geometric mean is a more general notion than the mean proportional. The geometric mean of n nonnegative real numbers is the nth root of their product: the geometric mean of $a_1, a_2, \ldots, a_n$ is $(a_1 a_2 \cdots a_n)^{1/n}$. Thus, the geometric mean of a and b is $\sqrt{ab}$.

We will generally use the term "mean proportional" in its historical context.

To be concrete, let $a = 100$ and $b = 15$, so that x satisfies the proportion

$$\frac{100}{x} = \frac{x}{15}.$$

Cross multiplying, we find that

$$x^2 = 1500, \quad \text{so that} \quad x = \sqrt{1500} \approx 38.729\,833\,46.$$

(Note that $x = -\sqrt{1500}$ is also a solution of $x^2 = 1500$, since $(-\sqrt{1500})^2 = 1500$ also, but we choose the positive square root because a and b are positive.) However, if that were it, then this would not be a very interesting example. Instead, let us find the mean proportional using the table for $N = 10^2(0.9)^L$.

First, note that

$$\frac{100}{x} = \frac{x}{15} \quad \Longrightarrow \quad \mathcal{L}(100) - \mathcal{L}(x) = \mathcal{L}(x) - \mathcal{L}(15),$$

so that

$$2\mathcal{L}(x) = \mathcal{L}(100) + \mathcal{L}(15) \quad \text{or} \quad \mathcal{L}(x) = \frac{1}{2}(\mathcal{L}(100) + \mathcal{L}(15)).$$

Thus, instead of having to multiply two numbers then extract a square root, which can be a lot of work if done by hand, now we only have to add once then divide by 2.

When we look at the table, we find $N = 100$, but we do not find $N = 15$; for the latter, the closest we find is $N = 15.009\,463\,53$. Hence, we may only approximate the mean proportional using the table:

$$\mathcal{L}(x) = \frac{1}{2}(\mathcal{L}(100) + \mathcal{L}(15))$$

$$\approx \frac{1}{2}(\mathcal{L}(100) + \mathcal{L}(15.009\,463\,53))$$

$$= \frac{1}{2}(0 + 18)$$

$$= 9.$$

Looking at the table again, we see that

$$9 = \mathcal{L}(38.742\,048\,9).$$

Therefore, we conclude that the mean proportional between 100 and 15 is

$$x \approx 38.742\,048\,9,$$

which agrees with the value $\sqrt{1500} \approx 38.729\,833\,46$ that we obtained earlier to one decimal place. This is not bad considering that, if done by hand, extracting the square root of 1500 would be considerably more work than dividing 18 by 2. This would be especially so without the use of a calculator, which was not available until the latter part of the twentieth century.   □

Now You Try 24.2 Use the table for $N = 10^2(0.9)^L$ to find or approximate the mean proportional between the numbers a and b, then compare the result with that obtained using a calculator to approximate the associated square root.

  1. $a = 60, b = 3$ 2. $a = 43, b = 8$

□

Think About It 24.1 How can you use a table for $N = ab^L$, so that $\mathcal{L}(N) = L$, to find or approximate $\sqrt{s}$, $\sqrt[3]{s}$, or in general $\sqrt[k]{s}$ for any positive integer k?   □

Example 24.3 In this example, to find or approximate the product of 12 and 5, we use a table for $N = ab^L$ with $a = 10^2$ again, but now with $b = 0.99$ instead of 0.9. Here is a portion of the table. Observe that we have "finer divisions" for N this time.

\multicolumn{6}{c}{$N = 10^2(0.99)^L$}					
L	N	L	N	L	N
0	100.	50	60.50060671	298	5.003662287
1	99.	51	59.89560065	299	4.953625664
2	98.01	⋮	.	⋮	.
3	97.0299	210	12.11688164	458	1.002118606
4	96.059601	211	11.99571282	459	0.99290742
⋮	.	⋮	.	⋮	.

To use the table, we recall that

$$\mathcal{L}(12 \cdot 5) = \mathcal{L}(12) + \mathcal{L}(5) - \mathcal{L}(1).$$

Looking at the table, however, we again do not find $N = 12$ nor $N = 5$ nor $N = 1$; the closest we find are

$$N = 11.995\,712\,82, \quad N = 5.003\,662\,287, \quad \text{and} \quad N = 1.002\,118\,606.$$

Hence, we may again only approximate the product using the table:

$$\begin{aligned}
\mathcal{L}(12 \cdot 5) &= \mathcal{L}(12) + \mathcal{L}(5) - \mathcal{L}(1) \\
&\approx \mathcal{L}(11.995\,712\,82) + \mathcal{L}(5.003\,662\,287) - \mathcal{L}(1.002\,118\,606) \\
&= 211 + 298 - 458 \\
&= 51.
\end{aligned}$$

Looking at the table once more, we see that

$$51 = \mathcal{L}(59.895\,600\,65).$$

Therefore, we conclude that

$$12 \cdot 5 \approx 59.895\,600\,65 \quad \text{or} \quad 60$$

when rounded to the nearest one.

Note that using $N = 10^2(0.99)^L$ yields a better approximation of $12 \cdot 5$ than using $N = 10^2(0.9)^L$.

$$\begin{aligned}
N = 10^2(0.99)^L : &\quad 12 \cdot 5 \approx 59.895\,600\,65 \\
N = 10^2(0.9)^L : &\quad 12 \cdot 5 \approx 59.049
\end{aligned}$$

☐

Now You Try 24.3 Create a table for $N = ab^L$, so that $\mathcal{L}(N) = L$, with $a = 1$, $b = 1.1$, and $L = 0, 1, 2, \ldots, 49$. Use the table to find or approximate the following products.

1. $12 \cdot 5$ 2. $15 \cdot 4$ 3. $2 \cdot 3 \cdot 10$ 4. 3^4

☐

With the ubiquity of handheld calculators today, it may seem that using TABLE 24.1, for example, to multiply, divide, or find roots is very cumbersome. This is true; however, imagine needing to multiply together two large numbers, say 4,030,185 and 5,518,112, before the invention of the common handheld calculator. In this case, since each number has seven digits, finding the product by hand would require 7×7 partial products that would then have to be summed. Not only would this be a considerable amount of work, but it would also provide considerable opportunities for mistakes. On the other hand, suppose that we had a table for $N = ab^L$, so that $\mathcal{L}(N) = L$, with $a = 10^7$ and $b = 0.999\,999\,9$. Here is the relevant part of such a table.[6]

$N = 10^7(0.9999999)^L$			
L	N	L	N
0	10000000.	9087727	4030185.2697937
$\vdots$	.	9087728	4030184.8667752
15033220	2223901.3487497	$\vdots$	.
15033221	2223901.1263596	5945492	5518112.4992145
15033222	2223900.9039694	5945493	5518111.9474032
$\vdots$	.	$\vdots$	.

Looking at the table, we see that

$$\mathcal{L}\left(\frac{4,030,185 \times 5,518,112}{10^7}\right) = \mathcal{L}(4,030,185) + \mathcal{L}(5,518,112)$$
$$- \mathcal{L}(10,000,000)$$
$$\approx 9,087,728 + 5,945,493 - 0$$
$$= 15,033,221.$$

Looking at the table again, we see that

$$2,223,901.126\,359\,6 = \mathcal{L}(15,033,221).$$

Therefore, we conclude that

$$\frac{4,030,185 \times 5,518,112}{10^7} \approx 2,223,901.126\,359\,6,$$

so that
$$4,030,185 \times 5,518,112 \approx 10^7 \times 2,223,901.126\,359\,6 = 22,239,011,263,596.$$

This approximate result actually compares very well with the actual product, namely,

$$4,030,185 \times 5,518,112 = 22,239,012,210,720.$$

The difference between the exact product and the approximate result is only 947,124, which, even though it appears to be a large number, is less than $0.000\,005\%$ of the exact product. Moreover, it was much easier to look up the table and add two seven-digit numbers than it would have been to multiply the two seven-digit numbers by hand.

---

[6]The computations in this example are carried out using the Wolfram Web resource, WolframAlpha computational knowledge engine <http://www.wolframalpha.com>.

Think About It 24.2 Let $\mathcal{L}(N) = L$ if $N = ab^L$. Then,

$$\mathcal{L}(XY) = \mathcal{L}(X) + \mathcal{L}(Y) - \mathcal{L}(1) \quad \text{and} \quad \mathcal{L}\!\left(\frac{X}{Y}\right) = \mathcal{L}(X) - \mathcal{L}(Y) + \mathcal{L}(1).$$

Explain why

$$\mathcal{L}\!\left(\frac{XY}{a}\right) = \mathcal{L}(X) + \mathcal{L}(Y).$$

$\square$

Of course, you may be familiar with the adage, "There is no free lunch." This is certainly true in this case, for while it may be an easy task to look up the table and to add two seven-digit numbers by hand, *constructing* such a table *by hand* would require a lot of work. On the other hand, such a table may be used over and over again.

Two men are credited to have come up with the idea of using tables similar to the ones in our examples to reduce the work required to multiply, divide, and find roots independently and around the same time. They are John Napier (1550–1617), a Scottish baron, and Joost Bürgi (or Jost Bürgi; 1552–1632), a Swiss craftsman [40]. Napier settled on a table that was essentially for $N = 10^7(0.999\,999\,9)^L$,[7] and Bürgi used a table for $N = 10^8(1.0001)^L$. Note that, as L increases, the values in Napier's table decrease, whereas the values in Bürgi's increase. What is truly remarkable is that both men invented logarithms without the advantage of our modern notation for exponents, b^L, that is generally credited to the French mathematician René Descartes (1596–1650).

It appears that Bürgi had developed his idea for logarithms by 1558, but that his work was not published until 1620, and then only anonymously [80, pp. 13–14]. In the meantime, Napier published his work on logarithms in 1614 and, because of that, it is Napier's work, and not Bürgi's, that had a profound impact on mathematics; and it is Napier's name, and not Bürgi's, that is now firmly attached to the conception of logarithms [80, pp. 11–14].

In his table for $N = 10^8(1.0001)^L$, Bürgi lists values of $10L$ and N, instead of L and N. The values of $10L$ were printed in red and the values of N were printed in black; consequently, Bürgi aptly called the values of $10L$ *red numbers* and the values of L *black numbers*. See FIGURE 24.1.

Napier's table lists values of L that he at first calls "artificial numbers"; however, he later calls the values of L *logarithms*of the corresponding values of N. He coined the term "logarithm," meaning "the number of the ratios" [37, p. 82], from the Greek words *logos* (ratio) and *arithmos* (number). In fact, this is how Napier originally defines his logarithms [87, pp. 16–19]:

> 22. *It remains, in the Third table at least, to place beside the sines or natural numbers decreasing geometrically their logarithms or artificial numbers increasing arithmetically.*

> 23. *To increase arithmetically is, in equal times, to be augmented by a quantity always the same.*

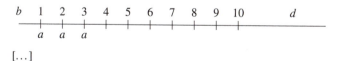

[...]

---

[7]Napier's entire construction was more elaborate. He actually constructed three tables: basically, the first table with 101 terms for $N = 10^7(0.999\,999\,9)^L$, that is, a geometric progression with ratio $r_1 = 0.999\,999\,9$; then, he constructed a second table with 51 terms for $N = 10^7(0.999\,99)^L$, where the ratio $r_2 = 0.999\,99$ is approximately the ratio of the last term to the first term of the first table; and, finally, he constructed a third table with 21 rows and 69 columns. In the third table, the first column contains 21 terms for $N = 10^7(0.9995)^L$, where the ratio $r_3 = 0.9995$ is approximately the ratio of the last term to the first term of the second table; the first row contains 69 terms for $N = 10^7(0.99)^L$, where the ratio $r_4 = 0.99$ is approximately the ratio of the last term to the first term of the first column; and the remaining 68 columns were then filled in with geometric progressions using the ratio r_3. Carslaw [37] explains the construction of Napier's three tables fully; he tells us that "Napier thus obtained in his Third Table a set of numbers lying between 10^7 and very nearly $\frac{1}{2}10^7$, and these numbers form a set dense enough to allow this Table to be used in dealing with the sines of angles from $90°$ to $30°$. The numbers are not exactly in a Geometrical Progression, but each of the 69 columns of this Table is a Geometrical Progression of 21 terms."

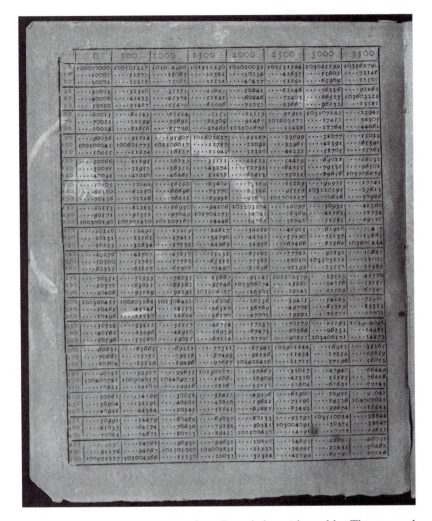

Figure 24.1: A portion of a page from Bürgi's logarithm table. The row and column headings appear in red, and the entries appear in black. For example, we see that the black number 101907877 is in the row with red number 390 and column with red number 1500 (third row from the bottom, last column), and 390 + 1500 = 1890. Therefore, the black number 101907877 has red number 1890, which is 10L, so that $L = 189$: $101{,}907{,}877 = 10^8(1.0001)^{189}$ rounded. (Source: Courtesy of the University Library of Graz.)

24. *To decrease geometrically is this, that in equal times, first the whole quantity then each of its successive remainders is diminished, always by a like proportional part.*

```
        1   2  3  4 5 6
    T ──┼───┼──┼──┼┼┼──────────────────────── S
        G   G  G
```

[…]

25. *Whence a geometrically moving point approaching a fixed one has its velocities proportionate to its distances from the fixed one.*

[…]

26. *The logarithm of a given sine is that number which has increased arithmetically with the same velocity throughout as that with which radius began to decrease geometrically, and in the same time as radius has decreased to the given sine.*

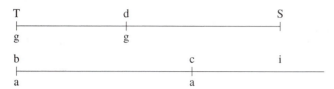

Let the line T S be radius, and d S a given sine in the same line; let g move geometrically from T to d in certain determinate moments of time. Again, let b i be another line, infinite towards i, along which, from b, let a move arithmetically with the same velocity as g had at first when at T; and from the fixed point b in the direction of i let a advance in just the same moments of time up to the point c. The number measuring the line b c is called the logarithm of the given sine d S.

Let us explain Napier's definition of the logarithm that is quoted above. Napier begins with a line segment TS and a ray or half-line bi (that begins at b and extends infinitely in the direction of i). Now, suppose that a point g moves from T towards S in a way that its speed is proportional to its distance from S; thus, the point g continually slows down as it moves from T towards S. At the same time, suppose that a point a moves from b in the direction of i at a constant speed that is equal to the initial speed of g. Assume at some time the point g is at position d on TS, and the point a is at position c on bi. If we let x denote the distance from d to S (a geometric progression), and let y denote the distance from b to c (an arithmetic progression), then Napier defines y to be the logarithm of x.

This is what Napier had in mind when he constructed his table of logarithms, which results in essentially that the *Naperian logarithm* (or Napierian logarithm)

$$\text{Nap}\log x = y \quad \text{if} \quad x = 10^7(0.999\,999\,9)^y.$$

It is worth noting again that it is remarkable that Napier and Bürgi invented logarithms without the advantage of our modern notation for exponents.

Remark 24.1 In his definition of logarithms, Napier refers to a "radius" and "sine." The radius is the radius of a circle and the sine is a trigonometric function. There are six trigonometric functions: sine, cosine, tangent, cotangent, secant, and cosecant. Trigonometry is an area of mathematics that has many applications, including astronomy, and it was principally to ease the calculations in astronomy that Napier invented logarithms. So, as Carslaw says [37, p. 77],

Before we can understand Napier's work, it is necessary to call to mind that his object was to render calculations with sines, cosines, etc., especially the calculations of the astronomer, an easier matter. To the mathematicians of his time, the trigonometrical functions were not *ratios*, but *lines*, or the measures of lines. The *sine of the arc AB* [accompanying figure], as they would put it, was the *line BM*....

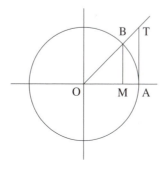

In the Trigonometrical Tables, which had been calculated by German mathematicians to an enormous degree of accuracy, these functions were given as integers, the radius being the sine of 90°, or the *sinus totus*. When additional accuracy was required, the radius was chosen proportionally large. In the Trigonometrical Tables used by Napier, it was 10^7....

Logarithms proved to be extremely helpful in reducing lengthy calculations, so much so that Laplace (1749–1827) is noted to have said that logarithms "by shortening the labors doubled the life of the astronomer" [56, p. 185]. □

24.1   LOGARITHMS TODAY

Gone are the days when logarithms and logarithm tables were used to facilitate multiplication, division, and the finding of roots, for today we have electronic calculators and computer programs to perform such computations. Instead, today one uses the logarithm *function* in mathematical modeling in a variety of areas, for example, in chemistry, seismology, psychology, and probability and statistics. Logarithms are also found to be used in finance, computer science, number theory, and music, for instance. We begin with a definition of the logarithm.[8]

Definition 24.1 Let $b > 1$ be a real number. A real number y is called the *logarithm* of x to the base b if $b^y = x$. We write

$$y = \log_b x \quad \text{if} \quad b^y = x.$$

The expression $\log_b x$ may be read, "log to the base b of x" or "log base b of x." □

Think About It 24.3

1. In the expression $\log_b x$, can $x \leq 0$?

2. How does the modern definition of logarithm compare with the function $\mathcal{L}(N) = L$ if $N = ab^L$?

□

To say it again, the logarithm of x to the base b is the exponent y such that $b^y = x$. We emphasize that this is not how either Napier or Bürgi thought about logarithms. It was not until John Wallis (1616–1703) that the possibility of defining logarithms as exponents was recognized, and not until about 1742 before there was a systematic explanation [23, p. 46].

Example 24.4

1. The logarithm of 8 to the base 2 is the exponent y such that $2^y = 8$:

$$\boxed{\log_2 8 = 3} \quad \text{because} \quad 2^3 = 8.$$

2. The logarithm of 25 to the base 5 is the exponent y such that $5^y = 25$:

$$\boxed{\log_5 25 = 2} \quad \text{because} \quad 5^2 = 25.$$

3. The logarithm of 5 to the base 25 is the exponent y such that $25^y = 5$:

$$\boxed{\log_{25} 5 = \tfrac{1}{2}} \quad \text{because} \quad 25^{1/2} = 5.$$

---

[8]There are other ways to define a logarithm that are equivalent.

4. The logarithm of 1 to the base 10 is the exponent y such that $10^y = 1$:

$$\boxed{\log_{10} 1 = 0} \quad \text{because} \quad 10^0 = 1.$$

5. The logarithm of $\frac{1}{36}$ to the base 6 is the exponent y such that $6^y = \frac{1}{36}$:

$$\boxed{\log_6 \frac{1}{36} = -2} \quad \text{because} \quad 6^{-2} = \frac{1}{6^2} = \frac{1}{36}.$$

$\square$

Now You Try 24.4 For each of the following, write the appropriate statement: "The logarithm of x to the base b is the exponent y such that $b^y = x$: $\log_b x = y$ because $b^y = x$."

1. $\log_2 32$ 2. $\log_{10} 0.01$ 3. $\log_{81} 3$ 4. $\log_{\sqrt{2}} 1$

$\square$

Example 24.5

1. From $2^3 = 8$, we have $\log_2 8 = 3$ and, thus,

$$2^{\log_2 8} = 2^3 = 8 : \quad \boxed{2^{\log_2 8} = 8.}$$

2. From $5^2 = 25$, we have $\log_5 25 = 2$ and, thus,

$$5^{\log_5 25} = 5^2 = 25 : \quad \boxed{5^{\log_5 25} = 25.}$$

3. From $25^{1/2} = 5$, we have $\log_{25} 5 = \frac{1}{2}$ and, thus,

$$25^{\log_{25} 5} = 25^{1/2} = 5 : \quad \boxed{25^{\log_{25} 5} = 5.}$$

4. From $10^0 = 1$, we have $\log_{10} 1 = 0$ and, thus,

$$10^{\log_{10} 1} = 10^0 = 1 : \quad \boxed{10^{\log_{10} 1} = 1.}$$

5. From $6^{-2} = \frac{1}{36}$, we have $\log_6 \frac{1}{36} = -2$ and, thus,

$$6^{\log_6(1/36)} = 6^{-2} = \frac{1}{36} : \quad \boxed{6^{\log_6(1/36)} = \frac{1}{36}.}$$

$\square$

$$\boxed{\text{From } b^y = x, \text{ we have } \log_b x = y \text{ and, thus, } b^{\log_b x} = x.}$$

Now You Try 24.5 Simplify the following.

1. $2^{\log_2 32}$ 2. $10^{\log_{10} 0.01}$ 3. $81^{\log_{81} 3}$ 4. $\sqrt{2}^{\log_{\sqrt{2}} 1}$

□

From $b^y = x$, we have $\log_b x = y$ and, thus, $\log_b b^y = y$.

Now You Try 24.6 Simplify the following.

1. $\log_2 2^5$ 2. $\log_2 32$ 3. $\log_2 8$ 4. $\log_2 2$

□

We also note that, because $b^0 = 1$ for *any* base $b > 1$, we have that the logarithm of 1 to the base b is zero for any base $b > 1$.

$\log_b 1 = 0$ for any base $b > 1$.

24.2   PROPERTIES OF LOGARITHMS

Now, because a logarithm is an exponent, there are rules for logarithms that correspond to the rules for exponents (page 338).

Rule 24.1 (RULES FOR LOGARITHMS) In the following, let $a > 1$, $b > 1$, $ab > 1$, or $\frac{a}{b} > 1$ as appropriate for the definition of the logarithm. Then,

	Exponents	Logarithms
(1)	$b^x b^y = b^{x+y}$	$\log_b(uv) = \log_b u + \log_b v$
(2)	$\dfrac{b^x}{b^y} = b^{x-y}$	$\log_b\left(\dfrac{u}{v}\right) = \log_b u - \log_b v$
(3)	$(b^x)^y = b^{xy}$	$\log_b(u^y) = y \log_b u$

□

Remark 24.2 It is important to note that

$$\log_b(u + v) \neq \log_b u + \log_b v, \quad \log_b(u - v) \neq \log_b u - \log_b v,$$

and

$$(\log_b u)^y \neq y \log_b u.$$

□

Example 24.6 We demonstrate rules (1) and (3) for logarithms.

(1) Let $u = b^x$ and $v = b^y$, so that

$$x = \log_b u \quad \text{and} \quad y = \log_b v.$$

Now,

$$uv = b^x b^y = b^{x+y},$$

so that

$$\log_b(uv) = x + y.$$

Thus,

$$\log_b(uv) = \log_b u + \log_b v.$$

(3) Let $u = b^x$, so that

$$x = \log_b u.$$

Now,

$$u^y = (b^x)^y = b^{xy},$$

so that

$$\log_b(u^y) = xy.$$

Thus

$$\log_b(u^y) = (\log_b u)y = y \log_b u.$$

$\square$

Now You Try 24.7 Demonstrate rule (2) for logarithms, namely,

$$\log_b\left(\frac{u}{v}\right) = \log_b u - \log_b v.$$

$\square$

Think About It 24.4 Let $\mathcal{L}(N) = L$ if $N = ab^L$. How do the relations

$$\mathcal{L}(XY) = \mathcal{L}(X) + \mathcal{L}(Y) - \mathcal{L}(1),$$

$$\mathcal{L}\left(\frac{X}{Y}\right) = \mathcal{L}(X) - \mathcal{L}(Y) + \mathcal{L}(1),$$

$$\mathcal{L}(X^k) = k\mathcal{L}(X) - (k-1)\mathcal{L}(1)$$

compare with the rules for logarithms (RULE 24.1)?

$\square$

For any two real numbers u and v,

$(uv)^{1/2}$ is called the *geometric mean* of u and v, and

$\dfrac{u+v}{2}$ is called the *arithmetic mean* of u and v.

(See the footnote on page 463.) Now, using the rules for logarithms, observe that

$$\log_b(uv)^{1/2} = \frac{1}{2}\log_b(uv) = \frac{1}{2}(\log_b u + \log_b v),$$

that is,

$$\log_b(uv)^{1/2} = \frac{\log_b u + \log_b v}{2}.$$

In words, this relation states that the logarithm of a geometric mean of two numbers equals the arithmetic mean of the respective logarithms of the numbers.

Think About It 24.5 For any three real numbers u, v, and w, the geometric mean is $(uvw)^{1/3}$ and the arithmetic mean is $\frac{u+v+w}{3}$. Is it true that the logarithm of the geometric mean of three numbers equals the arithmetic mean of the respective logarithms of the numbers? What about in general for any n real numbers? □

Example 24.7 Recall that $\log_b 1 = 0$ for any base b. Now, if we suppose that

$$\log_b 10 = 10{,}000{,}000{,}000,$$

it would then follow that

$$\log_b \sqrt{10} = \log_b (1 \cdot 10)^{1/2} = \frac{\log_b 1 + \log_b 10}{2} = 5{,}000{,}000{,}000,$$

$$\log_b \sqrt{10\sqrt{10}} = \log_b (10\sqrt{10})^{1/2} = \frac{\log_b 10 + \log_b \sqrt{10}}{2} = 7{,}500{,}000{,}000,$$

and so on. □

Now You Try 24.8

1. Given that $\log_b 10 = 10{,}000{,}000{,}000$, find

   a) $\log_b \sqrt{10\sqrt{10\sqrt{10}}}$.

   b) $\log_b \sqrt{10\sqrt{10\sqrt{10\sqrt{10}}}}$.

2. Use a calculator to approximate to 7 decimal places the quantities

$$\sqrt{10}, \qquad \sqrt{10\sqrt{10}}, \qquad \sqrt{10\sqrt{10\sqrt{10}}}, \qquad \sqrt{10\sqrt{10\sqrt{10\sqrt{10}}}}.$$

□

According to Carslaw [38, p. 118], using the geometric mean and the arithmetic mean in this way is the second of three methods that Napier gives for finding logarithms when one "adopts a cypher [zero] as the Logarithm of unity [one], and 10,000,000,000 as the Logarithm of either one-tenth of unity or ten times unity." Further, Carslaw tells us, "This is the method which Briggs followed in the calculation of his Tables, and it was adopted by Vlacq in his continuation of Briggs' work.....It must, however, be remembered that in Napier's time the index [that is, exponent] notation was unknown, and that these results are stated by him in the language of proportionals."

24.3 BASES OF A LOGARITHM

To find or approximate the value of $\log_b x$, we recall that $\log_b x = y$ if $b^y = x$. Thus,

$$\log_2 32 = 5 \quad \text{because} \quad 2^5 = 32.$$

In the same way,

$$\log_{10} 0.01 = -2 \quad \text{because} \quad 10^{-2} = 0.01,$$

and

$$\log_{81} 3 = \frac{1}{4} \quad \text{because} \quad 81^{1/4} = 3.$$

On the other hand, because there is no rational number y such that $2^y = 10$, we know that $\log_2 10$ is not a rational number.

To bound $\log_2 10$ between two integers, note that, because $2^3 = 8$ and $2^4 = 16$,

$$2^3 < 10 < 2^4.$$

Therefore,

$$3 < \log_2 10 < 4.$$

As another example, to bound $\log_{10} 749$ between two integers, note that, because $10^2 = 100$ and $10^3 = 1000$,

$$10^2 < 749 < 10^3.$$

Therefore,

$$2 < \log_{10} 749 < 3.$$

Now You Try 24.9 Find the logarithm or bound it between two integers.

1. $\log_2 50$ 2. $\log_4 64$ 3. $\log_5 300$ 4. $\log_{10} 8$

□

It is common to write "log" for "$\log_{10}$":

$$\log x = \log_{10} x.$$

The logarithm to the base 10 is called the *common logarithm function*.

Remark 24.3 Just as it is common to write "log" for "$\log_{10}$," it is also common to write "log" for "$\log_e$" (see SUBSECTION 24.4). Thus, one has to be cognizant of the notation being used at the moment. We shall follow the convention of writing "log" for "$\log_{10}$."

It is almost universal, however, that the $\boxed{\text{LOG}}$ key on a calculator refers to $\log_{10}$.   □

24.3.1   Using a Calculator

You may use the $\boxed{\text{LOG}}$ key on your calculator to find or approximate $\log_{10} x$.

Example 24.8 Find the exact value or approximate to 4 decimal places.[9]

1. $\log_{10} 0.01$

    Pressing the keys

    produces the result

    $$-2.$$

    Hence,

    $$\log_{10} 0.01 = -2.$$

2. $\log_{10} 749$

    Pressing the keys

    produces the result

    $$2.874481818.$$

    Hence,

    $$\log_{10} 749 \approx 2.8745.$$

    $\square$

Now You Try 24.10 Find the exact value or approximate to 4 decimal places.

1. $\log_{10} 8$ 2. $\log_{10} \dfrac{1}{2}$ 3. $\log_{10} 100,000,000$ 4. $\log_{10} 101,907,877$

    $\square$

Think About It 24.6 The $\boxed{\text{LOG}}$ key calculates the logarithm to the base 10. How may you use the $\boxed{\text{LOG}}$ key to calculate logarithms to other bases?

$\square$

With a little thought, we may also use a calculator to find or approximate the logarithm to any base using the $\boxed{\text{LOG}}$ key, that is, the key for $\log_{10} x$.

---

[9]We use a Texas Instruments TI-30X II S calculator here. The exact sequence of key presses on your particular calculator may differ from what is shown here. The TI-30X II S calculator inserts an open bracket automatically upon pressing the $\boxed{\text{LOG}}$ key and the $\boxed{\text{LN}}$ key, among others. If your calculator does not, then you may press the keys

for example.

Example 24.9 Consider $\log_2 15$. We see that

$$3 < \log_2 15 < 4 \quad \text{because} \quad 2^3 < 15 < 2^4.$$

To find the *exact* value of $\log_2 15$, let $\log_2 15 = y$. Then,

$$2^y = 15$$
$$\implies \quad \log_{10}(2^y) = \log_{10} 15$$
$$\implies \quad y \log_{10} 2 = \log_{10} 15 \qquad (\text{because } \log_b(u^y) = y \log_b u)$$
$$\implies \qquad y = \frac{\log_{10} 15}{\log_{10} 2}.$$

Therefore,

$$\log_2 15 = \frac{\log_{10} 15}{\log_{10} 2}$$

exactly. To approximate $\log_2 15$, we use a calculator to divide the two logarithms on the right-hand side:

$$\log_2 15 = \frac{\log_{10} 15}{\log_{10} 2} \approx 3.9069.$$

Note that this agrees with our bound on $\log_2 15$, that it is between 3 and 4. □

The preceding example, to find the exact value of $\log_2 15$, is completely general, that is to say, we may follow the same steps in the example to find the exact value of $\log_b x$ for any base b and positive real number x.

Let $\log_b x = y$. Then,

$$b^y = x$$
$$\implies \quad \log_{10}(b^y) = \log_{10} x$$
$$\implies \quad y \log_{10} b = \log_{10} x \qquad (\text{because } \log_b(u^y) = y \log_b u)$$
$$\implies \qquad y = \frac{\log_{10} x}{\log_{10} b}.$$

Therefore,

$$\boxed{\log_b x = \frac{\log_{10} x}{\log_{10} b}}$$

exactly. Observe that this relation says that, for any base b, $\log_b x$ is directly proportional to $\log_{10} x$,

$$\log_b x = k \log_{10} x,$$

with constant of proportionality $k = (\log_{10} b)^{-1}$.

Think About It 24.7 Show that, more generally,

$$\log_b x = \frac{\log_a x}{\log_a b}$$

for any base a, and not only for $a = 10$. This formula is called the *change of base* formula. Observe that this relation says that $\log_b x$ is directly proportional to $\log_a x$ for any base a; what is the constant of proportionality? □

Example 24.10 Write an expression for the exact value of the logarithm in terms of the logarithm to the base 10, then approximate the logarithm to 7 decimal places.

1. $\log_2 3$

   The exact value is

   $$\log_2 3 = \frac{\log_{10} 3}{\log_{10} 2}.$$

   Using a calculator, we find that

   $$\log_2 3 = \frac{\log_{10} 3}{\log_{10} 2} \approx 1.584\,962\,5.$$

2. $\log_2 10$

   The exact value is

   $$\log_2 10 = \frac{\log_{10} 10}{\log_{10} 2}.$$

   Using a calculator, we find that

   $$\log_2 10 = \frac{\log_{10} 10}{\log_{10} 2} \approx 3.321\,928\,1.$$

3. $\log_3 \dfrac{1}{2}$

   The exact value is

   $$\log_3 \frac{1}{2} = \frac{\log_{10} \frac{1}{2}}{\log_{10} 3}.$$

   Using a calculator, we find that

   $$\log_3 \frac{1}{2} = \frac{\log_{10} \frac{1}{2}}{\log_{10} 3} \approx -0.630\,929\,8.$$

   □

Now You Try 24.11 Write an expression for the exact value of the logarithm in terms of the logarithm to the base 10, then approximate the logarithm to 7 decimal places.

1. $\log_2 \dfrac{1}{3}$ 2. $\log_2 \dfrac{1}{10}$ 3. $\log_3 2$ 4. $\log_3 \dfrac{1}{10}$

   □

24.3.2 Comparing Logarithms

Let us consider the logarithm to the base 2, the logarithm to the base 3, and the logarithm to the base 10. To compare them, we compute a table of values for the three logarithms. Recall that $y = \log_b x$ if $b^y = x$ with $b > 1$; therefore, in $\log_b x$, we must have $x > 0$. The values shown in the table below are rounded to 7 decimal places.

x	$\log_2 x$	$\log_3 x$	$\log_{10} x$
1	0	0	0

2	1	0.630 929 8	0.301 030 0
3	1.584 962 5	1	0.477 121 3
10	3.321 928 1	2.095 903 3	1
$\frac{1}{2}$	−1	−0.630 929 8	−0.301 030 0
$\frac{1}{3}$	−1.584 962 5	−1	−0.477 121 3
$\frac{1}{10}$	−3.321 928 1	−2.095 903 3	−1

Observe that, for each logarithm, $\log_b \frac{1}{x} = -\log_b x$. For instance, $\log_2 \frac{1}{10} = -3.321\,928\,1$ and $\log_2 10 = 3.321\,928\,1$, so that $\log_2 \frac{1}{10} = -\log_2 10$. This should not be surprising because, by the rules for logarithms,

$$\log_b \frac{1}{x} = \log_b(x^{-1}) = -1 \cdot \log_b x = -\log_b x.$$

However, even with many more values of x, comparing the behaviors of the three logarithms by looking at a table of values is not so easy. An easier way to compare the three logarithms is to graph them together.

FIGURE 24.2 depicts the graphs of $y = \log_b x$ for $b = 2, 3$, and 10 using the seven values of x in the table above. We plot the pairs $(x, \log_b x)$ for each logarithm and connect the points with line segments, so that each graph is a broken line. Were we to allow x to vary over all positive real numbers instead, the graphs of the logarithms would be smooth curves. For reference, we also plot the line $y = x$.

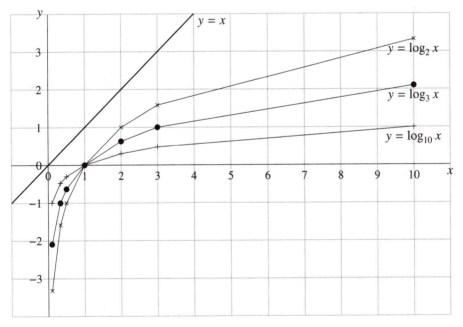

Figure 24.2: Comparing the graphs of the logarithms to the bases 2, 3, and 10.

Observe first that the graphs of the three logarithms are only to the right of the y axis. This is because $x > 0$ in $\log_b x$. However, the graphs of the logarithms are both above and below the x axis. This is because $\log_b x$ can be either positive (if $x > 1$), zero (if $x = 1$), or negative (if $x < 1$). Observe next that the graph of $y = \log_{10} x$ does not increase (rise) as rapidly as the graph of $y = \log_3 x$, and that the graph of $y = \log_3 x$ does not increase as rapidly as the graph of $y = \log_2 x$. This is true generally: The graph of $y = \log_b x$ does not increase as rapidly as the graph of $y = \log_a x$ if $b > a > 1$.

Now You Try 24.12 Sketch the graphs of $y = \log_4 x$, $y = \log_4 \frac{1}{x}$, and $\log_8 x$ together. Describe some similarities and differences among the graphs. □

Think About It 24.8

1. What point is common to the graph of all logarithm functions?

2. For a given base b, how are the graphs of $y = \log_b x$ and $y = b^x$ related?

□

There are two bases that are commonly used for logarithms: one is 10, and the other is the irrational number e.

24.4   LOGARITHM TO THE BASE e AND APPLICATIONS

To begin, we consider two sequences of numbers.

1. Consider a sequence of numbers that is obtained by raising numbers $1 + \frac{1}{n}$ (which are decreasing to 1 as the natural number n increases) to a fixed power, N. For example, let $N = 100$ and consider the sequence of numbers

$$\left(1 + \frac{1}{n}\right)^N = \left(1 + \frac{1}{n}\right)^{100} \quad \text{for} \quad n = 10, 100, 1000, \ldots,$$

that is, the sequence

$$1.1^{100}, \; 1.01^{100}, \; 1.001^{100}, \ldots .$$

2. Consider a sequence of numbers that is obtained by raising the fixed number $1 + \frac{1}{N}$ (which is a little greater than one), to increasingly greater powers n. For example, let $N = 100$ and consider the sequence of numbers

$$\left(1 + \frac{1}{N}\right)^n = \left(1 + \frac{1}{100}\right)^n \quad \text{for} \quad n = 10, 100, 1000, \ldots,$$

that is, the sequence

$$1.01^{10}, \; 1.01^{100}, \; 1.01^{1000}, \ldots .$$

We tabulate the first several terms in these two sequences.

n	$\left(1 + \dfrac{1}{n}\right)^{100}$	$\left(1 + \dfrac{1}{100}\right)^n$
10	13780.61234	1.104622125
100	2.704813829	2.704813829
1000	1.105115698	20959.15564
10,000	1.010049662	$1.635828711 \times 10^{43}$
100,000	1.001000495	OVERFLOW Error

The values shown in the table were copied from a calculator display. The result "OVERFLOW Error" means that the calculator was unable to complete the calculation.

Observe that

1. When the exponent is a fixed number, N, and the base, $1 + \frac{1}{n}$, is a number that is decreasing to 1, the terms in the sequence, $(1 + \frac{1}{n})^N$, appear to be decreasing to 1.

2. When the base is a fixed number, $1 + \frac{1}{N}$, which is a little greater than 1 and the exponents, n, are increasing, the terms in the sequence, $(1 + \frac{1}{N})^n$, appear to be continually increasing.

Thus, we expect that, in the sequence of numbers

$$\left(1 + \frac{1}{n}\right)^n \quad \text{for} \quad n = 10,\ 100,\ 1000, \ldots,$$

that is, in the sequence of numbers

$$1.1^{10},\ 1.01^{100},\ 1.001^{1000}, \ldots,$$

there will be a "competition" between the base, $1 + \frac{1}{n}$, and the exponent, n: The base, because it is decreasing to the number 1, will try to force the terms in the sequence to decrease to the number 1—in which case the base would "win"—and the exponent, because it is increasing, will try to force the terms in the sequence to increase continually—in which case the exponent would "win." Which would "win" ultimately? We tabulate the first several terms in the sequence.

n	10	100	1000
$(1 + \frac{1}{n})^n$	2.59374246	2.704813829	2.716923932

n	10,000	100,000	1,000,000
$(1 + \frac{1}{n})^n$	2.718145927	2.718268237	2.718280469

It turns out that neither the base nor the exponent "wins," but that the sequence of numbers approaches a constant that is greater than 1 as n continues to increase. This constant, denoted e, is an irrational number:

$$e = 2.718\,281\,828\,459\,045\,235\,360\,287\,471\,352\,662\,497\,757\ldots.$$

We say that the "limit" of the sequence $(1 + \frac{1}{n})^n$ is e,

$$\lim\left(1 + \frac{1}{n}\right)^n = e,$$

as n becomes increasingly greater.

Remark 24.4 It is important to note that it is the "limit" of the sequence of numbers $(1 + \frac{1}{n})^n$ that equals e, but that each term $(1 + \frac{1}{N})^N$ in the sequence, however, does not itself equal e:

$$\lim\left(1 + \frac{1}{n}\right)^n = e$$

as n becomes increasingly greater, but

$$\left(1 + \frac{1}{N}\right)^N \neq e$$

for any particular value N. □

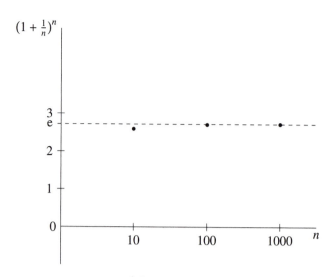

Figure 24.3: The value of $(1 + \frac{1}{n})^n$ approaches the constant irrational number e as n becomes increasingly greater.

It was apparently in 1690 that the constant e first appears fully recognized as a number in a letter that Gottfried Leibniz (1646–1716) had written to Christian Huygens (1629–1695). (See FIGURE 24.3.) In the letter, Leibniz called the constant b instead of e. It was Leonhard Euler (1707–1783), whom we credit for having invented or popularized many of the mathematical notations we use now, who popularized using the letter e for the constant.

From the time of Napier, the number e made veiled appearances, but was not recognized [107]. Although Napier did not realize this, his logarithm

$$\text{Nap}\log x = 10^7 \log_e\left(\frac{10^7}{x}\right),$$

where e is the base (see page 498). In the intervening years until 1683, the number e was also behind the work of others, but it was in 1683 that Jacob Bernoulli (1654–1705) recognized in his work on compound interest that the "limit" of the sequence $(1 + \frac{1}{n})^n$ as n becomes increasingly greater is a constant. He estimated the constant to be between 2 and 3, but did not make any connection between his work on compound interest and logarithms. (We introduce "compound interest" below.)

As Maor [80, p. 187] tells it,

> The history of π goes back to ancient times; that of e spans only about four centuries. The number π originated with a problem in geometry: how to find the circumference and area of a circle. The origins of e are less clear; they seem to go back to the sixteenth century, when it was noticed that the expression $(1 + 1/n)^n$ appearing in the formula for compound interest tends to a certain limit—about 2.71828—as n increases. Thus e became the first number to be *defined* by a limiting process....

Remark 24.5 In addition to christening the "limit" of the sequence $(1 + \frac{1}{n})^n$ as n becomes increasingly greater e, Euler also provided the following expressions for the constant.

- An "infinite sum."

$$e = 1 + \frac{1}{1} + \frac{1}{2} + \frac{1}{6} + \frac{1}{24} + \cdots .$$

This is often written using the *factorial* notation, namely,

$$e = 1 + \frac{1}{1!} + \frac{1}{2!} + \frac{1}{3!} + \frac{1}{4!} + \cdots ,$$

where

$$n! = n(n-1)(n-2)\cdots3\cdot2\cdot1 \quad \text{if } n > 0, \text{ and } \quad 0! = 1.$$

The expression $n!$ is read, "n factorial." For example, 3! is read, "3 factorial."

- A continued fraction.

$$e - 1 = 1 + \cfrac{1}{1 + \cfrac{1}{2 + \cfrac{1}{1 + \cfrac{1}{1 + \cfrac{1}{4 + \cfrac{1}{1 + \cfrac{1}{1 + \cfrac{1}{6 + \cdots}}}}}}}}.$$

(See SECTION 20.3 on continued fractions.)

□

Now You Try 24.13 Recall that

$$n! = n(n-1)(n-2)\cdots3\cdot2\cdot1 \quad \text{if } n > 0, \text{ and } \quad 0! = 1.$$

Thus, for example,

$$1! = 1,$$
$$2! = 2 \cdot 1 = 2,$$
$$3! = 3 \cdot 2 \cdot 1 = 6,$$
$$4! = 4 \cdot 3 \cdot 2 \cdot 1 = 24.$$

1. Evaluate the following.

   a) 5!   b) 6!   c) 10!

2. The first five terms of the "infinite sum" expression for the constant e are shown here:

$$e = 1 + \frac{1}{1!} + \frac{1}{2!} + \frac{1}{3!} + \frac{1}{4!} + \cdots = 1 + \frac{1}{1} + \frac{1}{2} + \frac{1}{6} + \frac{1}{24} + \cdots.$$

   a) Approximate e by adding the first five terms in the above expansion.
   b) Write the "infinite sum" expression for the constant e that shows the first ten terms of the expression.

□

> It is common to write "ln" for "$\log_e$":
>
> $$\ln x = \log_e x.$$
>
> The logarithm to the base e is called the *natural logarithm function*.

Remark 24.6

1. When one refers to *the* logarithm function, one usually is referring to the natural logarithm function.

2. It is just as common to write "log" for "$\log_e$" as it is to write "ln" for "$\log_e$." Thus, one has to be cognizant of the notation being used at the moment. We shall follow the convention of writing "ln" for "$\log_e$."

   It is almost universal, however, that the $\boxed{\text{LN}}$ key on a calculator refers to $\log_e$, and the $\boxed{\text{LOG}}$ key refers to $\log_{10}$.

□

You may use the $\boxed{\text{LN}}$ key on your calculator to approximate $\log_e x$ or $\ln x$ in the same way that you would use the $\boxed{\text{LOG}}$ key to find or approximate $\log_{10} x$.

Now You Try 24.14 Approximate to 4 decimal places.

1. $\ln 2$ 2. $\ln \dfrac{1}{2}$ 3. $\ln 100{,}000{,}000$ 4. $\ln 101{,}907{,}877$

□

24.4.1   Compound Interest

The calculation of interest can be traced as far back as to the Babylonians. Cooke [41, p. 403], referring to reports by Neugebauer, tells us that "Mesopotamian mathematicians moved beyond algebra proper and investigated the laws of exponents, compiling tables of successive powers of numbers and determining the power to which one number must be raised in order to yield another. Such problems occur in a comercial context, involving compound interest."

The amount of money in a bank savings account that earns interest is called the *principal*.[10] If the interest is calculated upon the original principal only, so that the principal remains the same no matter how long the money remains in the account, then the interest is called *simple interest*. On the other hand, if the principal increases by the amount of the interest every time, so that the interest is calculated upon an ever increasing principal, then the interest is called *compound interest*.

The basic formula for calculating interest is

$$I = Prt,$$

where I denotes the interest earned, P denotes the principal, r is the interest rate per time period, and t is the number of time periods. Note that the time period must be the same for both r and t. For example, if the interest rate is r per year, then t is the number of years that interest is earned. The time period may be fractional.

Example 24.11 Two persons, M and N, each deposit \$100 into savings accounts that pay an annual interest or annual percentage rate (APR) of 10%; however, M's account pays *simple* interest and N's account pays interest that is *compounded* annually. How much would M and N each have in their individual accounts at the end of 3 years?

Let T denote the number of years that have passed, P_T denote the principal at the beginning of year T, I_T denote the amount of interest earned at the end of year T, and A_T denote the total amount that is in the

---

[10]The practice of paying or charging interest for savings or loans is ancient, and a historically controversial idea.

account at the end of year T. Since the interest is calculated after each year, $t = 1$ in the formula $I = Prt$. Then, for M,

$$
\begin{aligned}
P_1 &= 100, & I_1 &= 100(10\%)(1) = 10, & A_1 &= 100 + 10 = 110, \\
P_2 &= 100, & I_2 &= 100(10\%)(1) = 10, & A_2 &= 110 + 10 = 120, \\
P_3 &= 100, & I_3 &= 100(10\%)(1) = 10, & A_3 &= 120 + 10 = 130.
\end{aligned}
$$

Thus, M would have \$130 in his account at the end of 3 years.

Now, for N,

$$
\begin{aligned}
P_1 &= 100, & I_1 &= 100(10\%)(1) = 10, & A_1 &= 100 + 10 = 110, \\
P_2 &= 110, & I_2 &= 110(10\%)(1) = 11, & A_2 &= 110 + 11 = 121, \\
P_3 &= 121, & I_3 &= 121(10\%)(1) = 12.10, & A_3 &= 121 + 12.10 = 133.10.
\end{aligned}
$$

Thus, N would have \$133.10 at in his account at the end of 3 years.

We see, therefore, that compound interest makes for a better investment than simple interest at the same APR. □

Example 24.12 Suppose that \$100 is deposited into a savings account that pays an APR of 10% that is compounded semiannually. How much would be in the account at the end of 3 years?

Interest that is compounded *semiannual* is calculated every half-a-year or every six months, so $t = 0.5$ in the formula $I = Prt$. Hence, using the same variables as we did in the last example, we find that

$$
\begin{aligned}
P_{1/2} &= 100, & I_{1/2} &= 100(10\%)(0.5) = 5 = 5, & A_{1/2} &= 105, \\
P_1 &= 105, & I_1 &= 105(10\%)(0.5) = 5.25, & A_1 &= 110.25, \\
P_{3/2} &= 110.25, & I_{3/2} &= 110.25(10\%)(0.5) = 5.5125, & A_{3/2} &= 116.7625, \\
P_2 &= 115.7625, & I_2 &= 115.7625(10\%)(0.5) & A_2 &= 121.550\,625, \\
& & &= 5.788\,125 & & \\
P_{5/2} &= 121.550\,625, & I_{5/2} &= 121.550\,625(10\%)(0.5) & A_{5/2} &= 127.628\,156\,25, \\
& & &= 6.077\,531\,25, & & \\
P_3 &= 127.628\,156\,25, & I_3 &= 127.628\,156\,25(10\%)(0.5) & A_3 &= 134.009\,564\,062\,5. \\
& & &= 6.381\,407\,812\,5, & &
\end{aligned}
$$

Thus, we would have about \$134.01 in the account at the end of 3 years.

We see, therefore, that interest that is compounded semiannually makes for a better investment than interest that is compounded annually at the same APR. □

In general, the greater the number of times that interest is compounded in a year, the better the investment at the same APR.

Let us now develop formulas for simple interest and compound interest. We begin with simple interest.

Suppose that P dollars are deposited into a savings account that pays $R\%$ simple interest annually. How much would be in the account at the end of T years? To determine this, we let as before

$$
\begin{aligned}
P_T &= \text{the principal at the beginning of year } T, \\
I_T &= \text{the amount of interest earned at the end of year } T, \\
A_T &= \text{the total amount that is in the account at the end of year } T, \\
r &= R\%.
\end{aligned}
$$

Since the interest is calculated after each year, $t = 1$ in the formula $I = Prt$. (We shall use these variables also when we develop a formula for compound interest below.) Then,

$$P_1 = P, \quad I_1 = Pr(1) = Pr, \quad A_1 = P + Pr,$$
$$P_2 = P, \quad I_2 = Pr(1) = Pr, \quad A_2 = (P + Pr) + Pr = P + 2Pr,$$
$$P_3 = P, \quad I_3 = Pr(1) = Pr, \quad A_3 = (P + 2Pr) + rP = P + 3Pr,$$
$$\vdots$$
$$A_T = P + TPr.$$

Thus, we would have $(P + TPr)$ dollars at the end of T years.

Simple Interest

If P dollars are deposited into a savings account that pays $r = R\%$ simple interest annually, then the amount that would be in the account at the end of T years is
$$A_T = P(1 + rT).$$
The number of years T may be any nonnegative real number.

Next, we consider compound interest.

Suppose that P dollars are deposited into a savings account that pays an APR of $R\%$ that is compounded annually. How much would be in the account at the end of T years? To determine this, we use the same variables, $P_T, I_T, A_T, r = R\%$, and $t = 1$ as before. Then,

$$P_1 = P, \qquad I_1 = Pr(1), \qquad A_1 = P + Pr = P(1 + r),$$
$$P_2 = P(1 + r), \qquad I_2 = [P(1 + r)](r)(1) \qquad A_2 = P(1 + r) + Pr(1 + r)$$
$$= Pr(1 + r), \qquad = P(1 + r)^2,$$
$$P_3 = P(1 + r)^2, \qquad I_3 = [P(1 + r)^2](r)(1), \qquad A_3 = P(1 + r)^2 + Pr(1 + r)^2$$
$$= Pr(1 + r)^2, \qquad = P(1 + r)^3,$$
$$\vdots$$
$$A_T = P(1 + r)^T.$$

Example 24.13 Suppose that \$100 is deposited into a savings account that pays an APR of 10% that is compounded annually. Then,
$$A_3 = 100(1 + 0.10)^3 = 133.10,$$
so that we would have \$133.10 in the account at the end of 3 years, as we found earlier. $\qquad\qquad\square$

Now suppose that P dollars are deposited into a savings account that pays an APR of $R\%$ that is compounded semiannually. How much would be in the account at the end of T years? Using the variables $P_T, I_T, A_T, r = R\%$, and $t = \frac{1}{2}$ because the interest is calculated every half-a-year, we find that

$$P_{1/2} = P, \qquad I_{1/2} = Pr\left(\frac{1}{2}\right) = P\left(\frac{r}{2}\right), \qquad A_{1/2} = P + P\frac{r}{2} = P\left(1 + \frac{r}{2}\right)$$

$$P_1 = P\left(1 + \frac{r}{2}\right), \qquad I_1 = \left[P\left(1 + \frac{r}{2}\right)\right](r)\left(\frac{1}{2}\right) \qquad A_1 = P\left(1 + \frac{r}{2}\right) + \left[P\left(1 + \frac{r}{2}\right)\right]\left(\frac{r}{2}\right)$$

$$= \left[P\left(1 + \frac{r}{2}\right)\right]\left(\frac{r}{2}\right), \qquad = P\left(1 + \frac{r}{2}\right)^2,$$

$$P_{3/2} = P\left(1 + \frac{r}{2}\right)^2, \qquad I_{3/2} = P\left(1 + \frac{r}{2}\right)^2 (r)\left(\frac{1}{2}\right) \qquad A_{3/2} = P\left(1 + \frac{r}{2}\right)^2 + \left[P\left(1 + \frac{r}{2}\right)^2\right]\left(\frac{r}{2}\right)$$

$$= \left[P\left(1 + \frac{r}{2}\right)^2\right]\left(\frac{r}{2}\right), \qquad = P\left(1 + \frac{r}{2}\right)^3,$$

$$P_2 = P\left(1 + \frac{r}{2}\right)^3, \qquad I_2 = P\left(1 + \frac{r}{2}\right)^3 (r)\left(\frac{1}{2}\right) \qquad A_2 = P\left(1 + \frac{r}{2}\right)^3 + \left[P\left(1 + \frac{r}{2}\right)^3\right]\left(\frac{r}{2}\right)$$

$$= \left[P\left(1 + \frac{r}{2}\right)^3\right]\left(\frac{r}{2}\right), \qquad = P\left(1 + \frac{r}{2}\right)^4,$$

$$P_{5/2} = P\left(1 + \frac{r}{2}\right)^4, \qquad I_{5/2} = P\left(1 + \frac{r}{2}\right)^4 (r)\left(\frac{1}{2}\right) \qquad A_{5/2} = P\left(1 + \frac{r}{2}\right)^4 + \left[P\left(1 + \frac{r}{2}\right)^4\right]\left(\frac{r}{2}\right)$$

$$= \left[P\left(1 + \frac{r}{2}\right)^4\right]\left(\frac{r}{2}\right), \qquad = P\left(1 + \frac{r}{2}\right)^5,$$

$$P_3 = P\left(1 + \frac{r}{2}\right)^5, \qquad I_3 = P\left(1 + \frac{r}{2}\right)^5 (r)\left(\frac{1}{2}\right) \qquad A_3 = P\left(1 + \frac{r}{2}\right)^5 + \left[P\left(1 + \frac{r}{2}\right)^5\right]\left(\frac{r}{2}\right)$$

$$= \left[P\left(1 + \frac{r}{2}\right)^5\right]\left(\frac{r}{2}\right), \qquad = P\left(1 + \frac{r}{2}\right)^6,$$

$$\vdots$$

$$A_T = P\left(1 + \frac{r}{2}\right)^{2T}.$$

Example 24.14 Suppose that $100 is deposited into a savings account that pays an APR of 10% that is compounded semiannually. Then,

$$A_3 = 100(1 + \tfrac{0.10}{2})^{2 \cdot 3} = 134.009\,564\,062\,5,$$

so that we would have about $134.01 in the account at the end of 3 years, as we found in EXAMPLE 24.12.
□

Now You Try 24.15 Suppose that P dollars are deposited into a savings account that pays an APR of $r = R\%$. How much would be in the account after T years if the interest is

1. compounded quarterly (every 3 months)?   2. compounded monthly?

□

Periodically Compound Interest

If P dollars are deposited into a savings account that pays an APR of $r = R\%$ that is compounded n times a year, then the amount that would be in the account at the end of T years is

$$A_T^{(n)} = P\left(1 + \frac{r}{n}\right)^{nT}.$$

The number of years T may be any nonnegative real number.

For interest that is compounded

- annually, $n = 1$.
- semiannually, $n = 2$.
- quarterly, $n = 4$.
- monthly, $n = 12$.

Think About It 24.9 What would be n for interest that is compounded weekly? What about interest that is compounded daily? What about interest that is compounded every hour or every minute or every second? $\square$

Now You Try 24.16 Suppose that $100 is deposited into a savings account that pays an APR of 8%. How much would be in the account after 3 years if the interest is

1. simple interest?
2. compounded annually?
3. compounded semiannually?
4. compounded quarterly?

$\square$

Think About It 24.10 Which is the better investment: a savings account that pays $R\%$ *simple* interest annually or one that pays $\frac{R}{2}\%$ simple interest semiannually? What about one that pays $\frac{R}{4}\%$ simple interest quarterly or, in general, one that pays $\frac{R}{n}\%$ simple interest n times a year? $\square$

As we mentioned, in general, the greater the number of times that interest is compounded in a year, the better the investment at the same APR. Thus, at the same APR, interest that is compounded semiannually makes a better investment than interest that is compounded annually; interest that is compounded quarterly makes a better investment than interest that is compounded semiannually; interest that is compounded monthly makes a better investment than interest that is compounded quarterly; and so on for interest that is compounded weekly, hourly, every minute, and every second. It would seem, then, that the *very best investment* would come from interest that is *compounded every moment of time* or *continuously*. So, we ask,

> If P dollars are deposited into a savings account that pays an APR of $r = R\%$ that is compounded *continuously*, how much would be in the account at the end of T years?

For interest that is compounded n times a year, the amount in the account at the end of T years would be

$$A_T^{(n)} = P\left(1 + \frac{r}{n}\right)^{nT}.$$

To find the amount that would be in the account if the interest were compounded continuously, we would let n become increasingly greater. Let us focus our attention on the factor

$$\left(1 + \frac{r}{n}\right)^{nT}.$$

If we choose m so that $n = mr$, then

$$\left(1 + \frac{r}{n}\right)^{nT} = \left(1 + \frac{1}{m}\right)^{mrT} = \left(\left(1 + \frac{1}{m}\right)^{m}\right)^{rT}.$$

Now, let us focus our attention on the base,

$$\left(1 + \frac{1}{m}\right)^{m},$$

and recall that the "limit" of the sequence $(1 + \frac{1}{m})^m$ is e:

$$\lim\left(1 + \frac{1}{m}\right)^m = e$$

as m becomes increasingly greater. (See page 481.) Consequently,

$$\lim\left(\left(1 + \frac{1}{m}\right)^m\right)^{rT} = e^{rT}$$

as m becomes increasingly greater.

Therefore,

$$\lim P\left(1 + \frac{r}{n}\right)^{nT} = Pe^{rT}$$

as n becomes increasingly greater.

Continuously Compounded Interest

If P dollars are deposited into a savings account that pays an APR of $r = R\%$ that is compounded continuously, then the amount that would be in the account at the end of T years is

$$A_T^{(\infty)} = Pe^{rT}.$$

The number of years T may be any nonnegative real number.

Remark 24.7 Notice that the exponent in the formula for interest that is compounded continuously is different from the exponent in the formula for interest that is compounded periodically: the exponent in the former is the interest rate r times the number of years t, whereas the exponent in the latter is the number of times a year n that the interest is compounded times the number of years t.

Compound Interest	
continuously	$A_t^{(\infty)} = Pe^{rt}$
n times a year	$A_t^{(n)} = P(1 + \frac{r}{n})^{nt}$

$\square$

Example 24.15 Suppose that $100 is deposited into a savings account that pays an APR of 10% that is compounded continuously. How much would be in the account at the end of 3 years?

We find that

$$A_3^{(\infty)} = 100e^{(0.10)(3)} = 134.9859\ldots,$$

so that we would have about $134.99 in the account at the end of 3 years. $\square$

Now You Try 24.17

1. Suppose that $100 is deposited into a savings account that pays an APR of 8%. How much would be in the account after 3 years if the interest is compounded continuously?

2. Suppose that $100 is deposited into a savings account that pays an APR of 6%. How much would be in the account after 3 years if the interest is

a) compounded semiannually?   c) compounded monthly?

b) compounded quarterly?   d) compounded continuously?

☐

Example 24.16 Suppose that $100 is deposited into a savings account that pays an APR of 10% that is compounded continuously. How many years would it take for the principal to double? Round to the nearest 0.1 year.

We note that the amount that would be in the account after t years is

$$A_t^{(\infty)} = 100e^{0.10t}.$$

Hence, the question is asking us to find t when $A_t^{(\infty)} = 200$ (double the principal of $100), that is, to solve the equation

$$200 = 100e^{0.10t}.$$

To begin, we divide through the equation by 100 to obtain

$$2 = e^{0.10t}.$$

Next, we recall that $x = b^y$ if and only if $y = \log_b x$. Thus,

$$2 = e^{0.10t} \quad \text{if and only if} \quad 0.10t = \log_e 2.$$

Solving for t, we find that

$$t = \frac{\log_e 2}{0.10} = 10 \log_e 2.$$

Hence, we find that the principal of $100 would take *exactly* $10 \log_e 2$ years to double. We use a calculator to approximate this.

Pressing the keys

$$\boxed{1} \quad \boxed{0} \quad \boxed{LN} \quad \boxed{2} \quad \boxed{)} \quad \boxed{=}$$

produces the result

$$6.931471806.$$

Therefore, we conclude that it would take about 6.9 years for the principal to double. This is called the *doubling time* of the investment.   ☐

Example 24.17 Suppose that $100 is deposited into a savings account that pays an APR of 10% that is compounded monthly. What is the doubling time of the investment? Round to the nearest 0.1 year.

We note that the amount that would be in the account after t years is

$$A_t^{(12)} = 100\left(1 + \frac{0.10}{12}\right)^{12t}.$$

Hence, the question is asking us to find t when $A_t^{(12)} = 200$ (double the principal of $100), that is, to solve the equation

$$200 = 100\left(1 + \frac{0.10}{12}\right)^{12t}.$$

To begin, let $b = 1 + \frac{0.10}{12}$ in the equation we wish to solve:

$$200 = 100b^{12t}.$$

Next, divide through the equation by 100 to obtain

$$2 = b^{12t},$$

and note that

$$2 = b^{12t} \quad \text{if and only if} \quad 12t = \log_b 2.$$

Solving for t, we find that

$$t = \frac{\log_b 2}{12}.$$

Hence, we find that the principal of $100 would take *exactly* $\frac{\log_b 2}{12}$ years to double. We use a calculator to approximate this.

First, using the change of base formula, we find that

$$\frac{\log_b 2}{12} = \frac{\frac{\ln 2}{\ln b}}{12} = \frac{\ln 2}{12 \ln b}.$$

Second, recalling that $b = 1 + \frac{0.01}{12}$, pressing the keys

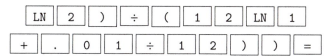

produces the result

$$6.960312992.$$

Therefore, we conclude that the doubling time of the investment is about 7.0 years, that is, it would take about 7.0 years for the principal to double. □

Now You Try 24.18 Suppose that P dollars are deposited into a savings account that pays an APR of 6%. What is the doubling time of the investment if the interest is

1. simple interest?

2. compounded annually?

3. compounded quarterly?

4. compounded continuously?

□

Think About It 24.11 Does the doubling time of an investment depend on the principal invested? On what does the doubling time depend? □

24.4.2 Amortization

Have you ever wondered how the monthly repayment of a loan is calculated? The amount of a loan is usually *amortized* over the life of the loan, a process called *amortization*, and the monthly repayment is a result of that. We present here the formula that may be used to calculate the monthly repayment of a loan that has a fixed APR.

To begin, recall that if P dollars are deposited into a savings account that pays an APR of $r = R\%$ that is compounded n times a year, then the amount that would be in the account at the end of T years is

$$A_T^{(n)} = P\left(1 + \frac{r}{n}\right)^{nT}.$$

Now, suppose that we wish to borrow $300,000 to purchase a house at a fixed APR of 5.5% that is to be repaid monthly over 30 years. We first calculate the amount of money a lender would have at the end of

30 years were the lender to deposit the $300,000 into a savings account that pays an APR of 5.5% that is compounded monthly instead of lending us the money.

$$A_{30}^{(12)} = 300,000\left(1 + \frac{0.055}{12}\right)^{12 \cdot 30} \approx 1,566,216.35.$$

Thus, the lender would have a total of $1,566,216.35 at the end of 30 years were the lender to deposit the $300,000 into a savings account instead of lending us the money—a tidy sum!

Let the monthly repayment for the loan be p dollars. To calculate p, we use the formula

$$A_T^{(n)} = p\left[\frac{\left(1 + \frac{r}{n}\right)^{nT} - 1}{r/n}\right].$$

(We shall not derive the formula, which requires using a geometric sum.) This gives

$$1,566,216.35 = p\left[\frac{\left(1 + \frac{0.055}{12}\right)^{12 \cdot 30} - 1}{0.055/12}\right].$$

Dividing both sides by the quantity in the square brackets, we find that

$$p \approx 1703.37.$$

Thus, our monthly repayment would be $1,703.37.

We summarize the two formulas that are needed to calculate the monthly repayment of a loan.

Repayment of a Loan

Suppose that P dollars are borrowed at a fixed APR of $r = R\%$ that is to be repaid n times a year over a period of T years. Then the amount of each repayment would be p dollars that is given in the formula

$$A_T^{(n)} = p\left[\frac{\left(1 + \frac{r}{n}\right)^{nT} - 1}{r/n}\right],$$

where

$$A_T^{(n)} = P\left(1 + \frac{r}{n}\right)^{nT}.$$

The number of years T may be any nonnegative real number.

Now You Try 24.19

1. Verify the calculations in the introductory example, in which we borrow $300,000 at a fixed APR of 5.5% that is to be repaid monthly over 30 years.

2. Find the monthly repayment of a $300,000 loan that is to be repaid over 15 years if the fixed APR is 6.75%.

☐

Think About It 24.12 Think about what the formula

$$A_T^{(n)} = p\left[\frac{(1 + \frac{r}{n})^{nT} - 1}{r/n}\right]$$

is saying, then try to derive the formula. In particular, why is there a minus 1 in the numerator? □

Example 24.18 Recall that the monthly repayment of a $300,000 loan at a fixed APR of 5.5% that is to be repaid over 30 years is $1,703.37. If we multiply the monthly repayment by 360 (30 years = 360 months), we find that we would have repaid a total of $613,213.20 at the end. This means that the total interest for the 30-year loan would be $313,213.20, the difference between $613,213.20 and $300,000. That is a lot in interest!

Now, suppose that our budget allows us to make a monthly repayment of $2,000, that is, $296.63 more per month. If we were to make that extra repayment against the principal of the loan every month, in how many months would we repay the loan and how much would we save in total interest?

Let $N = nt$ denote the total number of repayments; here, $N = 12t$. Then,

$$A_{30}^{(12)} = 300,000\left(1 + \frac{0.055}{12}\right)^N = 300,000(1.0046)^N.$$

Thus, if the monthly repayment is $2,000, so that $p = 2000$, it follows that

$$300,000(1.0046)^N = 2000\left[\frac{(1 + \frac{0.055}{12})^N - 1}{0.055/12}\right]$$

or

$$300,000(1.0046)^N = 2000\left[\frac{1.0046^N - 1}{0.0046}\right],$$

where we have rounded $\frac{0.055}{12} \approx 0.00458$. We solve this equation for N.

$$300,000(1.0046)^N = 2000\left[\frac{1.0046^N - 1}{0.0046}\right]$$

$$150(1.046)^N = \frac{1.0046^N - 1}{0.0046} \qquad \text{(divide by 2000)}$$

$$0.69(1.0046)^N = 1.0046^N - 1 \qquad \text{(multiply by 0.0046)}$$

$$-0.31(1.0046)^N = -1 \qquad \text{(subtract } 1.0046^N\text{)}$$

$$1.0046^N = \frac{1}{0.31} \qquad \text{(divide by } -0.31\text{)}$$

$$N = \log_{1.0046}\frac{1}{0.31} \qquad \text{(definition of } \log_b\text{)}$$

$$N = \frac{\log\frac{1}{0.31}}{\log 1.0046} \qquad \text{(change of base formula)}$$

$$N \approx 255.19$$

Thus, we would repay the loan in about 255.19 months or, dividing by 12, about 21.27 years. That is considerably less than the original 30 years for the length of the loan.

Finally, if we multiply the increased monthly repayment of \$2,000 by 255.19 months, we find that we would have repaid a total of \$510,380 at the end. This means that the total interest at the end would be \$210,380 (the difference between \$510,380 and \$300,000), which is about two-thirds of the total interest of \$313,213.20 were we not to make the extra monthly repayment of \$296.63.   □

Now You Try 24.20 You wish to borrow \$136,000 to purchase a car. You visit your friendly neighborhood loan shark, who provides you two options.

1. A fixed APR of 3.875% that is to be repaid monthly over 30 years.

2. A fixed APR of 4.25% that is to be repaid monthly over 15 years.

Find the monthly repayment and the total amount of interest you would have paid at the end for each option. Which is the better loan? Or should you instead increase the amount of the monthly repayment of the 30-year loan to equal the amount of the monthly repayment of the 15-year loan?   □

24.4.3   Exponential Growth and Decay

There are many instances in which the rate of growth or decay of a "substance" is directly proportional to the amount of the "substance" present at the time. For example, the number of microorganisms in a culture can be observed to increase for a time at a rate that is directly proportional to the number presently in the culture. As another example, an infectious disease may be observed to spread through a population at a rate that is directly proportional to the the number of persons who are presently infected. As a third example, a radioactive material may be observed to decay (lose its radioactivity) at a rate that is directly proportional to the amount of radioactive material present.

Let k be a positive constant (with units of per time), and let

$$t = \text{number of time units},$$
$$x_t = \text{amount of "substance" present at time } t,$$
$$x_0 = \text{initial amount of the "substance."}$$

For example, t may be the number of years ($t \geq 0$), x_t may be the number of grams of the "substance" at t years, and x_0 may be the initial number of grams of the "substance" (when $t = 0$); in this case, the constant k would have units of per year. For then, in the equation

$$x_t = x_0 e^{kt},$$

the quantity kt would be dimensionless:

$$(k \text{ per year})(t \text{ years}) = \left(k \, \frac{1}{\text{year}}\right)(t \text{ years}) = kt.$$

In this way, the units in the equation $x_t = x_0 e^{kt}$ match:

$$x_t \text{ grams} = (x_0 \text{ grams}) e^{kt} \quad \text{so that} \quad \text{grams} = \text{grams}.$$

Exponential Growth or Decay

Let $k > 0$ be a constant (the constant of proportionality).
If an amount x_t grows at the rate kx_t, then

$$x_t = x_0 e^{kt},$$

and we say that x_t grows exponentially.
If an amount x_t decays at the rate $-kx_t$, then

$$x_t = x_0 e^{-kt},$$

and we say that x_t decays exponentially.

Think About It 24.13 If P dollars are deposited into a savings account that pays an APR of $r = R\%$ that is compounded continuously, then the amount that would be in the account at the end of t years is $A_t^{(\infty)} = Pe^{rt}$. What does this say about the rate of growth of the amount in the account? □

Definition 24.2 Let $b > 1$ be a real number. The expression b^x, where x is any real number, is called the *exponential function* to the base b. The exponential function to the base e, e^x, is called the *natural exponential function*. □

Remark 24.8 When one refers to *the* exponential function, one usually is referring to the natural exponential function. □

Think About It 24.14 Recall that $y = e^x$ if and only if $x = \ln y$. For $b > 1$, it is common to define an exponential function to the base b to be $e^{x \ln b}$ instead of b^x. Why is this reasonable? Observe that, in this case, every exponential function $b^x = e^{kx}$, where $k = \ln b$. □

This allows us to use the e^x function on a calculator to approximate the exponential function when the value x is given. On some calculators, this may mean using the combination of keys $\boxed{\text{2nd}}$ $\boxed{\text{LN}}$.

Example 24.19 Approximate to $e^{0.14}$ to 4 decimal places.

Pressing the keys

$$\boxed{\text{2nd}}\ \boxed{\text{LN}}\ \boxed{.}\ \boxed{1}\ \boxed{4}\ \boxed{)}\ \boxed{=}$$

produces the result

$$1.150273799.$$

Hence, $e^{0.14} \approx 1.1503$. □

Example 24.20 The census records of a certain country indicate that the size of its population has been growing at the rate of 2.4% per year directly proportional to its present size. If the population now is 3.5 million people, what would be the population 5 years from now? Round to the nearest 0.1 million.

Since the growth rate has units of per year, let t denote the number of years from now, with now being $t = 0$; let x_t denote the size of the population at the end of t years; and let x_0 denote the size of the population now (when $t = 0$). Then, that the size of the population is growing at a rate that is directly proportional to its present size tells us that

$$x_t = x_0 e^{kt},$$

that is, the size of the population is growing exponentially.
    In this problem,

$$x_0 = 3.5 \text{ (million)} \quad \text{and} \quad k = 0.024 \text{ (per year)}.$$

Thus, the size of the population at the end of t years from now would be

$$x_t = 3.5 e^{0.024t}$$

in millions. In 5 years, $t = 5$:

$$x_5 = 3.5 e^{(0.024)(5)}.$$

We use a calculator to approximate this.
    Pressing the keys

produces the result

$$3.946238981.$$

    Therefore, we conclude that the size of the population would be about 3.9 million people 5 years from now.   □

Example 24.21 How many years would it take the size of the population described in the last example to double? Round to the nearest 0.1 year.

    We note that the size of the population at the end of t years from now would be

$$x_t = 3.5 e^{0.024t}$$

in millions. Hence, the question is asking us to find t when $x_t = 7$ (million; double the initial population of 3.5 million), that is, to solve the equation

$$7 = 3.5 e^{0.024t}.$$

    To begin, we divide through the equation by 3.5 to obtain

$$2 = e^{0.024t}.$$

Next, we recall that $x = b^y$ if and only if $y = \log_b x$. Thus,

$$2 = e^{0.024t} \quad \text{if and only if} \quad 0.024t = \log_e 2.$$

Solving for t, we find that

$$t = \frac{\ln 2}{0.024}.$$

Hence, according to the model, we find that the size of the population would take *exactly* $\frac{\ln 2}{0.024}$ years to double. We use a calculator to approximate this.
    Pressing the keys

produces the result

$$28.88113252.$$

Therefore, we conclude that it would take about 28.9 years for the size of the population to double, that is, the doubling time of the size of the population is about 28.9 years. □

Example 24.22 Experiments show that when 1 mg of a particular antibiotic is applied to a culture that contains a certain bacteria in a petri dish, the count of that bacteria in the dish decays at the rate of 25% per day directly proportional to its present count. If a petri dish presently has a culture that contains 140,000 of this bacteria and 1 mg of this antibiotic is applied to the culture, how many days would it take the number of bacteria in the dish to decay to half its initial count? Round to the nearest 0.1 day.

Since the decay rate has units of per day, let t denote the number of days from now, with now being $t = 0$; let x_t denote the bacteria count at the end of t days; and let x_0 denote the bacteria count now (when $t = 0$). Then, that the count of the bacteria decays at a rate that is directly proportional to its present count tells us that

$$x_t = x_0 e^{-kt},$$

that is, the count of the bacteria decays exponentially.
In this problem,

$$x_0 = 140,000 \quad \text{and} \quad k = 0.25 \text{ (per day)}.$$

Thus, the count of the bacteria at the end of t days from now would be

$$x_t = 140,000 e^{-0.25t}.$$

Hence, the question is asking us to find t when $x_t = 70,000$ (half the initial count), that is, to solve the equation

$$70,000 = 140,000 e^{-0.25t}.$$

To begin, we divide through the equation by 140,000 to obtain

$$\frac{1}{2} = e^{-0.25t},$$

and note that

$$\frac{1}{2} = e^{-0.25t} \quad \text{if and only if} \quad -0.25t = \log_e \frac{1}{2}.$$

Solving for t, we find that

$$t = \frac{\ln \frac{1}{2}}{-0.25} = -4 \ln \frac{1}{2}.$$

Hence, we find that the bacteria count would take *exactly* $-4 \ln \frac{1}{2}$ days to decay to half its initial count. We use a calculator to approximate this.
Pressing the keys

produces the result

$$2.772588722.$$

Therefore, we conclude that it would take about 2.8 days for the number of bacteria in the dish to decay to half its initial count. This is called the *half-life* of the bacteria. □

Now You Try 24.21

1. Suppose that 100 computers that are on a company network are infected by a certain computer virus today. If the virus spreads through the network at the rate of 8% per day directly proportional to the number of computers that are presently infected, how many computers on the network would be infected at the end of 3 days? Round to the nearest one computer.

2. A certain radioactive material decays exponentially at the rate of 3.5% per hour. If you have 100 g of the material on hand, how many grams would be left at the end of one week? Round to the nearest 0.1 g.

3. Suppose that a gas is leaking from a tank exponentially at the rate of 3% per minute. If there is presently M kilograms of the gas in the tank, when would half of the gas have leaked out?

4. If the half-life of a certain radioactive material that decays exponentially is 100 years, what is the rate of decay? Round to the nearest 0.1%.

□

Think About It 24.15 Can you develop a general formula for the doubling time or the half-life of an exponential growth or exponential decay process?   □

Remark 24.9 Recall that

$$\log_b x = y \quad \text{if} \quad x = b^y$$

with $b > 1$ (see page 470), and that

$$\text{Nap} \log x = y \quad \text{if} \quad x = 10^7 (0.999\,999\,9)^y$$

(see page 469). Thus, we see that

$$\log_b 1 = 0 \quad \text{for } any \text{ base } b, \text{ but} \quad \text{Nap} \log 1 \neq 0.$$

What, then, is the base of the Napierian logarithm?

A relatively simple exercise in calculus shows that, in fact,[11]

$$\text{Nap} \log x = 10^7 \log_e\left(\frac{10^7}{x}\right),$$

so that

$$\text{Nap} \log 1 = 10^7 \log_e 10^7 \neq 0,$$

but instead

$$\text{Nap} \log 10^7 = 10^7 \log_e 1 = 0.$$

In fact,

$$\text{Nap} \log x \neq \log_b x$$

for any base b. In particular, contrary to popular belief, the Napierian logarithm is not the natural logarithm.[12] Indeed, whereas $\log_b x$ for any base $b > 1$ increases as x increases, Nap $\log x$ *decreases* as x increases.

---

[11] In the definition of the logarithm to the base b (DEFINITION 24.1) we restricted the base $b > 1$. It is also common to define the logarithm to the base b with a looser restriction on the base b, namely, that $b > 0$, but $b \neq 1$. In this case, we find that $\log_{1/b} x = \log_b(\frac{1}{x})$, so that Nap $\log x = 10^7 \log_{1/e}(\frac{x}{10^7})$.

[12] However, according to Cajori [24, p. 8], "A readjustment of Napier's original logarithms was made in John Speidell's *New Logarithmes*,[†] published in 1619 in London, whereby the logarithms virtually became the so-called 'natural logarithms' of to-day."

[†] John Speidell's book is reprinted in Maseres' *Scriptores logarithmici*, Vol. 6, 1807, pp. 728–759.

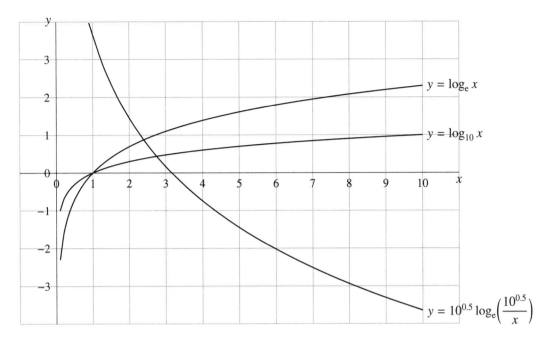

Figure 24.4: Comparing the graphs of $y = \log_e x$, $y = \log_{10} x$, and $y = 10^{0.5} \log_e(\frac{10^{0.5}}{x})$. The graph of $y = \text{Nap} \log x = 10^7 \log_e(\frac{10^7}{x})$ behaves similarly to the graph of $y = 10^{0.5} \log_e(\frac{10^{0.5}}{x})$, but would be off the scale shown here.

FIGURE 24.4 shows the graphs of $y = \log_e x$, $y = \log_{10} x$, and $y = 10^{0.5} \log_e(\frac{10^{0.5}}{x})$. The graph of $y = \text{Nap} \log x = 10^7 \log_e(\frac{10^7}{x})$ behaves similarly to the graph of $y = 10^{0.5} \log_e(\frac{10^{0.5}}{x})$, but would be off the scale shown in the figure. The following table illustrates this, where the values have been rounded; note that $10^{0.5} \approx 3.162\,277\,8$.

x	$y = 10^{0.5} \log_e(\frac{10^{0.5}}{x})$	$y = 10^7 \log_e(\frac{10^7}{x})$
1	3.640 706 7	161,180,957
$10^{0.5}$	0	149,668,031
10	−3.640 706 7	138,155,106
10^7	−47.329 187 1	0
10^{10}	−69.173 427 3	−69,077,553

Observe that the graphs of $y = \log_e x$ and $y = \log_{10} x$ are increasing, whereas the graph of $y = 10^{0.5} \log_e(\frac{10^{0.5}}{x})$, as would the graph of $y = \text{Nap} \log x$, is decreasing. □

24.5 LOGARITHM TO THE BASE 10 AND APPLICATION TO EARTHQUAKES

The choice of base 10 for the logarithm is substantially due to Henry Briggs (1561–1631), who was a professor of geometry at Gresham College, London. When he learned about Napier's logarithms, Briggs set off to meet Napier. E. T. Bell relates a very colorful account of Briggs's first encounter with Napier [9, p. 526]:

In the *History of his Life and Times* the astrologer William Lilly (1602–1681) records an amusing—
if incredible—account of the meeting between John Napier (1550–1617), of Merchiston, the inventor
of logarithms, and Henry Briggs (1561–1631) of Gresham College, London, who computed the first
table of common logarithms. One John Marr, "an excellent mathematician and geometrician," had gone
"into Scotland before Mr. Briggs, purposely to be there when these two learned persons should meet.
Mr. Briggs appoints a certain day when to meet in Edinburgh; but failing thereof, the lord Napier was
doubtful he would not come. It happened one day as John Marr and the lord Napier were speaking
of Mr. Briggs: 'Ah John (said Merchiston), Mr. Briggs will not now come.' At the very moment one
knocks at the gate; John Marr hastens down, and it proved Mr. Briggs to his great contentment. He
brings Mr. Briggs up into my lord's chamber, where almost *one quarter of an hour was spent*, each
beholding other with admiration, *before one word was spoke*."

Then Briggs broke the silence, saying ([56, p.182] and [80, p. 11]):

"My lord, I have undertaken this long journey purposely to see your person, and to learn by what
engine of wit or ingenuity you came first to think of this most excellent help in astronomy. But, my lord,
being by you found out, I wonder nobody found it out before when now known it appears so easy."

Recall that Nap $\log 10^7 = 0$. Briggs, during his visit, suggested two changes to Napier's logarithms
to make them easier to use, namely, to have the logarithm of 1, instead of 10^7, equal 0 and to have the
logarithm of 10 equal some power of 10 [80, p. 12]. In the end, both men decided to use a table for which
the logarithm of 1 equals 0 and the logarithm of 10 equals 1; in other words, they decided to use a table
for $N = 10^L$. Today we recognize this as the logarithm to the base 10, and for this reason the base 10 or
common logarithm is also called the *Briggsian logarithm* [56, p. 184].

Figure 24.5: A Gunter's scale. (Source: MIT Slide Rule Collection, MIT Museum,
Cambridge, Massachusetts. U.S. public domain.)

With Napier coming to the end of his life and lacking the energy, the task of creating a table for
common logarithms was left to Briggs. In 1620, Briggs, together with his colleague Edmund Gunter
(1581–1626), who was a professor of astronomy at Gresham College, published a table of the common
logarithms for the sine and tangent[13] accurate to 7 decimal places. Gunter also produced the predecessor

---

[13]The tangent is a trigonometric function. Recall that there are six trigonometric functions: sine, cosine, tangent, cotangent,
secant, and cosecant. Gunter is credited with coining the names "cosine" and "cotangent" [56, p. 185]. Gunter is also known for

to the slide rule (FIGURE 24.6) known as the *Gunter's scale* (FIGURE 24.5) for computing logarithms and for other calculations.

In 1624, Briggs published his *Arithmetica logarithmica* that contains the common logarithms of numbers from 1 to 20,000 and from 90,000 to 100,000 accurate to 14 decimal places. Brigg's table was filled in later with the help of the Dutch bookseller and publisher Adriaen Vlacq (1600–1666). According to Eves [56, p. 185], "Briggs and Vlacq published four fundamental tables of logarithms, which were not superseded until, between 1924 and 1949, extensive 20-place tables were calculated in England as part of the celebration of the tercentenary of the invention of logarithms."

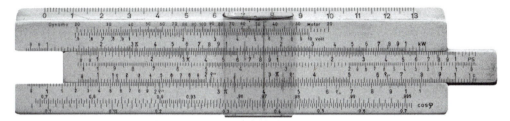

(a) Standard slide rule.

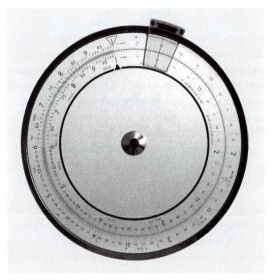

(b) Circular slide rule.

Figure 24.6: A standard slide rule and a circular slide rule. Before electronic calculators became commonplace, slide rules were used by scientists and engineers, and others, to make quick work of lengthy calculations. (Source: © Thinkstock.)

One application of the logarithm to the base 10 among very many is to earthquakes. The *Richter scale* is a typical way to measure the magnitude of an earthquake that was developed by Charles F. Richter of the

---

having "designed the logarithmic 'line of numbers,' which is simply a straight line, with the digits 1, 2, 3,..., 10 arranged upon it from one extremity to the other, in such a way that the distance on the line from the end marked 1, to the figure 2, is to the distance from 1 to any other number, as the logarithm of 2 is to the logarithm of that other numbers....Gunter mounted this line, together with other lines giving the logarithms of trigonometric functions, upon a ruler or scale, commonly called 'Gunter's scale' [see FIGURE 24.5], by means of which questions in navigation could be resolved with the aid of a pair of compasses....Gunter described his logarithmic 'line of numbers' in his *Canon Triangulorum*, London, 1620, as well as in his *Description and Use of the Sector, Cross-Staff and other Instruments*, London, 1624" [32, pp. 2–3]. Cajori [32] points out that Gunter's scale is not a slide rule (see FIGURE 24.6), and that Gunter has been frequently mistaken for having invented the slide rule.

Assuming that the Californian shocks are representative of general conditions, and attaching the results from California to those found directly for the whole world in the higher magnitude levels, the conclusions follow:

	Magnitude	Annual number
Great earthquakes	8 or more	1
Major earthquakes	7–7.9	10
Destructive shocks	6–6.9	100
Damaging shocks	5–5.9	1000
Minor strong shocks	4–4.9	10000
Generally felt	3–3.9	100000

Table 24.2: Gutenberg-Richter Table.

California Institute of Technology in 1935. According to the USGS,[14] "The magnitude of an earthquake is determined from the logarithm [to the base 10] of the amplitude of waves [that is, so to speak, the amount of shaking] recorded by seismographs.... Because of the logarithmic basis of the scale, each whole number increase in magnitude represents a tenfold increase in measured amplitude; as an estimate of energy, each whole number step in the magnitude scale corresponds to the release of about 31 times more energy than the amount associated with the preceding whole number value." Thus, for example, the amplitude of the ground waves of a magnitude 5.0 earthquake is 10 times that of a magnitude 4.0 earthquake ($10^{5.0-4.0} = 10^{1.0}$ or 10), and 100 times that of a magnitude 3.0 earthquake ($10^{5.0-3.0} = 10^{2.0}$ or 100). On the other hand, the amount of energy released by a magnitude 5.0 earthquake (or, equivalently, the amount of energy that had to build up prior to a magnitude 5.0 earthquake) is about 31 times that of a 4.0 magnitude earthquake, and about 961 ($= 31^2$) times that of a magnitude 3.0 earthquake.

Now You Try 24.22 The famed 1906 San Francisco earthquake is noted to have been magnitude 8.3. The 1989 Loma Prieta earthquake, about 56 miles south of San Francisco and that interrupted the World Series baseball game between the San Francisco Giants and the Oakland A's, is noted to have been magnitude 6.9. How many times greater was the amplitude of the ground wave of the San Francisco earthquake than that of the Loma Prieta earthquake? How many times greater was the amount of energy released? Round to the nearest one.   □

Walter [124, p. 1279] reproduces a table by Gutenberg and Richter[15] (shown here as TABLE 24.2) and notes: "We are told that the numbers in this first column are logarithms (presumably to base 10) of something we shall refer to for simplicity as earthquake *intensity*. If we take the logarithm, to base 10, say, of the second column, which is the number of earthquakes of a given intensity occurring in a given year, we notice a simple invariant. The sum of 'log of intensity of event' and 'log of the number of events of that intensity (per year)' is approximately constant, namely, 8." This is remarkable!

Think About It 24.16

1. The magnitude of an earthquake is determined from the logarithm to the base 10 of the amplitude of the waves recorded by seismographs. What should be the base if we were to determine the magnitude of an earthquake from a logarithm of the amount of energy released?

---

[14] United States Geological Survey. <http://earthquake.usgs.gov/learn/topics/richter.php>

[15] Beno Gutenberg and C. F. Richter, *Seismicity of the Earth*, The Geological Society of America, No. 34 (1941).

2. Verify what Walter notes about TABLE 24.2: that the "sum of 'log of intensity of event' [that is, the magnitude] and 'log of the number of events of that intensity [that is, the log of the magnitude] (per year) is approximately constant, namely, 8." What do you think this observation may imply about earthquakes?

☐

A formula for calculating the Richter scale magnitude M of an earthquake is

$$M = \log\left(\frac{I}{S}\right),$$

where I is the intensity of the earthquake that is determined from the amplitude of waves measured by a seismograph reading that is taken 100 kilometers from the epicenter of the earthquake, and S is the intensity of a "standard" earthquake whose amplitude of waves is 1 micron or 10^{-4} centimeters.[16] Recall that "log" means "$\log_{10}$."

Example 24.23 The 1906 San Francisco earthquake is noted to have been magnitude 8.3. How many times as intense was the San Francisco earthquake compared to a standard earthquake? Round to the nearest one.

We substitute $M = 8.3$ into the formula for the magnitude of an earthquake, and recall that

$$8.3 = \log\left(\frac{I}{S}\right) \quad \text{if and only if} \quad \frac{I}{S} = 10^{8.3}.$$

We use a calculator to approximate this.
    Pressing the keys

$$\boxed{1}\ \boxed{0}\ \boxed{\frown}\ \boxed{8}\ \boxed{.}\ \boxed{3}\ \boxed{=}$$

produces the result

$$199526231.5.$$

Thus,

$$\frac{I}{S} \approx 199{,}526{,}232 \quad \text{or} \quad I \approx 199{,}526{,}232S.$$

Therefore, we conclude that the San Francisco earthquake was about 199,526,232 as intense as a standard earthquake.
    Note that using logarithms makes dealing with such large numbers routine.   ☐

---

[16] According to C. M. R. Fowler, *The Solid Earth: An Introduction to Global Geophysics*, Cambridge University Press, Cambridge (1990), p. 88,

> A number of other logarithmic *magnitude scales* for earthquakes exist, all of which are based on measurements of the amplitude of the seismic waves. In addition there are *intensity scales* such as the Mercalli scale, which are subjective and are based on the shaking of buildings, breaking of glass, ground cracking, people running outside, and so on.
>   All the magnitude scales are of the form
>
> $$M = \log_{10}\left(\frac{A}{T}\right) + q(\Delta, h) + a$$
>
> where M is the magnitude, A the maximum amplitude of the waves (in 10^{-6} meters), T the period of the wave (in seconds), q a function correcting for the decrease in amplitude of the wave with distance from the epicentre and focal depth, Δ the angular distance from seismometer to epicenter, h the focal depth of the earthquake, and a an empirical constant.

Example 24.24 What is the magnitude of an earthquake whose intensity is 10,000,000 times that of a standard earthquake?

We substitute $I = 10{,}000{,}000S$ into the formula for the magnitude of an earthquake,

$$M = \log\left(\frac{I}{S}\right).$$

This yields

$$M = \log\left(\frac{10\,000\,000S}{S}\right) = \log 10{,}000{,}000 = 7.$$

Therefore, we conclude that an earthquake whose intensity is 10,000,000 times that of a standard earthquake is magnitude 7.   □

Now You Try 24.23

1. The 1989 Loma Prieta earthquake was magnitude 6.9. How many times as intense was the Loma Prieta earthquake compared to a standard earthquake? Round to the nearest one.

2. What is the magnitude of an earthquake whose intensity is three hundred million times that of a standard earthquake? Round to the nearest one.

3. What is the magnitude of a standard earthquake?

4. How many times greater was the amount of energy released by the Loma Prieta earthquake compared to a standard earthquake? Round to the nearest one.

□

Chapter 25

Exercises

Ex. 25.1 — *False position*. The figure on the left below shows the graph of $y = ax$ and the numbers ξ, b, g, and h. We remark that g may be to the right of ξ, that is, our guess, g, may be greater than the number we seek, ξ. Now, the graph admits the figure on the right that shows two similar triangles.

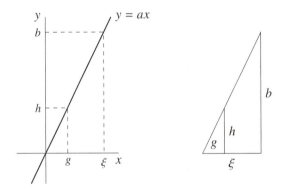

Use similar triangles to provide an argument for the method of false position for solving a linear equation in the form $ax = b$.

Ex. 25.2 — Provide an algebraic argument (without appealing to a graph or to similar triangles) for the method of false position for solving a linear equation in the form $ax = b$.

Ex. 25.3 — To introduce the method of false position, Chabert et al. [39] give as an example the following question posed by Francès Pellos, a fifteenth-century nobleman from Nice: "A lance has half and a third in the water and 9 palms outside. I ask you how long is it?" (A *palm* is the length measured by the width of a person's palm.)

1. Solve the problem using the method of false position. Hint: The problem is equivalent to: A quantity, its half and its third are removed from it so that 9 results. Begin with a guess that is a common multiple of 2 and 3. Pellos "invites us to 'put 12 for it as we please', that is, to carry out the calculations use 12 as a false position" [39].

2. Solve the problem using symbolic algebra: Let x denote the length of the lance, write an equation that describes the problem, and solve the equation.

Ex. 25.4 — Solve the following two problems found in Babylonian cuneiform tablets [39, p. 86] using the method of false position. Note that it is not known if the problems were originally solved in this way, for the answer given in the tablet to the first problem is incorrect, and the solution to the second is too imprecise for us to make out the method that was used.

1. (YBC 4669 problem 3)

> I have eaten two thirds of my provisions: here is left 7. What was the original (amount) of my provisions?

2. (AO 6770 problem 3)

> I took a stone: I did not know its weight. I took away 1/7, the third of a shekel and 15 grains. I put back 1/11 of what I had taken and five sixths of a shekel: my stone was restored to its original state. What was the original (weight) of my stone?

Ex. 25.5 — Solve the problem in EXAMPLE 19.8 using an algebraic equation: Let l denote the length of the rectangle, write an equation that describes the problem, and solve the equation.

Ex. 25.6 — To analyze the problem in EXAMPLE 23.6, Chabert et al. [39, p. 98] first describe the problem with the following equation:

$$\frac{1}{10}\left[5x - \frac{1}{3}(5x)\right] + \frac{1}{3}x + \frac{1}{2}x + \frac{1}{4}x = 70 - 2.$$

1. Show that this equation reduces to $ax = b$.

2. Solve the equation and compare the solution here to the solution provided in the example.

Ex. 25.7 — Solve (or attempt to solve) the equation $2(x + 3) + 4x = 15$

1. by false position.   2. by double false position.

* Ex. 25.8 — Write an algorithm for a method of double false position to solve linear equations that are equivalent to $ax + b = cx + d$ without transposition. Use the method to solve the equation $3(x - 1) = x + 7$.

Ex. 25.9 — Solve the problem in EXAMPLE 19.9 using an algebraic equation. Let x denote the amount of money that A should get.

Ex. 25.10 — Solve the problem in EXAMPLE 19.7 using a system of two linear equations. Let x denote the area of the first field and y denote the area of the second field.

Ex. 25.11 — Solve the problem in EXAMPLE 19.11 using a system of two linear equations. Let n denote the number of people and p denote the price of one chicken.

Ex. 25.12 — Solve the problem in EXAMPLE 19.12 using a system of two linear equations. Let n denote the number of people and p denote the price of one sheep.

Ex. 25.13 — Use the methods in chapter 7 of the *Nine Chapters* [44, p. 272] to solve the following problem.

> Now a total of 1 *qing* farmland is bought, the price is 10,000 coins. Tell: the good and poor farmland, how much each?

Ex. 25.14 — Solve the problem in EXAMPLE 19.7 using the method of double false position. For example, use $g_1 = 900$ and $g_2 = 1500$ as two guesses for the area of the first field.

Ex. 25.15 — (*Liber Abaci* chapter 13 problem "On Four Men Who Found a Purse" [90, p. 456])

> Four men having denari found a purse of denari; the first man said that if he would have the denari of the purse, then he would have twice as many as the second. The second, if he would have the purse, then he would have three times as many as the third, and the third, if he would have it, then he would have four times as many as the fourth. The fourth, five times as many as the first; it is sought how many denari each has.

1. Solve the problem using the second elchataym method. First, guess that the first man has 9 denari and the purse has 21 denari; for the second guess, either increase or decrease the amount that the first man has or the amount that is in the purse.

2. Solve the problem using a system algebraic equations. Let x denote the amount that the first man has and let y denote the amount in the purse.

Ex. 25.16 — (*Liber Abaci* chapter 13 problem "On Five Man Who Bought a Horse" [90, p. 457])

Five men having bezants wished to buy a horse; the first man takes from the second half of his bezants, and the second takes from the third a third, and the third takes from the fourth one fourth, and the fourth takes from the fifth one fifth, and the fifth similarly takes from the first one sixth of his bezants. And thus each of them proposes to buy the horse. It is sought how many bezants each of them had.

Ex. 25.17 — Solve the following problems using the rule of three or the inverse rule of three, as appropriate. Identify the "fruit of the desire," the "fruit quantity," the "desire quantity," and the "measure." (See SECTION 23.2.)

1. (*Līlāvatī* v. 78 [34, p. 222])

A heap of grain having been meted with a measure containing seven *ad'hacas*, if a hundred such measures were found, what would be the result with one containing five *ad'hacas*?

2. Boyle's law is a special case of the *ideal gas law* that relates the pressure and the volume of a gas that is kept at a constant temperature. According to Boyle's law, at a constant temperature, the absolute pressure and the volume of a gas are inversely proportional, meaning that a numerical increase in the pressure will result in an inversely similar numerical decrease in the volume. Suppose that a certain gas that occupies a 2-cubic-meter spherical chamber exerts a pressure of 1000 pascals on the chamber wall. If the gas is evacuated from this chamber into a 3-cubic-meter spherical chamber without any change in its temperature, how much pressure will the gas exert on the wall of the second (larger) chamber?

Ex. 25.18 — Solve the problem in EXAMPLE 23.3 using a proportion equation. Let x denote the "fruit of the desire."

Ex. 25.19 — Solve the problem in EXAMPLE 23.4 using a proportion equation. Let x denote the "fruit of the desire."

Ex. 25.20 — Solve the problem in EXAMPLE 23.5 using a proportion equation. Let x denote the "fruit of the desire."

Ex. 25.21 — FIGURE 19.2 shows the cuneiform tablet YBC 7289 from the Yale Babylonian Collection. It pictures a square with one side marked 30 and one diagonal marked 1 24 51 10 and, below that, 42 25 35.

1. Find the values of the sexagesimal numbers 30, 1 24 51 10, and 42 25 35.

2. How are the three numbers here related?

Ex. 25.22 — Refer to page 297. To arrive at the formula for double false position,

$$\xi = \frac{g_1 e_2 - g_2 e_1}{e_2 - e_1} \quad \text{or} \quad \frac{g_2 e_1 - g_1 e_2}{e_1 - e_2},$$

we computed the slope of the line $y = ax + b$ in two ways: First, we used the known points $P_1 = (g_1, h_1)$ and the unknown point $P_2 = (\xi, b)$; second, we used the known point $P_2 = (g_2, h_2)$ and the point P_3.

1. Derive a formula for double false position by computing the slope of the line in this way instead: First, use the two known points, P_1 and P_2; second, use the point P_1 and the unknown point P_3.

2. Use this formula to solve the problem in EXAMPLE 19.9 and the equation in EXAMPLE 19.10.

Ex. 25.23 — One of the best-known problems due to Leonardo of Pisa is the "rabbit problem" in *Liber Abaci* [90, p. 404]:

How Many Pairs of Rabbits Are Created by One Pair in One Year.

A certain man had one pair of rabbits together in a certain enclosed place, and one wishes to know how many are created from the pair in one year when it is the nature of them in a single month to bear another pair, and in the second month those born to bear also. Because the above written pair in the first month bore, you will double it; there will be two pairs in one month. One of these, namely the first, bears in the second month, and thus there are in the second month 3 pairs; of these in one month two are pregnant, and in the third month 2 pairs of rabbits are born, and thus there are 5 pairs in the month....

1. Complete the following table.

Month	Number of pairs born
beginning	1
first	2
second	3
third	5
fourth	
fifth	
sixth	
seventh	
eighth	
ninth	
tenth	
eleventh	
twelfth	

2. This sequence of numbers is generally written

$$1, 1, 2, 3, 5, \dots .$$

Continue this sequence of numbers to the twentieth term. This is called the *Fibonacci sequence*. (See SUBSECTION 20.3.3 on the Fibonacci sequence.)

3. Find the ratios of successive terms of the Fibonacci sequence,

$$\frac{1}{1}, \frac{2}{1}, \frac{3}{2}, \frac{5}{3}, \dots ,$$

using up to the twentieth term of the Fibonacci sequence. Compare each of these ratios to the golden ratio,

$$\phi = \frac{1 + \sqrt{5}}{2}$$

(see NOW YOU TRY 19.11). What do you notice?

* Ex. 25.24 — Find examples of how the Fibonacci sequence shows up in nature.

Ex. 25.25 — Build a new Fibonacci sequence by starting with the initial terms of 1 and 2 instead of 1 and 1. Now find the ratio of successive terms. Does it approach ϕ or some other number?

Ex. 25.26 — Create your own sequence like the Fibonacci sequence by changing the the initial terms, changing the rules, and so on.

* **Ex. 25.27** — J. W. L. Glaisher opines, "Without suitable notation it would have been impossible to express differential equations, or even to conceive of them if complicated, much less to deal with them…" (see page 268). What are "differential equations," and how are they different from "algebraic equations"?

* **Ex. 25.28** — Select and write about one of the dynasties in China's history.

* **Ex. 25.29** — Write about the history of the Chinese mathematics text, *Nine Chapters*.

* **Ex. 25.30** — Write about the life and work of the Indian mathematician Bhāskara II.

* **Ex. 25.31** — Write about the golden ratio (or divine proportion) and its applications.

* **Ex. 25.32** — The golden ratio has been claimed to be used in the building of the Great Pyramids and the Greek Parthenon, for example, as well as in many famous works of art. Does this sound credible? Analyze these claims.

Ex. 25.33 — Identify each of the problems in examples 20.2, 20.3, 20.4, and 20.5 as one of the five types of quadratic equations classified by al-Khwārizmī (page 314).

Ex. 25.34 — In EXAMPLE 20.7 on page 316, verify the steps by constructing the geometric interpretations as in EXAMPLE 20.6.

Ex. 25.35 — Fill in the missing algebra steps in parts 1 and 2 of RULE 20.3 on page 348.

Ex. 25.36 — In EXAMPLE 20.8 on page 317, we see that al-Khwārizmī finds that the solutions of the quadratic equation $x^2 + 21 = 10x$ are 3 and 7.

1. The following is the graph of the equation $xy = 21$ for $x > 0$ and $y > 0$. Graph the equation $x + y = 10$ on the same set of axes to solve the system of equations

$$x + y = 10, \quad xy = 21.$$

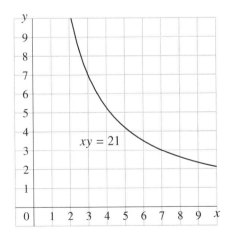

2. Show that the system of equations $\{x + y = 10, \ xy = 21\}$ leads to the quadratic equation $x^2 + 21 = 10x$. Thus, a Babylonian scribe may solve the single quadratic equation in one unknown here by considering, instead, a system of two equations in two unknowns.

Ex. 25.37 — We restate the problem in YBC 6967 in EXAMPLE 20.2 in this way:

A number exceeds another by 7, and the product of the two number is 60. Find the numbers.

Recall that, because the Babylonians used base 60, in YBC 6967 it was understood that two numbers a and b are reciprocals if their product $ab = 60$. Solve the problem with the understanding that a and b are reciprocals if $ab = 1$.

Ex. 25.38 — (Al-Khwārizmī [2, p. 10])

The proceeding will be the same if the instance be, "half of a square and five roots are equal to twenty-eight dirhams"; that is to say, what must be the amount of a square, the moiety [half] of which, when added to the equivalent of five of its roots, is equal to twenty-eight dirhams?

1. Which of al-Khwārizmī's five types of quadratic equations is this?

2. Al-Khwārizmī begins the solution by stating: "Your first business must be to complete your square, so that it amounts to one whole square. This you effect by doubling it. Therefore, double it, and double also that which is added to it, as well as what is equal to it. Then you have...." Complete the solution, writing the solution in words as al-Khwārizmī does.

3. Express the quadratic equation symbolically using x for the unknown. Check the solution that you found.

Ex. 25.39 — (Al-Khwārizmī [2, pp. 12–13])

Roots and Numbers are equal to Squares; for instance, "three roots and four of simple numbers are equal to a square." Solution: Halve the roots; the moiety is one and a half. Multiply this by itself; the product is two and a quarter. Add this to the four; the sum is six and a quarter. Extract its root; it is two and a half. Add this to the moiety of the roots, which was one and a half; the sum is four. This is the root of the square, and the square is sixteen.

1. Which of al-Khwārizmī's five types of quadratic equations is this?

2. Summarize the steps of al-Khwārizmī's solution.

3. Express the quadratic equation symbolically using x for the unknown. Check that 4 is a solution, but also that -1 is a solution.

4. How does this problem compare with the problem in BM 13901 (ii) (EXAMPLE 20.5 on page 309)?

5. Is it possible to solve the problem in BM 13901 (ii) using the steps for solving this problem, namely, "three roots and four of simple numbers are equal to a square"? If so, do it.

6. Is it possible to solve this problem using the steps for solving the problem in BM 13901 (ii)? If so, do it.

Ex. 25.40 — If we follow al-Khwārizmī's steps in EXAMPLE 20.8 to solve the general quadratic equation $x^2 + Bx + C = 0$, where B and C are any real constants, we find that the equation has two solutions, namely,

$$\frac{-B + \sqrt{B^2 - 4C}}{2} \quad \text{and} \quad \frac{-B - \sqrt{B^2 - 4C}}{2}.$$

Use the substitutions $B = \frac{b}{a}$ and $C = \frac{c}{a}$, where a is any nonzero real constant, to show that the quadratic equation $ax^2 + bx + c = 0$ has the two solutions

$$\frac{-b + \sqrt{b^2 - 4ac}}{2a} \quad \text{and} \quad \frac{-b - \sqrt{b^2 - 4ac}}{2a}$$

that are given by the quadratic formula (RULE 20.1 on page 330).

Ex. 25.41 — Factor the polynomial $x^2 + 6x + 7$ using the factor theorem (THEOREM 22.1 on page 424).

Ex. 25.42 — Factor the polynomial $2x^2 - 4x - 1$.

Ex. 25.43 — Verify that -4 is a zero of the polynomial $x^3 + 6x^2 + 9x + 4$, then factor the polynomial completely.

Ex. 25.44 — Verify that 3 is a zero of the polynomial $x^3 - 2x^2 - 4x + 3$, then factor the polynomial completely.

* **Ex. 25.45** — Research other methods of polynomial division and their history.

* **Ex. 25.46** — Research the geometric interpretation of the zeros of a polynomial or the roots of a polynomial equation.

Ex. 25.47 — Work through Harriot's method of solving a quadratic on page 335. Note that he is using a for the variable.

Ex. 25.48 — Express the following as simple continued fractions.

1. $\dfrac{47}{11}$ 2. $-\dfrac{47}{11}$ 3. $\dfrac{94}{22}$ 4. $\dfrac{11}{47}$

(Do not reduce $\frac{94}{22}$.)

Ex. 25.49 — Express the following as simple continued fractions. Check by determining the values of the continued fractions.

1. $-\sqrt{10}$

2. $\sqrt{53}$

3. $25 + \sqrt{53}$

4. $\dfrac{25 + \sqrt{53}}{22}$

* **Ex. 25.50** — Write about René Descartes and some of his accomplishments in mathematics.

* **Ex. 25.51** — Write about Pierre de Fermat and Blaise Pascal, and their contributions to the theory of probability.

* **Ex. 25.52** — Research the uses of finding roots. Why would this have been an important operation, especially in medieval and Renaissance Europe?

* **Ex. 25.53** — Write about the history of the golden ratio and the golden rectangle.

* **Ex. 25.54** — Explain why it could be argued that $1 - 1 + 1 - 1 + 1 + \cdots = \frac{1}{2}$.

* **Ex. 25.55** — Find a proof that the geometric sequence $1 + \frac{1}{2} + \frac{1}{4} + \cdots = 2$ that does not rely on a figure or on physical motion.

* **Ex. 25.56** — Write about some uses of continued fractions.

Ex. 25.57 — Verify that $(a - b)^3 = a^3 - 3a^2b + 3ab^2 - b^3$.

Ex. 25.58 — Verify that the three cube roots of -1 are

$$-1, \quad \frac{1}{2} + i\frac{\sqrt{3}}{2}, \quad \text{and} \quad \frac{1}{2} - i\frac{\sqrt{3}}{2}.$$

Identify the complex conjugate pairs.

Ex. 25.59 — Verify that the four fourth roots of -1 are

$$\frac{\sqrt{2}}{2} + i\frac{\sqrt{2}}{2}, \quad -\frac{\sqrt{2}}{2} + i\frac{\sqrt{2}}{2}, \quad -\frac{\sqrt{2}}{2} - i\frac{\sqrt{2}}{2}, \quad \frac{\sqrt{2}}{2} - i\frac{\sqrt{2}}{2}.$$

Identify the complex conjugate pairs.

Ex. 25.60 — Verify that the eight eighth roots of unity are

$$\omega_0 = 1, \quad \omega_1 = \frac{\sqrt{2}}{2} + i\frac{\sqrt{2}}{2}, \quad \omega_2 = i, \quad \omega_3 = -\frac{\sqrt{2}}{2} + i\frac{\sqrt{2}}{2},$$

$$\omega_4 = -1, \quad \omega_5 = -\frac{\sqrt{2}}{2} - i\frac{\sqrt{2}}{2}, \quad \omega_6 = -i, \quad \omega_7 = \frac{\sqrt{2}}{2} - i\frac{\sqrt{2}}{2}.$$

Identify the complex conjugate pairs.

Ex. 25.61 — Find the three cube roots of -8. Identify the complex conjugate pairs.

Ex. 25.62 — Find the four fourth roots of -16. Identify the complex conjugate pairs.

Ex. 25.63 — Find the eight eighth roots of 256. Identify the complex conjugate pairs.

Ex. 25.64 — The two square roots of i are

$$\frac{\sqrt{2}}{2} + i\frac{\sqrt{2}}{2} \quad \text{and} \quad -\frac{\sqrt{2}}{2} - i\frac{\sqrt{2}}{2}.$$

Find the following square roots.

1. $\sqrt{4i}$ 2. $\sqrt{10i}$ 3. $-\sqrt{16i}$ 4. $\sqrt{-i}$

* **Ex. 25.65** — What are some of the uses of complex numbers today?

* **Ex. 25.66** — What are the Gaussian integers? Give some examples. Where are they used?

* **Ex. 25.67** — Write about the history of numbers such as zero, negative numbers, and complex numbers that were slow to be accepted and even considered suspect.

Ex. 25.68 — Show that the change of variable $x = y - \frac{b}{3}$ in the general cubic equation

$$x^3 + bx^2 + cx + d = 0$$

leads to the reduced cubic equation

$$y^3 + py + q = 0,$$

where

$$p = c - \frac{b^2}{3} \quad \text{and} \quad q = \frac{2b^3}{27} - \frac{bc}{3} + d.$$

Ex. 25.69 — In chapter XV of *Ars magna*, "On the Cube and Square Equal to the Number" $[x^3 + bx^2 = d]$, Cardano provides the rule [36, p. 114]:

Cube one-third the coefficient of x^2, multiply the result by 2, and take the difference between this and the constant of the equation. [The result is the constant of a new equation.] Then multiply the square of one-third the coefficient of x^2 by 3 and you will have [the number of] y's which are equal to the cube and the constant, if twice the cube is greater than the constant of the equation, or [the number of] y's which equal the cube, if the difference between these two numbers is zero. Having derived the solution [for this equation], subtract one-third the coefficient of x^2 from it and the remainder is the value of x.

1. Show that Cardano's rule is the same as using the change of variable $x = y - \frac{b}{3}$ in the equation $x^3 + bx^2 - d = 0$.

2. Cardano illustrates the preceding rule by using it to solve the equation $x^3 + 6x^2 = 100$. Use Cardano's rule to solve the equation.

Ex. 25.70 — In chapter XVI, "On the Cube and Number Equal to the Square" $[x^3 + d = bx^2]$, Cardano provides the rule [36, p. 118]:

Multiply the cube root of the constant by the coefficient of x^2. This will produce a number of y's equal to the cube and the same constant $[py = y^3 + d$, where $p = b\sqrt[3]{d}\,]$. Having derived the solutions [for this equation], square the cube root of the constant and divide it by whatever values for y have been derived, and the results will be the values of x that are being sought.

Cardano illustrates the preceding rule by using it to solve the equation $x^3 + 64 = 18x^2$. Use Cardano's rule to solve the equation. Verify the solutions.

Ex. 25.71 — Complete the following problems.

1. (*Ars magna* chapter XVII problem III [36, p. 126])

    An oracle ordered a prince to build a sacred building whose space should be 400 cubits, the length being six cubits more than the width, and the width three cubits more than the height. These quantities are to be found.
    Let the altitude be x; the width be $x + 3$; and the length be $x + 9$. Multiplied in turn, you will have

$$x^3 + 12x^2 + 27x = 400.$$

    Put $x = y - \beta$ into the equation to obtain a new equation in the variable y. Determine the value of β that will eliminate the y^2 term in the new equation. Rearrange the terms in the equation after eliminating the y^2 term so that all of the coefficients are positive, and then solve the resulting equation using one of Cardan's methods. Finally, obtain the solution x.

2. (*Ars magna* chapter XXXVIII problem I [36, pp. 222–223])

    [F]ind two numbers the difference between which is 8 and the sum of the cube of one of which and the square of the other is 100.... [L]et the part to be cubed be $x + 2$ and the part to be squared $x - 6$.

    Write an equation for x, and then proceed as in the last problem to solve for x.

3. (*Ars magna* chapter XXXVIII problem II [36, pp. 223–224])

    Find two numbers the difference between which is 8 and the difference between the cube of one of which and the square of the other is 100.... [P]ostulate $x - 2$ and $x + 6$. Cube $x - 2$ and square $x + 6$, and take the difference.

Ex. 25.72 — "Vectors" provide us another geometric representation of complex numbers. If z is a complex number, then a *vector* that represents z may be seen as an arrow that is drawn from 0 to z in the complex plane. Here are a few examples.

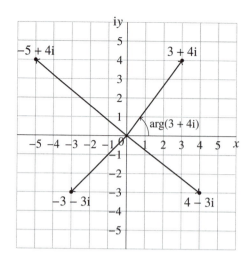

1. Let $z = 4 + 3i$ and $w = -3 - 5i$.

   a) Draw vectors for z, $-z$, $2z$, and $\frac{1}{2}z$. How are the vectors related?

   b) Draw vectors for z, w, $\bar{z}$, and $\bar{w}$. How are the vectors for z and $\bar{z}$ related? How are the vectors for w and $\bar{w}$ related? (Recall that a bar denotes the complex conjugate.)

2. Draw vectors for $z = 4 + 2i$ and $w = 1 + 3i$. Show that the vector for $z + w$ is the diagonal of a parallelogram with adjacent sides that are the vectors for z and w.

3. For a complex number z, the *argument* of z, denoted arg(z), is the angle measured between the vector for z and the positive real axis (x axis), with $0° \leq \arg(z) < 360°$. In general, angles measured in the counterclockwise direction are positive, and angles measured in the clockwise direction are negative.
   Draw vectors for $z = 4 + 3i$, $w = -\frac{3}{2} + 2i$, and zw. Verify that

$$|zw| = |z||w|$$

   and, using a protractor, that

$$\arg(zw) = \arg(z) + \arg(w).$$

   This is true in general provided we subtract an appropriate multiple of 360° to make the sum less than 360°.

4. Locate on the unit circle (the circle of radius one centered at the origin) in the complex plane

   a) the two square roots of unity.   d) the six sixth roots of unity.

   b) the three cube roots of unity.   e) the eight eighth roots of unity.

   c) the four fourth roots of unity.   f) the two square roots of i.

5. Draw the vector for i in the complex plane.

   a) Find $|i|$ and arg(i).

   b) By repeatedly multiplying $|i|$ and repeatedly adding arg(i), locate i^2, i^3, i^4, i^5, i^6, i^7, and i^8 in the complex plane. For example,

$$|i^2| = |i||i|, \qquad \arg(i^2) = \arg(i) + \arg(i),$$
$$|i^3| = |i^2||i|, \qquad \arg(i^3) = \arg(i^2) + \arg(i),$$

and so on. Describe what is happening geometrically as you find succeeding positive integer powers of i.

   c) Let n be a positive integer. Determine the location of i^n in the complex plane.

Ex. 25.73 — In chapter XXV of the *Ars magna* [36, p. 160], Cardano gives this rule:

> 2. When the cube is equal to the first power and constant, find two numbers whose product is the constant and one of which is the square root of the sum of the other and the coefficient of x. The square root is the value.

Consider the equation $x^3 = px + q$, with p and q positive.

1. Suppose that $ab = q$ and that $\sqrt{a + p} = b$. Show that b is a solution of the equation.

2. Use the rule to solve the equation $x^3 = 15x + 4$.

Ex. 25.74 — In EXAMPLE 21.10, Cardano shows that $\sqrt[3]{\sqrt{108} + 10} - \sqrt[3]{\sqrt{108} - 10}$ is a root of the equation $x^3 + 6x = 20$. Show that

$$\sqrt[3]{\sqrt{108} + 10} = 1 + \sqrt{3} \quad \text{and} \quad \sqrt[3]{\sqrt{108} - 10} = -1 + \sqrt{3},$$

so that, in fact,

$$\sqrt[3]{\sqrt{108} + 10} - \sqrt[3]{\sqrt{108} - 10} = 2.$$

Here are two suggestions to do this:

1. Note that $108 = 6\sqrt{3}$ and let

$$\sqrt[3]{\sqrt{108} + 10} = a + b\sqrt{3} \quad \text{and} \quad \sqrt[3]{\sqrt{108} - 10} = -a + b\sqrt{3}.$$

Taking the first equation, cube both sides and collect like terms on the right-hand side; then, equate coefficients of like terms to obtain the system

$$a(a^2 + 9b^2) = 10,$$
$$3b(a^2 + b^2) = 6.$$

Now let $a = 1$ and solve for b; then do the same for the second equation.

2. Let

$$\sqrt[3]{\sqrt{108} + 10} = 1 + x \quad \text{and} \quad \sqrt[3]{\sqrt{108} - 10} = -1 + x;$$

cube both sides of each equation; subtract the two resulting equations to obtain a quadratic equation in x; and solve for x.

Ex. 25.75 — In EXAMPLE 21.13, Cardano shows that $\sqrt[3]{20 + \sqrt{392}} + \sqrt[3]{20 - \sqrt{392}}$ is a root of the equation $x^3 = 6x + 40$. Show that

$$\sqrt[3]{20 + \sqrt{392}} = 2 + \sqrt{2} \quad \text{and} \quad \sqrt[3]{20 - \sqrt{392}} = 2 - \sqrt{2},$$

so that, in fact,

$$\sqrt[3]{20 + \sqrt{392}} + \sqrt[3]{20 - \sqrt{392}} = 4.$$

Here are two suggestions to do this:

1. Note that $\sqrt{392} = 14\sqrt{2}$ and let

$$\sqrt[3]{20 + \sqrt{392}} = a + b\sqrt{2} \quad \text{and} \quad \sqrt[3]{20 - \sqrt{392}} = a - b\sqrt{2}.$$

Taking the first equation, cube both sides and collect like terms on the right-hand side; then, equate coefficients of like terms to obtain a system of equations in a and b. Now let $a = 2$ and solve for b; then do the same for the second equation.

2. Let

$$\sqrt[3]{20 + \sqrt{392}} = 2 + x \quad \text{and} \quad \sqrt[3]{20 - \sqrt{392}} = 2 - x;$$

cube both sides of each equation; add the two resulting equations to obtain a quadratic equation in x; and solve for x.

* **Ex. 25.76 —** The term "sophistry" appears in several of the original sources we have cited. What does sophistry mean? What is the philosophy of sophists?

* **Ex. 25.77 —** Write about Girolamo Cardano: his life and his impact on mathematics.

* **Ex. 25.78 —** Write about Rafael Bombelli: his life and his impact on mathematics.

* **Ex. 25.79 —** Write about Leonhard Euler: his life and his impact on mathematics.

* **Ex. 25.80 —** Write about Carl Friedric Gauss: his life and his impact on mathematics.

* **Ex. 25.81 —** Write about the solution of the general quartic equation.

* **Ex. 25.82 —** Write about the solution of the general quintic equation.

Ex. 25.83 — Solve the quadratic equation $x^2 - 2x - 2 = 0$, then express the coefficients of the linear term and the constant term in the equation in terms of its roots. Show that the square of the difference between the two roots is the discriminant given in the quadratic formula on page 330.

Ex. 25.84 — Suppose that $r_1, r_2, \ldots, r_n$ are zeros of the polynomial

$$x^n + a_{n-1}x^{n-1} + a_{n-2}x^{n-2} + \cdots + a_2x^2 + a_1x + a_0,$$

and let

$$\delta = (r_1 - r_2)(r_1 - r_3) \cdots (r_1 - r_n)(r_2 - r_3)(r_2 - r_4) \cdots (r_2 - r_n) \cdots (r_{n-1} - r_n).$$

Then, $\Delta = \delta^2$ is called the *discriminant* of the polynomial.

1. Suppose that r_1 and r_2 are zeros of the quadratic polynomial $x^2 + bx + c$, and let $\delta = r_1 - r_2$. Show that $\Delta = \delta^2$ is the discriminant given in the quadratic formula on page 330.

2. Suppose that r_1, r_2, and r_3 are zeros of the cubic polynomial $x^3 + sx^2 + px + q$, and let $\delta = (r_1 - r_2)(r_1 - r_3)(r_2 - r_3)$. Verify that $\Delta = \delta^2 = s^2p^2 - 4s^3q + 18spq - 4p^3 - 27q^2$.

3. Suppose that r_1, r_2, and r_3 are zeros of the cubic polynomial $x^3 + px + q$, and let $\delta = (r_1 - r_2)(r_1 - r_3)(r_2 - r_3)$. How does $\Delta = \delta^2$ compare to the discriminant given in the cubic formula on page 405?

4. Find the discriminant of $x^4 + 7x^3 + 5x^2 - 31x - 30$. The zeros are 2, -1, -3, and -5.

5. Consider the cubic polynomial $x^3 - 3x^2 - 18x + 40$. The zeros are -4, 2, and 5.

   a) Find the discriminant of $x^3 - 3x^2 - 18x + 40$.

   b) Use the change of variable $x = y + 1$ to obtain the reduced cubic equation $x^3 + px + q = 0$. Show that -5, 1, and 4 are also the zeros of the reduced cubic. How does the discriminant of the reduced cubic compare to the discriminant of $x^3 - 3x^2 - 18x + 40$?

Ex. 25.85 — Suppose that a polynomial equation has roots a, b, c, and d. What may be the degree of the equation, and what are the coefficients of the terms in the equation in terms of its roots?

Ex. 25.86 — The roots of $x^4 + 7x^3 + 5x^2 - 31x - 30 = 0$ are 2, -1, -3, and -5. Express the coefficients of the cubic, quadratic, and linear terms and the constant term in the equation in terms of its roots.

Ex. 25.87 — Verify that 2 is a zero of the polynomial $x^3 - 3x^2 - 18x + 40$, then factor the polynomial completely. Express the coefficients of the quadratic and linear terms and the constant term in the equation $x^3 - 3x^2 - 18x + 40 = 0$ in terms of its roots.

Ex. 25.88 — Find a polynomial with zeros $-3, 4$, and 9. Use this polynomial to find a different polynomial with zeros 3, -4, and -9.

Ex. 25.89 — Suppose that $x^2 + Bx + C = (x - a)(x - b)$, where a and b are real numbers. Show that Descartes's rule of signs is true for the equation $x^2 + Bx + C = 0$.

Ex. 25.90 — Show that if r is a root of the equation

$$ax^4 + bx^3 + cx^2 + dx + e = 0,$$

then $-r$ is a root of

$$ax^4 - bx^3 + cx^2 - dx + e = 0.$$

Ex. 25.91 — Show that if r is a root of

$$ax^5 + bx^4 + cx^3 + dx^2 + ex + f = 0,$$

then $-r$ is a root of

$$ax^5 - bx^4 + cx^3 - dx^2 + ex - f = 0,$$

Ex. 25.92 — For each of the following, write the appropriate statement: "The logarithm of x to the base b is the exponent y such that $b^y = x$: $\log_b x = y$ because $b^y = x$."

1. $\log_3 81$
2. $\log_{2/5} \dfrac{4}{25}$
3. $\log_8 2$
4. $\log_4 \dfrac{1}{64}$

Ex. 25.93 — Simplify the following.

1. $2^{\log_2 x}$
2. $2^{3 \log_2 x}$
3. $2^{\frac{1}{2} \log_2 x^2}$
4. $2^{3 + \log_2 x}$

Ex. 25.94 — Use the rules for logarithms (RULE 24.1) to rewrite each expression using a single logarithm.

1. $\log_3 x + \log_3(x + 5)$

2. $\log_3 x - \log_3(x + 5)$

3. $\log_3 10 - \log_3 x - \log_3(x + 5)$

4. $2 \log_3 x - \dfrac{1}{2} \log_3(x + 5) + \log_3(x - 1)$

Ex. 25.95 — Express the given logarithm in terms of the logarithm to the base a for the specified base a.

1. $\log_2 5$; $a = 8$

2. $\log_{12} 100$; $a = \pi$

3. $\log 32$; $a = 4$

4. $\ln 3$; $a = 12$

Ex. 25.96 — Write an expression for the exact value of the logarithm in terms of the logarithm to the base e, then approximate the logarithm to 7 decimal places.

1. $\log_2 e$
2. $\log_2 \dfrac{1}{e}$
3. $\log_3 243$
4. $\log 1000$

Ex. 25.97 — Is there a flaw in the following argument?

Let a and b be positive real numbers, and let $u = a^x$ and $v = b^x$, so that

$$x = \log_a u \quad \text{and} \quad x = \log_b v.$$

Then,

$$uv = (a^x)(b^x) = (ab)^x,$$

so that

$$\log_{ab}(uv) = x.$$

Thus,

$$\log_{ab}(uv) = \log_a u \quad \text{or} \quad \log_{ab}(uv) = \log_b v.$$

Ex. 25.98 — Is there a flaw in the following argument?

Let a and b be positive real numbers, and let $u = a^x$ and $v = b^x$, so that

$$x = \log_a u \quad \text{and} \quad x = \log_b v.$$

Then,

$$\frac{u}{v} = \frac{a^x}{b^x} = \left(\frac{a}{b}\right)^x,$$

so that

$$\log_{a/b}\left(\frac{u}{v}\right) = x.$$

Thus,

$$\log_{a/b}\left(\frac{u}{v}\right) = \log_a u \quad \text{or} \quad \log_{a/b}\left(\frac{u}{v}\right) = \log_b v.$$

Ex. 25.99 — Recall that the "limit" of the sequence $(1 + \frac{1}{n})^n$ is e:

$$\lim\left(1 + \frac{1}{n}\right)^n = e$$

as n becomes increasingly greater. Find the "limit" of the sequence

$$\left(1 - \frac{1}{n}\right)^n$$

as n becomes increasingly greater. Hint: Let $m = -n$.

Ex. 25.100 — Recall that in the definition of the logarithm to the base b (DEFINITION 24.1) we restricted the base $b > 1$. Now, if $b > 1$, then $0 < \frac{1}{b} < 1$. It is also common, therefore, to define the logarithm to the base b with a weaker restriction on the base b, namely, that $b > 0$, but $b \neq 1$.
Let A be a positive real number. Show that

$$\log_b\left(\frac{A}{x}\right) = \log_{1/b}\left(\frac{x}{A}\right).$$

Ex. 25.101 — Recall that

$$\text{Nap} \log x = 10^7 \log_e\left(\frac{10^7}{x}\right).$$

Show that equivalently

$$\text{Nap} \log x = 10^7 \log_{1/e}\left(\frac{x}{10^7}\right).$$

Ex. 25.102 — You borrow \$5,000 at a fixed APR of 18% that is to be repaid fortnightly or semimonthly over 6 months (12 repayments). Find the amount of each repayment and the total amount in interest you would have paid at the end.

Ex. 25.103 — Which is a better investment: 4.9% APR compounded monthly or 5% APR compounded quarterly?

Ex. 25.104 — What APR compounded quarterly would be the same as 5% APR compounded monthly?

Ex. 25.105 — How much should you deposit into a bank account that pays 3.125% APR compounded quarterly so that you would have accumulated \$18,000 in five years? The amount that you should deposit is called the *present value*, and the amount that you wish to accumulate (\$18,000 here) is called the *future value*.

* **Ex. 25.106** — Research the history of (compound) interest. Give examples from at least two different civilizations or societies.

* **Ex. 25.107** — Write about the history of amortization. Besides loans, what other amounts may be amortized today?

* **Ex. 25.108** — Look up "annual percentage yield" (APY). Calculate the APY for an investment at 4.9% APR compounded monthly, and for an investment at 5% APR compounded quarterly. Which is the better investment?

Ex. 25.109 — The acidity of an aqueous solution is determined by the molar concentration of hydrogen ions in the solution: the greater the concentration of hydrogen ions, the more acidic is the solution.[1] Now, the concentration of hydrogen ions in an aqueous solution can range from more than 1 M (mole per liter) to less than 0.000 000 000 000 01 M. To make manageable dealing with such small numbers routinely, we use logarithms. Specifically, the "pH" (potential of hydrogen) of a solution is defined to be the negative logarithm to the base 10 of its molar hydrogen ion concentration:

$$\text{pH} = -\log[H^+],$$

where $[H^+]$ is the molar hydrogen concentration. Using this scale, a neutral solution (one that is neither an acid nor a base) has pH equal to 7; acids have pH less than 7; and bases have pH greater than 7.

1. Find the pH of the solution given its molar hydrogen ion concentration. State if the solution is neutral, an acid, or a base.

   a) $[H^+] = 0.001\ M$ b) $[H^+] = 0.000\,000\,001\ M$

2. Find the molar hydrogen ion concentration given the pH. State if the solution is neutral, an acid, or a base.

   a) pH 3.2   b) pH 9

---

[1]Charles H. Corwin, *Introductory Chemistry: Concepts and Critical Thinking*, 6th ed., Prentice Hall, Boston (2011).

3. Explain why a smaller molar hydrogen ion concentration corresponds to a larger pH, and inversely.

4. Can a pH be zero? Negative?

* **Ex. 25.110** — We provided examples of how exponential functions are used, such as seismology. Research other areas that use exponential functions.

* **Ex. 25.111** — Research radiocarbon dating and how it uses exponential functions. Give an example.

* **Ex. 25.112** — Look up the intensity of sound and explain the decibel scale, including the history of its development.

* **Ex. 25.113** — Write about Newton's law of cooling. Provide an example of its use.

* **Ex. 25.114** — Explain how Napier constructed his table of logarithms.

* **Ex. 25.115** — Explain why Napier chose to use 10^7 in his definition of his logarithms. What was the geometrical interpretation he was working in?

* **Ex. 25.116** — Explain how Bürgi constructed his table of logarithms.

* **Ex. 25.117** — What is a "logarithmic curve"? Is it different from a "logarithmic spiral"?

* **Ex. 25.118** — Explain what is a slide rule and how one is used to perform calculations.

* **Ex. 25.119** — Investigate and report on why Euler may have chosen to use the letter e for the constant that is the "limit" of the sequence $\left(1 + \frac{1}{n}\right)^n$ as n becomes increasingly greater.

Bibliography

[1] Michael Adams, editor. *The Middle East*, volume 1 of *Handbooks to the Modern World*. Facts on File Publications, New York, 1988. General Editor: Andrew C. Kimmens.

[2] Muḥammad ibn-Mūsā al-Khwārizmī. *The Algebra of Mohammed ben Musa*. J. Murray, London, 1831. Edited and translated by Frederic Rosen.

[3] Gerald L. Alexanderson. About the cover: Euler and Königsberg's bridges: a historical view. *Bull. Amer. Math. Soc.*, 43(4):567–573, October 2006.

[4] Marlow Anderson, Victor Katz, and Robin Wilson, editors. *Sherlock Holmes in Babylon and Other Tales of Mathematical History*. The Mathematical Association of America, Washington, D.C., 2004.

[5] Apollonius. *Apollonius of Perga, Treatise on Conic Sections*. Cambridge University Press, Cambridge, 1896. Edited in modern notation with introductions by T. L. Heath.

[6] Apollonius. *Apollonius, Conics, Books V to VII: The Arabic Translation of the Lost Greek Original in the Version of the Banū Mūsā*. Springer, New York, 1990. Edited with translation and commentary by G. J. Toomer.

[7] Marcia Ascher. Before the conquest. *Math. Mag.*, 65:211–218, 1992. In Marlow Anderson, Victor Katz, and Robin Wilson, editors, *Sherlock Holmes in Babylon and Other Tales of Mathematical History*. The Mathematical Association of America, Washington, D.C., 2004.

[8] I. G. Bashmakova and G. S. Smirnova. *The Beginnings and Evolution of Algebra*. Number 23 in Dolciani Mathematical Expositions. The Mathematical Association of America, Washington, D.C., 2000. Translated from the Russian by Abe Shenitzer with the editorial assistance of David A. Cox.

[9] E. T. Bell. *Men of Mathematics*. Simon and Schuster, New York, 1937.

[10] J. Lennart Berggren. Mathematics in Medieval Islam. In Victor J. Katz, editor, *The Mathematics of Egypt, Mesopotamia, China, India, and Islam: A Sourcebook*. Princeton University Press, Princeton, 2007.

[11] William P. Berlinghoff and Fernando Q. Gouvêa. *Math through the Ages: A Gentle History for Teachers and Others*. Oxton House Publishers, Farmington, 2002.

[12] Norman Biggs. Mathematics of currency and exchange: arithmetic at the end of the thirteenth century. *Brit. Soc. Hist. Math. Bull.*, 24(2):67–77, 2009.

[13] Elizabeth Boag. Lattice multiplication. *Brit. Soc. Hist. Math. Bull.*, 22(3):182, 2007.

[14] Rafael Bombelli. *L'Algebra, prima edizione integrale*. Feltrinelli, Milano, 1966. Italian. Preface by Ettore Bortolotti and Umberto Forti. A reprint of the 1579 edition in Italian is at <http://mathematica.sns.it/volume.asp?Id=8&PrevId=9>.

[15] Carl B. Boyer. *A History of Mathematics*. John Wiley & Sons, Inc., Hoboken, 2nd edition, 1991. Revised by Uta C. Merzbach.

[16] Brahmagupta and Bhāskara. *Algebra, with Arithmetic and Mensuration, from the Sanscrĭt of Brahmegupta and Bhascara*. John Murray, London, 1817. Translated by Henry Thomas Colebrooke, Esq. <http://books.google.com/books?id=ebZIAAAAcAAJ>.

[17] R. Creighton Buck. Sherlock Holmes in Babylon. *Amer. Math. Monthly*, 87:335–345, 1980. In Marlow Anderson, Victor Katz, and Robin Wilson, editors, *Sherlock Holmes in Babylon and Other Tales of Mathematical History*. The Mathematical Association of America, Washington, D.C., 2004.

[18] Charles Burnett. Why We Read Arabic Numerals Backwards. In Patrick Suppes, Julius M. Moravscik, and Henry Mendell, editors, *Ancient & Medieval Traditions in the Exact Sciences: Essays in Memory of Wilbur Knorr*. CSLI Publications, Stanford, 2000.

[19] Charles Burnett. The semantics of Indian numerals in Arabic, Greek and Latin. *J. Indian Philosophy*, 34:15–30, 2006.

[20] E. A. Burt. *The Metaphysical Foundations of Modern Science*. Dover Publications, Inc., New York, 2003.

[21] David M. Burton. *The History of Mathematics: An Introduction*. McGraw-Hill, Boston, 6th edition, 2007.

[22] David M. Burton. *The History of Mathematics: An Introduction*. McGraw-Hill, New York, 7th edition, 2011.

[23] Florian Cajori. History of the exponential and logarithmic concepts. *Amer. Math. Monthly*, 20(2):35–47, Feb. 1913. Part 2 of 7.

[24] Florian Cajori. History of the exponential and logarithmic concepts. *Amer. Math. Monthly*, 20(1):5–14, Jan. 1913. Part 1 of 7.

[25] Florian Cajori. History of the exponential and logarithmic concepts. *Amer. Math. Monthly*, 20(3):75–84, Mar. 1913. Part 3 of 7.

[26] Florian Cajori. History of the exponential and logarithmic concepts. *Amer. Math. Monthly*, 20(4):107–117, Apr. 1913. Part 4 of 7.

[27] Florian Cajori. History of the exponential and logarithmic concepts. *Amer. Math. Monthly*, 20(5):148–151, May 1913. Part 5 of 7.

[28] Florian Cajori. History of the exponential and logarithmic concepts. *Amer. Math. Monthly*, 20(6):173–182, Jun. 1913. Part 6 of 7.

[29] Florian Cajori. History of the exponential and logarithmic concepts. *Amer. Math. Monthly*, 20(7):205–210, Sep. 1913. Part 7 of 7.

[30] Florian Cajori. *A History of Elementary Mathematics with Hints on Methods of Teaching*. Macmillan Company, London, 1921.

[31] Florian Cajori. *A History of Mathematical Notation: Two Volumes Bound As One*. Dover Publications, Inc., New York, 1993.

[32] Florian Cajori. *A History of the Logarithmic Slide Rule and Allied Instruments*. Astragal Press, Mendham, 1994. Includes *On the History of Gunter's Scale and the Slide Rule during the Seventeenth Century* (begins on page 137).

[33] Florian Cajori. *A History of Elementary Mathematics*. Cosimo Classics, New York, 2007. Originally published in 1896.

[34] Ronald Calinger, editor. *Classics of Mathematics*. Prentice Hall, Upper Saddle River, 1995.

[35] Ronald Calinger. *A Contextual History of Mathematics*. Prentice Hall, Upper Saddle River, 1999.

[36] Girolamo Cardano. *The Rules of Algebra (Ars Magna)*. Dover Publications, Inc., Mineola, 2007. Translated and edited by T. Richard Witmer. Latin text at <http://www.filosofia.unimi.it/cardano/testi/operaomnia/vol_4_s_4.pdf> or <http://daten.digitale-sammlungen.de/~db/0002/bsb00029375/images/>.

[37] H. S. Carslaw. The discovery of logarithms by Napier. *Math. Gazette*, 8(117):76–84, May 1915. Part 1 of 2.

[38] H. S. Carslaw. The discovery of logarithms by Napier (concluded). *Math. Gazette*, 8(118):115–119, Jul. 1915. Part 2 of 2.

[39] Jean-Luc Chabert (editor), Évelyne Barbin, Jacques Borowczyk, Michel Guillemot, and Anne Michel-Pajus. *A History of Algorithms: From the Pebble to the Microchip*. Springer-Verlag, Berlin, 1999. Translator of the English Edition: Chris Weeks.

[40] Kathleen Clark and Clemency Montelle. Logarithms: the early history of a familiar function. *Loci*, DOI: 10.4169/loci003495:1–11, Jun. 2010. <http://www.maa.org/publications/periodicals/convergence/logarithms-the-early-history-of-a-familiar-function>.

[41] Roger Cooke. *The History of Mathematics: A Brief Course*. Wiley-Interscience. John Wiley & Sons, Inc., Hoboken, 2nd edition, 2005.

[42] Michael J. Crowe. *A History of Vector Analysis: The Evolution of the Idea of a Vectorial System*. Dover Publications, Inc., New York, 1967.

[43] Nathan Daboll. *Daboll's Schoolmaster's Assistant: Improved and Enlarged, Being a Plain Practical System of Arithmetic Adapted to the United States*. E. & E. Hosford, Albany, 1825. With the addition of *The Practical Accountant; or, Farmers' and Mechanics' Best Method of Book-Keeping; for the Easy Instrction of Youth* by Samuel Green. <http://books.google.com/books?id=rWEVAAAAYAAJ>.

[44] Joseph W. Dauben. Chinese Mathematics. In Victor J. Katz, editor, *The Mathematics of Egypt, Mesopotamia, China, India, and Islam: A Sourcebook*. Princeton University Press, Princeton, 2007.

[45] H. Davenport. *The Higher Arithmetic*. Cambridge University Press, Cambridge, 5th edition, 1982.

[46] Rene Descartes. *The Geometry of Rene Descartes*. Cosimo Classics [Science]. Cosimo, Inc., New York, 2007. David Eugene Smith and Marcia Latham, translators.

[47] William Dunham. *Journey through Genius: The Great Theorems of Mathematics*. Penguin Books, New York, 1990.

[48] W. C. Eells. Number systems of the North American Indians. *Amer. Math. Monthly*, 20:263–272, 293–299, 1913. In Marlow Anderson, Victor Katz, and Robin Wilson, editors, *Sherlock Holmes in Babylon and Other Tales of Mathematical History*. The Mathematical Association of America, Washington, D.C., 2004.

[49] Anislie T. Embree, editor. *Encyclopedia of Asian History*, volume 2. Charles Scribner's Sons, New York, 1988. Prepared under the auspices of The Asia Society.

[50] Anislie T. Embree, editor. *Encyclopedia of Asian History*, volume 1. Charles Scribner's Sons, New York, 1988. Prepared under the auspices of The Asia Society.

[51] Herbert B. Enderton. *A Mathematical Introduction to Logic*. Academic Press, San Diego, 1972.

[52] Euclid. *The Bones*. Green Lion Press, Santa Fe, 2002. A handy where-to-find-it pocket reference to Euclid's *Elements*.

[53] Euclid. *Elements: All Thirteen Books Complete in One Volume*. Green Lion Press, Santa Fe, 2003. The Thomas L. Heath Translation. Dana Densmore, editor.

[54] Leonhard Euler. *Elements of Algebra*. Tarquin Publications, St. Albans, 2006. Edited by Christopher James Sangwin, this book is an edited reprint of Part I of J. Hewlett's 1822 English translation from the French of Leonhard Euler's *Elements of algebra*. Hewlett's translation is in the public domain at <http://books.google.com/books?id=X8yvOsj4_1YC>.

[55] Howard Eves. *Great Moments in Mathematics After 1650*. Number 7 in Dolciani Mathematical Expositions. The Mathematical Association of America, Washington, D.C., 1983.

[56] Howard Eves. *Great Moments in Mathematics Before 1650*. Number 5 in Dolciani Mathematical Expositions. The Mathematical Association of America, Washington, D.C., 1983.

[57] John Fauvel and Jeremy Gray, editors. *The History of Mathematics: A Reader*. Macmillan Press, London, 1987.

[58] Midhat J. Gazalé. *Number: From Ahmes to Cantor*. Princeton University Press, Princeton, 1999.

[59] W. Gellert, H. Kustner, M. Hellwish, and H. Kastner, editors. *The VNR Concise Encyclopedia of Mathematics*. Van Norstrand Reinhold, New York, 1975.

[60] Owen Gingerich. *The Book Nobody Read: Chasing the Revolutions of Nicolaus Copernicus*. Walker & Company, New York, 2004.

[61] William N. Goetzmann. Fibonacci and the Financial Revolution. In William N. Goetzmann and K. Geert Rouwenhorst, editors, *The Origins of Value: The Financial Innovations That Created Modern Capital Markets*. Oxford University Press, Oxford, 2005.

[62] Hardy Grant. Mathematics and the liberal arts. *Coll. Math. Journal*, 30(2):96–105, March 1999.

[63] Hardy Grant. Mathematics and the liberal arts—II. *Coll. Math. Journal*, 30(3):197–203, May 1999.

[64] A. Rupert Hall. *Philosophers at War: The Quarrel between Newton and Leibniz*. Cambridge University Press, Cambridge, 1980.

[65] Rachel W. Hall. A course in multicultural mathematics. *PRIMUS*, XVII(3):209–227, 2007.

[66] G. H. Hardy. *A Mathematician's Apology*. Cambridge University Press, Cambridge, 1992.

[67] Cynthia Hay, editor. *Mathematics from Manuscript to Print: 1300–1600*. Clarendon Press, Oxford, 1988.

[68] Sir Thomas Heath. *A History of Greek Mathematics Volume I: From Thales to Euclid*. Dover Publications, Inc., New York, 1981.

[69] Sir Thomas Heath. *A History of Greek Mathematics Volume II: From Aristarchus to Diophantus*. Dover Publications, Inc., New York, 1981.

[70] Annette Imhausen. Egyptian Mathematics. In Victor J. Katz, editor, *The Mathematics of Egypt, Mesopotamia, China, India, and Islam: A Sourcebook*. Princeton University Press, Princeton, 2007.

[71] Thomas Jech. Set Theory. In Edward N. Zalta, editor, *The Stanford Encyclopedia of Philosophy*. Stanford University, winter 2011 edition, 2011.

[72] George Gheverghese Joseph. *The Crest of the Peacock: Non-European Roots of Mathematics*. I. B. Tauris, London, 1991.

[73] Victor J. Katz. *A History of Mathematics: An Introduction*. Addison Wesley Longman, Reading, 2nd edition, 1998.

[74] Victor J. Katz. *A History of Mathematics: Brief Edition*. Pearson Addison Wesley, Boston, 2004.

[75] Victor J. Katz. *A History of Mathematics: An Introduction*. Pearson Education, Inc., Boston, 3rd edition, 2009.

[76] Federica La Nave and Barry Mazur. Reading Bombelli. *The Math. Intellingencer*, 24(1):12–21, 2002.

[77] Lay Yong Lam and Tian Se Ang. *Fleeting Footsteps (Revised Edition): Tracing the Conception of Arithmetic and Algebra in Ancient China*. World Scientific Publishing Company, New Jersey, 2004.

[78] Elisha S. Loomis. *The Pythagorean Proposition: Its Demonstrations Analyzed and Classified, and Biography of Sources for Data of the Four Kinds of "Proofs"*. Edwards Brothers, Inc., Ann Arbor, 2nd edition, 1940.

[79] Lǐ Yǎn and Dù Shírán. *Chinese Mathematics: A Concise History*. Clarendon Press, Oxford, 1987. Translated by John N. Crossley and Anthony W.-C. Lun.

[80] Eli Maor. *e: The Story of a Number*. Princeton University Press, Princeton, 1994.

[81] Eli Maor. *The Pythagorean Theorem: A 4,000-Year History*. Princeton University Press, Princeton, 2007.

[82] Karl Menninger. *Number Words and Number Symbols*. The MIT Press, Cambridge, 1977. Translated by Paul Broneer from the revised German edition. *Zahlwort und Ziffer*, Vandenhoeck & Ruprecht, Göttingen, 1958.

[83] Yoshio Mikami. *The Development of Mathematics in China and Japan*. Chelsea Publishing Company, New York, 1913.

[84] Sylvanus Griswold Morley. *An Introduction to the Study of the Maya Hieroglyphs*. Number 57 in Smithsonian Institution Bureau of American Ethnology. Government Printing Office, Washington, 1915. In the public domain at <http://books.google.com/books?id=MV4SAAAAYAAJ>.

[85] Trevor Mostyn and Albert Hourani, editors. *The Cambridge Encyclopedia of the Middle East and North Africa*. Cambridge University Press, Cambridge, 1988.

[86] Paul J. Nahin. *An Imaginary Tale: The Story of $\sqrt{-1}$*. Princeton University Press, Princeton, 1998.

[87] John Napier. *The Construction of the Wonderful Canon of Logarithms*. William Blackwood and Sons, Edinburgh and London, 1889. Translated from Latin into English with notes and a catalogue of the various editions of Napier's works by William Rae Macdonald. In the public domain at <http://books.google.com/books?id=Zlu4AAAAIAAJ>.

[88] O. (Otto) Neugebauer. *The Exact Sciences in Antiquity*. Princeton University Press, Princeton, 1952.

[89] Ivan Niven. *Numbers: Rational and Irrational*. Number 1 in New Mathematical Library. The Mathematical Association of America, Washington, D.C., 1961.

[90]   Leonardo of Pisa (Fibonacci). *Fibonacci's Liber Abaci: A Translation into Modern English of Leonardo Pisano's Book of Calculation*. Springer, New York, 2002. Translated L. E. Sigler.

[91]   C. D. Olds. *Continued Fractions*. Number 9 in New Mathematical Library. The Mathematical Association of America, Washington, D.C., 1963.

[92]   Oystein Ore. *Number Theory and Its History*. McGraw-Hill Book Company, Inc., New York, 1948.

[93]   Lynn M. Osen. *Women in Mathematics*. The MIT Press, Cambridge, 1988.

[94]   Victor Larios Osorio. Sistemas numéricos en el México prehispánico (Number systems in the prehispanic Mexico; in Spanish). *Xixím*, 1(3):24–34, April 2001. <http://www.uaq.mx/matematicas/redm/articulos.html?0303>.

[95]   Kim Plofker. Mathematics in India. In Victor J. Katz, editor, *The Mathematics of Egypt, Mesopotamia, China, India, and Islam: A Sourcebook*. Princeton University Press, Princeton, 2007.

[96]   Kim Plofker. *Mathematics in India*. Princeton University Press, Princeton, 2009.

[97]   Helena M. Pycior. *Symbols, Impossible Numbers, and Geometric Entanglements: British Algebra through the Commentaries on Newton's* Universial Arithmetick. Cambridge University Press, Cambridge, 1997.

[98]   Robert Recorde. *The Whetstone of Witte*. Open Source Books, www.archive.org, 1557 (2009). In the public domain at <http://www.archive.org/details/TheWhetstoneOfWitte>.

[99]   Barbara E. Reynolds. The algorists vs. the abacists: An ancient controversy on the use of calculators. *Coll. Math. Journal*, 24:218–223, 1993. In Marlow Anderson, Victor Katz, and Robin Wilson, editors, *Sherlock Holmes in Babylon and Other Tales of Mathematical History*. The Mathematical Association of America, Washington, D.C., 2004.

[100] A. W. Richeson. The number system of the Mayas. *Amer. Math. Monthly*, 40:542–546, 1933. In Marlow Anderson, Victor Katz, and Robin Wilson, editors, *Sherlock Holmes in Babylon and Other Tales of Mathematical History*. The Mathematical Association of America, Washington, D.C., 2004.

[101] Eleanor Robson. Words and pictures: new light on Plimpton 322. *Amer. Math. Monthly*, 109:105–120, 2002. In Marlow Anderson, Victor Katz, and Robin Wilson, editors, *Sherlock Holmes in Babylon and Other Tales of Mathematical History*. The Mathematical Association of America, Washington, D.C., 2004.

[102] Eleanor Robson. Mesopotamian Mathematics. In Victor J. Katz, editor, *The Mathematics of Egypt, Mesopotamia, China, India, and Islam: A Sourcebook*. Princeton University Press, Princeton, 2007.

[103] Eleanor Robson. *Mathematics in Ancient Iraq: A Social History*. Princeton University Press, Princeton, 2008.

[104] Steven Schwartzman. *The Words of Mathematics: An Etymological Dictionary of Mathematical Terms Used in English*. MAA Spectrum. The Mathematical Association of America, Washington, D.C., 1994.

[105] Muriel Seltman. The *Artis analyticae praxis* of Harriot and Warner in focus. *BSHM Bull.*, 23(2):73–80, 2008.

[106] Muriel Seltman and Robert Goulding, editors. *Thomas Harriot's* Artis analyticae praxis: *An English Translation with Commentary*. Sources and Studies in the History of Mathematics and Physical Sciences. Springer, New York, 2007.

[107] Amy Shell-Gellasch. Napier's e. *Loci*, DOI: 10.4169/loci003209:1–9, Dec. 2008. <http://mathdl.maa.org/mathDL/46/?pa=content&sa=viewDocument&nodeId=3209>.

[108] Kangshen Shen, John N. Crossley, Anthony W.-C. Lun, and Liu Hui. *The Nine Chapters on the Mathematical Art: Companion and Commentary*. Oxford University Press, Oxford, 1999.

[109] Jacqueline A. Stedall. *A Discourse Concerning Algebra: English Algebra to 1685*. Oxford University Press, Oxford, 2002.

[110] Jacqueline A. Stedall. *The Greate Invention of Algebra: Thomas Harriot's Treatise on Equations*. Oxford University Press, Oxford, 2003.

[111] D. J. Struik, editor. *A Source Book in Mathematics, 1200–1800*. Harvard University Press, Cambridge, 1969.

[112] Jeff Suzuki. *A History of Mathematics*. Prentice Hall, Upper Saddle River, 2002.

[113] Frank J. Swetz, editor. *From Five Fingers to Infinity: A Journey through the History of Mathematics*. Open Court, Chicago, 1994.

[114] Frank J. Swetz. Similarity vs. the "in-and-out complementary principle": a cultural faux pas. *Math. Mag.*, 85(1):3–11, Feb. 2012.

[115] Frank J. Swetz and T. I. Kao. *Was Pythagoras Chinese? An Examination of Right Triangle Theory in Ancient China*. Pennsylvania State University Studies No. 40. The Pennsylvania State University Press, University Park, 1977.

[116] John E. Teeple. Maya Astronomy. In *Contributions to American Anthropology and History*, volume 1 (number 2), pages 29–116. Carnegie Institution of Washington, New York, 1931.

[117] Sir J. Eric S. Thompson. *A Commentary on the Dresden Codex: A Maya Hieroglyphic Book*, volume 93 of *Memoirs of the American Philosophical Society*. Amer. Phil. Soc., Philadelphia, 1972.

[118] Sir J. Eric S. Thompson. *Maya Hieroglyphic Writing: An Introduction*. University of Oklahoma Press, Norman, 3rd edition, 1978.

[119] J. Hilton Turner. Roman elementary mathematics: the operations. *The Classical J.*, 47(2):63–74 and 106–108, Nov. 1951. The article has been reproduced at <http://penelope.uchicago.edu/Thayer/E/Roman/Texts/secondary/journals/CJ/47/2/Roman_Elementary_Mathematics*.html>.

[120] François Viète. *In artem analyticem Isagoge, Seorsim excussa ab Opere restitutæ Mathematicæ Analyseos, Seu, Algebrâ novâ*. apud J. Mettayer (Turonis), bnf.fr, 1591. Francisci Vietæ. In the public domain at <http://gallica2.bnf.fr/ark:/12148/bpt6k108865t>.

[121] François Viète. De numerosa potestatum purarum, atque adfectarum ad exegesin resolutione tractus. In Francisci a Schooten, editor, *Francisci Vietæ Opera mathematica, In unum Volumen congesta, ac recognita*, pages 163–228. ex officina B. et A. Elzeviriorum (Lugduni Batavorum), bnf.fr, 1646. In the public domain at <http://gallica2.bnf.fr/ark:/12148/bpt6k107597d>.

[122] François Viete. *The Analytic Art: Nine Studies in Algebra, Geometry and Trigonometry from the Opus Restitutae Mathematicae Analyseos, seu Algebrâ Novâ*. The Kent State University Press, Kent, 1983. Translated by T. Richard Witmer.

[123] Kurt Von Fritz. The discovery of incommensurability by Hippasus of Metapontum. *The Annals of Mathematics*, 46(2):242–264, Apr. 1945.

Index